Mathematics Benchmarking Report
TIMSS 1999 – Eighth Grade

Achievement for U.S. States and Districts in an International Context

Ina V.S. Mullis
Michael O. Martin
Eugenio J. Gonzalez
Kathleen M. O'Connor
Steven J. Chrostowski
Kelvin D. Gregory
Robert A. Garden
Teresa A. Smith

ISC BOSTON COLLEGE
The International Study Center
Lynch School of Education

IEA
The International Association
for the Evaluation of
Educational Achievement

A Bridge to School Improvement

April 2001

Mathematics Benchmarking Report: TIMSS 1999 – Eighth Grade / by Ina V.S. Mullis, Michael O. Martin, Eugenio J. Gonzalez, Kathleen M. O'Connor, Steven J. Chrostowski, Kelvin D. Gregory, Robert A. Garden, Teresa A. Smith

Publisher: International Study Center
Lynch School of Education
Boston College

Library of Congress
Catalog Card Number: 2001087804

ISBN 1-889938-19-X

For more information about TIMSS contact:

The International Study Center
Lynch School of Education
Manresa House
Boston College
Chestnut Hill, MA 02467
United States

For information on ordering this report, write to the above address or call +1-617-552-1600

This report also is available on the World Wide Web: http://www.timss.org

Funding for the TIMSS 1999 Benchmarking Study was provided by the National Center for Education Statistics and the Office of Educational Research and Improvement of the U.S. Department of Education, the U.S. National Science Foundation, and participating jurisdictions.

Boston College is an equal opportunity, affirmative action employer.

Printed and bound in the United States

CONTENTS

Contents

CONTENTS

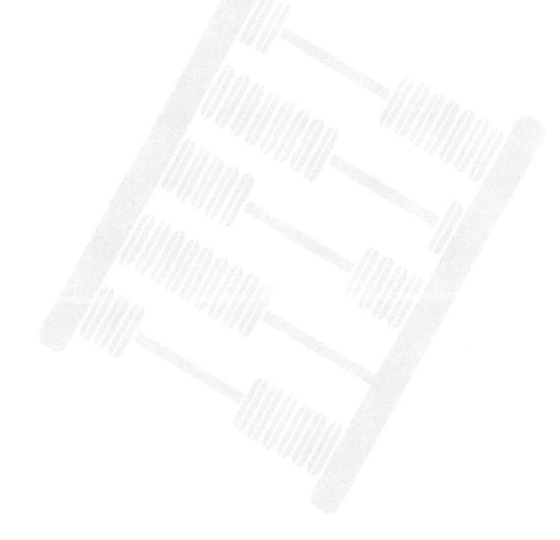

CONTENTS

Executive Summary

TIMSS 1999 Benchmarking

A Bridge to School Improvement

S

Executive Summary

TIMSS 1999, a successor to the acclaimed 1995 Third International Mathematics and Science Study (TIMSS), focused on the mathematics and science achievement of eighth-grade students. Thirty-eight countries including the United States participated in TIMSS 1999 (also known as TIMSS-Repeat or TIMSS-R).[1] Even more significantly for the United States, however, TIMSS 1999 included a voluntary Benchmarking Study. Twenty-seven jurisdictions from all across the nation, including 13 states and 14 districts or consortia (see below), participated in the Benchmarking Study.

TIMSS 1999 Benchmarking Participants

States	Districts and Consortia
Connecticut	Academy School District #20, Colorado Springs, CO
Idaho	Chicago Public Schools, IL
Illinois	Delaware Science Coalition, DE
Indiana	First in the World Consortium, IL
Maryland	Fremont/Lincoln/Westside Public Schools, NE
Massachusetts	Guilford County, NC
Michigan	Jersey City Public Schools, NJ
Missouri	Miami-Dade County Public Schools, FL
North Carolina	Michigan Invitational Group, MI
Oregon	Montgomery County, MD
Pennsylvania	Naperville School District #203, IL
South Carolina	Project SMART Consortium, OH
Texas	Rochester City School District, NY
	Southwest Pennsylvania Math and Science Collaborative, PA

Each jurisdiction had its own reasons for taking part in the TIMSS 1999 Benchmarking Study. In general, participation provided an unprecedented opportunity for jurisdictions to assess the comparative international standing of their students' achievement and to evaluate their mathematics and science programs in an international context. Participants were also able to compare their achievement with that of the United States as a whole,[2] and in the cases where they both participated, school districts could compare with the performance of their states.

Each participating entity invested valuable resources in this effort, primarily for data collection and team building, but also for staff development to facilitate use of the TIMSS 1999 results as an effective tool for school improvement. Despite each participant's deep commitment to educational improvement by virtue of its participation in such a venture, it took courage and initiative to join such a high profile enterprise as the TIMSS 1999 Benchmarking Study. Whether students' achievement fell at the top, middle, or bottom of the range of results for countries internationally, each participant will be asked to explain the results to its parents and communities.

1 IEA's International Study Center at Boston College reported the international results for TIMSS 1999 as well as trends between 1995 and 1999 in two companion volumes – the *TIMSS 1999 International Mathematics Report* and the *TIMSS 1999 International Science Report.* Performance in the United States relative to that of other nations was reported by the U.S. National Center for Education Statistics in *Pursuing Excellence: Comparisons of International Eighth-Grade Mathematics and Science Achievement from a U.S. Perspective, 1995 and 1999.* (See the Introduction for full citations.)

2 For the most part, the U.S. TIMSS national sample was separate from the students assessed in each of the Benchmarking jurisdictions. Each Benchmarking participant had its own sample to provide comparisons to each of the TIMSS 1999 countries including the United States. Collectively, the Benchmarking participants are not representative of the United States even though the effort was substantial in scope.

This report provides a preliminary overview of the results for the Benchmarking Study in mathematics. The real work will take place as each participating entity begins to examine its curriculum, teaching force, instructional approaches, and school environment in an international context. As those working on school improvement know full well, there is no "silver bullet" or single factor that is the answer to higher achievement in mathematics or any other school subject. Making strides in raising student achievement requires tireless diligence, as policy makers, administrators, teachers, and communities work to make improvements in a number of important areas related to educational quality.

Unlike in many countries around the world where educational decision making is highly centralized, in the United States the opportunities to learn mathematics derive from an educational system that operates through states and districts, allocating opportunities through schools and then through classrooms. Improving students' opportunities to learn requires examining every step of the educational system, including the curriculum, teacher quality, availability and appropriateness of resources, student motivation, instructional effectiveness, parental support, and school safety.

Particularly since *A Nation at Risk*[3] was issued eighteen years ago, many states and school districts have been working on the arduous task of improving education in their jurisdictions. During the past decade, content-driven systemic school reform has emerged as a promising model for school improvement.[4] That is, curriculum frameworks establishing what students should know and be able to do provide a coherent direction for improving the quality of instruction. Teacher preparation, instructional materials, and other aspects of the system are then aligned to reflect the content of the frameworks in an integrated way to reinforce and sustain high-quality teaching and learning in schools and classrooms.

There has been concerted effort across the nation in writing and revising academic standards that has very much included attention to mathematics. All states except Iowa (which as a matter of policy publishes no state standards) now have content or curriculum standards in mathematics, and many educational jurisdictions have worked successfully to improve their initial standards in clarity and content.[5] Forty-three states also have some type of criterion-referenced mathematics assessment aligned to state standards.[6] Much of this effort has been based on work done at the national level over the past decade to develop standards

3 *A Nation at Risk: The Imperative for Education Reform* (1983), Washington, DC: National Commission on Excellence in Education.

4 O'Day, J.A. and Smith, M.S. (1993), "Systemic Reform and Educational Opportunity" in S.H. Fuhrman (ed.), *Designing Coherent Education Policy: Improving the System*, San Francisco, CA: Jossey-Bass, Inc.

5 Raimi, R.A. (2000), "The State of State Standards in Mathematics" in C.E. Finn and M.J. Petrilli (eds.), *The State of State Standards*, Washington, DC: Thomas B. Fordham Foundation; Glidden, H. (1999), *Making Standards Matter 1999*, Washington, DC: American Federation of Teachers.

6 Orlofsky, G.F. and Olson, L. (2001), "The State of the States" in *Quality Counts 2001, A Better Balance: Standards, Tests, and the Tools to Succeed*, Education Week, 20(17).

aimed at increasing the mathematics competencies of all students. Since 1989, when the National Council of Teachers of Mathematics (NCTM) published *Curriculum and Education Standards for School Mathematics*, the mathematics education community has had the benefit of a unified set of goals for mathematics teaching and learning. The NCTM standards have been a springboard for state and local efforts to focus and improve mathematics education.[7]

Despite considerable energy devoted to educational improvement, achievement in mathematics has shown only modest gains since 1983.[8] The TIMSS results show little change in eighth-grade mathematics achievement between 1995 and 1999. In 1999, the U.S. eighth graders performed significantly above the TIMSS international average in mathematics, but about in the middle of the achievement distribution of the 38 participating countries (above 17 countries, similar to 6, and below 14). In TIMSS 1999, the world class performance levels in mathematics were set essentially by five Asian countries. Singapore, the Republic of Korea, Chinese Taipei, and Hong Kong SAR had the highest average performance, with Singapore and Korea having significantly higher achievement than all other participating countries. Japan, the fifth, also performed very well, as did Belgium (Flemish)[9] (see Exhibits 1.1 and 1.2 in Chapter 1).

7 Kelly, D.L., Mullis, I.V.S., and Martin, M.O. (2000), *Profiles of Student Achievement in Mathematics at the TIMSS International Benchmarks: U.S. Performance and Standards in an International Context*, Chestnut Hill, MA: Boston College.

8 Campbell, J.R., Hombo, C.M., and Mazzeo, J. (2000), *NAEP 1999 Trends in Academic Progress: Three Decades of Student Performance*, NCES 2000-469, Washington, DC: National Center for Education Statistics.

9 Belgium has two separate educational systems, Flemish and French. The Flemish system participated in TIMSS 1999.

Major Findings from the TIMSS 1999 Benchmarking Study

- Average mathematics performance for the 13 Benchmarking states was clustered in the middle of the international distribution of results for the 38 countries. All of the Benchmarking states performed either significantly above or similar to the international average, yet significantly below the high-performing Asian countries.

- The Benchmarking Study underscores the extreme importance of looking beyond the averages to the range of performance found across the nation. Performance across the participating school districts and consortia reflected nearly the full range of achievement internationally. Although achievement was not as high as Singapore, Korea, and Chinese Taipei, the top-performing Benchmarking jurisdictions of the Naperville School District and the First in the World Consortium (both in Illinois) performed similarly to Hong Kong, Japan, Belgium (Flemish), and the Netherlands. At the other end of the continuum, urban districts with high percentages of students from low-income families, such as the Chicago Public Schools, the Rochester City School District, and the Miami-Dade County Public Schools, performed more similarly to lower-performing countries such as Thailand, Macedonia, and Iran, respectively, but significantly higher than the lowest-scoring countries.

- The TIMSS 1999 Benchmarking Study provides evidence that some schools in the U.S. are among the best in the world, but that a world-class education is not available to all children across the nation. The TIMSS index of home educational resources (based on books in the home, availability of study aids, and parents' education level) shows that students with more home resources have higher mathematics achievement. Furthermore, the Benchmarking jurisdictions with the greatest percentages of students with high levels of home resources were among the top-performing jurisdictions, and those with the lowest achievement were four urban districts that also had the lowest percentages of students with high levels of home resources. These and other TIMSS 1999 Benchmarking results support research indicating that students in urban districts with a high proportion of low-income families and minorities often attend schools with fewer resources than in non-urban districts, including less experienced teachers, fewer appropriate instructional materials, more emphasis on lower-level content, less access to gifted and talented programs, higher absenteeism, more inadequate buildings, and more discipline problems.

- It is good news that in mathematics at the eighth grade, the TIMSS 1999 Benchmarking Study shows relatively equivalent average achievement for girls and boys in each of the Benchmarking jurisdictions. This follows the national and international pattern where the United States was one of 34 countries in 1999 with girls and boys performing similarly.

- Of the five mathematics content areas assessed by TIMSS, U.S. eighth graders performed higher than the international average in fractions and number sense; data representation, analysis, and probability; and algebra; but only at the international average in measurement and geometry. Despite the major differences among the Benchmarking participants geographically, economically, and culturally, most to some extent followed the national pattern. It will be important, however, for each participant to determine its specific relative strengths and weaknesses in mathematics achievement.

- The Benchmarking results indicate that students' relatively lower achievement in geometry is most likely related to less coverage of geometry topics in mathematics classrooms. Teachers also expressed the least confidence in their preparation to teach geometry.

- The content area emphasis differed dramatically from jurisdiction to jurisdiction, however. For example, teachers in Naperville reported emphasizing algebra for nearly all their students (91 percent), and those in the Academy School District, the Michigan Invitational Group, and Montgomery County for about half. In contrast, about 70 percent of the students in Jersey City and Rochester received a combined emphasis on algebra, geometry, number, etc., and nearly half the students in Chicago had an emphasis mainly on number.

- Research shows that higher achievement in mathematics is associated with teachers having a bachelor's and/or master's degree in mathematics.[10] According to their teachers, however, U.S. eighth-grade students were less likely than those in other countries to be taught mathematics by teachers with a major area of study in mathematics (41 percent in the U.S. compared with 71 percent internationally, on average). Among the Benchmarking jurisdictions, the percentages of students taught by teachers with mathematics as a major area of study varied dramatically from 70 to 73 percent in the First in the World Consortium, Naperville, and Rochester, to less than one-quarter in the Delaware Science Coalition and Jersey City.

10 Goldhaber, D.D. and Brewer, D.J. (1997), "Evaluating the Effect of Teacher Degree Level on Educational Performance" in W. Fowler (ed.), *Developments in School Finance, 1996*, NCES 97-535, Washington DC: National Center for Education Statistics; Darling-Hammond, L. (2000), *Teacher Quality and Student Achievement: A Review of State Policy Evidence*, Education Policy Analysis Archives, 8(1).

In general, teachers in many Benchmarking entities and in the United States overall may be overconfident about their preparation to teach eighth-grade mathematics. More teachers in the Benchmarking jurisdictions and in the U.S. nationally reported feeling very well prepared to teach mathematics compared with their counterparts in other countries. In half the Benchmarking jurisdictions, 90 percent of the students had teachers who felt "very well prepared" to teach across a range of 12 general mathematics topics covered by TIMSS. Across the Benchmarking entities, the smallest percentage of students with teachers highly confident in their preparation to teach mathematics was 75 percent, which was higher than the international average of 63 percent. The comparable figure for the U.S. was 87 percent.

- Since entering teachers make up a relatively small percentage of the teaching force, improving teacher quality depends on providing opportunities for professional development. Across the Benchmarking participants, there was considerable variation in the type of professional development that teachers engaged in. For example, only in the First in the World Consortium and Montgomery County did more than half the students have mathematics teachers who reported both observing and being observed by other teachers. In many of the Benchmarking entities, half or more of the students had teachers who reported that their professional development activities emphasized curriculum, but only about one-quarter had teachers who reported that their professional development activities emphasized content knowledge.

- The choices teachers make determine, to a large extent, what students learn. In effective teaching, worthwhile mathematical problems are used to introduce important ideas and engage students' thinking. The Benchmarking results show that higher achievement is related to the emphasis that teachers place on reasoning and problem-solving activities. This finding is consistent with the video study component of TIMSS conducted in 1995.[11] Analyses of videotapes of mathematics classes revealed that in the typical mathematics lesson in Japan students worked on developing solution procedures to report to the class that were often expected to be original constructions. In contrast, in the typical U.S. lesson students essentially practiced procedures that had been demonstrated by the teacher.

[11] Stigler, J.W., Gonzales, P., Kawanaka, T., Knoll S., and Serrano, A. (1999), *The TIMSS Videotape Classroom Study: Methods and Findings from an Exploratory Research Project on Eighth-Grade Mathematics Instruction in Germany, Japan, and the United States,* NCES 1999-074, Washington, DC: National Center for Education Statistics.

- In TIMSS 1999, about half the Japanese students had teachers who reported a high degree of emphasis on reasoning activities in their mathematics classes, more than in any other country. The degree of emphasis on reasoning and problem-solving varied dramatically among Benchmarking participants. At the top end, between 41 and 46 percent of the students in Jersey City, the First in the World Consortium, and the Michigan Invitational Group had teachers who reported a high degree of emphasis on mathematics reasoning and problem-solving. Oregon and Chicago had the smallest percentages of students (eight and nine percent, respectively) with teachers reporting this degree of emphasis.

- In general, the TIMSS 1999 data reveal that in most mathematics classes teachers do not focus on mathematics reasoning. Just as was found in the 1995 videotapes, it appears that usually the teacher states the problem, demonstrates the solution, and then asks the students to practice. Ninety-four percent of U.S. eighth graders reported that their teachers showed them how to do mathematics problems almost always or pretty often during mathematics lessons, and 86 percent reported working from worksheets or textbooks on their own this frequently. According to U.S. mathematics teachers, class time is spent as follows: 15 percent on homework review; 20 percent on lecture style teacher presentation; 35 percent on teacher-guided or independent student practice; 12 percent on re-teaching and clarification; 11 percent on tests and quizzes, six percent on administrative tasks; and four percent on other activities.

- The TIMSS 1999 data indicate that the instructional time for learning mathematics, beyond being spent primarily on demonstrations of procedures and repeated practice, becomes further eroded by non-instructional tasks. In Japan and Korea, more than half the students were in classes that never had interruptions for announcements or administrative tasks. Among the Benchmarking participants, the results ranged from 22 percent of the eighth graders in such classes in Naperville to only five percent in Jersey City. Also, 74 percent of the U.S. students reported that they began their mathematics homework during class almost always or pretty often, well above the international average of 42 percent. In most Benchmarking jurisdictions, the results followed the national pattern, although the percentage varied from 43 to 90 percent.

- The Benchmarking Study shows that students in schools that are well-resourced have higher mathematics achievement. Among the Benchmarking participants, three-fourths or more of the students in the Academy School District, the First in the World Consortium, and Naperville were in schools where the capacity to provide mathematics instruction was largely unaffected by shortages or inadequacies in instructional materials, supplies, buildings, space, computers and computer software, calculators, library materials and audio-visual resources. These high percentages exceeded those of all the TIMSS 1999 countries, with the highest percentages (about 50 percent) reported by Belgium (Flemish), Singapore, and the Czech Republic.

- Discipline that maintains a safe and orderly atmosphere conducive to learning is very important to school quality, and research indicates that urban schools have conditions less conducive to learning than non-urban schools.[12] For example, urban schools report more crime against students and teachers at school and that physical conflict among students is a serious or moderate problem. Among the Benchmarking participants there was considerable variation in principals' reports about the seriousness of a variety of potential discipline problems. In several of the urban districts, however, 10 percent or more of the students were in schools where absenteeism, classroom disturbances, and physical injury to students were felt to be serious problems. Also in several of these districts, 20 percent or more of the students were in schools where intimidation or verbal abuse among students was a serious problem.

12 Mayer, D.P., Mullens, J.E., and Moore, M.T. (2000), *Monitoring School Quality: An Indicators Report*, NCES 2001-030, Washington, DC: National Center for Education Statistics; Kaufman, P., Chen, X., Choy, S.P., Ruddy, S.A., Miller, A.K., Fleury, J.K., Chandler, K.A., Rand, M.R., Klaus, P., and Planty, M.G. (2000), *Indicators of School Crime and Safety, 2000*, NCES 2001-017/NCJ-184176, Washington, DC: U.S. Departments of Education and Justice.

Among the 27 participants in the TIMSS 1999 Benchmarking Study, there was particularly extreme variation in mathematics achievement among the school districts and consortia, but less among the states. Several districts in relatively wealthy communities had comparatively high achievement in mathematics, while others in urban areas with high percentages of students from low-income families had relatively low achievement, compared with the TIMSS 1999 results internationally. Regardless of its performance, however, each state, district, and consortium now has a better idea of the challenges ahead and access to a rich array of data about various facets of its educational system. The TIMSS 1999 data provide an excellent basis for examining how best to move from developing a curriculum framework or standards in mathematics to meeting the extraordinary challenge of actually implementing the standards in schools and classrooms often characterized by considerable cultural, social, and experiential diversity.

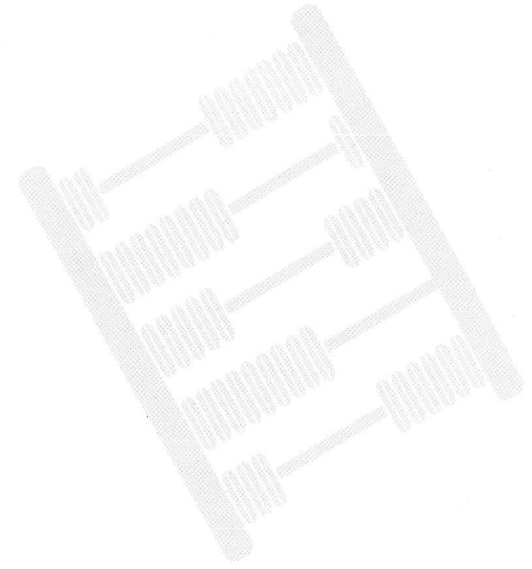

1 Introduction

TIMSS 1999 BENCHMARKING

I

Over the last decade, many states and school districts have created content and performance standards targeted at improving students' achievement in mathematics and science. In mathematics, in particular, most states are in the process of updating and revising their standards. All states except Iowa (which as a matter of policy publishes no state standards) now have content or curriculum standards in mathematics, and many educational jurisdictions have worked successfully to improve their initial standards in clarity and content.[1] Much of this effort has been based on work done at the national level during this period to develop standards aimed at increasing the mathematics competencies of all students. Since 1989, when the National Council of Teachers of Mathematics (NCTM) published *Curriculum and Education Standards for School Mathematics,* the mathematics education community has had the benefit of a unified set of goals for mathematics teaching and learning. The NCTM standards have been a springboard for state and local efforts to focus and improve mathematics education.[2]

Particularly during the past decade, there has been an enormous amount of energy expended in states and school districts not only on developing mathematics content standards but also on improving teacher quality and school environments as well as on developing assessments and accountability measures.[3] Participating in an international assessment provides states and school districts a global context for evaluating the success of their policies and practices aimed at raising students' academic achievement.

What Is TIMSS 1999 Benchmarking?

TIMSS 1999, a successor to the 1995 Third International Mathematics and Science Study (TIMSS), focused on the mathematics and science achievement of eighth-grade students. Thirty-eight countries including the United States participated in TIMSS 1999 (also known as TIMSS-Repeat or TIMSS-R). Even more significantly for the United States, however, TIMSS 1999 included a voluntary Benchmarking Study. Participation in the TIMSS 1999 Benchmarking Study at the eighth grade provided states, districts, and consortia an unprecedented opportunity to assess the comparative international standing of their students' achievement and evaluate their mathematics and science programs in an international context. Participants were also able to

1 Raimi, R.A. (2000), "The State of State Standards in Mathematics" in C.E. Finn and M.J. Petrilli (eds.), *The State of State Standards,* Washington, DC: Thomas B. Fordham Foundation; Glidden, H. (1999), *Making Standards Matter 1999,* Washington, DC: American Federation of Teachers.

2 Kelly, D.L., Mullis, I.V.S., and Martin, M.O. (2000), *Profiles of Student Achievement in Mathematics at the TIMSS International Benchmarks: U.S. Performance and Standards in an International Context,* Chestnut Hill, MA: Boston College.

3 Orlofsky, G.F. and Olson, L. (2001), "The State of the States" in *Quality Counts 2001, A Better Balance: Standards, Tests, and the Tools to Succeed,* Education Week, 20(17).

compare their achievement with that of the United States as a whole, and in the cases where they both participated, school districts could compare with the performance of their states.

Originally conducted in 1994-1995,[4] TIMSS compared the mathematics and science achievement of students in 41 countries at five grade levels. Using questionnaires, videotapes, and analyses of curriculum materials, TIMSS also investigated the contexts for learning mathematics and science in the participating countries. TIMSS results, which were first reported in 1996, have stirred debate, spurred reform efforts, and provided important information to educators and decision makers around the world. The findings from TIMSS 1999, a follow-up to the earlier study, add to the richness of the TIMSS data and their potential to have an impact on policy and practice in mathematics and science teaching and learning.

Twenty-seven jurisdictions from all across the nation, including 13 states and 14 districts or consortia, participated in the Benchmarking Study (see Exhibit 1). To conduct the Benchmarking Study, the TIMSS 1999 assessments were administered to representative samples of eighth-grade students in each of the participating districts and states in the spring of 1999, at the same time and following the same guidelines as those established for the 38 countries.

In addition to testing achievement in mathematics and science, the TIMSS 1999 Benchmarking Study involved administering a broad array of questionnaires. TIMSS collected extensive information from students, teachers, and school principals as well as system-level information from each participating entity about mathematics and science curricula, instruction, home contexts, and school characteristics and policies. The TIMSS data provide an abundance of information making it possible to analyze differences in current levels of performance in relation to a wide variety of factors associated with classroom, school, and national contexts within which education takes place.

Why Did Countries, States, Districts, and Consortia Participate?

The decision to participate in any cycle of TIMSS is made by each country according to its own data needs and resources. Similarly, the states, districts, and consortia that participated in the Benchmarking Study decided to do so for various reasons.

4 TIMSS was administered in the spring of 1995 in northern hemisphere countries and in the fall of 1994 in southern hemisphere countries, both at the end of the school year.

Primarily, the Benchmarking participants are interested in building educational capacity and looking at their own situations in an international context as a way of improving mathematics and science teaching and learning in their jurisdictions. International assessments provide an excellent basis for gaining multiple perspectives on educational issues and examining a variety of possible reasons for observed differences in achievement. While TIMSS helps to measure progress towards learning goals in mathematics and science, it is much more than an educational Olympics. It is a tool to help examine such questions as:

- How demanding are our curricula and expectations for student learning?
- Is our classroom instruction effective? Is the time provided for instruction being used efficiently?
- Are our teachers well prepared to teach mathematics concepts? Can they help students understand mathematics?
- Do our schools provide an environment that is safe and conducive to learning?

Unlike in many countries around the world where educational decision making is highly centralized, in the United States the opportunities to learn mathematics derive from an educational system that operates through states and districts, allocating opportunities through schools and then through classrooms. Improving students' opportunities to learn requires examining every step of the educational system, including the curriculum, teacher quality, availability and appropriateness of resources, student motivation, instructional effectiveness, parental support, and school safety.

Which Countries, States, Districts, and Consortia Participated?

Exhibit 1 shows the 38 countries, 13 states, and the 14 districts and consortia that participated in TIMSS 1999 and the Benchmarking Study.

The consortia consist of groups of entire school districts or individual schools from several districts that organized together either to participate in the Benchmarking Study or to collaborate across a range of educational issues. Descriptions of the consortia that participated in the project follow.

Delaware Science Coalition. The Delaware Science Coalition (DSC) is a coalition of 15 school districts working in partnership with the Delaware Department of Education and the business-based Delaware Foundation for Science and Mathematics Education. The mission of the DSC is to improve the teaching and learning of science for all students in grades K-8. The Coalition includes more that 2,200 teachers who serve more than 90 percent of Delaware's public school students.

First in the World Consortium. The First in the World Consortium consists of a group of 18 districts from the North Shore of Chicago that have joined forces to bring a world-class education to the region's students and to improve mathematics and science achievement in their schools. Resulting from meetings of district superintendents in 1995, the consortium decided to focus on three main goals: benchmarking their performance to educational standards through participating in the original TIMSS in 1996 and again in 1999; creating a forum to share the vision with businesses and the community of benchmarking to world-class standards; and establishing a network of learning communities of teachers, researchers, parents, and community members to conduct the work needed to achieve their goal.

Fremont/Lincoln/Westside Public Schools. The Fremont/Lincoln/Westside consortium is comprised of three public school districts in Nebraska. These districts joined together specifically to participate in the TIMSS 1999 Benchmarking Study.

Michigan Invitational Group. The Michigan Invitational Group is a heterogeneous and socioeconomically diverse group composed of urban, suburban, and rural schools across Michigan. Schools invited to participate as part of this consortia were those that were using National Science Foundation (NSF) materials, well-developed curricula, and provided staff development to teachers.

Project SMART Consortium. SMART (Science & Mathematics Achievement Required For Tomorrow) is a consortium of 30 diverse school districts in northeast Ohio committed to continuous improvement, long term systemic change, and improved student learning in science and mathematics in grades K-12. It is jointly funded by the Ohio Department of Education and the Martha Holden Jennings Foundation. The schools that participated in the project represent 17 of the 30 districts.

Southwest Pennsylvania Math and Science Collaborative. The Southwest Pennsylvania Math and Science Collaborative, established in 1994, coordinates efforts and focuses resources on strengthening math and science education in the entire southwest Pennsylvania workforce region that has Pittsburgh as its center. Committed to gathering and using good information that can help prepare its students to be productive citizens, the Collaborative is composed of all 118 "local control" public districts, as well as the parochial and private schools in the nine-county region. Several of these districts are working together in selecting exemplary materials, developing curriculum frameworks, and building sustained professional development strategies to strengthen math and science instruction.

ISC
TIMSS 1999
Benchmarking
Boston College

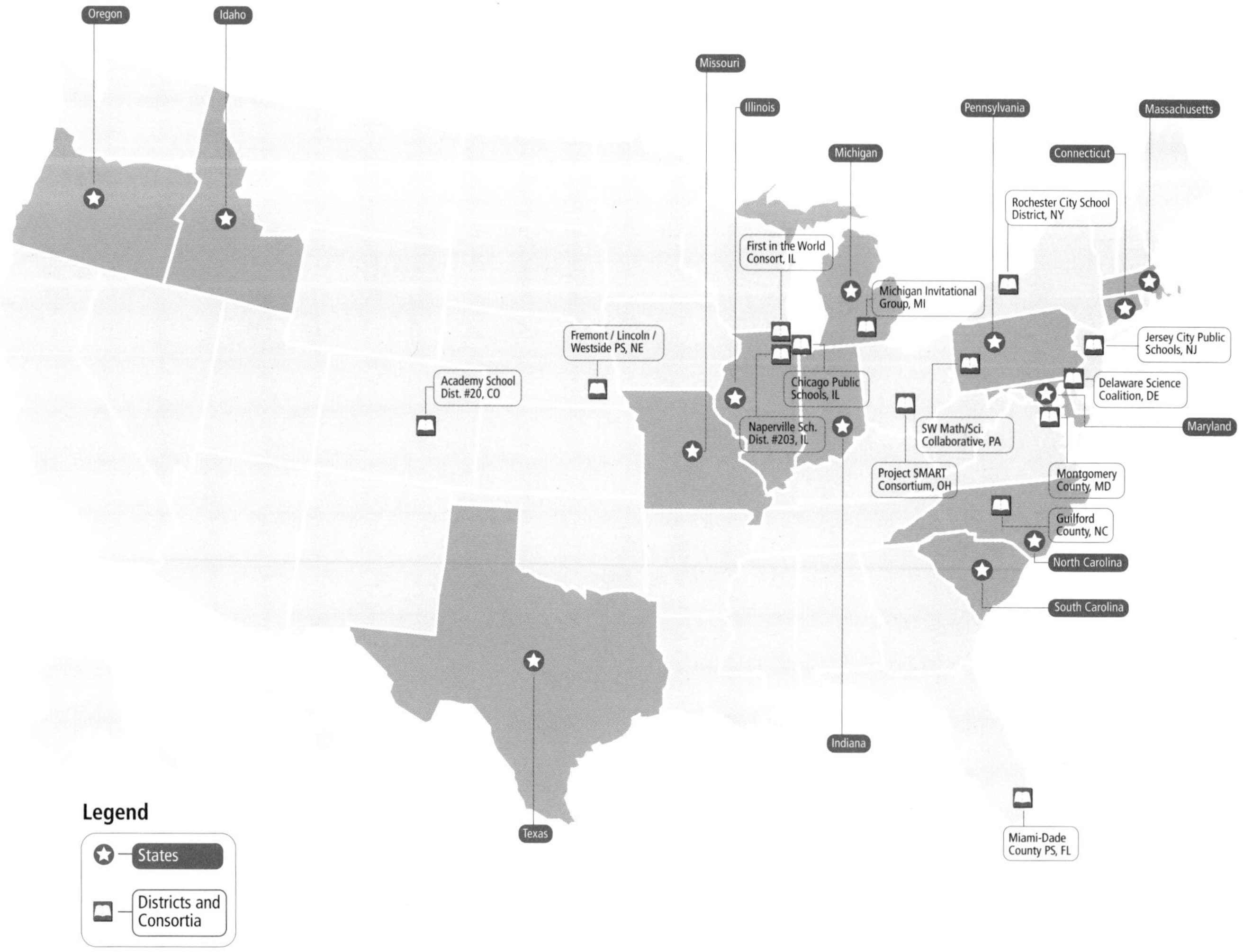

Oregon
Idaho
Missouri
Illinois
Michigan
Pennsylvania
Massachusetts
Connecticut
Rochester City School District, NY
First in the World Consort, IL
Michigan Invitational Group, MI
Fremont / Lincoln / Westside PS, NE
Jersey City Public Schools, NJ
Academy School Dist. #20, CO
Chicago Public Schools, IL
Delaware Science Coalition, DE
Naperville Sch. Dist. #203, IL
SW Math/Sci. Collaborative, PA
Maryland
Project SMART Consortium, OH
Montgomery County, MD
Guilford County, NC
North Carolina
South Carolina
Indiana
Texas
Miami-Dade County PS, FL
Legend
States
Districts and Consortia

States

Connecticut
Idaho
Illinois
Indiana
Maryland
Massachusetts
Michigan
Missouri
North Carolina
Oregon
Pennsylvania
South Carolina
Texas

Districts and Consortia

Academy School District #20, Colorado Springs, CO

Chicago Public Schools, IL

Delaware Science Coalition, DE

First in the World Consortium, IL

Fremont/Lincoln/Westside Public Schools, NE

Guilford County, NC

Jersey City Public Schools, NJ

Miami-Dade County Public Schools, FL

Michigan Invitational Group, MI

Montgomery County, MD

Naperville Community Unit School District #203, IL

Project SMART Consortium, OH

Rochester City School District, NY

Southwest Pennsylvania Math and Science Collaborative, PA

Countries

Australia
Belgium (Flemish)
Bulgaria
Canada
Chile
Chinese Taipei
Cyprus
Czech Republic
England
Finland
Hong Kong, SAR
Hungary
Indonesia
Iran, Islamic Republic
Israel
Italy
Japan
Jordan
Korea, Republic of
Latvia (LSS)
Lithuania
Macedonia, Republic of
Malaysia
Moldova
Morocco
Netherlands
New Zealand
Philippines
Romania
Russian Federation
Singapore
Slovak Republic
Slovenia
South Africa
Thailand
Tunisia
Turkey
United States

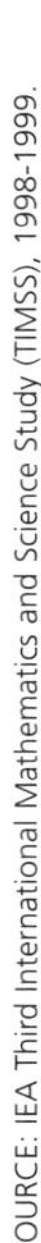

Countries Participating in TIMSS 1999

What Is the Relationship Between the TIMSS 1999 Data for the United States and the Data for the Benchmarking Study?

The results for the 38 countries participating in TIMSS 1999, including those for the United States, were reported in December 2000 in two companion reports – the *TIMSS 1999 International Mathematics Report* and the *TIMSS 1999 International Science Report.*[5] Performance in the United States relative to that of other nations was reported by the U.S. National Center for Education Statistics in *Pursuing Excellence.*[6] The results for the United States in those reports, as well as in this volume and its companion science report,[7] were based on a nationally representative sample of eighth-grade students drawn in accordance with TIMSS guidelines for all participating countries.

Because having valid and efficient samples in each country is crucial to the quality and integrity of TIMSS, procedures and guidelines have been developed to ensure that the national samples are of the highest quality possible. Following the TIMSS guidelines, representative samples were also drawn for the Benchmarking entities. Sampling statisticians at Westat, the organization responsible for sampling and data collection for the United States, worked in accordance with TIMSS standards to design procedures that would coordinate the assessment of separate representative samples of students within each Benchmarking entity.

For the most part, the U.S. TIMSS 1999 national sample was separate from the students assessed in each of the Benchmarking jurisdictions. Each Benchmarking participant had its own sample to provide comparisons with each of the TIMSS 1999 countries including the United States. In general, the Benchmarking samples were drawn in accordance with the TIMSS standards, and achievement results can be compared with confidence. Deviations from the guidelines are noted in the exhibits in the reports. The TIMSS 1999 sampling requirements and the outcomes of the sampling procedures for the participating countries and Benchmarking jurisdictions are described in Appendix A. Although taken collectively the Benchmarking participants are not representative of the United States, the effort was substantial in scope involving approximately 1,000 schools, 4,000 teachers, and 50,000 students.

5 Mullis, I.V.S., Martin, M.O., Gonzalez, E.J., Gregory, K.D., Garden, R.A., O'Connor, K.M., Chrostowski, S.J., and Smith, T.A. (2000), *TIMSS 1999 International Mathematics Report: Findings from IEA's Repeat of the Third International Mathematics and Science Study at the Eighth Grade*, Chestnut Hill, MA: Boston College; Martin, M.O., Mullis, I.V.S., Gonzalez, E.J., Gregory, K.D., Smith, T.A., Chrostowski, S.J., Garden, R.A., and O'Connor, K.M. (2000), *TIMSS 1999 International Science Report: Findings from IEA's Repeat of the Third International Mathematics and Science Study at the Eighth Grade*, Chestnut Hill, MA: Boston College.

6 Gonzales, P., Calsyn, C., Jocelyn, L., Mak, K., Kastberg, D., Arafeh, S., Williams, T., and Tsen, W. (2000), *Pursuing Excellence: Comparisons of International Eighth-Grade Mathematics and Science Achievement from a U.S. Perspective, 1995 and 1999*, NCES 2001-028, Washington, DC: National Center for Education Statistics.

7 Martin, M.O., Mullis, I.V.S., Gonzalez, E.J., O'Connor, K.M., Chrostowski, S.J., Gregory, K.D., Smith, T.A., and Garden, R.A. (2001), *Science Benchmarking Report, TIMSS 1999 – Eighth Grade: Achievement for U.S. States and Districts in an International Context*, Chestnut Hill, MA: Boston College.

How Was the TIMSS 1999 Benchmarking Study Conducted?

The TIMSS 1999 Benchmarking Study was a shared venture. In conjunction with the Office of Educational Research and Improvement (OERI) and the National Science Foundation (NSF), the National Center for Education Statistics (NCES) worked with the International Study Center at Boston College to develop the study. Each participating jurisdiction invested valuable resources in the effort, primarily for data collection including the costs of administering the assessments at the same time and using identical procedures as for TIMSS in the United States. Many participants have also devoted considerable resources to team building as well as to staff development to facilitate use of the TIMSS 1999 results as an effective tool for school improvement.

The TIMSS studies are conducted under the auspices of the International Association for the Evaluation of Educational Achievement (IEA), an independent cooperative of national and governmental research agencies with a permanent secretariat based in Amsterdam, the Netherlands. Its primary purpose is to conduct large-scale comparative studies of educational achievement to gain a deeper understanding of the effects of policies and practices within and across systems of education.

TIMSS is part of a regular cycle of international assessments of mathematics and science that are planned to chart trends in achievement over time, much like the regular cycle of national assessments in the U.S. conducted by the National Assessment of Educational Progress (NAEP). Work has begun on TIMSS 2003, and a regular cycle of studies is planned for the years beyond.

The IEA delegated responsibility for the overall direction and management of TIMSS 1999 to the International Study Center in the Lynch School of Education at Boston College, headed by Michael O. Martin and Ina V.S. Mullis. In carrying out the project, the International Study Center worked closely with the IEA Secretariat, Statistics Canada in Ottawa, the IEA Data Processing Center in Hamburg, Germany, and Educational Testing Service in Princeton, New Jersey. Westat in Rockville, Maryland, was responsible for sampling and data collection for the Benchmarking Study as well as the U.S. component of TIMSS 1999 so that procedures would be coordinated and comparable.

Funding for TIMSS 1999 was provided by the United States, the World Bank, and the participating countries. Within the United States, funding agencies included NCES, NSF, and OERI, the same group of organizations supporting major components of the TIMSS 1999 Benchmarking Study for states, districts, and consortia, including overall coordination as well as data analysis, reporting, and dissemination.

What Was the Nature of the Mathematics Test?

The TIMSS curriculum frameworks developed for 1995 were also used for 1999. They describe the content dimensions for the TIMSS tests as well as the performance expectations (behaviors that might be expected of students in school mathematics).[8] Five content areas were covered in the TIMSS 1999 mathematics test. These areas and the percentage of the test items devoted to each are fractions and number sense (38 percent), measurement (15 percent), data representation, analysis, and probability (13 percent), geometry (13 percent), and algebra (22 percent). The performance expectations include knowing (19 percent), using routine procedures (23 percent), using complex procedures (24 percent), investigating and solving problems (31 percent), and communicating and reasoning (two percent).

The test items were developed through a cooperative and iterative process involving the National Research Coordinators (NRCs) of the participating countries. All of the items were reviewed thoroughly by subject matter experts and field tested. Nearly all the TIMSS 1999 countries participated in field testing with nationally representative samples, and the NRCs had several opportunities to review the items and scoring criteria. The TIMSS 1999 mathematics test contained 162 items representing a range of mathematics topics and skills.

About one-fourth of the questions were in the free-response format, requiring students to generate and write their answers. These questions, some of which required extended responses, were allotted about one-third of the testing time. Responses to the free-response questions were evaluated to capture diagnostic information, and some were scored using procedures that permitted partial credit. Chapter 2 of this report contains 16 example items illustrating the range of mathematics concepts and processes covered in the TIMSS 1999 test. Appendix D contains descriptions of the topics and skills assessed by each item.

8 Robitaille, D.F., McKnight, C.C., Schmidt, W.H., Britton, E.D., Raisen, S.A., and Nicol, C. (1993), *TIMSS Monograph No. 1: Curriculum Frameworks for Mathematics and Science*, Vancouver, BC: Pacific Educational Press.

Testing was designed so that no one student took all the items, which would have required more than three hours of testing time. Instead, the test was assembled in eight booklets, each requiring 90 minutes to complete. Each student took only one booklet, and the items were rotated through the booklets so that each item was answered by a representative sample of students.

How Does TIMSS 1999 Compare with NAEP?

The National Assessment of Educational Progress (NAEP) is an ongoing program that has reported the mathematics achievement of U.S. students for some 30 years. TIMSS and NAEP were designed to serve different purposes, and this is evident in the types of assessment items as well as the content areas and topics covered in each assessment. TIMSS and NAEP both assess students at the eighth grade, and both tend to focus on mathematics as it is generally presented in classrooms and textbooks. However, TIMSS is based on the curricula that students in the participating countries are likely to have encountered by the eighth grade, while NAEP is based on an expert consensus of what students in the United States should know and be able to do in mathematics and other academic subjects at that grade. For example, TIMSS 1999 appears to place more emphasis on number sense, properties, and operations than NAEP. NAEP appears to distribute its focus more equally across the content areas included in the assessment frameworks.[9]

Whereas NAEP is designed to provide comparisons among and between states and the nation as a whole, the major purpose of the TIMSS 1999 Benchmarking Study was to provide entities in the United States with a way to compare their achievement and instructional programs in an international context. Thus, the point of comparison or "benchmark" consists primarily of the high-performing TIMSS 1999 countries. The sample sizes were designed to place participants near the top, middle, or bottom of the TIMSS continuum of performance internationally, but not necessarily to detect differences in performance among different Benchmarking participants. For example, all 13 of the participating states performed similarly in mathematics in relation to the TIMSS countries – near the middle. As findings from the NAEP assessment in 2000 are released, it is important to understand the differences and similarities in the assessments to be able to make sense of the findings in relation to each other.

9 Nohara, D. (working paper 2001), *A Comparison of Three Educational Assessments: NAEP, TIMSS-R, and PISA*, Washington, DC: National Center for Education Statistics.

How Do Country Characteristics Differ?

International studies of student achievement provide valuable comparative information about student performance, instructional practice, and curriculum. Accompanying the benefits of international studies, though, are challenges associated with making comparisons across countries, cultures, and languages. TIMSS attends to these issues through careful planning and documentation, cooperation among the participating countries, standardized procedures, and rigorous attention to quality control throughout.[10]

It is extremely important, nevertheless, to consider the TIMSS 1999 results in light of countrywide demographic and economic factors. Some selected demographic characteristics of the TIMSS 1999 countries are presented in Exhibit 2. Countries ranged widely in population, from almost 270 million in the United States to less than one million in Cyprus, and in size, from almost 17 million square kilometers in the Russian Federation to less than one thousand in Hong Kong SAR and Singapore. Countries also varied widely on indicators of health, such as life expectancy at birth and infant mortality rate, and of literacy, including adult literacy rate and daily newspaper circulation. Exhibit 3 shows information for selected economic indicators, such as gross national product (GNP) per capita, expenditure on education and research, and development aid. The data reveal that there is great disparity in the economic resources available to participating countries.

10 Appendix A contains an overview of the procedures used. More detailed information is provided in Martin, M.O., Gregory, K.A., and Stemler, S.E., eds., (2000), *TIMSS 1999 Technical Report*, Chestnut Hill, MA: Boston College.

Exhibit 2 Selected Characteristics of TIMSS 1999 Countries

	Population Size (in millions)[1]	Area of Country (1000 square kilometers)[2]	Life Expectancy at Birth[3]	Infant Mortality Rate (per 1000 live births)[4]	Adult Literacy Rate (%)[5]	Daily Newspaper Circulation (per 1000)[6]
United States	267.6	9159	76	7	99.0	212
Australia	18.5	7682	78	5	99.0	296
Belgium (Flemish) [7]	10.2	33	77	6	99.0	161
Bulgaria	8.3	111	71	18	98.2	254
Canada	30.3	9221	79	6	99.0	158
Chile	14.6	749	75	11	95.2	98
Chinese Taipei [8]	22.1	36	75	8	–	–
Cyprus [9]	0.8	9	–	6	95.9	111
Czech Republic	10.3	77	74	6	99.0	254
England [10]	50.0	130	–	–	99.0	–
Finland	5.1	305	77	4	99.0	455
Hong Kong	6.5	1	79	5	92.4	786
Hungary	10.2	92	71	10	99.0	186
Indonesia	200.4	1812	65	47	85.0	23
Iran, Islamic Rep.	60.9	1622	69	32	73.3	26
Israel [11]	6.1	21	78	7	95.4	288
Italy	57.5	294	78	5	98.3	104
Japan	126.1	377	80	4	99.0	578
Jordan	4.4	89	71	29	87.2	42
Korea, Rep.	46.0	99	72	9	97.2	394
Latvia	2.5	62	69	15	99.0	247
Lithuania	3.7	65	71	10	99.0	93
Macedonia	2.0	25	72	16	94.0	21
Malaysia	21.7	329	72	11	85.7	163
Moldova	4.3	33	67	20	98.3	60
Morocco [12]	27.3	711	67	51	45.9	27
Netherlands	15.6	34	78	5	99.0	306
New Zealand	3.8	268	77	7	99.0	216
Philippines	73.5	298	68	35	94.6	82
Romania	22.6	230	69	22	97.8	298
Russian Federation	147.3	16889	67	17	99.0	105
Singapore	3.1	1	76	4	91.4	324
Slovak Republic	5.4	48	73	9	99.0	184
Slovenia	2.0	20	75	5	99.0	199
South Africa	40.6	1221	65	48	84.0	34
Thailand	60.6	511	69	33	94.7	64
Tunisia	9.2	155	70	30	67.0	31
Turkey [13]	62.5	815	69	40	83.2	110

SOURCE: IEA Third International Mathematics and Science Study (TIMSS), 1998-1999.

1 Estimates for 1997 based, in most cases, on a de facto definition. Refugees not permanently settled in the country of asylum are generally considered to be part of their country of origin. World Bank (1999) World Development Indicators, p. 42-44.

2 Area is the total surface area in square kilometers, comprising all land area and inland waters. World Bank (1999) World Development Indicators, p. 120-122

3 Number of years a newborn infant would live if prevailing patterns of mortality at its birth were to stay the same throughout its life. World Bank (1999) World Development Indicators, p. 110-112.

4 Infant mortality rate is the number of deaths of infants under one year of age during 1997 per 1,000 live births in the same year. World Bank (1999) World Development Indicators, p.16-18.

5 Population aged 15 years and over. UNDP (1999) Human Development Report 1999 (134-137).

6 A newspaper issued at least four times a week is considered to be a daily newspaper. Circulation figures show the average circulation. UNESCO (1999) Statistical Yearbook, IV (106-133).

7 Figures for Belgium (Flemish) are for the whole country of Belgium.

8 Data provided by Department of Statistics, Ministry of Interior, Republic of China.

9 Data for population, area, and infant mortality provided by Cypriot Government Statistics

10 The Statesman's Yearbook, 1998-99. Edited by Barry Turner, p.1411.

11 Data provided by Israel's Central Bureau of Statistics, publication no. 1133.

12 Data provided by Ministere du plan et de l'initiation economique: Annuaire de Maroc, 1999.

13 Data provided by Turkey's State Institute of Statistics.

A dash (–) indicates data are not available.

	Gross National Product per Capita (in US dollars)[1]	GNP per Capita (Purchasing Power Parity)[2]	Expenditure on Education as % of Gross National Product[3]	Expenditure on Research and Development as % of Gross National Product[4]	Total Unemployment (% of total labor force)[5]	Aid per Capita[6]
United States	29080	29080	5.4	2.6	5.0	–
Australia	20650	19510	5.5	1.8	8.4	–
Belgium (Flemish) [7]	26730	23090	3.1	1.6	12.7	–
Bulgaria	1170	3870	3.2	0.6	11.1	25
Canada	19640	21750	6.9	1.7	9.4	0
Chile	4820	12240	3.6	0.6	5.3	9
Chinese Taipei [8]	13235	–	4.9	2.0	2.9	–
Cyprus	–	–	4.5	0.2	–	–
Czech Republic	5240	10380	5.1	1.2	3.1	10
England	–	–	–	–	–	–
Finland	24790	19660	7.5	2.8	14.7	–
Hong Kong	25200	24350	2.9	0.3	2.2	–
Hungary	4510	6970	4.6	0.7	10.5	16
Indonesia	1110	3390	1.4	0.1	–	4
Iran, Islamic Rep.	1780	5690	4.0	0.5	–	3
Israel [9]	16180	17680	10.1	2.4	7.7	204
Italy	20170	20100	4.9	2.2	12.1	–
Japan	38160	24400	3.6	2.8	3.2	–
Jordan	1520	3350	7.9	0.3	–	104
Korea, Rep.	10550	13430	3.7	2.8	2.7	-3
Latvia	2430	3970	6.3	0.4	7.0	33
Lithuania	2260	4140	5.5	0.7	7.1	27
Macedonia	1100	3180	5.1	–	38.8	75
Malaysia	4530	7730	4.9	0.2	2.5	-11
Moldova	460	1450	10.6	0.9	1.6	15
Morocco	1260	3210	5.3	–	17.8	17
Netherlands	25830	21300	5.1	2.1	6.2	–
New Zealand	15830	15780	7.3	1.0	6.0	–
Philippines	1200	3670	3.4	0.2	7.9	9
Romania	1410	4270	3.6	0.7	6.3	9
Russian Federation	2680	4280	3.5	0.9	3.4	5
Singapore	32810	29230	3.0	1.1	2.4	0
Slovak Republic	3680	7860	5.0	1.1	12.6	13
Slovenia	9840	11880	5.7	1.5	13.9	49
South Africa	3210	7190	8.0	0.7	–	12
Thailand	2740	6490	4.8	0.1	0.9	10
Tunisia	2110	5050	7.7	0.3	–	21
Turkey	3130	6470	2.2	0.5	6.6	0

SOURCE: IEA Third International Mathematics and Science Study (TIMSS), 1998-1999.

1 World Bank (1999) World Development Indicators, p. 12-14.

2 An international dollar has the same purchasing power over GNP as a U.S. dollar in the United States. World Bank (1999) World Development Indicators, p. 12-14.

3 UNESCO (1999) Statistical Yearbook, p.II-(490-513); Belgium figure is for the Flemish community only; Cyprus is for Greek section only.

4 UNESCO (1999) Statistical Yearbook, p.III-(6-17); Belgium figure is for the Flemish community only; Cyprus is for Greek section only.

5 Unemployment is the share of the labor force that is without work but available for and seeking employment. Definitions of labor force and unemployment differ by country. World Bank (1999) World Development Indicators, p. 58-60.

6 World Bank (1999) World Development Indicators, p. 352-355. Aid per capita includes official development assistance, which consists of disbursement of loans and grants, and official aid, which consists of capital projects, budget and balance of payments support, food and other commodity services, technical co-operation and emergency relief. A negative value indicates repayments exceed aid payments.

7 Figures for Belgium (Flemish) are for the whole country of Belgium.

8 Data provided by Department of Statistics, Ministry of Interior, Republic of China.

9 Data Provided by Israel's Central Bureau of Statistics, publication no. 1133.

A dash (–) indicates data are not available or that aggregates cannot be calculated because of missing data in year shown.

How Do the Benchmarking Jurisdictions Compare on Demographic Indicators?

Together, the indicators in Exhibits 2 and 3 highlight the diversity of the TIMSS 1999 countries. Although the factors the indicators reflect do not necessarily determine high or low performance in mathematics, they do provide a context for considering the challenges involved in the educational task from country to country. Similarly, there was great diversity among the TIMSS 1999 Benchmarking participants. Exhibit 4 presents information about selected characteristics of the states, districts, and consortia that took part in the TIMSS 1999 Benchmarking Study.

As illustrated previously in Exhibit 1, geographically the Benchmarking jurisdictions were from all across the United States, although there was a concentration of east coast participants with six of the states and several of the districts and consortia from the eastern seaboard. Illinois was well represented, by the state as a whole and by three districts or consortia – the Chicago Public Schools, the Naperville School District, and the First in the World Consortium. Several other districts and consortia also had the added benefit of a state comparison – the Michigan Invitational Group and Michigan, Guilford County and North Carolina, Montgomery County and Maryland, and the Southwest Pennsylvania Math and Science Collaborative and Pennsylvania.

As shown in Exhibit 4, demographically the Benchmarking participants varied widely. They ranged greatly in the size of their total public school enrollment, from about 244,000 in Idaho to nearly four million in Texas among states, and from about 11,000 in the Michigan Invitational Group to about 430,000 in the Chicago Public Schools among districts and consortia.

It is extremely important to note that the Benchmarking jurisdictions had widely differing percentages of limited English proficient and minority student populations. They also had widely different percentages of students from low-income families (based on the percentage of students eligible to receive free or reduced-price lunch). Among states, Texas had more than half minority students compared with less than one-fifth in Idaho, Indiana, and Michigan. Among the school districts, those in urban areas had more than four-fifths minority students, including the Chicago Public Schools (89 percent), the Jersey City Public Schools (93 percent), the Miami-Dade County Public Schools (93 percent), and the Rochester City School District (84 percent).

These four districts also had very high percentages of students from low-income families. In comparison, Naperville and the Academy School District had less than one-fifth minority students and less than five percent of their students from low-income families.

Research on disparities between urban and non-urban schools reveals a combination of factors, often interrelated, that all mesh to lessen students' opportunities to learn in urban schools. Students in urban districts with high percentages of low-income families and minorities often attend schools with higher proportions of inexperienced teachers.[11] Urban schools also have fewer qualified teachers than non-urban schools. In reviewing the U.S. Department of Education's 1994 Schools and Staffing Survey, *Education Week* prepared a 1998 study on urban education that found that urban school districts experience greater difficulty filling teacher vacancies, particularly for certain fields including mathematics, and that they are more likely than non-urban schools to hire teachers who have an emergency or temporary license.[12] Studies of under-prepared teachers indicate that such teachers have more difficulty with classroom management, teaching strategies, curriculum development, and student motivation.[13] Teacher absenteeism is also a more serious problem in urban districts. An NCES report on urban schools found they have fewer resources, such as textbooks, supplies, and copy machines, available for their classrooms.[14] It also found that urban students had less access to gifted and talented programs than suburban students. Additionally, several large studies have found urban school facilities to be functionally older and in worse condition than non-urban ones.[15]

[11] Mayer, D.P., Mullens, J.E., and Moore, M.T. (2000), *Monitoring School Quality: An Indicators Report*, NCES 2001-030, Washington, DC: National Center for Education Statistics.

[12] *Quality Counts 1998, The Urban Challenge: Public Education in the 50 States*, Education Week, 17(17).

[13] Darling-Hammond, L. and Post, L. (2000), "Inequality in Teaching and Schooling: Supporting High Quality Teaching and Leadership in Low-Income Schools" in R. Kahlenberg (ed.), *A Notion at Risk: Preserving Public Education as an Engine for Social Mobility*, Century Foundation Press.

[14] Lippman, L., Burns, S., and McArthur, E. (1996), *Urban Schools: The Challenge of Location and Poverty*, NCES 96-184, Washington, DC: National Center for Education Statistics.

[15] Lewis, L., Snow, K., Farris, E., Smerdon, B., Cronen, S., Kaplan, J., and Greene, B. (2000), *Condition of America's Public School Facilities: 1999*, NCES 2000-032, Washington, DC: National Center for Education Statistics; *School Facilities: America's Schools Report Differing Conditions* (1996), GAO/HEHS-96-103, Washington, DC: U.S. General Accounting Office.

Exhibit 4 Selected Characteristics of States, Districts and Consortia*

8th Grade Mathematics

	Total Public Enrollment (All Grades)	Percentage of Students				Per Pupil Expenditure[3]
		Special Needs	Limited English Proficient	Minority[1]	Low Income[2]	
States						
Connecticut	544698	14	4	26	20	8827
Idaho	244722	11	7	17	37	4808
Illinois	2011530	14	6	35	31	6481
Indiana	988094	15	3	17	25	6420
Maryland	841671	13	2	45	28	7412
Massachusetts	962317	18	13	26	28	8064
Michigan	1720266	5	–	18	17	7330
Missouri	912445	14	1	22	34	5663
North Carolina	1254821	13	2	38	44	5367
Oregon	542809	11	7	20	33	6920
Pennsylvania	1816414	11	–	22	30	7409
South Carolina	664592	13	0	37	45	5204
Texas	3945367	12	14	53	48	5567
Districts and Consortia						
Academy School Dist. #20, CO	15821	7	–	18	4	4767
Chicago Public Schools, IL	430914	12	16	89	71	5784
Delaware Science Coalition, DE [4]	19830			37	40	
First in the World Consort., IL	35802	13	8	26	14	8924
Fremont/Lincoln/WestSide PS, NE	40769	15	2	17	23	5915
Guilford County, NC	61154	14	3	43	37	5431
Jersey City Public Schools, NJ	32505	9	–	93	89	9653
Miami-Dade County PS, FL	352536	11	14	93	59	5845
Michigan Invitational Group, MI [4]	10947			12	22	
Montgomery County, MD	127933	12	6	50	25	8223
Naperville Sch. Dist. #203, IL	18473	11	1	18	2	5988
Project SMART Consortium, OH [4]	15266			21	22	
Rochester City Sch. Dist., NY	38121	17	–	84	73	8490
SW Math/Sci. Collaborative, PA [5]	403347	11	–	13	33	6858

SOURCE: IEA Third International Mathematics and Science Study (TIMSS), 1998-1999.

* All data except percent minority and percent low income are from the Common Core of Data (CCD) published by the National Center for Education Statistics (NCES) of the U.S. Department of Education. The nonfiscal data are from School Year 1998-99; the state fiscal data are from Fiscal Year 1997-98, and the district/consortium fiscal data are from Fiscal Year 1996-97. A dash (–) indicates data were not reported to NCES; a blank indicates data are not available for a consortium. All percentages are rounded to the nearest whole number.

1 Percent minority is the percentage of non-white students as reported by participating schools (also shown in Exhibit 4.4, which provides the breakdown by race/ethnicity).

2 Percent low income is the percentage of students eligible to receive free or reduced-price lunch through the National School Lunch Program as of October 1, 1998, as reported by participating schools (also shown in Exhibit 7.1). Because school response data were available for less than 50% of students in Miami-Dade, its low-income figure shown is that reported by the Florida Department of Education's Bureau of Education Information and Accountability Services.

3 Per pupil expenditure is net current expenditures as defined by Hawkins-Stafford Education Amendments of 1988 (P.L. 100-297), divided by average daily attendance for states and by total enrollment for districts/consortia.

4 Data shown are for participating schools only.

5 Enrollment includes students attending private schools that are part of the consortium.

How Is the Report Organized?

This report provides a preliminary overview of the mathematics results for the Benchmarking Study. The real work will take place as policy makers, administrators, and teachers in each participating entity begin to examine the curriculum, teaching force, instructional approaches, and school environment in an international context. As those working on school improvement know full well, there is no "silver bullet" or single factor that is the answer to higher achievement in mathematics or any other school subject. Making strides in raising student achievement requires tireless diligence in all of the various areas related to educational quality.

The report is in two sections. Chapters 1 through 3 present the achievement results. Chapter 1 presents overall achievement results. Chapter 2 shows international benchmarks of mathematics achievement illustrated by results for individual mathematics questions. Chapter 3 gives results for the five mathematics content areas. Chapters 4 through 7 focus on the contextual factors related to teaching and learning mathematics. Chapter 4 examines student factors including the availability of educational resources in the home, how much time they spend studying mathematics outside of school, and their attitudes towards mathematics. Chapter 5 provides information about the curriculum, such as the mathematics included in participants' content standards and curriculum frameworks as well as the topics covered and emphasized by teachers in mathematics lessons. Chapter 6 presents information on mathematics teacher preparation and professional development activities as well as on classroom practices. Chapter 7 focuses on school factors, including the availability of resources for teaching mathematics and school safety.

Each of chapters 4 through 7 is accompanied by a set of reference exhibits in the reference section of the report, following the main chapters. Appendices at the end of the report summarize the procedures used in the Benchmarking Study, present the multiple comparisons for the mathematics content areas, provide the achievement percentiles, list the topics and processes measured by each item in the assessment, and acknowledge the numerous individuals responsible for implementing the TIMSS 1999 Benchmarking Study.

CHAPTER

1 Student Achievement in Mathematics

Chapter 1 summarizes eighth-grade achievement on the TIMSS 1999 mathematics assessment for each of the Benchmarking states, districts, and consortia, as well as for each participating country. Comparisons of participants' performance against international benchmarks, as well as gender differences in performance, are also provided.

1

How Do Participants Differ in Mathematics Achievement?

Exhibit 1.1 presents the distribution of student achievement for the 38 TIMSS 1999 countries and the 27 Benchmarking participants in a two-page display.[1] The left-hand page shows countries and Benchmarking participants together, in decreasing order of average (mean) scale score, and indicates whether the average for each participant is significantly higher or lower than the international average of 487. The international average was obtained by averaging across the mean scores for each of the 38 participating countries. On the right-hand page is a tabular display of average achievement, along with the number of years of formal schooling and the average age of students tested.

Many of the Benchmarking participants performed fairly well on the TIMSS 1999 mathematics assessment. Average performance for the 13 Benchmarking states was clustered in the middle of the international distribution of results for the 38 countries. All of the Benchmarking states performed either significantly above or similar to the international average. The United States as a whole also had average mathematics achievement just above the international average.

The Benchmarking Study underscores the extreme importance of looking beyond the averages to the range of performance found across the nation. Performance across the participating school districts and consortia reflected nearly the full range of achievement internationally. The two highest-achieving Benchmarking participants were the Naperville School District and the First in the World Consortium. These were two of the Benchmarking participants with the lowest percentages of students from low-income families (Naperville, 2 percent; First in the World, 14 percent).[2] Benchmarking participants with the lowest average mathematics achievement included four urban school districts with high percentages of students from low-income families – the Jersey City Public Schools (89 percent), the Chicago Public Schools (71 percent), the Rochester City School District (73 percent), and the Miami-Dade County Public Schools (59 percent). Although not quite as high as Singapore, Korea, and Chinese Taipei nor as low as the lowest-scoring countries in TIMSS 1999, the range of average performance across the Benchmarking districts and consortia was almost as broad as across all the TIMSS 1999 countries.

1 TIMSS used item response theory (IRT) methods to summarize the achievement results on a scale with a mean of 500 and a standard deviation of 100. Given the matrix-sampling approach, scaling averages students' responses in a way that accounts for differences in the difficulty of different subsets of items. It allows students' performance to be summarized on a common metric even though individual students responded to different items in the test. For more detailed information, see the "IRT Scaling and Data Analysis" section of Appendix A.

2 Low-income figures are percentages of students eligible to receive free or reduced-price lunch through the National School Lunch Program, as reported by participating schools.

That achievement is distributed broadly within as well as across participating entities is graphically illustrated in Exhibit 1.1 showing the distribution of student performance within each entity. Achievement for each participant is shown for the 25th and 75th percentiles as well as for the 5th and 95th percentiles.[3] Each percentile point indicates the percentages of students performing below and above that point on the scale. For example, 25 percent of the eighth-grade students in each participating entity performed below the 25th percentile for that entity, and 75 percent performed above the 25th percentile. The range between the 25th and 75th percentiles represents performance by the middle half of students. In most entities, the range of performance for the middle group was between 100 and 150 scale-score points. Performance at the 5th and 95th percentiles represents the extremes in both lower and higher achievement. The range of performance between these two score points, which includes 90 percent of the population, is between 250 and 300 points for most participants. The dark boxes at the midpoints of the distributions show the 95 percent confidence intervals around the average achievement in each entity.[4]

As well as showing the wide spread of student achievement within each entity, the percentiles also provide a perspective on the size of the differences among entities. Even though performance generally differed very little between one participant and the next higher- or lower-performing one, the range across participants was very large. For example, average performance in Singapore was comparable to or even exceeded performance at the 95th percentile in the lower-performing countries such as Chile, the Philippines, Morocco, and South Africa. This means that only the most proficient students in the lower-performing countries approached the level of achievement of Singaporean students of average proficiency.

Exhibit 1.2 compares overall mean achievement in mathematics among individual entities. This figure shows whether or not the differences in average achievement between pairs of participants are statistically significant. Selecting a participant of interest and reading across the exhibit, a triangle pointing up indicates significantly higher performance than the comparison participant listed across the top; a circle indicates no significant difference in performance; and a triangle pointing down indicates significantly lower performance.

The data in Exhibit 1.2 reinforce the point that, when ordered by average achievement, adjacent participants usually did not significantly differ from each other, although the differences in achievement between the high-performing and low-performing participants were very large.

3 Tables of the percentile values and standard deviations for all participants are presented in Appendix C.

4 See the "IRT Scaling and Data Analysis" section of Appendix A for more details about calculating standard errors and confidence intervals for the TIMSS statistics.

Singapore, Korea, Chinese Taipei, and Hong Kong had the highest performance, closely followed by Japan, the Naperville School District, the First in the World Consortium, and Belgium (Flemish).[5] Naperville and First in the World both performed similarly to Hong Kong, Japan, and Belgium (Flemish), but significantly below Singapore, Korea, and Chinese Taipei. The difference in performance from one participant to the next was often negligible. Montgomery County, the Michigan Invitational Group, the Academy School District, the Project SMART Consortium, the Southwest Pennsylvania Math and Science Collaborative, Michigan, Texas, Indiana, Oregon, Guilford County, Massachusetts, Connecticut, and Illinois were outperformed by only the top-performing eight or nine entities. These Benchmarking jurisdictions had average achievement most similar to the Netherlands, the Slovak Republic, Hungary, Canada, Slovenia, the Russian Federation, Australia, Finland, the Czech Republic, and Malaysia. Pennsylvania and South Carolina had achievement similar to that of Latvia (LSS),[6] the United States, and England, closely followed by North Carolina, Idaho, Maryland, Missouri, and the Fremont/Lincoln/Westside Public Schools. The Delaware Science Coalition and the Jersey City Public Schools had average achievement similar to that of Italy, outperforming eleven and nine of the TIMSS 1999 countries, respectively. The Chicago Public Schools had average achievement close to that in Moldova, Thailand, and Israel. The Rochester City School District and the Miami-Dade County Public Schools had average eighth-grade mathematics performance lower than most of the TIMSS 1999 countries. Rochester had performance similar to the Republic of Macedonia, but significantly higher than Indonesia and Chile. Miami-Dade had average achievement about the same as the Islamic Republic of Iran, but significantly higher than the three lowest-scoring countries (the Philippines, Morocco, and South Africa).

5 Belgium has two separate educational systems, Flemish and French. The Flemish system participated in TIMSS 1999.

6 Because coverage of its eighth-grade population falls below 65%, Latvia is annotated LSS for Latvian-Speaking Schools only.

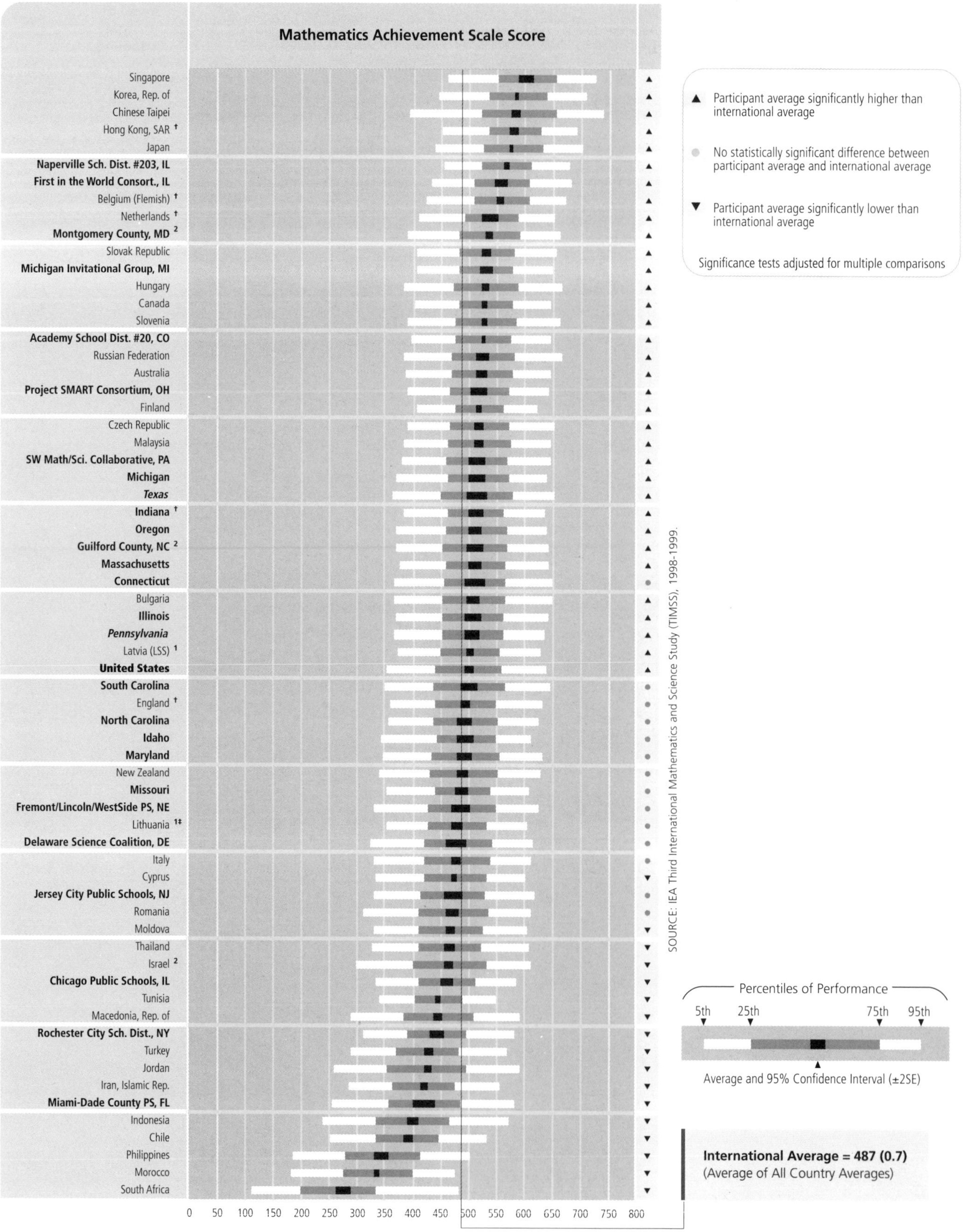

Mathematics Achievement Scale Score

Participant	
Singapore	▲
Korea, Rep. of	▲
Chinese Taipei	▲
Hong Kong, SAR †	▲
Japan	▲
Naperville Sch. Dist. #203, IL	▲
First in the World Consort., IL	▲
Belgium (Flemish) †	▲
Netherlands †	▲
Montgomery County, MD 2	▲
Slovak Republic	▲
Michigan Invitational Group, MI	▲
Hungary	▲
Canada	▲
Slovenia	▲
Academy School Dist. #20, CO	▲
Russian Federation	▲
Australia	▲
Project SMART Consortium, OH	▲
Finland	▲
Czech Republic	▲
Malaysia	▲
SW Math/Sci. Collaborative, PA	▲
Michigan	▲
Texas	▲
Indiana †	▲
Oregon	▲
Guilford County, NC 2	▲
Massachusetts	▲
Connecticut	●
Bulgaria	▲
Illinois	▲
Pennsylvania	▲
Latvia (LSS) 1	▲
United States	▲
South Carolina	●
England †	●
North Carolina	●
Idaho	●
Maryland	●
New Zealand	●
Missouri	●
Fremont/Lincoln/WestSide PS, NE	●
Lithuania 1‡	●
Delaware Science Coalition, DE	●
Italy	●
Cyprus	▼
Jersey City Public Schools, NJ	●
Romania	●
Moldova	▼
Thailand	▼
Israel 2	▼
Chicago Public Schools, IL	▼
Tunisia	▼
Macedonia, Rep. of	▼
Rochester City Sch. Dist., NY	▼
Turkey	▼
Jordan	▼
Iran, Islamic Rep.	▼
Miami-Dade County PS, FL	▼
Indonesia	▼
Chile	▼
Philippines	▼
Morocco	▼
South Africa	▼

0 50 100 150 200 250 300 350 400 450 500 550 600 650 700 750 800

▲ Participant average significantly higher than international average

● No statistically significant difference between participant average and international average

▼ Participant average significantly lower than international average

Significance tests adjusted for multiple comparisons

Percentiles of Performance: 5th, 25th, 75th, 95th

Average and 95% Confidence Interval (±2SE)

International Average = 487 (0.7)
(Average of All Country Averages)

SOURCE: IEA Third International Mathematics and Science Study (TIMSS), 1998-1999.

Countries

		Average Scale Score	Years of Formal Schooling	Average Age
United States	▲	502 (4.0)	8	14.2
Australia	▲	525 (4.8)	8 or 9	14.3
Belgium (Flemish) †	▲	558 (3.3)	8	14.1
Bulgaria	▲	511 (5.8)	8	14.8
Canada	▲	531 (2.5)	8	14.0
Chile	▼	392 (4.4)	8	14.4
Chinese Taipei	▲	585 (4.0)	8	14.2
Cyprus	▼	476 (1.8)	8	13.8
Czech Republic	▲	520 (4.2)	8	14.4
England †	●	496 (4.1)	9	14.2
Finland	▲	520 (2.7)	7	13.8
Hong Kong, SAR †	▲	582 (4.3)	8	14.2
Hungary	▲	532 (3.7)	8	14.4
Indonesia	▼	403 (4.9)	8	14.6
Iran, Islamic Rep.	▼	422 (3.4)	8	14.6
Israel [2]	▼	466 (3.9)	8	14.1
Italy	●	479 (3.8)	8	14.0
Japan	▲	579 (1.7)	8	14.4
Jordan	▼	428 (3.6)	8	14.0
Korea, Rep. of	▲	587 (2.0)	8	14.4
Latvia (LSS) [1]	▲	505 (3.4)	8	14.5
Lithuania [1‡]	●	482 (4.3)	8.5	15.2
Macedonia, Rep. of	▼	447 (4.2)	8	14.6
Malaysia	▲	519 (4.4)	8	14.4
Moldova	▼	469 (3.9)	9	14.4
Morocco	▼	337 (2.6)	7	14.2
Netherlands †	▲	540 (7.1)	8	14.2
New Zealand	●	491 (5.2)	8.5 to 9.5	14.0
Philippines	▼	345 (6.0)	7	14.1
Romania	●	472 (5.8)	8	14.8
Russian Federation	▲	526 (5.9)	7 or 8	14.1
Singapore	▲	604 (6.3)	8	14.4
Slovak Republic	▲	534 (4.0)	8	14.3
Slovenia	▲	530 (2.8)	8	14.8
South Africa	▼	275 (6.8)	8	15.5
Thailand	▼	467 (5.1)	8	14.5
Tunisia	▼	448 (2.4)	8	14.8
Turkey	▼	429 (4.3)	8	14.2
International Avg. (All Countries)		487 (0.7)		

States

		Average Scale Score	Years of Formal Schooling	Average Age
Connecticut	●	512 (9.1)	8	14.0
Idaho	●	495 (7.4)	8	14.2
Illinois	▲	509 (6.7)	8	14.2
Indiana †	▲	515 (7.2)	8	14.4
Maryland	●	495 (6.2)	8	13.9
Massachusetts	▲	513 (5.9)	8	14.1
Michigan	▲	517 (7.5)	8	14.1
Missouri	●	490 (5.3)	8	14.3
North Carolina	●	495 (7.0)	8	14.2
Oregon	▲	514 (6.0)	8	14.2
Pennsylvania	▲	507 (6.3)	8	14.2
South Carolina	●	502 (7.4)	8	14.2
Texas	▲	516 (9.1)	8	14.3

Districts and Consortia

		Average Scale Score	Years of Formal Schooling	Average Age
Academy School Dist. #20, CO	▲	528 (1.8)	8	14.2
Chicago Public Schools, IL	▼	462 (6.1)	8	14.2
Delaware Science Coalition, DE	●	479 (8.9)	8	14.1
First in the World Consort., IL	▲	560 (5.8)	8	14.2
Fremont/Lincoln/WestSide PS, NE	●	488 (8.2)	8	14.2
Guilford County, NC [2]	▲	514 (7.7)	8	14.2
Jersey City Public Schools, NJ	●	475 (8.6)	8	14.3
Miami-Dade County PS, FL	▼	421 (9.4)	8	14.3
Michigan Invitational Group, MI	▲	532 (5.8)	8	14.1
Montgomery County, MD [2]	▲	537 (3.5)	8	14.0
Naperville Sch. Dist. #203, IL	▲	569 (2.8)	8	14.1
Project SMART Consortium, OH	▲	521 (7.5)	8	14.2
Rochester City Sch. Dist., NY	▼	444 (6.5)	8	14.2
SW Math/Sci. Collaborative, PA	▲	517 (7.5)	8	14.2

SOURCE: IEA Third International Mathematics and Science Study (TIMSS), 1998-1999.

▲ Participant average significantly higher than international average

● No statistically significant difference between participant average and international average

▼ Participant average significantly lower than international average

Significance tests adjusted for multiple comparisons

States in *italics* did not fully satisfy guidelines for sample participation rates (see Appendix A for details).

† Met guidelines for sample participation rates only after replacement schools were included (see Exhibit A.6).

1 National Desired Population does not cover all of International Desired Population (see Exhibit A.3). Because coverage falls below 65%, Latvia is annotated LSS for Latvian-Speaking Schools only.

2 National Defined Population covers less than 90 percent of National Desired Population (see Exhibit A.3).

‡ Lithuania tested the same cohort of students as other countries, but later in 1999, at the beginning of the next school year.

( ) Standard errors appear in parentheses. Because results are rounded to the nearest whole number, some totals may appear inconsistent.

Exhibit 1.2	Multiple Comparisons of Average Mathematics Achievement

8th Grade Mathematics

Instructions: Read across the row for a participant to compare performance with the participants listed along the top of the chart. The symbols indicate whether the average achievement of the participant in the row is significantly lower than that of the comparison participant, significantly higher than that of the comparison participant, or if there is no statistically significant difference between the average achievement of the two participants.

	Singapore	Korea, Rep. of	Chinese Taipei	Hong Kong, SAR	Japan	**Naperville Sch. Dist. #203, IL**	**First in the World Consort., IL**	Belgium (Flemish)	Netherlands	**Montgomery County, MD**	Slovak Republic	**Michigan Invitational Group, MI**	Hungary	Canada	Slovenia	**Academy School Dist. #20, CO**	Russian Federation	Australia	**Project SMART Consortium, OH**	Finland	Czech Republic	Malaysia	**SW Math/Sci. Collaborative, PA**	**Michigan**	***Texas***	**Indiana**	**Oregon**	**Guilford County, NC**	**Massachusetts**	**Connecticut**	Bulgaria	**Illinois**
Singapore		●	●	●	▲	▲	▲	▲	▲	▲	▲	▲	▲	▲	▲	▲	▲	▲	▲	▲	▲	▲	▲	▲	▲	▲	▲	▲	▲	▲		▲
Korea, Rep. of	●		●	●	▲	▲	▲	▲	▲	▲	▲	▲	▲	▲	▲	▲	▲	▲	▲	▲	▲	▲	▲	▲	▲	▲	▲	▲	▲	▲	▲	
Chinese Taipei	●	●		●	●	▲	▲	▲	▲	▲	▲	▲	▲	▲	▲	▲	▲	▲	▲	▲	▲	▲	▲	▲	▲	▲	▲	▲	▲	▲	▲	▲
Hong Kong, SAR	●	●	●		●	●	●	▲	▲	▲	▲	▲	▲	▲	▲	▲	▲	▲	▲	▲	▲	▲	▲	▲	▲	▲	▲	▲	▲	▲	▲	▲
Japan	▼	▼	●	●		●	●	▲	▲	▲	▲	▲	▲	▲	▲	▲	▲	▲	▲	▲	▲	▲	▲	▲	▲	▲	▲	▲	▲	▲	▲	▲
Naperville Sch. Dist. #203, IL	▼	▼	▼	●	●		●	●	▲	▲	▲	▲	▲	▲	▲	▲	▲	▲	▲	▲	▲	▲	▲	▲	▲	▲	▲	▲	▲	▲	▲	▲
First in the World Consort., IL	▼	▼	▼	●	●	●		●	●	▲	▲	▲	▲	▲	▲	▲	▲	▲	▲	▲	▲	▲	▲	▲	▲	▲	▲	▲	▲	▲	▲	▲
Belgium (Flemish)	▼	▼	▼	▼	▼	●	●		●	▲	▲	▲	▲	▲	▲	▲	▲	▲	▲	▲	▲	▲	▲	▲	▲	▲	▲	▲	▲	▲	▲	▲
Netherlands	▼	▼	▼	▼	▼	▼	●	●		●	●	●	●	●	●	●	●	●	●	●	●	●	●	●	●	●	●	●	●	●	●	●
Montgomery County, MD	▼	▼	▼	▼	▼	▼	▼	▼	●		●	●	●	●	●	●	●	●	●	▲	●	▲	●	●	●	●	▲	●	▲	●	▲	▲
Slovak Republic	▼	▼	▼	▼	▼	▼	▼	▼	●	●		●	●	●	●	●	●	●	●	●	●	●	●	●	●	●	●	●	●	●	▲	●
Michigan Invitational Group, MI	▼	▼	▼	▼	▼	▼	▼	▼	●	●	●		●	●	●	●	●	●	●	●	●	●	●	●	●	●	●	●	●	●	●	●
Hungary	▼	▼	▼	▼	▼	▼	▼	▼	●	●	●	●		●	●	●	●	●	●	●	●	●	●	●	●	●	●	●	●	●	●	●
Canada	▼	▼	▼	▼	▼	▼	▼	▼	●	●	●	●	●		●	●	●	●	●	●	●	●	●	●	●	●	●	●	●	●	●	●
Slovenia	▼	▼	▼	▼	▼	▼	▼	▼	●	●	●	●	●	●		●	●	●	●	●	●	●	●	●	●	●	●	●	●	●	●	●
Academy School Dist. #20, CO	▼	▼	▼	▼	▼	▼	▼	▼	●	●	●	●	●	●	●		●	●	●	●	●	●	●	●	●	●	●	●	●	●	●	●
Russian Federation	▼	▼	▼	▼	▼	▼	▼	▼	●	●	●	●	●	●	●	●		●	●	●	●	●	●	●	●	●	●	●	●	●	●	●
Australia	▼	▼	▼	▼	▼	▼	▼	▼	●	●	●	●	●	●	●	●	●		●	●	●	●	●	●	●	●	●	●	●	●	●	●
Project SMART Consortium, OH	▼	▼	▼	▼	▼	▼	▼	▼	●	●	●	●	●	●	●	●	●	●		●	●	●	●	●	●	●	●	●	●	●	●	●
Finland	▼	▼	▼	▼	▼	▼	▼	▼	●	▼	●	●	●	●	●	●	●	●	●		●	●	●	●	●	●	●	●	●	●	●	●
Czech Republic	▼	▼	▼	▼	▼	▼	▼	▼	●	●	●	●	●	●	●	●	●	●	●	●		●	●	●	●	●	●	●	●	●	●	●
Malaysia	▼	▼	▼	▼	▼	▼	▼	▼	●	▼	●	●	●	●	●	●	●	●	●	●	●		●	●	●	●	●	●	●	●	●	●
SW Math/Sci. Collaborative, PA	▼	▼	▼	▼	▼	▼	▼	▼	●	●	●	●	●	●	●	●	●	●	●	●	●	●		●	●	●	●	●	●	●	●	●
Michigan	▼	▼	▼	▼	▼	▼	▼	▼	●	●	●	●	●	●	●	●	●	●	●	●	●	●	●		●	●	●	●	●	●	●	●
Texas	▼	▼	▼	▼	▼	▼	▼	▼	●	●	●	●	●	●	●	●	●	●	●	●	●	●	●	●		●	●	●	●	●	●	●
Indiana	▼	▼	▼	▼	▼	▼	▼	▼	●	●	●	●	●	●	●	●	●	●	●	●	●	●	●	●	●		●	●	●	●	●	●
Oregon	▼	▼	▼	▼	▼	▼	▼	▼	●	▼	●	●	●	●	●	●	●	●	●	●	●	●	●	●	●	●		●	●	●	●	●
Guilford County, NC	▼	▼	▼	▼	▼	▼	▼	▼	●	●	●	●	●	●	●	●	●	●	●	●	●	●	●	●	●	●	●		●	●	●	●
Massachusetts	▼	▼	▼	▼	▼	▼	▼	▼	●	▼	●	●	●	●	●	●	●	●	●	●	●	●	●	●	●	●	●	●		●	●	●
Connecticut	▼	▼	▼	▼	▼	▼	▼	▼	●	●	●	●	●	●	●	●	●	●	●	●	●	●	●	●	●	●	●	●	●		●	●
Bulgaria	▼	▼	▼	▼	▼	▼	▼	▼	●	▼	▼	●	●	●	●	●	●	●	●	●	●	●	●	●	●	●	●	●	●	●		●
Illinois	▼	▼	▼	▼	▼	▼	▼	▼	●	▼	●	●	●	●	●	●	●	●	●	●	●	●	●	●	●	●	●	●	●	●	●	
Pennsylvania	▼	▼	▼	▼	▼	▼	▼	▼	▼	▼	▼	●	▼	▼	▼	●	●	●	●	●	●	●	●	●	●	●	●	●	●	●	●	●
Latvia (LSS)	▼	▼	▼	▼	▼	▼	▼	▼	▼	▼	▼	▼	▼	▼	▼	▼	●	▼	●	▼	●	●	●	●	●	●	●	●	●	●	●	●
United States	▼	▼	▼	▼	▼	▼	▼	▼	▼	▼	▼	▼	▼	▼	▼	▼	▼	▼	●	▼	●	●	●	●	●	●	●	●	●	●	●	●
South Carolina	▼	▼	▼	▼	▼	▼	▼	▼	▼	▼	▼	●	▼	▼	▼	▼	●	●	●	●	●	●	●	●	●	●	●	●	●	●	●	●
England	▼	▼	▼	▼	▼	▼	▼	▼	▼	▼	▼	▼	▼	▼	▼	▼	▼	▼	●	▼	▼	▼	●	●	●	●	●	●	●	●	●	●
North Carolina	▼	▼	▼	▼	▼	▼	▼	▼	▼	▼	▼	▼	▼	▼	▼	▼	▼	▼	●	▼	●	●	●	●	●	●	●	●	●	●	●	●
Idaho	▼	▼	▼	▼	▼	▼	▼	▼	▼	▼	▼	▼	▼	▼	▼	▼	▼	▼	●	▼	●	●	●	●	●	●	●	●	●	●	●	●
Maryland	▼	▼	▼	▼	▼	▼	▼	▼	▼	▼	▼	▼	▼	▼	▼	▼	▼	▼	●	▼	▼	▼	●	●	●	●	●	●	●	●	●	●
New Zealand	▼	▼	▼	▼	▼	▼	▼	▼	▼	▼	▼	▼	▼	▼	▼	▼	▼	▼	▼	▼	▼	▼	●	●	●	●	●	●	●	●	●	●
Missouri	▼	▼	▼	▼	▼	▼	▼	▼	▼	▼	▼	▼	▼	▼	▼	▼	▼	▼	▼	▼	▼	▼	●	●	●	●	●	●	●	●	●	●
Fremont/Lincoln/WestSide PS, NE	▼	▼	▼	▼	▼	▼	▼	▼	▼	▼	▼	▼	▼	▼	▼	▼	▼	▼	●	▼	▼	▼	●	●	●	●	●	●	●	●	●	●
Lithuania	▼	▼	▼	▼	▼	▼	▼	▼	▼	▼	▼	▼	▼	▼	▼	▼	▼	▼	▼	▼	▼	▼	▼	▼	▼	▼	▼	▼	▼	●	▼	▼
Delaware Science Coalition, DE	▼	▼	▼	▼	▼	▼	▼	▼	▼	▼	▼	▼	▼	▼	▼	▼	▼	▼	▼	▼	▼	▼	●	●	●	●	▼	●	●	●	●	●
Italy	▼	▼	▼	▼	▼	▼	▼	▼	▼	▼	▼	▼	▼	▼	▼	▼	▼	▼	▼	▼	▼	▼	▼	▼	▼	▼	▼	▼	▼	▼	▼	▼
Cyprus	▼	▼	▼	▼	▼	▼	▼	▼	▼	▼	▼	▼	▼	▼	▼	▼	▼	▼	▼	▼	▼	▼	▼	▼	▼	▼	▼	▼	▼	▼	▼	▼
Jersey City Public Schools, NJ	▼	▼	▼	▼	▼	▼	▼	▼	▼	▼	▼	▼	▼	▼	▼	▼	▼	▼	▼	▼	▼	▼	▼	▼	▼	▼	▼	▼	▼	●	▼	●
Romania	▼	▼	▼	▼	▼	▼	▼	▼	▼	▼	▼	▼	▼	▼	▼	▼	▼	▼	▼	▼	▼	▼	▼	▼	▼	▼	▼	▼	▼	▼	▼	▼
Moldova	▼	▼	▼	▼	▼	▼	▼	▼	▼	▼	▼	▼	▼	▼	▼	▼	▼	▼	▼	▼	▼	▼	▼	▼	▼	▼	▼	▼	▼	▼	▼	▼
Thailand	▼	▼	▼	▼	▼	▼	▼	▼	▼	▼	▼	▼	▼	▼	▼	▼	▼	▼	▼	▼	▼	▼	▼	▼	▼	▼	▼	▼	▼	▼	▼	▼
Israel	▼	▼	▼	▼	▼	▼	▼	▼	▼	▼	▼	▼	▼	▼	▼	▼	▼	▼	▼	▼	▼	▼	▼	▼	▼	▼	▼	▼	▼	▼	▼	▼
Chicago Public Schools, IL	▼	▼	▼	▼	▼	▼	▼	▼	▼	▼	▼	▼	▼	▼	▼	▼	▼	▼	▼	▼	▼	▼	▼	▼	▼	▼	▼	▼	▼	▼	▼	▼
Tunisia	▼	▼	▼	▼	▼	▼	▼	▼	▼	▼	▼	▼	▼	▼	▼	▼	▼	▼	▼	▼	▼	▼	▼	▼	▼	▼	▼	▼	▼	▼	▼	▼
Macedonia, Rep. of	▼	▼	▼	▼	▼	▼	▼	▼	▼	▼	▼	▼	▼	▼	▼	▼	▼	▼	▼	▼	▼	▼	▼	▼	▼	▼	▼	▼	▼	▼	▼	▼
Rochester City Sch. Dist., NY	▼	▼	▼	▼	▼	▼	▼	▼	▼	▼	▼	▼	▼	▼	▼	▼	▼	▼	▼	▼	▼	▼	▼	▼	▼	▼	▼	▼	▼	▼	▼	▼
Turkey	▼	▼	▼	▼	▼	▼	▼	▼	▼	▼	▼	▼	▼	▼	▼	▼	▼	▼	▼	▼	▼	▼	▼	▼	▼	▼	▼	▼	▼	▼	▼	▼
Jordan	▼	▼	▼	▼	▼	▼	▼	▼	▼	▼	▼	▼	▼	▼	▼	▼	▼	▼	▼	▼	▼	▼	▼	▼	▼	▼	▼	▼	▼	▼	▼	▼
Iran, Islamic Rep.	▼	▼	▼	▼	▼	▼	▼	▼	▼	▼	▼	▼	▼	▼	▼	▼	▼	▼	▼	▼	▼	▼	▼	▼	▼	▼	▼	▼	▼	▼	▼	▼
Miami-Dade County PS, FL	▼	▼	▼	▼	▼	▼	▼	▼	▼	▼	▼	▼	▼	▼	▼	▼	▼	▼	▼	▼	▼	▼	▼	▼	▼	▼	▼	▼	▼	▼	▼	▼
Indonesia	▼	▼	▼	▼	▼	▼	▼	▼	▼	▼	▼	▼	▼	▼	▼	▼	▼	▼	▼	▼	▼	▼	▼	▼	▼	▼	▼	▼	▼	▼	▼	▼
Chile	▼	▼	▼	▼	▼	▼	▼	▼	▼	▼	▼	▼	▼	▼	▼	▼	▼	▼	▼	▼	▼	▼	▼	▼	▼	▼	▼	▼	▼	▼	▼	▼
Philippines	▼	▼	▼	▼	▼	▼	▼	▼	▼	▼	▼	▼	▼	▼	▼	▼	▼	▼	▼	▼	▼	▼	▼	▼	▼	▼	▼	▼	▼	▼	▼	▼
Morocco	▼	▼	▼	▼	▼	▼	▼	▼	▼	▼	▼	▼	▼	▼	▼	▼	▼	▼	▼	▼	▼	▼	▼	▼	▼	▼	▼	▼	▼	▼	▼	▼
South Africa	▼	▼	▼	▼	▼	▼	▼	▼	▼	▼	▼	▼	▼	▼	▼	▼	▼	▼	▼	▼	▼	▼	▼	▼	▼	▼	▼	▼	▼	▼	▼	▼

States in *italics* did not fully satisfy guidelines for sample participation rates (see Appendix A for details).

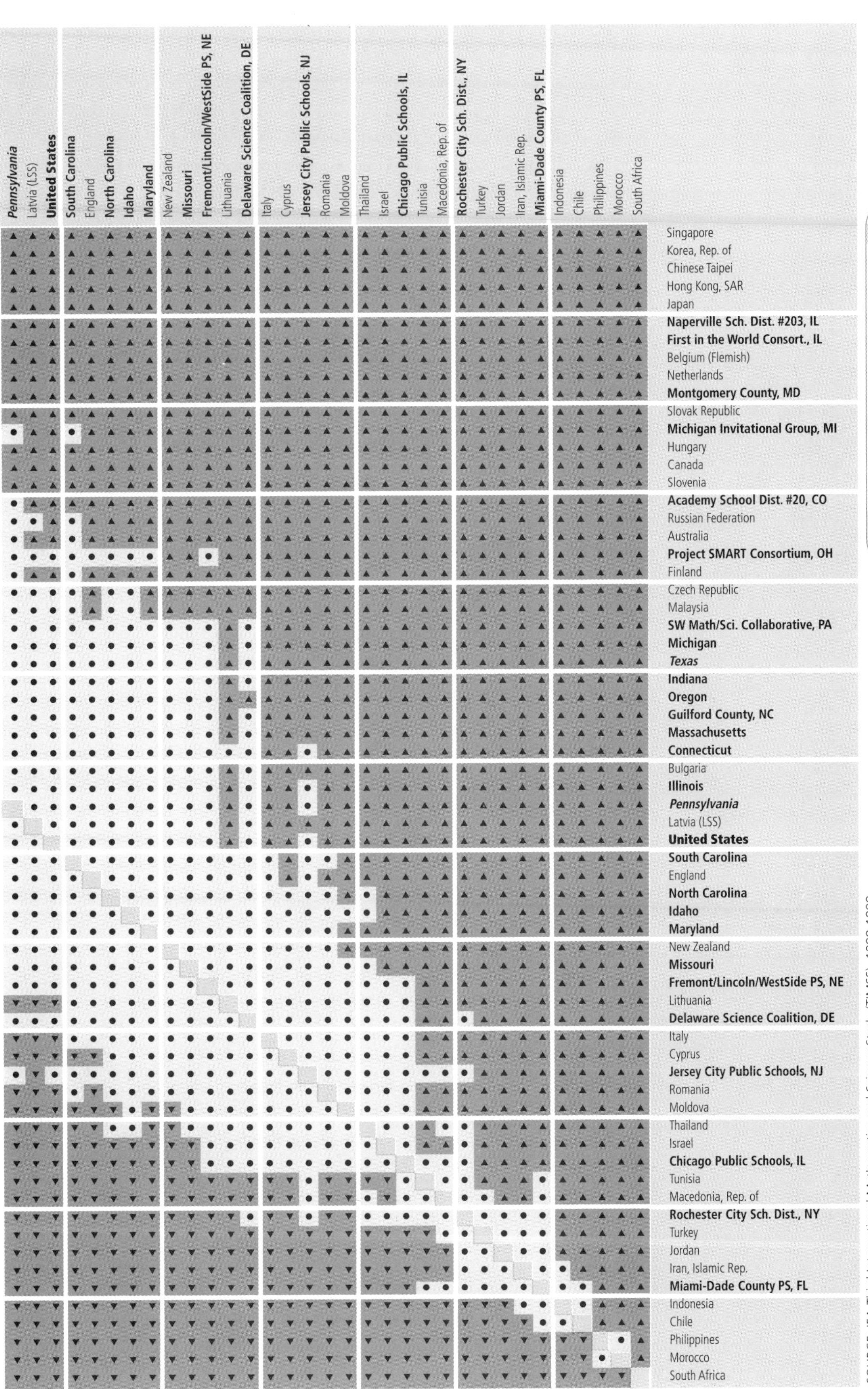

▲ Average achievement significantly higher than comparison participant

● No statistically significant difference from comparison participant

▼ Average achievement significantly lower than comparison participant

Significance tests adjusted for multiple comparisons

SOURCE: IEA Third International Mathematics and Science Study (TIMSS), 1998-1999.

How Do Benchmarking Participants Compare with International Benchmarks of Mathematics Achievement?

The TIMSS mathematics achievement scale summarizes student performance on test items designed to measure a wide range of student knowledge and proficiency. In order to provide descriptions of what performance could mean in terms of the mathematics that students know and can do, TIMSS identified four points on the scale for use as international benchmarks[7] or reference points, and conducted an ambitious scale anchoring exercise to describe students' performance at these benchmarks. Exhibit 1.3 shows the four international benchmarks of mathematics achievement and briefly describes what students scoring at these benchmarks typically know and can do. More detailed descriptions appear in Chapter 2, together with example test items illustrating performance at each benchmark.

The Top 10% Benchmark is defined at the 90th percentile on the TIMSS mathematics scale, taking into account the performance of all students in all countries participating in 1999. It corresponds to a scale score of 616 and is the point above which the top 10 percent of students in the TIMSS 1999 assessment scored. Students performing at this level demonstrated that they could organize information, make generalizations, and explain solution strategies in non-routine problem-solving situations.

The Upper Quarter Benchmark is the 75th percentile on the mathematics scale. This point, corresponding to a scale score of 555, is the point above which the top 25 percent of students scored. Students scoring at this benchmark demonstrated that they could apply their mathematical understanding and knowledge in a wide variety of relatively complex situations involving fractions, decimals, geometric properties, and algebraic expressions.

The Median Benchmark, with a score of 479, corresponds to the 50th percentile, or median. This is the point above which the top half of students scored on the TIMSS 1999 assessment. Students performing at this level showed that they could apply basic mathematical knowledge in straightforward situations, such as one-step word problems involving addition and subtraction or computational problems based on basic properties of geometric figures and simple algebraic relationships.

7 Readers should be careful not to confuse the international benchmarks, which are points on the international mathematics achievement scale chosen to describe specific achievement levels, with the benchmarking exercise itself, which is a process by which participants compare their achievement, curriculum, and instructional practices with those of the best in the world.

The Lower Quarter Benchmark is the 25th percentile and corresponds to a scale score of 396. This score point is reached by the top 75 percent of students and may be used as a benchmark of performance for lower-achieving students. Students scoring at this level typically demonstrated computational facility with whole numbers.

Exhibit 1.4 displays the percentage of students in each participating entity that reached each international benchmark, in decreasing order by the percentage reaching the Top 10% Benchmark. If student achievement in mathematics were distributed alike in every entity, then each entity would be expected to have about 10 percent of its students reaching the Top 10% Benchmark, 25 percent the Upper Quarter Benchmark, 50 percent the Median Benchmark, and 75 percent the Lower Quarter Benchmark. Although countries such as New Zealand, and Benchmarking participants such as Maryland, North Carolina, and the Delaware Science Coalition, came fairly close, no entity followed this pattern exactly. Instead, the high-performing entities generally had greater percentages of students reaching each benchmark, and the low-performing entities had lesser percentages.

Among the high performers, for example, Singapore, Chinese Taipei, Korea, Hong Kong, and Japan had one-third or more of their students reaching the Top 10% Benchmark, about two-thirds reaching the Upper Quarter Benchmark, around 90 percent reaching the Median Benchmark, and almost all (95 to 99 percent) reaching the Lower Quarter Benchmark. In comparison, the Naperville School District and the First in the World Consortium had 24 and 22 percent of their students, respectively, reaching the Top 10% Benchmark and 59 and 56 percent, respectively, reaching the Upper Quarter Benchmark, somewhat less than in the high-performing Asian countries. More like the top-performing Asian countries, these two high-performing districts had close to 90 percent of their students reaching the Median Benchmark (91 and 87 percent, respectively) and nearly all of their students reaching the Lower Quarter Benchmark (99 and 98 percent, respectively).

In contrast, the three lowest-performing Benchmarking participants, all urban districts, had two percent of their students reaching the Top 10% Benchmark, 9 to 12 percent reaching the Upper Quarter Benchmark, and from 29 to 41 percent reaching the Median Benchmark. The lowest-performing countries of South Africa, the

Philippines, and Morocco had almost no students reaching the Top 10% Benchmark, no more than one percent reaching the Upper Quarter Benchmark, less than 10 percent reaching the Median Benchmark, and no more than 31 percent reaching the Lower Quarter Benchmark.

Although Exhibit 1.4 is organized to draw particular attention to the percentage of high-achieving students in each entity, it conveys information about the distribution of middle and low performers also. For example, Canada, Australia, and Malaysia had 12 percent of their students reaching the Top 10% Benchmark, as might be expected, but 94 to 96 percent (rather than 75 percent) reaching the Lower Quarter Benchmark. Similarly, the Academy School District, the Michigan Invitational Group, and the Project SMART Consortium had 11 to 12 percent of their students reaching the Top 10% Benchmark but 95 to 96 percent reaching the Lower Quarter Benchmark.

• Top 10% Benchmark

Students can organize information, make generalizations, and explain solution strategies in non-routine problem solving situations. They can organize information and make generalizations to solve problems; apply knowledge of numeric, geometric, and algebraic relationships to solve problems (e.g., among fractions, decimals, and percents; geometric properties; and algebraic rules); and find the equivalent forms of algebraic expressions.

90th Percentile: 616

• Upper Quarter Benchmark

Students can apply their understanding and knowledge in a wide variety of relatively complex situations. They can order, relate and compute with fractions and decimals to solve word problems; solve multi-step word problems involving proportions with whole numbers; solve probability problems; use knowledge of geometric properties to solve problems; identify and evaluate algebraic expressions and solve equations with one variable.

75th Percentile: 555

• Median Benchmark

Students can apply basic mathematical knowledge in straightforward situations. They can add or subtract to solve one-step word problems involving whole numbers and decimals; identify representations of common fractions and relative sizes of fractions; solve for missing terms in proportions; recognize basic notions of percents and probability; use basic properties of geometric figures; read and interpret graphs, tables, and scales; and understand simple algebraic relationships.

50th Percentile: 479

• Lower Quarter Benchmark

Students can do basic computations with whole numbers. The few items that anchor at this level provide some evidence that students can add, subtract, and round with whole numbers. When there are the same number of decimal places, they can subtract with multiple regrouping. Students can round whole numbers to the nearest hundred. They recognize some basic notation and terminology.

25th Percentile: 396

SOURCE: IEA Third International Mathematics and Science Study (TIMSS), 1998-1999.

The international benchmarks are based on the combined data from the countries participating in 1999.

Exhibit 1.4 Percentages of Students Reaching TIMSS 1999 International Benchmarks of Mathematics Achievement

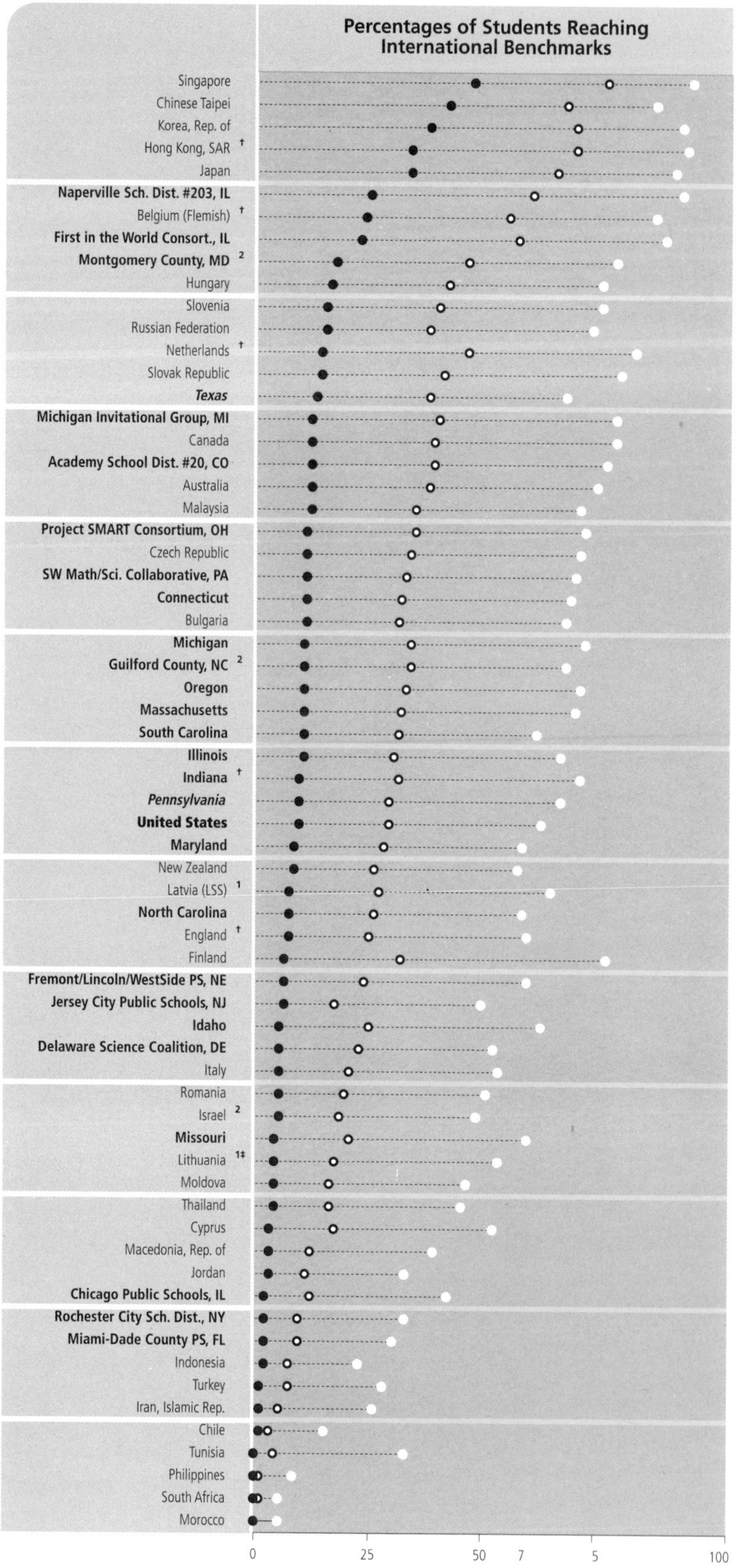

Percentages of Students Reaching International Benchmarks

Singapore
Chinese Taipei
Korea, Rep. of
Hong Kong, SAR †
Japan
Naperville Sch. Dist. #203, IL
Belgium (Flemish) †
First in the World Consort., IL
Montgomery County, MD 2
Hungary
Slovenia
Russian Federation
Netherlands †
Slovak Republic
Texas
Michigan Invitational Group, MI
Canada
Academy School Dist. #20, CO
Australia
Malaysia
Project SMART Consortium, OH
Czech Republic
SW Math/Sci. Collaborative, PA
Connecticut
Bulgaria
Michigan
Guilford County, NC 2
Oregon
Massachusetts
South Carolina
Illinois
Indiana †
Pennsylvania
United States
Maryland
New Zealand
Latvia (LSS) 1
North Carolina
England †
Finland
Fremont/Lincoln/WestSide PS, NE
Jersey City Public Schools, NJ
Idaho
Delaware Science Coalition, DE
Italy
Romania
Israel 2
Missouri
Lithuania 1‡
Moldova
Thailand
Cyprus
Macedonia, Rep. of
Jordan
Chicago Public Schools, IL
Rochester City Sch. Dist., NY
Miami-Dade County PS, FL
Indonesia
Turkey
Iran, Islamic Rep.
Chile
Tunisia
Philippines
South Africa
Morocco

0 25 50 7 5 100

SOURCE: IEA Third International Mathematics and Science Study (TIMSS), 1998-1999.

Top 10% Benchmark (90th Percentile) = 616

Upper Quarter Benchmark (75th Percentile) = 555

Median Benchmark (50th Percentile) = 479

Lower Quarter Benchmark (25th Percentile) = 396

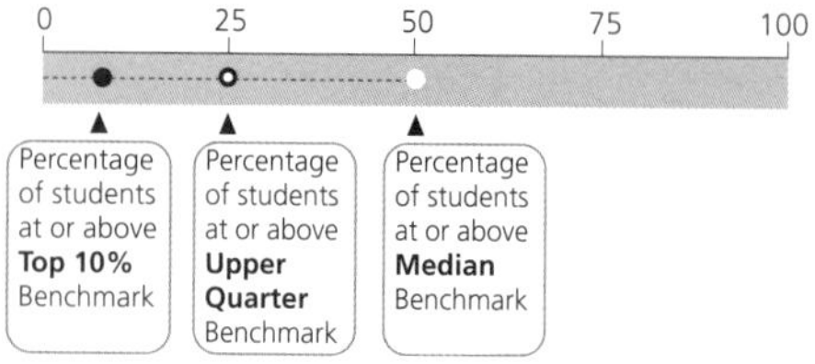

0 25 50 75 100

Percentage of students at or above **Top 10%** Benchmark

Percentage of students at or above **Upper Quarter** Benchmark

Percentage of students at or above **Median** Benchmark

Countries	Top 10%	Upper Quarter	Median	Lower Quarter
United States	9 (1.0)	28 (1.6)	61 (1.9)	88 (1.0)
Australia	12 (1.8)	37 (2.7)	73 (2.4)	94 (0.8)
Belgium (Flemish) †	23 (1.5)	54 (1.7)	85 (1.2)	98 (0.6)
Bulgaria	11 (2.3)	30 (3.0)	66 (2.6)	91 (1.3)
Canada	12 (1.1)	38 (1.5)	77 (1.3)	96 (0.6)
Chile	1 (0.5)	3 (1.1)	15 (1.8)	48 (2.0)
Chinese Taipei	41 (1.7)	66 (1.5)	85 (1.0)	95 (0.6)
Cyprus	3 (0.4)	17 (0.8)	51 (1.1)	84 (0.8)
Czech Republic	11 (1.4)	33 (2.1)	69 (2.3)	94 (1.1)
England †	7 (0.9)	24 (1.9)	58 (2.1)	89 (1.3)
Finland	6 (0.9)	31 (1.7)	75 (1.5)	96 (0.5)
Hong Kong, SAR †	33 (2.3)	68 (2.4)	92 (1.5)	99 (0.6)
Hungary	16 (1.2)	41 (1.9)	74 (1.6)	94 (1.0)
Indonesia	2 (0.4)	7 (0.9)	22 (1.4)	52 (2.2)
Iran, Islamic Rep.	1 (0.2)	5 (0.8)	25 (1.7)	63 (1.5)
Israel [2]	5 (0.6)	18 (1.3)	47 (1.8)	77 (1.9)
Italy	5 (0.7)	20 (1.4)	52 (2.1)	83 (1.4)
Japan	33 (1.1)	64 (0.9)	89 (0.5)	98 (0.3)
Jordan	3 (0.5)	11 (0.9)	32 (1.5)	62 (1.4)
Korea, Rep. of	37 (1.0)	68 (0.9)	91 (0.5)	99 (0.2)
Latvia (LSS) [1]	7 (0.9)	26 (1.8)	63 (2.0)	92 (1.0)
Lithuania [1]‡	4 (0.7)	17 (2.0)	52 (2.4)	86 (1.8)
Macedonia, Rep. of	3 (0.4)	12 (1.0)	38 (1.9)	72 (1.8)
Malaysia	12 (1.4)	34 (2.4)	69 (2.2)	94 (0.8)
Moldova	4 (0.7)	16 (1.5)	45 (2.2)	81 (1.7)
Morocco	0 (0.0)	0 (0.2)	5 (0.4)	27 (1.1)
Netherlands †	14 (2.3)	45 (4.1)	81 (3.5)	96 (1.3)
New Zealand	8 (1.2)	25 (2.4)	56 (2.5)	85 (1.5)
Philippines	0 (0.1)	1 (0.5)	8 (1.4)	31 (2.5)
Romania	5 (1.1)	19 (1.9)	49 (2.6)	80 (2.1)
Russian Federation	15 (1.8)	37 (2.8)	72 (2.7)	94 (1.2)
Singapore	46 (3.5)	75 (2.7)	93 (1.3)	99 (0.3)
Slovak Republic	14 (1.4)	40 (2.3)	78 (1.8)	96 (0.6)
Slovenia	15 (1.2)	39 (1.4)	74 (1.4)	95 (0.7)
South Africa	0 (0.2)	1 (0.4)	5 (1.0)	14 (2.0)
Thailand	4 (0.8)	16 (1.8)	44 (2.6)	81 (1.6)
Tunisia	0 (0.1)	4 (0.5)	32 (1.6)	80 (1.3)
Turkey	1 (0.3)	7 (1.0)	27 (1.9)	65 (2.0)

	Top 10%	Upper Quarter	Median	Lower Quarter
States				
Connecticut	11 (2.5)	31 (3.9)	67 (4.4)	91 (1.9)
Idaho	5 (1.1)	24 (2.9)	61 (3.5)	88 (2.2)
Illinois	10 (1.6)	29 (2.9)	65 (3.3)	92 (1.5)
Indiana †	9 (1.9)	30 (3.9)	69 (3.6)	94 (1.2)
Maryland	8 (1.4)	27 (2.5)	57 (3.2)	87 (2.0)
Massachusetts	10 (1.6)	31 (2.6)	68 (3.0)	92 (1.6)
Michigan	10 (2.0)	33 (3.7)	70 (3.3)	92 (1.7)
Missouri	4 (0.9)	20 (2.4)	58 (2.9)	89 (1.5)
North Carolina	7 (1.6)	25 (3.1)	57 (3.3)	88 (2.0)
Oregon	10 (1.8)	32 (2.8)	69 (2.8)	91 (1.4)
Pennsylvania	9 (1.3)	28 (2.6)	65 (3.0)	91 (1.8)
South Carolina	10 (2.0)	30 (3.2)	60 (3.5)	88 (1.8)
Texas	13 (2.2)	37 (3.8)	66 (4.3)	90 (2.1)
Districts and Consortia				
Academy School Dist. #20, CO	12 (0.8)	38 (1.5)	75 (1.5)	95 (0.7)
Chicago Public Schools, IL	2 (0.9)	12 (1.7)	41 (4.3)	81 (2.5)
Delaware Science Coalition, DE	5 (1.8)	22 (4.1)	51 (4.5)	83 (2.4)
First in the World Consort., IL	22 (3.2)	56 (3.3)	87 (2.1)	98 (0.6)
Fremont/Lincoln/WestSide PS, NE	6 (2.3)	23 (4.1)	58 (4.0)	84 (2.7)
Guilford County, NC [2]	10 (2.2)	33 (3.5)	66 (4.1)	91 (1.6)
Jersey City Public Schools, NJ	6 (1.9)	17 (3.4)	48 (3.9)	82 (2.9)
Miami-Dade County PS, FL	2 (0.9)	9 (2.4)	29 (3.6)	61 (3.5)
Michigan Invitational Group, MI	12 (2.4)	39 (3.4)	77 (3.0)	96 (1.3)
Montgomery County, MD [2]	17 (2.2)	45 (1.8)	77 (1.4)	95 (1.1)
Naperville Sch. Dist. #203, IL	24 (1.7)	59 (2.2)	91 (1.1)	99 (0.4)
Project SMART Consortium, OH	11 (2.9)	34 (4.7)	70 (3.1)	95 (1.0)
Rochester City Sch. Dist., NY	2 (0.9)	9 (2.5)	32 (3.2)	73 (2.9)
SW Math/Sci. Collaborative, PA	11 (2.7)	32 (3.9)	68 (3.1)	93 (1.6)

Top 10% Benchmark (90th Percentile) = 616

Upper Quarter Benchmark (75th Percentile) = 555

Median Benchmark (50th Percentile) = 479

Lower Quarter Benchmark (25th Percentile) = 396

SOURCE: IEA Third International Mathematics and Science Study (TIMSS), 1998-1999.

States in *italics* did not fully satisfy guidelines for sample participation rates (see Appendix A for details).

† Met guidelines for sample participation rates only after replacement schools were included (see Exhibit A.6).

1 National Desired Population does not cover all of International Desired Population (see Exhibit A.3). Because coverage falls below 65%, Latvia is annotated LSS for Latvian-Speaking Schools only.

2 National Defined Population covers less than 90 percent of National Desired Population (see Exhibit A.3).

‡ Lithuania tested the same cohort of students as other countries, but later in 1999, at the beginning of the next school year.

( ) Standard errors appear in parentheses. Because results are rounded to the nearest whole number, some totals may appear inconsistent.

What Are the Gender Differences in Mathematics Achievement?

Exhibit 1.5 presents average mathematics achievement separately for girls and boys for each of the participating entities, as well as the difference between the means, in increasing order of the difference. The gender difference for each entity is shown by a bar indicating the amount of the difference, whether its direction favored girls or boys, and whether it is statistically significant (a darkened bar).

It is good news that in mathematics at the eighth grade, the TIMSS 1999 Benchmarking Study shows relatively equivalent average achievement for girls and boys in each of the Benchmarking jurisdictions. The United States as well as a number of other countries around the world appear to be making progress towards gender equity in mathematics education. On average across all TIMSS 1999 countries, there was a modest but significant difference favoring boys, although this varied considerably from country to country. The only countries with differences large enough to be statistically significant were Israel, the Czech Republic, Iran, and Tunisia.

Although achievement differences between the genders are becoming smaller in mathematics, research indicates that they still exist in those areas involving the most complex mathematical tasks, particularly as students progress to middle and secondary schools.[8] Thus, Exhibit 1.6 provides information on gender differences in mathematics achievement among students with high performance compared with those in the middle of the achievement distribution. For each entity, score levels were computed for the highest-scoring 25 percent of students, called the upper quarter level, and for the highest-scoring 50 percent, called the median level. The percentages of girls and boys in each entity reaching each of the two levels were computed. For equitable performance, 25 percent each of girls and boys should have reached the upper quarter level, and 50 percent the median level.

On average across countries, 23 percent of girls compared with 27 percent of boys reached the upper quarter level, and 49 percent of girls compared with 51 percent of boys reached the median level. These gender differences, although small, were statistically significant. In all but four countries, however, the percentages reaching the upper quarter and median levels were not significantly different, indicating

8 Fennema, E. (1996), "Mathematics, Gender, and Research" in G. Hanna (ed.), *Towards Equity in Mathematics Education*, Dordrecht, the Netherlands: Kluwer Academic Publishers.

that gender equity exists in most countries at these levels. Even though the four countries with significant differences did include the United States (as well as Israel, the Philippines, and Tunisia), this was not reflected in the results for the Benchmarking jurisdictions. Michigan was the only Benchmarking jurisdiction to show a significant gender difference favoring males among high-performing students.

Exhibit 1.5 Gender Differences in Average Mathematics Achievement

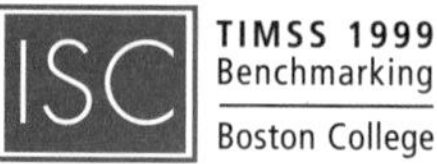

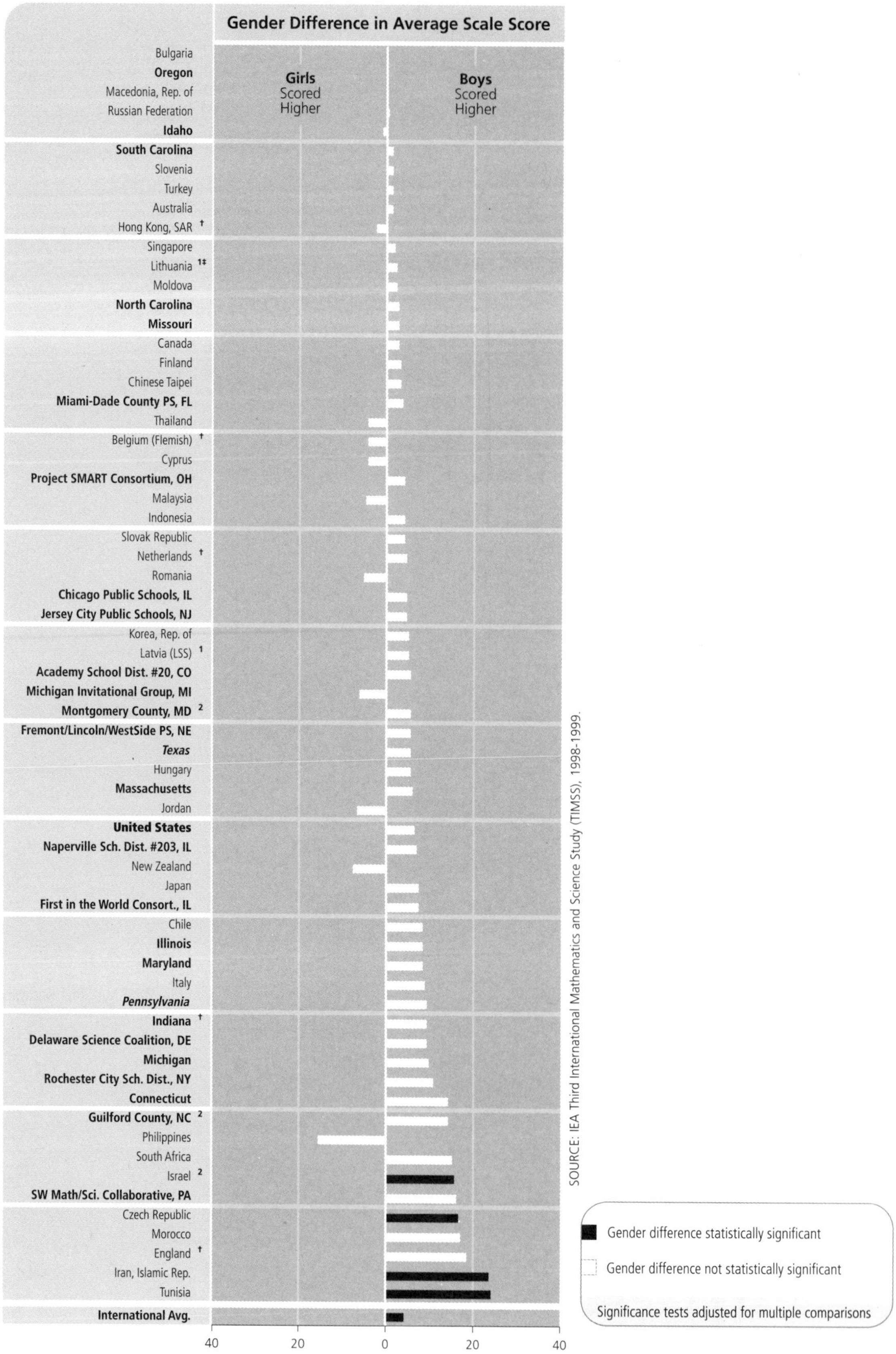
Gender Difference in Average Scale Score
Girls Scored Higher
Boys Scored Higher
Bulgaria
Oregon
Macedonia, Rep. of
Russian Federation
Idaho
South Carolina
Slovenia
Turkey
Australia
Hong Kong, SAR †
Singapore
Lithuania 1‡
Moldova
North Carolina
Missouri
Canada
Finland
Chinese Taipei
Miami-Dade County PS, FL
Thailand
Belgium (Flemish) †
Cyprus
Project SMART Consortium, OH
Malaysia
Indonesia
Slovak Republic
Netherlands †
Romania
Chicago Public Schools, IL
Jersey City Public Schools, NJ
Korea, Rep. of
Latvia (LSS) 1
Academy School Dist. #20, CO
Michigan Invitational Group, MI
Montgomery County, MD 2
Fremont/Lincoln/WestSide PS, NE
Texas
Hungary
Massachusetts
Jordan
United States
Naperville Sch. Dist. #203, IL
New Zealand
Japan
First in the World Consort., IL
Chile
Illinois
Maryland
Italy
Pennsylvania
Indiana †
Delaware Science Coalition, DE
Michigan
Rochester City Sch. Dist., NY
Connecticut
Guilford County, NC 2
Philippines
South Africa
Israel 2
SW Math/Sci. Collaborative, PA
Czech Republic
Morocco
England †
Iran, Islamic Rep.
Tunisia
International Avg.
40
20
0
20
40
SOURCE: IEA Third International Mathematics and Science Study (TIMSS), 1998-1999.
Gender difference statistically significant
Gender difference not statistically significant
Significance tests adjusted for multiple comparisons

Countries	Girls' Average Scale Score	Boys' Average Scale Score		Difference (Absolute Value)
United States	498 (3.9)	505 (4.8)		7 (3.4)
Australia	524 (5.7)	526 (5.7)		2 (6.0)
Belgium (Flemish) †	560 (7.2)	556 (8.3)		4 (14.2)
Bulgaria	510 (5.9)	511 (6.9)		0 (5.5)
Canada	529 (2.5)	533 (3.2)		3 (2.9)
Chile	388 (4.3)	397 (5.8)		9 (5.5)
Chinese Taipei	583 (3.9)	587 (5.3)		4 (4.6)
Cyprus	479 (2.1)	474 (2.7)		4 (3.3)
Czech Republic	512 (4.0)	528 (5.8)	▲	17 (5.0)
England †	487 (5.4)	505 (5.0)		19 (6.5)
Finland	519 (3.0)	522 (3.5)		3 (3.6)
Hong Kong, SAR †	583 (4.7)	581 (5.9)		2 (6.5)
Hungary	529 (4.0)	535 (4.3)		6 (3.7)
Indonesia	401 (5.4)	405 (5.0)		5 (3.3)
Iran, Islamic Rep.	408 (4.2)	432 (4.8)	▲	24 (6.5)
Israel [2]	459 (4.2)	474 (4.8)	▲	16 (4.6)
Italy	475 (4.5)	484 (4.3)		9 (4.2)
Japan	575 (2.4)	582 (2.3)		8 (3.3)
Jordan	431 (4.7)	425 (5.9)		7 (8.1)
Korea, Rep. of	585 (3.1)	590 (2.2)		5 (3.7)
Latvia (LSS) [1]	502 (3.8)	508 (4.4)		5 (4.5)
Lithuania [1‡]	480 (4.7)	483 (4.8)		3 (4.0)
Macedonia, Rep. of	446 (5.3)	447 (4.3)		0 (4.5)
Malaysia	521 (4.7)	517 (6.0)		5 (6.1)
Moldova	468 (4.1)	471 (4.7)		3 (4.1)
Morocco	326 (5.3)	344 (4.1)		17 (7.7)
Netherlands †	538 (7.6)	542 (7.0)		5 (3.0)
New Zealand	495 (5.5)	487 (7.6)		7 (8.3)
Philippines	352 (6.9)	337 (6.5)		15 (6.1)
Romania	475 (6.3)	470 (6.2)		5 (4.7)
Russian Federation	526 (6.0)	526 (6.4)		1 (3.3)
Singapore	603 (6.1)	606 (7.5)		2 (5.7)
Slovak Republic	532 (4.2)	536 (4.5)		5 (3.6)
Slovenia	529 (3.0)	531 (3.6)		1 (3.6)
South Africa	267 (7.5)	283 (7.3)		16 (5.9)
Thailand	469 (5.7)	465 (5.5)		4 (4.9)
Tunisia	436 (2.4)	460 (2.9)	▲	25 (2.2)
Turkey	428 (4.7)	429 (4.4)		2 (2.8)
International Avg. (All Countries)	485 (0.8)	489 (0.9)	▲	4 (1.1)

States	Girls' Average Scale Score	Boys' Average Scale Score	Difference (Absolute Value)
Connecticut	506 (8.9)	520 (9.8)	14 (5.3)
Idaho	495 (7.1)	495 (8.2)	1 (4.3)
Illinois	505 (8.0)	514 (6.1)	9 (4.5)
Indiana †	510 (6.8)	519 (8.0)	10 (3.9)
Maryland	490 (6.4)	499 (6.8)	9 (4.2)
Massachusetts	510 (6.4)	517 (6.0)	6 (3.5)
Michigan	512 (7.2)	522 (8.1)	10 (3.9)
Missouri	488 (5.9)	491 (5.6)	3 (4.5)
North Carolina	494 (7.9)	497 (6.9)	3 (4.9)
Oregon	514 (6.6)	514 (6.9)	0 (6.0)
Pennsylvania	503 (6.2)	512 (7.2)	10 (4.2)
South Carolina	501 (8.0)	502 (7.6)	1 (5.0)
Texas	513 (8.2)	519 (10.7)	6 (5.7)

Districts and Consortia	Girls' Average Scale Score	Boys' Average Scale Score	Difference (Absolute Value)
Academy School Dist. #20, CO	526 (2.9)	531 (3.4)	6 (5.2)
Chicago Public Schools, IL	460 (6.3)	465 (6.7)	5 (4.5)
Delaware Science Coalition, DE	475 (8.9)	485 (11.1)	10 (9.2)
First in the World Consort., IL	556 (6.7)	564 (6.8)	8 (7.1)
Fremont/Lincoln/WestSide PS, NE	485 (8.3)	491 (10.2)	6 (8.7)
Guilford County, NC [2]	507 (8.3)	521 (8.2)	14 (5.8)
Jersey City Public Schools, NJ	472 (8.8)	478 (9.2)	5 (4.6)
Miami-Dade County PS, FL	419 (9.3)	423 (12.1)	4 (10.3)
Michigan Invitational Group, MI	535 (5.4)	529 (7.4)	6 (5.8)
Montgomery County, MD [2]	534 (5.5)	540 (4.4)	6 (7.0)
Naperville Sch. Dist. #203, IL	566 (3.3)	573 (3.3)	7 (3.3)
Project SMART Consortium, OH	518 (7.8)	523 (8.1)	4 (5.0)
Rochester City Sch. Dist., NY	439 (7.8)	450 (6.6)	11 (6.2)
SW Math/Sci. Collaborative, PA	509 (7.5)	525 (8.5)	16 (5.3)

SOURCE: IEA Third International Mathematics and Science Study (TIMSS), 1998-1999.

▲ Significantly higher than other gender

Significance tests adjusted for multiple comparisons

States in *italics* did not fully satisfy guidelines for sample participation rates (see Appendix A for details).

† Met guidelines for sample participation rates only after replacement schools were included (see Exhibit A.6).

1 National Desired Population does not cover all of International Desired Population (see Exhibit A.3). Because coverage falls below 65%, Latvia is annotated LSS for Latvian-Speaking Schools only.

2 National Defined Population covers less than 90 percent of National Desired Population (see Exhibit A.3)

‡ Lithuania tested the same cohort of students as other countries, but later in 1999, at the beginning of the next school year.

( ) Standard errors appear in parentheses. Because results are rounded to the nearest whole number, some totals may appear inconsistent.

Percentages of Girls and Boys Reaching Each Participant's Own Upper Quarter and Median Levels of Mathematics Achievement

8th Grade Mathematics

Countries	Upper Quarter: Percent of Girls	Upper Quarter: Percent of Boys	Median: Percent of Girls	Median: Percent of Boys
United States	23 (1.3)	27 (1.9) ▲	49 (2.0)	51 (2.3)
Australia	24 (2.8)	26 (2.6)	49 (3.2)	51 (3.0)
Belgium (Flemish) †	25 (2.5)	25 (2.5)	50 (3.1)	50 (3.5)
Bulgaria	24 (3.1)	26 (3.5)	51 (3.0)	49 (3.2)
Canada	24 (1.2)	26 (1.4)	49 (1.3)	51 (1.9)
Chile	23 (1.9)	27 (2.6)	48 (2.2)	52 (2.4)
Chinese Taipei	22 (1.5)	28 (1.9)	49 (1.9)	51 (2.1)
Cyprus	24 (1.4)	26 (1.4)	50 (1.4)	50 (1.5)
Czech Republic	22 (1.6)	28 (2.5)	46 (2.4)	54 (2.9)
England †	20 (2.7)	30 (2.4)	46 (3.0)	54 (2.7)
Finland	23 (1.8)	27 (2.2)	49 (1.9)	51 (2.2)
Hong Kong, SAR †	24 (2.5)	26 (2.4)	50 (2.9)	50 (3.1)
Hungary	24 (1.9)	26 (1.8)	48 (2.2)	52 (2.1)
Indonesia	25 (1.6)	25 (1.7)	49 (2.1)	52 (2.1)
Iran, Islamic Rep.	19 (2.0)	29 (2.2)	43 (2.5)	55 (2.5)
Israel 2	21 (1.5)	29 (1.7) ▲	47 (2.0)	53 (2.2)
Italy	23 (1.8)	28 (1.7)	47 (2.2)	53 (2.2)
Japan	23 (1.3)	27 (1.1)	47 (1.5)	53 (1.3)
Jordan	24 (1.7)	26 (2.1)	51 (2.0)	49 (2.2)
Korea, Rep. of	24 (1.1)	26 (1.0)	48 (1.5)	52 (1.3)
Latvia (LSS) 1	24 (1.9)	27 (2.1)	49 (2.2)	52 (2.2)
Lithuania 1‡	24 (2.5)	26 (2.3)	50 (2.5)	50 (2.5)
Macedonia, Rep. of	26 (1.8)	24 (1.6)	51 (2.4)	49 (2.0)
Malaysia	26 (2.3)	24 (2.9)	52 (2.6)	48 (3.4)
Moldova	24 (1.6)	27 (2.1)	50 (2.1)	51 (2.2)
Morocco	21 (1.7)	28 (1.5)	45 (2.2)	54 (1.7)
Netherlands †	24 (3.6)	26 (3.2)	48 (4.2)	52 (4.4)
New Zealand	26 (2.6)	24 (3.5)	52 (3.0)	48 (3.5)
Philippines	27 (2.7)	23 (2.5)	53 (2.7) ▲	46 (2.5)
Romania	25 (2.3)	25 (2.4)	51 (2.8)	49 (2.8)
Russian Federation	24 (2.4)	26 (2.5)	49 (2.9)	51 (3.2)
Singapore	23 (3.1)	26 (3.4)	49 (3.6)	51 (4.2)
Slovak Republic	23 (2.0)	27 (2.2)	48 (2.6)	52 (2.7)
Slovenia	24 (1.6)	26 (1.5)	49 (1.7)	51 (2.0)
South Africa	23 (2.7)	27 (2.3)	47 (2.5)	53 (2.1)
Thailand	25 (2.6)	24 (2.4)	50 (2.9)	50 (2.7)
Tunisia	19 (1.4)	31 (1.6) ▲	42 (1.7)	59 (1.6) ▲
Turkey	25 (1.8)	25 (1.9)	50 (2.2)	50 (1.8)
International Avg. (All Countries)	23 (0.4)	27 (0.4) ▲	49 (0.4)	51 (0.4) ▲

SOURCE: IEA Third International Mathematics and Science Study (TIMSS), 1998-1999.

▲ Significantly greater percentage than other gender

Significance tests adjusted for multiple comparisons

States in *italics* did not fully satisfy guidelines for sample participation rates (see Appendix A for details).

† Met guidelines for sample participation rates only after replacement schools were included (see Exhibit A.6).

1 National Desired Population does not cover all of International Desired Population (see Exhibit A.3). Because coverage falls below 65%, Latvia is annotated LSS for Latvian-Speaking Schools only.

2 National Defined Population covers less than 90 percent of National Desired Population (see Exhibit A.3).

‡ Lithuania tested the same cohort of students as other countries, but later in 1999, at the beginning of the next school year.

( ) Standard errors appear in parentheses. Because results are rounded to the nearest whole number, some totals may appear inconsistent.

	Upper Quarter			Median	
	Percent of Girls	Percent of Boys		Percent of Girls	Percent of Boys
States					
Connecticut	21 (3.1)	29 (3.9)		47 (4.7)	53 (4.4)
Idaho	24 (3.0)	26 (3.0)		49 (3.5)	51 (4.1)
Illinois	23 (3.1)	27 (2.9)		48 (3.7)	52 (3.1)
Indiana †	22 (3.6)	28 (3.7)		47 (4.1)	53 (5.1)
Maryland	22 (2.6)	28 (2.6)		48 (3.4)	52 (3.2)
Massachusetts	23 (2.7)	27 (2.7)		48 (3.4)	52 (3.0)
Michigan	22 (3.3)	29 (3.6)	▲	48 (4.3)	52 (3.6)
Missouri	23 (2.7)	27 (2.7)		49 (3.3)	51 (2.5)
North Carolina	24 (3.5)	26 (2.8)		49 (3.6)	51 (3.5)
Oregon	24 (2.7)	27 (2.8)		49 (3.2)	51 (3.5)
Pennsylvania	22 (3.0)	28 (2.9)		48 (3.2)	52 (3.6)
South Carolina	24 (3.2)	27 (3.2)		49 (3.8)	51 (3.3)
Texas	22 (3.1)	28 (3.7)		48 (4.4)	52 (4.7)
Districts and Consortia					
Academy School Dist. #20, CO	22 (1.6)	28 (1.9)		48 (2.3)	52 (2.1)
Chicago Public Schools,	23 (2.9)	27 (3.6)		50 (4.3)	51 (3.5)
Delaware Science Coalition, DE	22 (4.3)	29 (5.2)		47 (4.9)	53 (5.1)
First in the World Consort., IL	22 (3.8)	28 (3.7)		49 (3.6)	51 (3.9)
Fremont/Lincoln/WestSide PS, NE	24 (3.7)	26 (4.7)		50 (4.0)	50 (4.1)
Guilford County, NC ²	22 (3.0)	28 (4.2)		47 (4.6)	54 (4.3)
Jersey City Public Schools, NJ	24 (3.8)	26 (4.7)		49 (4.6)	51 (3.5)
Miami-Dade County PS, FL	23 (4.1)	27 (3.5)		50 (3.9)	50 (5.0)
Michigan Invitational Group, MI	25 (3.6)	25 (3.6)		51 (4.2)	49 (4.5)
Montgomery County, MD ²	24 (2.3)	26 (2.2)		48 (2.8)	52 (2.0)
Naperville Sch. Dist. #203, IL	23 (1.9)	27 (2.1)		49 (2.6)	51 (2.7)
Project SMART Consortium, OH	24 (4.5)	26 (4.4)		49 (4.8)	51 (5.0)
Rochester City Sch. Dist., NY	22 (3.9)	29 (3.0)		48 (4.4)	52 (3.7)
SW Math/Sci. Collaborative, PA	22 (3.1)	29 (4.2)		47 (4.3)	54 (4.3)

SOURCE: IEA Third International Mathematics and Science Study (TIMSS), 1998-1999.

▲ Significantly greater percentage than other gender

Significance tests adjusted for multiple comparisons

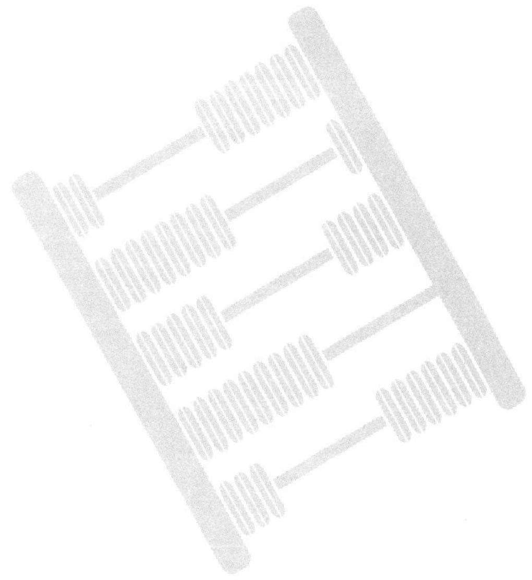

CHAPTER 2

Performance at International Benchmarks

The TIMSS 1999 international benchmarks delineate performance of the top 10 percent, top quarter, top half, and lower quarter of students in the entities participating in the study. To help interpret the achievement results, Chapter 2 describes eighth-grade mathematics achievement at each of these benchmarks together with examples of the types of items typically answered correctly by students performing at the benchmark.

2

To provide an idea of the mathematics understandings and skills displayed by students performing at different levels on the TIMSS mathematics achievement scale, TIMSS described performance at four international benchmarks. The TIMSS 1999 international benchmarks delineate performance of the top 10 percent, top quarter, top half, and lower quarter of students in the countries participating in the TIMSS 1999 study. (The benchmarks were set at the 90th, 75th, 50th, and 25th percentiles, respectively.)

As states and school districts spend time and energy on improving students' mathematics achievement, it is important that educators, curriculum developers, and policy makers understand what students know and can do in mathematics, and what areas, concepts, and topics need more focus and effort. To help interpret the range of achievement results for the TIMSS 1999 Benchmarking participants presented in Chapter 1, this chapter describes eighth-grade mathematics achievement at each of the TIMSS 1999 international benchmarks, explaining the types of mathematics understandings and skills typically displayed by students performing at the benchmarks. The benchmark descriptions are presented together with examples of the types of mathematics test questions typically answered correctly by students reaching the benchmark. Appendix D contains the descriptions of the understandings and skills assessed by each item in the TIMSS 1999 assessment at each benchmark.[1]

For each of the example test questions, the percentages of correct responses are provided for selected countries as well as for the jurisdictions participating in the TIMSS 1999 Benchmarking project. The countries and Benchmarking jurisdictions are presented in descending order, with those performing highest shown first. The countries included for purposes of comparison are the United States as well as a dozen European and Asian countries of interest. These include several high-performing European countries (Belgium (Flemish), the Czech Republic, the Netherlands, and the Russian Federation), countries that are major economic trading partners of the United States (Canada, England, and Italy), and the top-scoring Asian countries of Chinese Taipei, Hong Kong, Japan, Korea, and Singapore.

Presented previously in Chapter 1, Exhibit 1.4 shows the percentages of students in each participating entity reaching each international benchmark – Top 10%, Upper Quarter, Median, and Lower Quarter. If an entity had high average achievement in mathematics and a large percentage of its students at or above the upper benchmarks, this indicates that the students are concentrated among the highest-achieving

1 For a detailed description of the items and benchmarks for TIMSS 1995 at fourth and eighth grades and how they compare to the National Council of Teachers of Mathematics' (NCTM) *Principles and Standards for School Mathematics*, see Kelly, D.L., Mullis, I.V.S., and Martin, M.O., *Profiles of Student Achievement in Mathematics at the TIMSS International Benchmarks: U.S. Performance and Standards in an International Context*, Chestnut Hill, MA: Boston College.

students internationally. For example, top-performing Singapore had nearly half (46 percent) of its students reaching the Top 10% Benchmark and three-fourths (75 percent) reaching the Upper Quarter Benchmark – the point on the scale that typically only 25 percent of the students would be expected to reach if achievement were distributed equally from country to country. Most of the Singaporean students (93 percent) reached the Median Benchmark. Performance in the United States was closer to the distribution that might be expected if achievement were distributed the same from country to country: nine percent of the students reached the Top 10% Benchmark, 28 percent reached the Top Quarter Benchmark, and 61 percent reached the Median Benchmark.

The analysis of performance at these benchmarks in mathematics suggests that three primary factors appeared to differentiate performance at the four levels:

- The mathematical operation required
- The complexity of the numbers or number system
- The nature of the problem situation.

For example, there is evidence that students performing at the lower end of the scale could add, subtract, and multiply whole numbers. In contrast, students performing at the higher end of the scale solved non-routine problems involving relationships among fractions, decimals, and percents; various geometric properties; and algebraic rules.

How Were the Benchmark Descriptions Developed?

To develop descriptions of achievement at the TIMSS 1999 international benchmarks, the International Study Center used the scale anchoring method. Scale anchoring is a way of describing students' performance at different points on the TIMSS 1999 achievement scale in terms of the types of items they answered correctly. It involves an empirical component in which items that discriminate between successive points on the scale are identified, and a judgmental component in which subject-matter experts examine the content of the items and generalize to students' knowledge and understandings.

For the scale anchoring analysis, the results of students from all the TIMSS 1999 countries were pooled, so that the benchmark descriptions refer to all students achieving at that level. (That is, it does not matter which country the students are from, only how they performed on the test.) Certain criteria were applied to the TIMSS 1999 achievement scale results to identify the sets of items that students reaching each international benchmark were likely to answer correctly and those at the next lower benchmark were unlikely to answer correctly.[2] The sets of items thus produced represented the accomplishments of students reaching each benchmark and were used by a panel of subject-matter experts from the TIMSS countries to develop the benchmark descriptions.[3] The work of the panel involved developing a short description for each item of the mathematical understandings demonstrated by students answering it correctly, summarizing students' knowledge and understandings across the set of items for each benchmark to provide more general statements of achievement, and selecting example items illustrating the descriptions.

How Should the Descriptions Be Interpreted?

In general, the parts of the descriptions that relate to the understanding of mathematical concepts or familiarity with procedures are relatively straightforward. It needs to be acknowledged, however, that the cognitive behavior necessary to answer some items correctly may vary according to students' experience. An item may require only simple recall for a student familiar with the item's content and context, but necessitate problem-solving strategies from one unfamiliar with the material. Nevertheless, the descriptions are based on what the panel believed to be the way the great majority of eighth-grade students could be expected to perform.

It also needs to be emphasized that the descriptions of achievement characteristic of students at the international benchmarks are based solely on student performance on the TIMSS 1999 items. Since those items were developed in particular to sample the mathematics domains prescribed for this study, neither the set of items nor the descriptions based on them purport to be comprehensive. There are undoubtedly other mathematics curriculum elements on which students at the various benchmarks would have been successful if they had been included in the assessment.

Please note that students reaching a particular benchmark demonstrated the knowledge and understandings characterizing that benchmark as well as those characterizing the lower benchmarks. The description of achievement at each benchmark is cumulative, building on the description of achievement demonstrated by students at the lower benchmarks.

2 For example, for the Top 10% Benchmark, an item was included if at least 65 percent of students scoring at the scale point corresponding to this benchmark answered the item correctly and less than 50 percent of students scoring at the Upper Quarter Benchmark answered it correctly. Similarly, for the Upper Quarter Benchmark, an item was included if at least 65 percent of students scoring at that point answered the item correctly and less than 50 percent of students at the Median Benchmark answered it correctly.

3 The participants in the scale anchoring process are listed in Appendix E.

Finally, it must be emphasized that the descriptions of the international benchmarks are one possible way of beginning to examine student performance. Some students scoring below a benchmark may indeed know or understand some of the concepts that characterize a higher level. Thus, it is important to consider performance on the individual items and clusters of items in developing a profile of student achievement in each participating entity.

Several example items are included for each benchmark to complement the descriptions by giving a more concrete notion of the abilities students demonstrated. Each example item is accompanied by the percentage of correct responses for each TIMSS 1999 Benchmarking participant. Percentages are also provided for selected countries, as is the international average for all 38 countries that participated in TIMSS 1999. In general, the several entities scoring highest on the overall test also scored highest on many of the example items. Not surprisingly, this was true for items assessing a range of performance expectations – recall, ability to carry out routine procedures, and ability to solve routine and non-routine problems. The TIMSS 1999 results support the premise that successful problem solving is grounded in mastery of more fundamental knowledge and skills.

Item Examples and Student Performance

The remainder of this chapter describes each benchmark and presents three to five example items illustrating what students know and can do at that level. The correct answer is circled for multiple-choice items. For open-ended items, the answers shown exemplify the types of student responses that were given full credit. The example items are ones that students reaching each benchmark were likely to answer correctly, and they represent the types of items used to develop the description of achievement at that benchmark.[4]

4 Some of the items used to develop the benchmark descriptions are being kept secure to measure achievement trends in future TIMSS assessments and are not available for publication.

Achievement at the Top 10% Benchmark

Exhibit 2.1 describes performance at the Top 10% Benchmark. Students reaching this benchmark demonstrated the ability to organize information in problem-solving situations and to apply their understanding of mathematical relationships. They typically demonstrated success on the knowledge and skills represented by this benchmark, as well as those demonstrated at the three lower benchmarks.

Example Item 1 in Exhibit 2.2 illustrates the type of measurement item a student performing at the Top 10% Benchmark generally answered correctly. As can be seen, students had to apply their knowledge of the area of rectangles and inscribed shapes to solve a two-step problem about the area of a garden path. The international average for this item was 42 percent correct, indicating that this was a relatively difficult item for eighth graders around the world. Nevertheless, more than two-thirds of the students answered the item correctly in Hong Kong, Singapore, Japan, Chinese Taipei, and Korea. Among the Benchmarking participants, eighth graders in the Naperville School District did as well as their counterparts in the high-performing Asian countries, with 69 percent answering correctly. Generally, however, students in the United States – in the country as a whole and in the Benchmarking entities – performed relatively less well than students internationally on measurement questions involving relationships between shapes. No other Benchmarking entity performed significantly above the international average on this test question, and students in six Benchmarking entities and in the United States overall performed significantly below the international average. On average internationally, more than 20 percent of students chose Option A, solving for the area of the larger rectangle rather than that of the path. Option C was an equally popular distracter, selected by more than 20 percent of students internationally.

Unlike students performing at lower benchmarks, students reaching the Top 10% Benchmark typically could correctly answer multistep word problems. Example Item 2 in Exhibit 2.3 requires students to select relevant information from two advertisements to solve a complex multistep word problem involving decimals. Given the price for each issue of a magazine and a certain number of free issues, students were asked to calculate which of the two magazine subscriptions was the less expensive for 24 issues. Students received full credit if they showed correct calculations for at least one of the subscriptions, identified the less expensive magazine, and calculated the difference between the two

• Top 10% Benchmark

Summary

Students can organize information, make generalizations, and explain solution strategies in non-routine problem solving situations. They can organize information and make generalizations to solve problems; apply knowledge of numeric, geometric, and algebraic relationships to solve problems (e.g., among fractions, decimals, and percents; geometric properties; and algebraic rules); and find the equivalent forms of algebraic expressions.

Students can organize information in problem-solving situations. They can select and organize information from two sources to solve a complex word problem involving decimals and organize information to solve a multi-step word problem involving whole numbers.

Students can correctly order the four basic operations in computing with decimals and fractions. Students use their understanding of fractions and decimals in multi-step problem situations. They can solve a problem involving both addition and subtraction of simple common fractions and a problem involving multiplication and subtraction of decimals. They can solve word problems involving fractions and decimals which require analysis of the verbal relations described. They can order a set of decimal fractions of up to three decimal places and can identify the pair of numbers satisfying given conditions involving ordering integers, decimals, and fractions. They can solve a time-distance-rate problem involving decimals and the conversion of minutes to seconds. They can work with part-whole ratios and can solve word problems to find the percent change.

Students can apply their knowledge of measurement in more complex problem situations. They can solve problems involving area and perimeter of rectangles and area of inscribed triangles. They apply knowledge of properties of squares to solve multi-step word problems and draw a new rectangle based on a given rectangle and express the ratio of their areas. They can relate different units of time and apply their knowledge of the number of milliliters in a liter to solve a word problem. They recognize that precision of measurement is related to the size of the unit of measurement.

Students can use their knowledge of angles – overlapping and measures of angles in quadrilaterals – to solve problems. They can use their knowledge of congruent and similar triangles to solve problems concerning corresponding parts. They can identify the coordinates of a point on a line given the coordinates of two other points on the line and locate a point on a number line given its distance from two other points on the line. They can identify the image of a triangle under a rotation in a plane.

Students can use proportion to find missing values in a table. Students can identify an equivalent form of a linear inequality involving a fraction. Students can recognize properties of number operations represented in symbolic form. They can solve a multi-step word problem in which there are two unknowns.

Given the first several terms in pictorial form, that grow in either one or two dimensions, students can make generalizations to find terms in the sequences (e.g. 51st), and they can explain the process used to find those terms.

90th Percentile: 616

SOURCE: IEA Third International Mathematics and Science Study (TIMSS), 1998-1999.

subscriptions. With an international average of 24 percent correct (for full credit), this item was among the most difficult in TIMSS 1999. Singapore, Korea, and Chinese Taipei were the only countries where the majority of the students answered correctly. The best performance by a Benchmarking entity was in Naperville, where 41 percent of the eighth graders answered correctly. Students in the First of World Consortium (36 percent) and Montgomery County (35 percent) also performed significantly above the international average.

Students reaching the Top 10% Benchmark exhibited an understanding of the properties of similar triangles, as shown by Example Item 3 (see Exhibit 2.4). Given two angle measurements, the length of a side of a triangle, and the dimensions of a second similar triangle, students needed to find the length of an unlabeled side of the first triangle. Internationally, most eighth-grade students had not mastered the concept of proportionality of corresponding sides or could not solve the resulting equation; only 37 percent, on average, answered the question correctly. In comparison, top-performing Korea had 70 percent correct responses. Among the TIMSS 1999 countries, only in Korea, Japan, Singapore, Hong Kong, Chinese Taipei, and Belgium (Flemish) did at least half the students answer correctly. In the Benchmarking jurisdictions, correct responses were provided by more than half the eighth graders in Naperville (56 percent) and the First in the World Consortium (52 percent).

The eighth-grade students reaching the Top 10% Benchmark typically were able to apply a generalization to solve a sequence problem like the one shown in Example Item 4 in Exhibit 2.5. In this algebra problem, given the initial terms in a sequence and the 50th term of that sequence, students generalized to find the 51st term. Even though results are presented only for Part C, this problem was presented in three parts, A, B, and C. To provide some scaffolding, parts A and B asked students to indicate how many circles would be in the 5th and 7th figures, respectively, if the pattern were extended. On average internationally, 65 percent of the students answered Part A correctly and 54 percent successfully extended the sequence to the 7th figure in Part B.

To receive full credit for Part C, students had to show or explain how they arrived at their answer by providing a general expression or an equation and by calculating the correct number of circles for the 51st figure. Internationally on average, 30 percent of the students received full credit for their responses. In comparison, about two-thirds of the students in Korea, Chinese Taipei, Japan, and Singapore received full credit. Although eighth graders in six Benchmarking entities – First in the World, Naperville, the Michigan Invitational Group, Montgomery County, the Academy School District, and Oregon – performed significantly above the international average, their performance was below that of the top performers, ranging from 54 to 39 percent correct. Most students added the sequence number to the number of circles in the preceding figure: 1275 + 51 = 1326. Very few calculated the answer by a general expression: n(n+1)/2 or 51(52)/2 (although 13 percent of the Dutch students did so).

Content Area: Measurement

Description: Finds the area between two rectangles when one is inside the other and their sides are parallel.

A rectangular garden that is next to a building has a path around the other three sides, as shown.

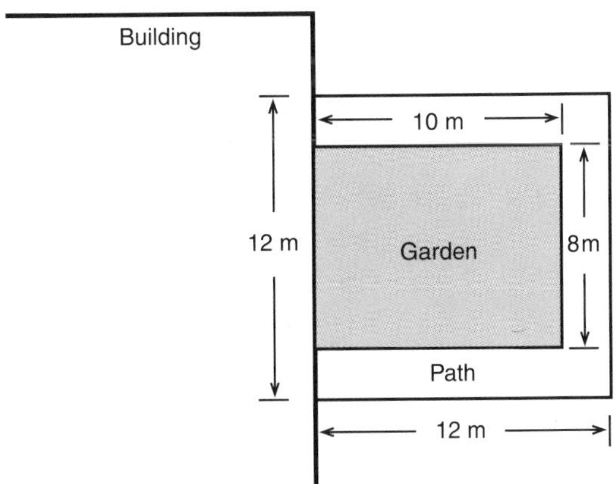

What is the area of the path?

A. 144 m^2

(B.) 64 m^2

C. 44 m^2

D. 16 m^2

	Overall Percent Correct	
Hong Kong, SAR †	79 (2.0)	▲
Singapore	78 (2.6)	▲
Japan	74 (1.9)	▲
Chinese Taipei	73 (2.1)	▲
Naperville Sch. Dist. #203, IL	69 (4.0)	▲
Korea, Rep. of	67 (1.7)	▲
Netherlands †	57 (4.4)	▲
First in the World Consort., IL	56 (5.9)	●
Canada	51 (3.0)	●
Belgium (Flemish) †	51 (2.2)	▲
Montgomery County, MD 2	46 (4.6)	●
Italy	45 (2.7)	●
Oregon	42 (3.9)	●
Michigan Invitational Group, MI	41 (4.4)	●
Czech Republic	40 (3.5)	●
England †	40 (3.3)	●
Illinois	40 (3.4)	●
Project SMART Consortium, OH	39 (5.2)	●
Maryland	38 (2.8)	●
Russian Federation	38 (3.2)	●
Fremont/Lincoln/WestSide PS, NE	38 (3.6)	●
SW Math/Sci. Collaborative, PA	38 (4.1)	●
Texas	38 (3.8)	●
Massachusetts	35 (3.0)	●
Academy School Dist. #20, CO	35 (4.4)	●
Guilford County, NC 2	35 (5.7)	●
Indiana †	34 (3.1)	●
Idaho	34 (2.8)	●
Connecticut	33 (3.9)	●
Michigan	33 (3.3)	●
United States	33 (1.6)	▼
Delaware Science Coalition, DE	32 (3.1)	●
Missouri	31 (3.0)	▼
Pennsylvania	30 (2.8)	▼
South Carolina	30 (4.1)	●
North Carolina	29 (3.2)	▼
Rochester City Sch. Dist., NY	27 (4.4)	▼
Chicago Public Schools, IL	26 (5.9)	●
Jersey City Public Schools, NJ	25 (3.5)	▼
Miami-Dade County PS, FL	21 (3.0)	▼
International Avg. (All Countries)	42 (0.4)	

▲ Participant average significantly higher than international average

● No statistically significant difference between participant average and international average

▼ Participant average significantly lower than international average

Significance tests adjusted for multiple comparisons

SOURCE: IEA Third International Mathematics and Science Study (TIMSS), 1998-1999.

* The item was answered correctly by a majority of students reaching this benchmark.

States in *italics* did not fully satisfy guidelines for sample participation rates (see Appendix A for details).

† Met guidelines for sample participation rates only after replacement schools were included (see Exhibit A.6).

2 National Defined Population covers less than 90 percent of National Desired Population (see Exhibit A.3).

( ) Standard errors appear in parentheses. Because results are rounded to the nearest whole number, some totals may appear inconsistent.

Exhibit 2.3 | **Top 10% TIMSS International Benchmark – Example Item 2**
An Item That Students Reaching the Top 10% International Benchmark Are Likely to Answer Correctly*

Content Area: Data Representation, Analysis and Probability

Description: Selects relevant information from two advertisements to solve a complex word problem involving decimals.

Chris plans to order 24 issues of a magazine. He reads the following advertisements for two magazines. *Ceds* are the units of currency in Chris' country.

Teen Life Magazine	**Teen News Magazine**
24 issues First four issues FREE The rest 3 *ceds* each.	24 issues First six issues FREE The rest 3.5 *ceds* each.

Which magazine is the least expensive for 24 issues? How much less expensive? Show your work.

Teen Life = 20
× 3
60 ceds

24 = 60 ceds

Teen News = 18
×3.5
90
540
63.0 ceds

24 = 63 ceds

Teen Life is less expensive by 3 ceds.

The answer shown illustrates the type of student response that was given full credit.

	Overall Percent Correct	
Singapore	57 (2.1)	▲
Korea, Rep. of	52 (1.5)	▲
Chinese Taipei	50 (1.8)	▲
Belgium (Flemish) †	42 (1.7)	▲
Naperville Sch. Dist. #203, IL	41 (2.6)	▲
Japan	39 (1.5)	▲
First in the World Consort., IL	36 (2.9)	▲
Montgomery County, MD 2	35 (2.8)	▲
Hong Kong, SAR †	34 (1.8)	▲
Czech Republic	34 (2.5)	▲
Canada	32 (1.8)	▲
Connecticut	32 (2.7)	●
Texas	31 (4.0)	●
Russian Federation	30 (2.4)	●
Project SMART Consortium, OH	30 (3.5)	●
Indiana †	29 (3.5)	●
Massachusetts	29 (2.7)	●
Michigan Invitational Group, MI	29 (2.2)	●
Academy School Dist. #20, CO	27 (2.5)	●
Italy	27 (1.7)	●
Jersey City Public Schools, NJ	27 (4.4)	●
SW Math/Sci. Collaborative, PA	27 (3.2)	●
Guilford County, NC 2	26 (2.4)	●
Pennsylvania	26 (2.9)	●
United States	26 (1.4)	●
Michigan	26 (2.2)	●
Illinois	25 (3.1)	●
Netherlands †	25 (2.7)	●
South Carolina	25 (2.2)	●
Idaho	25 (2.8)	●
North Carolina	23 (2.2)	●
Maryland	23 (2.1)	●
Oregon	22 (2.5)	●
Delaware Science Coalition, DE	22 (3.8)	●
Missouri	21 (1.6)	●
Fremont/Lincoln/WestSide PS, NE	20 (3.7)	●
Chicago Public Schools, IL	19 (3.4)	●
England †	17 (1.9)	▼
Rochester City Sch. Dist., NY	15 (2.3)	▼
Miami-Dade County PS, FL	11 (2.3)	▼
International Avg. (All Countries)	24 (0.3)	

SOURCE: IEA Third International Mathematics and Science Study (TIMSS), 1998-1999.

▲ Participant average significantly higher than international average

● No statistically significant difference between participant average and international average

▼ Participant average significantly lower than international average

Significance tests adjusted for multiple comparisons

* The item was answered fully correctly by a majority of students reaching this benchmark.

States in *italics* did not fully satisfy guidelines for sample participation rates (see Appendix A for details).

† Met guidelines for sample participation rates only after replacement schools were included (see Exhibit A.6).

2 National Defined Population covers less than 90 percent of National Desired Population (see Exhibit A.3).

( ) Standard errors appear in parentheses. Because results are rounded to the nearest whole number, some totals may appear inconsistent.

Top 10% TIMSS International Benchmark – Example Item 3

An Item That Students Reaching the Top 10% International Benchmark Are Likely to Answer Correctly*

8th Grade Mathematics

Content Area: Geometry

Description: Uses properties of similar triangles to find the length of a corresponding side.

The figure represents two similar triangles. The triangles are not drawn to scale.

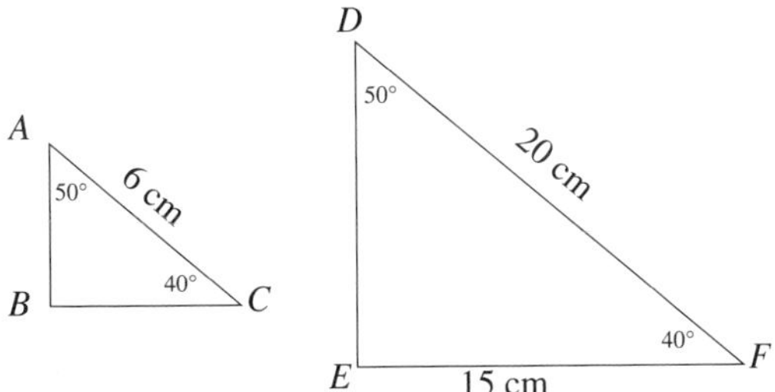

In the actual triangle *ABC*, what is the length of side *BC*?

A 3.5 cm

 B. 4.5 cm

C. 5 cm

D. 5.5 cm

E. 8 cm

	Overall Percent Correct	
Korea, Rep. of	70 (1.9)	▲
Japan	68 (1.9)	▲
Singapore	64 (2.7)	▲
Hong Kong, SAR †	56 (2.2)	▲
Naperville Sch. Dist. #203, IL	56 (3.6)	▲
First in the World Consort., IL	52 (4.7)	▲
Chinese Taipei	52 (2.3)	▲
Belgium (Flemish) †	50 (3.2)	▲
Academy School Dist. #20, CO	46 (4.2)	●
Guilford County, NC [2]	45 (5.4)	●
Netherlands †	44 (3.1)	●
SW Math/Sci. Collaborative, PA	43 (2.9)	●
Texas	43 (5.0)	●
Montgomery County, MD [2]	42 (3.6)	●
Russian Federation	41 (2.7)	●
Connecticut	40 (3.8)	●
Illinois	40 (2.2)	●
Idaho	39 (4.2)	●
Massachusetts	38 (2.8)	●
North Carolina	38 (3.4)	●
Indiana †	38 (3.7)	●
Michigan	37 (3.3)	●
South Carolina	37 (2.6)	●
Delaware Science Coalition, DE	37 (4.1)	●
Oregon	36 (3.9)	●
United States	36 (1.6)	●
Maryland	35 (2.5)	●
Michigan Invitational Group, MI	35 (4.0)	●
Canada	35 (2.2)	●
England †	34 (2.7)	●
Chicago Public Schools, IL	32 (4.5)	●
Jersey City Public Schools, NJ	32 (3.7)	●
Miami-Dade County PS, FL	32 (3.2)	●
Czech Republic	32 (2.5)	●
Pennsylvania	32 (2.8)	●
Project SMART Consortium, OH	31 (4.4)	●
Italy	29 (2.4)	▼
Fremont/Lincoln/WestSide PS, NE	29 (5.7)	●
Missouri	27 (3.0)	●
Rochester City Sch. Dist., NY	26 (4.0)	●
International Avg. (All Countries)	37 (0.4)	

▲ Participant average significantly higher than international average

● No statistically significant difference between participant average and international average

▼ Participant average significantly lower than international average

Significance tests adjusted for multiple comparisons

SOURCE: IEA Third International Mathematics and Science Study (TIMSS), 1998-1999.

* The item was answered correctly by a majority of students reaching this benchmark.

States in *italics* did not fully satisfy guidelines for sample participation rates (see Appendix A for details).

† Met guidelines for sample participation rates only after replacement schools were included (see Exhibit A.6).

2 National Defined Population covers less than 90 percent of National Desired Population (see Exhibit A.3).

( ) Standard errors appear in parentheses. Because results are rounded to the nearest whole number, some totals may appear inconsistent.

Top 10% TIMSS International Benchmark – Example Item 4

An Item That Students Reaching the Top 10% International Benchmark Are Likely to Answer Correctly*

8th Grade Mathematics

Content Area: Algebra

Description: Given the initial terms in a sequence and, for example, the 50th term of that sequence, generalizes to find the next term.

The figures show four sets consisting of circles.

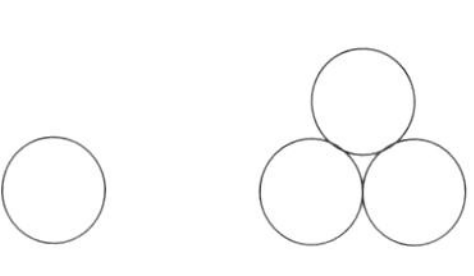

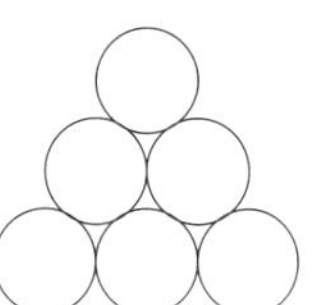

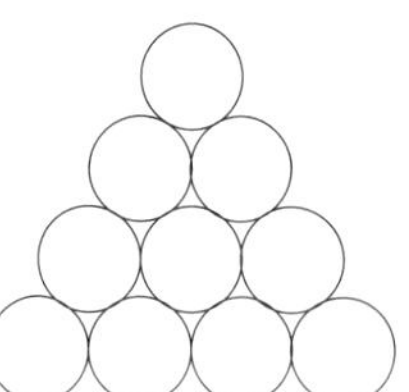

Figure 1 Figure 2 Figure 3 Figure 4

a) Complete the table below. First, fill in how many circles make up Figure 4. Then, find the number of circles that would be needed for the 5th figure if the sequence of figures is extended.

Figure	Number of circles
1	1
2	3
3	6
4	10
5	15

b) The sequence of figures is extended to the 7th figure. How many circles would be needed for Figure 7?

Answer: 28

c) The 50th figure in the sequence contains 1275 circles. Determine the number of circles in the 51st figure. Without drawing the 51st figure, explain or show how you arrived at your answer.

Because it is the 51st figure you have to add 51 to the base of the figure before it

1275
+ 51
1326

The answer shown illustrates the type of student response that was given full credit.

	Overall Percent Correct	
Korea, Rep. of	70 (1.2)	▲
Chinese Taipei	68 (1.5)	▲
Japan	66 (1.6)	▲
Singapore	65 (2.4)	▲
Hong Kong, SAR †	57 (2.0)	▲
First in the World Consort., IL	54 (4.2)	▲
Naperville Sch. Dist. #203, IL	53 (3.6)	▲
Michigan Invitational Group, MI	53 (3.7)	▲
Netherlands †	48 (3.0)	▲
Montgomery County, MD 2	46 (3.8)	▲
Belgium (Flemish) †	44 (1.7)	▲
Academy School Dist. #20, CO	44 (4.1)	▲
Canada	43 (2.2)	▲
Massachusetts	40 (3.5)	●
Connecticut	39 (3.5)	●
Oregon	39 (2.7)	▲
Michigan	39 (3.2)	●
Project SMART Consortium, OH	39 (4.5)	●
Indiana †	37 (3.8)	●
Guilford County, NC 2	36 (2.8)	●
Texas	36 (4.0)	●
SW Math/Sci. Collaborative, PA	35 (3.5)	●
Fremont/Lincoln/WestSide PS, NE	35 (3.6)	●
Delaware Science Coalition, DE	35 (3.3)	●
England †	35 (2.5)	●
United States	34 (1.3)	▲
Pennsylvania	34 (2.7)	●
South Carolina	34 (2.7)	●
Czech Republic	34 (2.5)	●
Jersey City Public Schools, NJ	33 (4.8)	●
Idaho	32 (2.9)	●
North Carolina	32 (2.6)	●
Maryland	30 (2.8)	●
Missouri	30 (1.8)	●
Illinois	30 (2.4)	●
Russian Federation	27 (2.0)	●
Italy	24 (1.8)	▼
Miami-Dade County PS, FL	20 (2.5)	▼
Chicago Public Schools, IL	19 (3.2)	▼
Rochester City Sch. Dist., NY	17 (3.1)	▼
International Avg. (All Countries)	30 (0.3)	

SOURCE: IEA Third International Mathematics and Science Study (TIMSS), 1998-1999.

▲ Participant average significantly higher than international average

● No statistically significant difference between participant average and international average

▼ Participant average significantly lower than international average

Significance tests adjusted for multiple comparisons

* The item was answered fully correctly by a majority of students reaching this benchmark.

States in *italics* did not fully satisfy guidelines for sample participation rates (see Appendix A for details

† Met guidelines for sample participation rates only after replacement schools were included (see Exhibit A.6).

2 National Defined Population covers less than 90 percent of National Desired Population (see Exhibit A.3).

( ) Standard errors appear in parentheses. Because results are rounded to the nearest whole number, some totals may appear inconsistent.

Achievement at the Upper Quarter Benchmark

Exhibit 2.6 describes performance at the Upper Quarter Benchmark. Eighth-grade students performing at this level applied their mathematical knowledge and understandings in a wide variety of relatively complex problem situations. For example, they demonstrated facility with fractions in various formats, as illustrated by Example Item 5 shown in Exhibit 2.7. This item required students to shade squares in a rectangular grid to represent a given fraction. Since the grid is divided into squares that are a multiple of the fraction's denominator, more than one step is required to solve the problem. Internationally, about half the students (49 percent on average) were able to shade in nine of the 24 squares to represent 3/8 of the region. Eighty percent or more of the students in Singapore, Hong Kong, Belgium (Flemish), Korea, and Chinese Taipei answered the question correctly. No Benchmarking entities performed that well, but students in the First in World Consortium, Naperville, the Michigan Invitational Group, and Massachusetts performed significantly above the international average.

Example Item 6 is a proportional reasoning word problem that students at the Upper Quarter Benchmark typically answered correctly (see Exhibit 2.8). Given the number of magazines sold by each of two boys and the total amount of money made from the sales, students were to calculate how much money one of the boys made by selling his 80 magazines. On average, 44 percent of students internationally answered this question correctly. In Singapore and Chinese Taipei at least three-quarters of the students answered correctly. No Benchmarking participant performed significantly above the international average, and students in Maryland, the Michigan Invitational Group, the Chicago Public Schools, the Rochester City School District, and the Miami-Dade County Public Schools performed significantly below the international average.

Students reaching the Upper Quarter Benchmark generally were able to apply knowledge of geometric properties. In Example Item 7 in Exhibit 2.9, students needed to use their knowledge of the properties of parallelograms and rectangles to solve for the area of the rectangle (dimensions not labeled) that was part of a different figure with given dimensions. Three-quarters or more of the students in Singapore, Japan, Hong Kong, Korea, and Chinese Taipei answered the item correctly. Internationally, however, less than half the eighth-grade students (43 percent on average) did so. The United States performed significantly below the international average, as did eight of the Benchmarking entities: North Carolina, South Carolina, Missouri, the Delaware Science Coalition, and the public school systems in Jersey City, Chicago, Miami-Dade, and Rochester.

Example Item 8 shown in Exhibit 2.10 asks students for the number of triangles of a given dimension needed to cover a rectangle of given dimensions. The international average on this item was 46 percent correct. Many students (approximately 29 percent internationally) incorrectly chose Option A, which is half the number of required triangles needed to fill the rectangle but just enough to cover the perimeter. Japanese students had the highest performance on this item, with 80 percent answering correctly. About two-thirds or more of the students in Korea, Hong Kong, Singapore, Belgium (Flemish), and the Netherlands answered the item correctly. Performance among the Benchmarking participants ranged from 62 percent correct responses in Naperville to 30 percent in Miami-Dade. The United States as a whole performed at about the international average, and most of the Benchmarking jurisdictions performed similarly.

Unlike students at lower benchmarks, those reaching the Upper Quarter Benchmark typically could solve simple linear equations. As illustrated by Example Item 9 in Exhibit 2.11, for example, students successfully solved for the value of x in a linear equation involving the variable on both sides of the equation. Eighty percent or more of the students in Japan, Hong Kong, and Korea answered this item correctly. Even though the United States did relatively well in algebra (see Chapter 3), this problem posed difficulties for students in the Benchmarking entities. Naperville (72 percent) and First in the World (61 percent) were the only Benchmarking participants that performed significantly above the international average of 44 percent correct responses. The United States performed below average (34 percent) on this question, as did students in 11 of the Benchmarking entities.

• Upper Quarter Benchmark

Summary

Students can apply their understanding and knowledge in a wide variety of relatively complex situations. They can order, relate and compute with fractions and decimals to solve word problems; solve multi-step word problems involving proportions with whole numbers; solve probability problems; use knowledge of geometric properties to solve problems; identify and evaluate algebraic expressions and solve equations with one variable.

Students demonstrate some facility with fractions and decimals through computation, ordering, rounding, and use in word problems. They can recognize equivalent fractions, add, subtract, multiply and divide fractions with unlike denominators, and correctly order operations. They can identify the smallest decimal from a set of decimals with differing number of places and provide a fraction that is less than a given fraction. They can solve word problems involving multiplication and division of whole numbers and fractions and use pictorial representations of fractions in solving problems. They can identify the fraction of an hour representing a given time interval and identify fractions representing the comparison of part to whole, given each of two parts in a word problem setting.

Students can select the correct rounding of a number involving four decimal places, identify the decimal that is between two decimals given in hundredths, and solve a word problem that involves multiplying a decimal in thousandths by a multiple of a hundred. They can produce an example of a number that would round to a given value. Given a length rounded to the nearest centimeter, they can identify an example of the actual length expressed to one decimal place. Students can identify the ratio expressing a given whole number comparison in a word problem and recognize the effect of adding the same amount to both terms of a ratio. They can estimate products of whole numbers to solve problems. They can solve multi-step word problems involving proportions with whole numbers.

Students demonstrate their understanding of measurement in several settings. They can compare volumes by visualizing and counting cubes. They can calculate the areas of rectangles contained in diagrams of combined shapes. Given the start time and the duration of an event expressed as a fraction of an hour, they can determine the end time. They can estimate the distance between two points on a map, given the scale, and can read unlabeled tick marks on a scale.

Students can use basic properties of triangles, properties of angles on a straight line, and knowledge of symmetry to find the measures of angles. They can identify the angle in a diagram that represents the best estimate of a given measure and recognize that internal angles on a transversal are supplementary. They can visualize the center of a rotation for a two-dimensional figure, the arrangement of faces of a cube when shown its net, and the number of triangles of given dimensions needed to cover a given rectangle. They can identify false statements about congruent triangles and the properties of rectangles.

Students understand elementary concepts of probability, including independent events. They can solve simple problems involving the relationship between successful and unsuccessful outcomes and probabilities. They also recognize that when outcomes are expressed as fractions of a whole, the least likely outcome corresponds to the smallest fraction. They can extrapolate from a graph and determine the number of values on the horizontal axis of a line graph that correspond to a given value on the vertical axis. On a given graph, students can interpolate to find a value between gradations on one axis matching a given value on the other axis.

Students can recognize that multiplication can represent repeated addition. They can identify the algebraic equation corresponding to a verbal description. They can select a simple, multiplicative expression in one variable that is positive for all negative values of the variable. They can substitute numbers for variables to evaluate an expression, and subtract fractions represented algebraically with the same numeric denominator.

Students can solve a linear equation with or without parentheses. They can identify the linear equation that describes the relationship between two variables given in a table of values and select the formula satisfied by the given values of the variables. They can identify the relationship between the first and second terms in a set of ordered pairs.

Given the first several terms of a sequence in pictorial form, growing in either one or two dimensions, they can find specified terms to extend the sequence.

75th Percentile: 555

SOURCE: IEA Third International Mathematics and Science Study (TIMSS), 1998-1999.

Exhibit 2.7 Upper Quarter TIMSS International Benchmark – Example Item 5

An Item That Students Reaching the Upper Quarter International Benchmark Are Likely to Answer Correctly*

8th Grade Mathematics

Content Area: Fractions and Number Sense

Description: Shades squares in a rectangular grid to represent a given fraction.

Shade in $\frac{3}{8}$ of the unit squares in the grid.

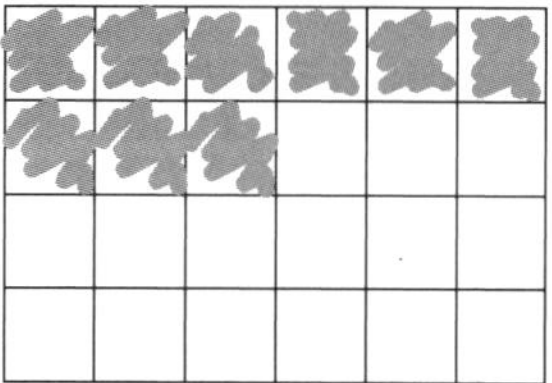

8 | 24

3×3 = 9

The answer shown illustrates the type of student response that was given credit.

	Overall Percent Correct	
Singapore	89 (1.7)	▲
Hong Kong, SAR †	87 (1.7)	▲
Belgium (Flemish) †	87 (1.8)	▲
Korea, Rep. of	81 (1.4)	▲
Chinese Taipei	80 (1.9)	▲
Japan	78 (1.9)	▲
First in the World Consort., IL	71 (5.6)	▲
Canada	68 (2.6)	▲
Naperville Sch. Dist. #203, IL	67 (3.6)	▲
Michigan Invitational Group, MI	65 (5.0)	▲
Netherlands †	61 (4.7)	●
Fremont/Lincoln/WestSide PS, NE	59 (5.2)	●
Massachusetts	59 (3.1)	▲
Montgomery County, MD 2	59 (4.7)	●
Texas	58 (4.6)	●
Academy School Dist. #20, CO	57 (4.2)	●
Indiana †	55 (4.9)	●
Michigan	54 (3.8)	●
Pennsylvania	53 (4.0)	●
England †	52 (2.9)	●
Russian Federation	52 (3.2)	●
Connecticut	52 (5.6)	●
Guilford County, NC 2	51 (4.8)	●
Project SMART Consortium, OH	51 (5.6)	●
Illinois	50 (4.2)	●
Oregon	49 (3.2)	●
SW Math/Sci. Collaborative, PA	49 (3.7)	●
United States	49 (1.9)	●
Missouri	47 (4.2)	●
Idaho	46 (4.1)	●
Italy	46 (2.6)	●
North Carolina	44 (4.5)	●
Delaware Science Coalition, DE	43 (5.4)	●
South Carolina	43 (3.3)	●
Czech Republic	42 (3.2)	●
Maryland	42 (4.1)	●
Jersey City Public Schools, NJ	38 (4.1)	●
Chicago Public Schools, IL	37 (3.8)	●
Rochester City Sch. Dist., NY	32 (5.0)	▼
Miami-Dade County PS, FL	20 (3.6)	▼
International Avg. (All Countries)	49 (0.4)	

▲ Participant average significantly higher than international average

● No statistically significant difference between participant average and international average

▼ Participant average significantly lower than international average

Significance tests adjusted for multiple comparisons

* The item was answered correctly by a majority of students reaching this benchmark.

States in *italics* did not fully satisfy guidelines for sample participation rates (see Appendix A for details).

† Met guidelines for sample participation rates only after replacement schools were included (see Exhibit A.6).

2 National Defined Population covers less than 90 percent of National Desired Population (see Exhibit A.3).

( ) Standard errors appear in parentheses. Because results are rounded to the nearest whole number, some totals may appear inconsistent.

Upper Quarter TIMSS International Benchmark – Example Item 6

An Item That Students Reaching the Upper Quarter International Benchmark Are Likely to Answer Correctly*

8th Grade Mathematics

Content Area: Fractions and Number Sense

Description: Solves a multi-step word problem that involves dividing a quantity in a given ratio.

John sold 60 magazines and Mark sold 80 magazines. The magazines were all sold for the same price. The total amount of money received for the magazines was $700. How much money did Mark receive?

Answer: $400

80
+60
140 TOTAL

MARK = 80/140 = 8/14 = 4/7

7) 700 = 100

100
x4
$400

The answer shown illustrates the type of student response that was given credit.

	Overall Percent Correct	
Singapore	84 (2.0)	▲
Chinese Taipei	75 (1.8)	▲
Hong Kong, SAR †	72 (2.1)	▲
Korea, Rep. of	69 (1.4)	▲
Japan	67 (2.0)	▲
Belgium (Flemish) †	60 (3.7)	▲
First in the World Consort., IL	55 (6.1)	●
Montgomery County, MD 2	54 (4.1)	●
Czech Republic	54 (3.8)	●
Netherlands †	53 (4.5)	●
Russian Federation	52 (3.1)	●
Naperville Sch. Dist. #203, IL	49 (3.9)	●
Massachusetts	46 (4.0)	●
Canada	46 (2.4)	●
Illinois	44 (2.5)	●
Oregon	43 (4.2)	●
Texas	42 (4.7)	●
South Carolina	42 (3.0)	●
SW Math/Sci. Collaborative, PA	42 (2.8)	●
Michigan	42 (3.0)	●
Jersey City Public Schools, NJ	41 (5.6)	●
Academy School Dist. #20, CO	41 (4.1)	●
United States	41 (2.0)	●
Indiana †	40 (4.4)	●
Pennsylvania	39 (3.5)	●
Guilford County, NC 2	38 (4.8)	●
Connecticut	38 (4.3)	●
North Carolina	36 (3.8)	●
Italy	36 (2.6)	●
Fremont/Lincoln/WestSide PS, NE	36 (6.5)	●
Missouri	35 (4.6)	●
Idaho	35 (3.0)	●
Delaware Science Coalition, DE	34 (5.0)	●
Project SMART Consortium, OH	34 (4.3)	●
Maryland	33 (2.4)	▼
Michigan Invitational Group, MI	32 (2.9)	▼
Chicago Public Schools, IL	32 (3.8)	▼
England †	31 (2.6)	▼
Rochester City Sch. Dist., NY	19 (3.1)	▼
Miami-Dade County PS, FL	18 (4.1)	▼
International Avg. (All Countries)	44 (0.4)	

▲ Participant average significantly higher than international average

● No statistically significant difference between participant average and international average

▼ Participant average significantly lower than international average

Significance tests adjusted for multiple comparisons

SOURCE: IEA Third International Mathematics and Science Study (TIMSS), 1998-1999.

* The item was answered correctly by a majority of students reaching this benchmark.

States in *italics* did not fully satisfy guidelines for sample participation rates (see Appendix A for details).

† Met guidelines for sample participation rates only after replacement schools were included (see Exhibit A.6).

2 National Defined Population covers less than 90 percent of National Desired Population (see Exhibit A.3).

( ) Standard errors appear in parentheses. Because results are rounded to the nearest whole number, some totals may appear inconsistent.

Upper Quarter TIMSS International Benchmark – Example Item 7

An Item That Students Reaching the Upper Quarter International Benchmark Are Likely to Answer Correctly*

8th Grade Mathematics

Content Area: Measurement

Description: Finds the area of a rectangle contained in a parallelogram of given dimensions.

The figure shows a shaded rectangle inside a parallelogram.

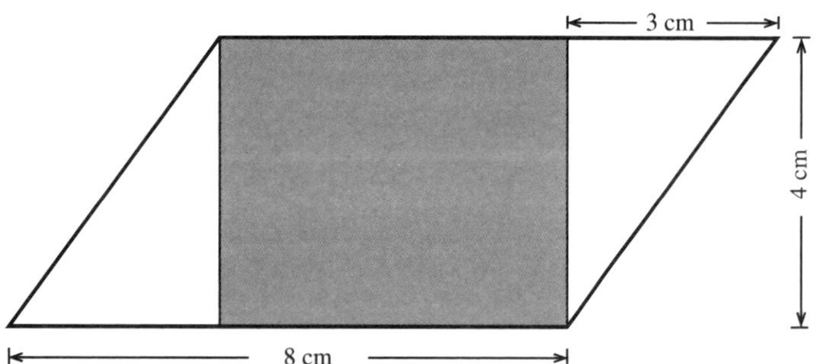

What is the area of the shaded rectangle?

Answer: 20

8-3=5

5
x 4
20

The answer shown illustrates the type of student response that was given credit.

	Overall Percent Correct	
Singapore	83 (1.5)	▲
Japan	80 (1.2)	▲
Hong Kong, SAR †	78 (1.6)	▲
Korea, Rep. of	78 (1.3)	▲
Chinese Taipei	75 (1.4)	▲
Naperville Sch. Dist. #203, IL	65 (2.8)	▲
Belgium (Flemish) †	65 (2.0)	▲
First in the World Consort., IL	62 (4.3)	▲
Canada	58 (1.6)	▲
Netherlands †	55 (4.7)	●
Academy School Dist. #20, CO	49 (3.4)	●
Russian Federation	49 (2.8)	●
Italy	48 (2.1)	●
England †	48 (2.3)	●
Czech Republic	46 (2.9)	●
Oregon	46 (4.0)	●
Michigan Invitational Group, MI	46 (3.9)	●
Montgomery County, MD [2]	45 (3.9)	●
Project SMART Consortium, OH	44 (4.5)	●
Massachusetts	44 (2.8)	●
Illinois	41 (2.9)	●
Idaho	41 (3.8)	●
Connecticut	40 (4.2)	●
SW Math/Sci. Collaborative, PA	40 (3.6)	●
Texas	40 (4.1)	●
Michigan	39 (2.9)	●
Fremont/Lincoln/WestSide PS, NE	38 (3.5)	●
Indiana †	38 (3.9)	●
Pennsylvania	34 (2.9)	▼
Maryland	34 (2.5)	▼
Guilford County, NC [2]	34 (4.6)	●
United States	34 (1.4)	▼
North Carolina	33 (2.9)	▼
South Carolina	32 (3.2)	▼
Missouri	30 (2.5)	▼
Delaware Science Coalition, DE	24 (3.6)	▼
Jersey City Public Schools, NJ	22 (4.1)	▼
Chicago Public Schools, IL	18 (4.4)	▼
Miami-Dade County PS, FL	14 (2.4)	▼
Rochester City Sch. Dist., NY	12 (1.9)	▼
International Avg. (All Countries)	43 (0.3)	

▲ Participant average significantly higher than international average

● No statistically significant difference between participant average and international average

▼ Participant average significantly lower than international average

Significance tests adjusted for multiple comparisons

SOURCE: IEA Third International Mathematics and Science Study (TIMSS), 1998-1999.

* The item was answered correctly by a majority of students reaching this benchmark.

States in *italics* did not fully satisfy guidelines for sample participation rates (see Appendix A for details).

† Met guidelines for sample participation rates only after replacement schools were included (see Exhibit A.6).

2 National Defined Population covers less than 90 percent of National Desired Population (see Exhibit A.3).

( ) Standard errors appear in parentheses. Because results are rounded to the nearest whole number, some totals may appear inconsistent.

Exhibit 2.10 Upper Quarter TIMSS International Benchmark – Example Item 8

An Item That Students Reaching the Upper Quarter International Benchmark Are Likely to Answer Correctly*

8th Grade Mathematics

Content Area: Geometry

Description: Determines the number of triangles of given dimensions needed to cover a given rectangle.

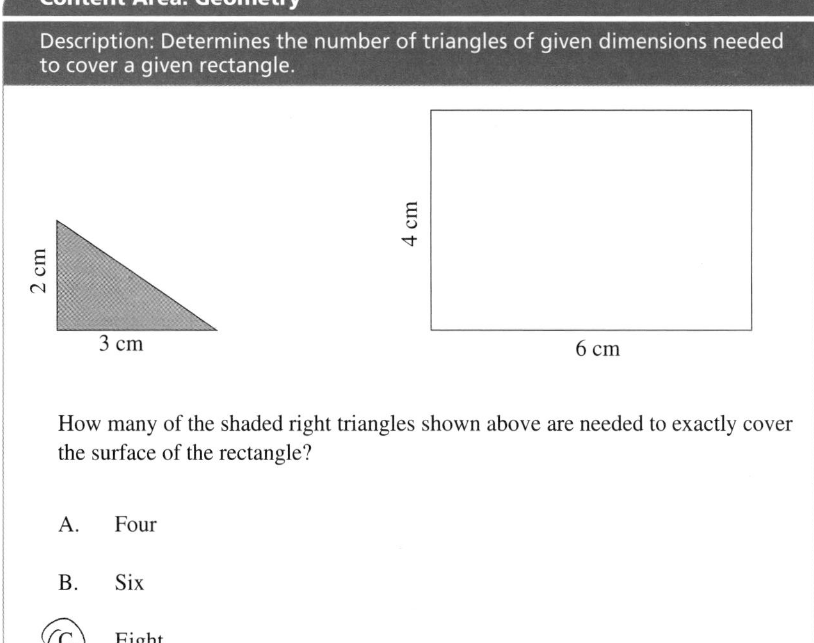

How many of the shaded right triangles shown above are needed to exactly cover the surface of the rectangle?

A. Four

B. Six

(C.) Eight

D. Ten

	Overall Percent Correct	
Japan	80 (1.8)	▲
Korea, Rep. of	76 (1.7)	▲
Hong Kong, SAR †	75 (2.0)	▲
Singapore	72 (2.2)	▲
Belgium (Flemish) †	68 (2.7)	▲
Netherlands †	66 (3.8)	▲
Naperville Sch. Dist. #203, IL	62 (3.5)	▲
Chinese Taipei	60 (1.8)	▲
Michigan Invitational Group, MI	57 (6.8)	●
Montgomery County, MD 2	57 (3.9)	●
Guilford County, NC 2	56 (5.0)	●
First in the World Consort., IL	56 (5.1)	●
Czech Republic	55 (3.6)	●
South Carolina	53 (2.9)	●
Michigan	52 (3.7)	●
Oregon	50 (4.0)	●
Canada	50 (2.4)	●
Texas	50 (3.3)	●
Italy	49 (2.7)	●
England †	48 (2.6)	●
Academy School Dist. #20, CO	48 (4.3)	●
United States	47 (2.0)	●
Indiana †	47 (2.4)	●
Delaware Science Coalition, DE	46 (4.2)	●
Idaho	46 (3.7)	●
Illinois	44 (2.5)	●
Connecticut	44 (4.1)	●
SW Math/Sci. Collaborative, PA	44 (2.8)	●
Russian Federation	44 (2.8)	●
Project SMART Consortium, OH	44 (4.3)	●
Pennsylvania	42 (3.1)	●
Fremont/Lincoln/WestSide PS, NE	42 (5.5)	●
North Carolina	42 (3.2)	●
Massachusetts	41 (2.0)	●
Chicago Public Schools, IL	40 (5.4)	●
Missouri	39 (2.6)	●
Jersey City Public Schools, NJ	39 (4.4)	●
Maryland	38 (2.8)	●
Rochester City Sch. Dist., NY	31 (5.2)	●
Miami-Dade County PS, FL	30 (4.4)	▼
International Avg. (All Countries)	46 (0.4)	

▲ Participant average significantly higher than international average

● No statistically significant difference between participant average and international average

▼ Participant average significantly lower than international average

Significance tests adjusted for multiple comparisons

SOURCE: IEA Third International Mathematics and Science Study (TIMSS), 1998-1999.

* The item was answered correctly by a majority of students reaching this benchmark.

States in *italics* did not fully satisfy guidelines for sample participation rates (see Appendix A for details).

† Met guidelines for sample participation rates only after replacement schools were included (see Exhibit A.6).

2 National Defined Population covers less than 90 percent of National Desired Population (see Exhibit A.3).

( ) Standard errors appear in parentheses. Because results are rounded to the nearest whole number, some totals may appear inconsistent.

Exhibit 2.11

Upper Quarter TIMSS International Benchmark – Example Item 9

An Item That Students Reaching the Upper Quarter International Benchmark Are Likely to Answer Correctly*

TIMSS 1999 Benchmarking
Boston College

8th Grade Mathematics

Content Area: Algebra

Description: Solves a linear equation involving transposing.

Find the value of x if $12x - 10 = 6x + 32$

Answer: 7

$12x - 6x - 10 = 32$

$\Rightarrow 6x = 42$

$\Rightarrow \frac{6x}{6} = \frac{42}{6}$

$\Rightarrow x = 7.$

The answer shown illustrates the type of student response that was given credit.

	Overall Percent Correct	
Japan	85 (1.4)	▲
Hong Kong, SAR †	80 (1.9)	▲
Korea, Rep. of	80 (1.5)	▲
Russian Federation	77 (3.1)	▲
Singapore	75 (2.8)	▲
Chinese Taipei	73 (2.0)	▲
Naperville Sch. Dist. #203, IL	72 (3.6)	▲
Czech Republic	66 (2.8)	▲
First in the World Consort., IL	61 (5.0)	▲
Belgium (Flemish) †	58 (1.9)	▲
Academy School Dist. #20, CO	57 (5.1)	●
Montgomery County, MD 2	55 (4.1)	●
Italy	46 (2.8)	●
Indiana †	44 (5.7)	●
Michigan	40 (3.7)	●
Guilford County, NC 2	40 (6.1)	●
Massachusetts	39 (3.7)	●
South Carolina	39 (3.9)	●
Texas	38 (5.3)	●
SW Math/Sci. Collaborative, PA	38 (3.8)	●
Oregon	37 (3.9)	●
Maryland	35 (3.7)	●
Idaho	34 (5.2)	●
United States	34 (1.8)	▼
Michigan Invitational Group, MI	33 (6.5)	●
Canada	33 (3.1)	▼
Project SMART Consortium, OH	32 (5.3)	●
Connecticut	32 (3.7)	▼
Illinois	32 (4.1)	▼
Pennsylvania	31 (2.6)	▼
North Carolina	27 (3.9)	▼
Rochester City Sch. Dist., NY	27 (5.2)	▼
Jersey City Public Schools, NJ	26 (5.0)	▼
England †	26 (2.7)	▼
Delaware Science Coalition, DE	25 (5.2)	▼
Missouri	24 (3.1)	▼
Fremont/Lincoln/WestSide PS, NE	22 (4.1)	▼
Netherlands †	19 (2.9)	▼
Miami-Dade County PS, FL	17 (4.8)	▼
Chicago Public Schools, IL	10 (2.3)	▼
International Avg. (All Countries)	44 (0.4)	

SOURCE: IEA Third International Mathematics and Science Study (TIMSS), 1998-1999.

▲ Participant average significantly higher than international average

● No statistically significant difference between participant average and international average

▼ Participant average significantly lower than international average

Significance tests adjusted for multiple comparisons

* The item was answered correctly by a majority of students reaching this benchmark.

States in *italics* did not fully satisfy guidelines for sample participation rates (see Appendix A for details).

† Met guidelines for sample participation rates only after replacement schools were included (see Exhibit A.6).

2 National Defined Population covers less than 90 percent of National Desired Population (see Exhibit A.3).

( ) Standard errors appear in parentheses. Because results are rounded to the nearest whole number, some totals may appear inconsistent.

Achievement at the Median Benchmark

Students at the Median Benchmark demonstrated the ability to apply basic mathematical knowledge in straightforward situations (see Exhibit 2.12). For example, as shown by Example Item 10 in Exhibit 2.13, students showed that they understand rounding and can use it to estimate the results of computations. Given the number of rows of cars in a parking lot and the number of cars in each row, students chose the number sentence that would give the best estimate of the total number of cars. While students at the Lower Quarter Benchmark rounded to the nearest hundred, students at the Median Benchmark successfully rounded numbers to get the best estimate for a product. Moreover, middle-performing students demonstrated greater competence with word problems than did those at the Lower Quarter Benchmark. The Benchmarking participants performed particularly well on this test question involving rounding. The international average percent correct for this item was 65 percent, and all except five Benchmarking entities performed significantly above the international average. Among the high-achieving countries, Singapore outperformed other countries with 94 percent correct, followed by 85 percent in Hong Kong. More than 85 percent of students answered correctly in Naperville, the First in the World Consortium, Guilford County, the Academy School District, the Southwest Pennsylvania Math and Science Collaborative, Indiana, North Carolina, and Connecticut.

In geometry, students at the Median Benchmark were able to locate a point on a grid with five-unit divisions that lies between the grid lines (see Example Item 11 in Exhibit 2.14). Fifty-eight percent of students on average internationally correctly chose Point S as the point on the grid that could have the coordinates (7,16). In Japan, Korea, Chinese Taipei, Hong Kong, and Singapore, 80 percent or more of the students answered correctly, as did students in Naperville and First in the World. Generally, the Benchmarking participants performed relatively well on this question, with 13 of them performing significantly above the international average. As might be anticipated, students answering incorrectly most commonly chose Point Q (16,7).

Example Item 12 shown in Exhibit 2.15 illustrates students' emerging familiarity with algebraic representation. Internationally on average, nearly two-thirds of students correctly identified the linear equation corresponding to a given verbal statement involving a variable. In Hong Kong, Singapore, Japan, and Korea, 85 percent or more of the students answered correctly, and eighth graders in several Benchmarking districts and consortia performed similarly. Naperville (94 percent) topped the chart on this item, and 85 percent or more of the students in the First in the World Consortium, Montgomery County, and the Academy School District answered correctly.

• Median Benchmark

Summary

Students can apply basic mathematical knowledge in straightforward situations. They can add or subtract to solve one-step word problems involving whole numbers and decimals; identify representations of common fractions and relative sizes of fractions; solve for missing terms in proportions; recognize basic notions of percents and probability; use basic properties of geometric figures; read and interpret graphs, tables, and scales; and understand simple algebraic relationships.

Students can apply basic mathematical knowledge in straightforward situations. They are able to use addition and subtraction to solve one-step word problems involving whole numbers and decimals. They can round whole numbers to the nearest hundred and identify the number sentence that gives the best estimate for the product of two numbers after rounding. Students can arrange four given digits in descending and ascending order to form the largest and smallest possible numbers, and find the difference between those two numbers. Students can approximate the quantity remaining after an amount is reduced by a given percent.

Students demonstrate an understanding of place value in decimal numbers. They can estimate the location of a point representing a decimal number in tenths on a number line marked in whole numbers and identify an unlabeled midway point on a number line marked in tenths. They can set up and solve one-step problems involving addition and subtraction of numbers having up to three decimal places, including situations where the numbers have a different number of decimal places. Given an object of one length, to one decimal place, they can estimate the length of another object.

Students can select the smallest fraction from a list of fractions and can recognize models representing fractions as shaded regions. They can find the missing term in a proportion in word problems and number sentences. Students can solve a simple word problem involving the likelihood of a successful outcome.

Students are able to select the appropriate metric unit to measure the mass of an object. They recognize the inverse relationship between the length of a unit and the number of units required to cover a distance.

Students can locate and interpret data presented in bar graphs, pictographs, pie graphs, and line graphs. Given a table of values for two variables, they can select the graph that represents the given data.

Students can solve problems involving the properties of congruent figures and can select a pair of similar triangles from a set of triangles. They can visualize a rotation of a three-dimensional figure made of cubes. They can locate points in the first quadrant of the Cartesian plane.

Students can select an expression to represent a situation involving multiplication, and identify a linear equation corresponding to a verbal statement. They can find a missing value in a table of values relating x and y values. Using the properties of a balance, they can reason to find an unknown weight. Given diagrams representing the first few terms of a sequence, growing in one dimension, and a partially completed table, they can find the next two terms.

50th Percentile: 479

SOURCE: IEA Third International Mathematics and Science Study (TIMSS), 1998-1999.

Exhibit 2.13 Median TIMSS International Benchmark – Example Item 10

An Item That Students Reaching the Median International Benchmark Are Likely to Answer Correctly*

8th Grade Mathematics

Content Area: Fractions and Number Sense

Description: In a word problem, uses rounding to identify the number sentence that gives the best estimate for the product.

There are 68 rows of cars in a parking lot. Each row has 92 cars. Which of these would give the closest estimate of the total number of cars in the parking lot?

A. 60 × 90 = 5400

B. 60 × 100 = 6000

(C.) 70 × 90 = 6300

D. 70 × 100 = 7000

	Overall Percent Correct	
Naperville Sch. Dist. #203, IL	95 (2.1)	▲
Singapore	94 (1.0)	▲
First in the World Consort., IL	93 (3.2)	▲
Guilford County, NC [2]	87 (3.4)	▲
Academy School Dist. #20, CO	87 (3.0)	▲
SW Math/Sci. Collaborative, PA	87 (3.1)	▲
Indiana †	86 (2.6)	▲
North Carolina	86 (1.9)	▲
Connecticut	86 (3.6)	▲
Michigan Invitational Group, MI	85 (3.8)	▲
Illinois	85 (2.2)	▲
Hong Kong, SAR †	85 (1.7)	▲
Montgomery County, MD [2]	85 (3.2)	▲
Michigan	85 (2.6)	▲
Chicago Public Schools, IL	84 (2.1)	▲
Oregon	84 (2.1)	▲
Belgium (Flemish) †	83 (3.0)	▲
Japan	82 (1.4)	▲
Korea, Rep. of	82 (1.2)	▲
Chinese Taipei	81 (1.5)	▲
South Carolina	81 (2.9)	▲
Texas	81 (3.5)	▲
Netherlands †	81 (3.1)	▲
Idaho	81 (3.6)	▲
Pennsylvania	80 (3.9)	▲
Project SMART Consortium, OH	80 (4.7)	▲
United States	79 (1.8)	▲
Canada	78 (2.1)	▲
Czech Republic	78 (2.3)	▲
Massachusetts	76 (2.8)	▲
Missouri	75 (2.6)	▲
Fremont/Lincoln/WestSide PS, NE	75 (4.0)	●
Delaware Science Coalition, DE	74 (3.2)	●
England †	74 (2.8)	▲
Maryland	74 (1.9)	▲
Jersey City Public Schools, NJ	71 (3.2)	●
Rochester City Sch. Dist., NY	67 (3.8)	●
Russian Federation	65 (2.7)	●
Miami-Dade County PS, FL	60 (3.5)	●
Italy	52 (2.5)	▼
International Avg. (All Countries)	65 (0.4)	

▲ Participant average significantly higher than international average

● No statistically significant difference between participant average and international average

▼ Participant average significantly lower than international average

Significance tests adjusted for multiple comparisons

SOURCE: IEA Third International Mathematics and Science Study (TIMSS), 1998-1999.

* The item was answered correctly by a majority of students reaching this benchmark.

States in *italics* did not fully satisfy guidelines for sample participation rates (see Appendix A for details).

† Met guidelines for sample participation rates only after replacement schools were included (see Exhibit A.6).

2 National Defined Population covers less than 90 percent of National Desired Population (see Exhibit A.3).

( ) Standard errors appear in parentheses. Because results are rounded to the nearest whole number, some totals may appear inconsistent.

Median TIMSS International Benchmark – Example Item 11

An Item That Students Reaching the Median International Benchmark Are Likely to Answer Correctly*

8th Grade Mathematics

Content Area: Geometry

Description: Locates the point on a grid with 5-unit divisions when the point lies between the grid lines.

Which point on the graph could have coordinates (7,16)?

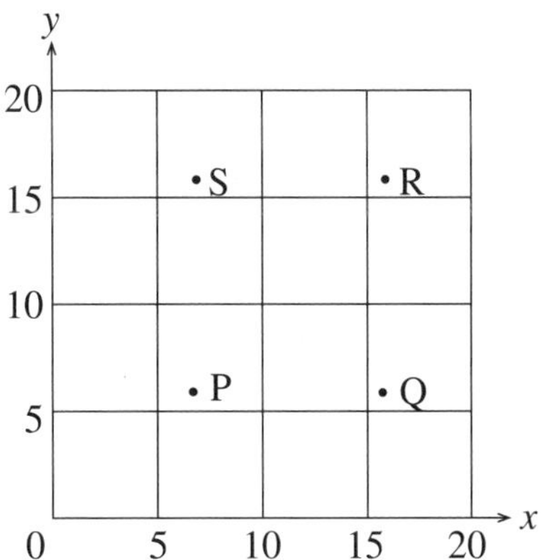

A. Point P

B. Point Q

C. Point R

(D) Point S

	Overall Percent Correct	
Naperville Sch. Dist. #203, IL	88 (2.9)	▲
Japan	84 (1.7)	▲
Korea, Rep. of	84 (1.4)	▲
Chinese Taipei	83 (1.5)	▲
First in the World Consort., IL	82 (3.2)	▲
Hong Kong, SAR †	81 (1.7)	▲
Singapore	80 (2.3)	▲
Netherlands †	78 (2.5)	▲
North Carolina	78 (3.2)	▲
Jersey City Public Schools, NJ	76 (4.4)	▲
Guilford County, NC 2	75 (4.2)	▲
England †	75 (3.2)	▲
SW Math/Sci. Collaborative, PA	74 (3.3)	▲
Texas	74 (3.4)	▲
South Carolina	73 (3.5)	▲
Academy School Dist. #20, CO	73 (3.3)	▲
Montgomery County, MD 2	73 (3.0)	▲
Michigan	72 (2.9)	▲
Pennsylvania	71 (2.0)	▲
Russian Federation	71 (2.2)	▲
Belgium (Flemish) †	71 (2.5)	▲
Oregon	70 (5.3)	●
Michigan Invitational Group, MI	69 (3.8)	●
Illinois	69 (3.3)	▲
Project SMART Consortium, OH	68 (4.8)	●
Canada	67 (2.6)	▲
Indiana †	67 (3.2)	●
United States	67 (1.6)	▲
Maryland	67 (3.7)	●
Massachusetts	64 (3.0)	●
Italy	62 (2.2)	●
Connecticut	61 (4.5)	●
Delaware Science Coalition, DE	60 (4.6)	●
Missouri	60 (3.0)	●
Czech Republic	58 (3.2)	●
Chicago Public Schools, IL	57 (5.3)	●
Miami-Dade County PS, FL	56 (4.2)	●
Idaho	56 (5.8)	●
Rochester City Sch. Dist., NY	55 (4.7)	●
Fremont/Lincoln/WestSide PS, NE	54 (6.1)	●
International Avg. (All Countries)	58 (0.4)	

▲ Participant average significantly higher than international average

● No statistically significant difference between participant average and international average

▼ Participant average significantly lower than international average

Significance tests adjusted for multiple comparisons

SOURCE: IEA Third International Mathematics and Science Study (TIMSS), 1998-1999.

* The item was answered correctly by a majority of students reaching this benchmark.

States in *italics* did not fully satisfy guidelines for sample participation rates (see Appendix A for details).

† Met guidelines for sample participation rates only after replacement schools were included (see Exhibit A.6).

2 National Defined Population covers less than 90 percent of National Desired Population (see Exhibit A.3).

( ) Standard errors appear in parentheses. Because results are rounded to the nearest whole number, some totals may appear inconsistent.

Median TIMSS International Benchmark – Example Item 12

An Item That Students Reaching the Median International Benchmark Are Likely to Answer Correctly*

8th Grade Mathematics

Content Area: Algebra

Description: Identifies the linear equation corresponding to a given verbal statement involving a variable.

n is a number. When n is multiplied by 7, and 6 is then added, the result is 41. Which of these equations represents this relation?

A. $7n + 6 = 41$

B. $7n - 6 = 41$

C. $7n \times 6 = 41$

D. $7(n + 6) = 41$

	Overall Percent Correct	
Naperville Sch. Dist. #203, IL	94 (1.4)	▲
Hong Kong, SAR †	93 (0.9)	▲
First in the World Consort., IL	90 (1.4)	▲
Singapore	89 (1.7)	▲
Montgomery County, MD 2	87 (1.4)	▲
Japan	86 (0.8)	▲
Academy School Dist. #20, CO	85 (1.6)	▲
Korea, Rep. of	85 (0.7)	▲
Chinese Taipei	84 (1.1)	▲
Michigan	82 (1.6)	▲
Canada	82 (1.0)	▲
Russian Federation	82 (1.6)	▲
Project SMART Consortium, OH	82 (2.1)	▲
Pennsylvania	81 (1.8)	▲
Belgium (Flemish) †	81 (1.2)	▲
Massachusetts	80 (2.1)	▲
Netherlands †	80 (2.5)	▲
Connecticut	80 (2.7)	▲
Indiana †	80 (2.3)	▲
Michigan Invitational Group, MI	80 (2.3)	▲
SW Math/Sci. Collaborative, PA	80 (2.5)	▲
Illinois	80 (2.3)	▲
Guilford County, NC 2	79 (2.3)	▲
Texas	78 (3.6)	▲
United States	77 (1.3)	▲
Oregon	77 (2.0)	▲
Idaho	77 (2.7)	▲
South Carolina	77 (2.3)	▲
North Carolina	75 (2.4)	▲
Missouri	73 (1.7)	▲
Maryland	72 (2.6)	●
Fremont/Lincoln/WestSide PS, NE	72 (4.1)	●
Czech Republic	72 (1.7)	▲
Delaware Science Coalition, DE	71 (3.3)	●
Chicago Public Schools, IL	71 (2.3)	●
Jersey City Public Schools, NJ	69 (2.6)	●
Rochester City Sch. Dist., NY	68 (3.4)	●
Miami-Dade County PS, FL	64 (2.3)	●
England †	62 (2.1)	●
Italy	58 (1.6)	▼
International Avg. (All Countries)	65 (0.3)	

▲ Participant average significantly higher than international average

● No statistically significant difference between participant average and international average

▼ Participant average significantly lower than international average

Significance tests adjusted for multiple comparisons

SOURCE: IEA Third International Mathematics and Science Study (TIMSS), 1998-1999.

* The item was answered correctly by a majority of students reaching this benchmark.

States in *italics* did not fully satisfy guidelines for sample participation rates (see Appendix A for details).

† Met guidelines for sample participation rates only after replacement schools were included (see Exhibit A.6).

2 National Defined Population covers less than 90 percent of National Desired Population (see Exhibit A.3).

( ) Standard errors appear in parentheses. Because results are rounded to the nearest whole number, some totals may appear inconsistent.

Achievement at the Lower Quarter Benchmark

As shown in Exhibit 2.16, the few items anchoring at the Lower Quarter Benchmark provided evidence that students performing at this level can add, subtract, and round with whole numbers. For example, students answering Example Item 13 correctly rounded 691 and 208 to estimate their sum as close to the sum of 700 and 200 (see Exhibit 2.17). The international average was 80 percent correct, and 27 countries had three-quarters or more of their students choosing the correct answer. In four countries – Singapore, Belgium (Flemish), Japan, and the Netherlands – 95 percent or more of the students gave the correct response. That level of performance was attained by students in twelve Benchmarking entities: Naperville, Indiana, the Michigan Invitational Group, the Southwest Pennsylvania Math and Science Collaborative, Montgomery County, the Project SMART Consortium, Connecticut, Pennsylvania, Illinois, Missouri, Texas, and the First in the World Consortium. Again, the Benchmarking participants did comparatively well on this rounding item. In all, students in every Benchmarking entity except the Miami-Dade County Public Schools achieved significantly above the international average.

As illustrated by Example Item 14 in Exhibit 2.18, students at the Lower Quarter Benchmark generally could subtract one three-decimal-place number from another with multiple regrouping. Internationally on average, 77 percent of the eighth-grade students selected the correct response to this item. Students in Texas (89 percent) performed significantly above the international average and similarly to students in Singapore, Korea, and the Russian Federation (88 to 90 percent). All of the other Benchmarking participants performed near the international average except the Michigan Invitational Group (60 percent), whose students performed below it.

Students at this level could subtract one four-digit integer from another involving multiple regrouping with zeroes (see Example Item 15 in Exhibit 2.19). On this subtraction item also, students in Texas (90 percent) performed similarly to those in Singapore, Chinese Taipei, and Hong Kong (90 to 92 percent). Students in the Naperville School District (88 percent), the Academy School District (84 percent), and Massachusetts (82 percent) also performed significantly above the international average of 74 percent.

In addition, Example Item 16 in Exhibit 2.20 shows that students at this level could read a thermometer and locate the correct reading in a table. Internationally on average, 79 percent of students answered the item correctly. Students in the Benchmarking entities performed comparatively well on this question. Sixteen of the Benchmarking participants performed significantly above the international average and none below it. Essentially all of the students in Naperville (99 percent) responded correctly, and 90 percent or more did so in First in the World, the Academy School District, Illinois, Project SMART, Indiana, the Southwest Pennsylvania Math and Science Collaborative, and Massachusetts.

• Lower Quarter Benchmark

Summary

Students can do basic computations with whole numbers.

The few items at this level provide some evidence that students can add, subtract, and round with whole numbers. When there are the same number of decimal places, they can subtract with multiple regrouping. Students can round whole numbers to the nearest hundred. They can read a thermometer and locate the reading in a table. Students recognize some basic notation.

25th Percentile: 396

SOURCE: IEA Third International Mathematics and Science Study (TIMSS), 1998-1999.

Lower Quarter TIMSS International Benchmark – Example Item 13

An Item That Students Reaching the Lower Quarter International Benchmark Are Likely to Answer Correctly*

Content Area: Fractions and Number Sense

Description: Rounds to estimate the sum of two three-digit numbers.

The sum 691 + 208 is closest to the sum

A. 600 + 200

(B.) 700 + 200

C. 700 + 300

D. 900 + 200

	Overall Percent Correct	
Naperville Sch. Dist. #203, IL	99 (0.5)	▲
Indiana †	97 (0.7)	▲
Singapore	97 (0.5)	▲
Michigan Invitational Group, MI	97 (1.0)	▲
Belgium (Flemish) †	96 (0.7)	▲
SW Math/Sci. Collaborative, PA	96 (1.1)	▲
Montgomery County, MD 2	95 (0.9)	▲
Project SMART Consortium, OH	95 (1.0)	▲
Japan	95 (0.5)	▲
Connecticut	95 (1.1)	▲
Pennsylvania	95 (1.1)	▲
Illinois	95 (0.9)	▲
Missouri	95 (0.8)	▲
Netherlands †	95 (0.8)	▲
Texas	95 (0.8)	▲
First in the World Consort., IL	95 (1.6)	▲
Academy School Dist. #20, CO	94 (1.2)	▲
Michigan	94 (1.0)	▲
North Carolina	94 (1.2)	▲
Oregon	94 (1.2)	▲
Idaho	93 (1.2)	▲
Massachusetts	93 (0.8)	▲
Hong Kong, SAR †	93 (0.7)	▲
Jersey City Public Schools, NJ	93 (1.5)	▲
Canada	93 (0.7)	▲
United States	93 (0.7)	▲
South Carolina	93 (1.1)	▲
Maryland	93 (1.0)	▲
Chicago Public Schools, IL	93 (1.6)	▲
Korea, Rep. of	93 (0.6)	▲
England †	92 (1.0)	▲
Guilford County, NC 2	92 (1.6)	▲
Delaware Science Coalition, DE	92 (1.4)	▲
Czech Republic	91 (1.0)	▲
Chinese Taipei	89 (0.7)	▲
Fremont/Lincoln/WestSide PS, NE	89 (1.5)	▲
Rochester City Sch. Dist., NY	88 (2.1)	▲
Russian Federation	83 (1.9)	●
Miami-Dade County PS, FL	83 (3.2)	●
Italy	77 (1.9)	●
International Avg. (All Countries)	80 (0.2)	

SOURCE: IEA Third International Mathematics and Science Study (TIMSS), 1998-1999.

Participant average significantly higher than international average ▲

No statistically significant difference between participant average and international average ●

Participant average significantly lower than international average ▼

Significance tests adjusted for multiple comparisons

* The item was answered correctly by a majority of students reaching this benchmark.

States in *italics* did not fully satisfy guidelines for sample participation rates (see Appendix A for details).

† Met guidelines for sample participation rates only after replacement schools were included (see Exhibit A.6).

2 National Defined Population covers less than 90 percent of National Desired Population (see Exhibit A.3).

( ) Standard errors appear in parentheses. Because results are rounded to the nearest whole number, some totals may appear inconsistent.

Lower Quarter TIMSS International Benchmark – Example Item 14

An Item That Students Reaching the Lower Quarter International Benchmark Are Likely to Answer Correctly*

8th Grade Mathematics

Content Area: Fractions and Number Sense

Description: Subtracts a three-decimal-place number from another with multiple regrouping.

Subtract: $4.722 - 1.935 =$

A. 2.787

B. 2.797

C. 2.887

D. 2.897

	Overall Percent Correct	
Singapore	90 (1.4)	▲
Texas	89 (2.1)	▲
Korea, Rep. of	88 (1.2)	▲
Russian Federation	88 (1.9)	▲
Japan	86 (1.3)	▲
Czech Republic	85 (2.8)	●
Chinese Taipei	84 (1.5)	▲
Naperville Sch. Dist. #203, IL	84 (2.9)	●
Chicago Public Schools, IL	83 (2.8)	●
Hong Kong, SAR †	83 (1.8)	▲
Indiana †	82 (2.7)	●
Montgomery County, MD 2	82 (3.4)	●
South Carolina	81 (2.6)	●
Academy School Dist. #20, CO	81 (3.3)	●
Canada	80 (1.8)	●
Illinois	78 (2.2)	●
Guilford County, NC 2	78 (4.0)	●
Pennsylvania	78 (2.8)	●
Project SMART Consortium, OH	78 (3.3)	●
Rochester City Sch. Dist., NY	77 (3.9)	●
Massachusetts	77 (2.6)	●
Maryland	77 (2.2)	●
United States	77 (1.7)	●
Italy	77 (2.3)	●
Connecticut	77 (4.0)	●
SW Math/Sci. Collaborative, PA	76 (3.4)	●
Jersey City Public Schools, NJ	76 (5.2)	●
North Carolina	76 (2.6)	●
Idaho	75 (3.9)	●
Michigan	74 (3.1)	●
Oregon	73 (3.6)	●
Belgium (Flemish) †	73 (2.0)	●
First in the World Consort., IL	73 (3.5)	●
Miami-Dade County PS, FL	71 (4.0)	●
Netherlands †	69 (4.3)	●
Missouri	68 (4.2)	●
Delaware Science Coalition, DE	68 (3.5)	●
Fremont/Lincoln/WestSide PS, NE	61 (5.6)	●
Michigan Invitational Group, MI	60 (4.4)	▼
England †	59 (2.7)	▼
International Avg. (All Countries)	77 (0.4)	

SOURCE: IEA Third International Mathematics and Science Study (TIMSS), 1998-1999.

▲ Participant average significantly higher than international average

● No statistically significant difference between participant average and international average

▼ Participant average significantly lower than international average

Significance tests adjusted for multiple comparisons

* The item was answered correctly by a majority of students reaching this benchmark.

States in *italics* did not fully satisfy guidelines for sample participation rates (see Appendix A for details).

† Met guidelines for sample participation rates only after replacement schools were included (see Exhibit A.6).

2 National Defined Population covers less than 90 percent of National Desired Population (see Exhibit A.3).

( ) Standard errors appear in parentheses. Because results are rounded to the nearest whole number, some totals may appear inconsistent.

Exhibit 2.19

Lower Quarter TIMSS International Benchmark – Example Item 15

An Item That Students Reaching the Lower Quarter International Benchmark Are Likely to Answer Correctly*

8th Grade Mathematics

Content Area: Fractions and Number Sense

Description: Subtracts a four-digit number from another involving zeroes.

Subtract:

$$\begin{array}{r} 7003 \\ -\ 4078 \\ \hline \end{array}$$

A. 2035

(B.) 2925

C. 3005

D. 3925

	Overall Percent Correct	
Singapore	92 (1.3)	▲
Chinese Taipei	90 (1.2)	▲
Texas	90 (1.9)	▲
Hong Kong, SAR †	90 (1.3)	▲
Korea, Rep. of	88 (1.2)	▲
Naperville Sch. Dist. #203, IL	88 (2.7)	▲
Japan	86 (1.4)	▲
Belgium (Flemish) †	85 (2.1)	▲
Academy School Dist. #20, CO	84 (2.8)	▲
Indiana †	84 (3.3)	●
Canada	83 (1.4)	▲
Massachusetts	82 (2.3)	▲
Montgomery County, MD ²	82 (4.3)	●
Illinois	82 (2.4)	●
Czech Republic	82 (2.4)	●
Jersey City Public Schools, NJ	81 (2.8)	●
Idaho	81 (2.8)	●
United States	81 (1.6)	▲
Oregon	80 (2.1)	●
Guilford County, NC ²	80 (4.5)	●
Chicago Public Schools, IL	80 (4.9)	●
Russian Federation	79 (2.2)	●
Netherlands †	79 (3.4)	●
SW Math/Sci. Collaborative, PA	79 (2.9)	●
Michigan Invitational Group, MI	78 (4.8)	●
Missouri	77 (3.3)	●
Pennsylvania	77 (2.4)	●
Connecticut	77 (3.8)	●
South Carolina	77 (2.6)	●
Project SMART Consortium, OH	76 (4.0)	●
Rochester City Sch. Dist., NY	76 (4.5)	●
Maryland	76 (2.2)	●
North Carolina	75 (2.7)	●
Delaware Science Coalition, DE	75 (3.2)	●
First in the World Consort., IL	74 (4.0)	●
Michigan	73 (3.2)	●
Miami-Dade County PS, FL	72 (2.8)	●
Fremont/Lincoln/WestSide PS, NE	68 (6.4)	●
Italy	67 (2.7)	●
England †	51 (3.1)	▼
International Avg. (All Countries)	74 (0.4)	

▲ Participant average significantly higher than international average

● No statistically significant difference between participant average and international average

▼ Participant average significantly lower than international average

Significance tests adjusted for multiple comparisons

SOURCE: IEA Third International Mathematics and Science Study (TIMSS), 1998-1999.

* This item was answered correctly by a majority of students reaching this benchmark.

States in *italics* did not fully satisfy guidelines for sample participation rates (see Appendix A for details).

† Met guidelines for sample participation rates only after replacement schools were included (see Exhibit A.6).

2 National Defined Population covers less than 90 percent of National Desired Population (see Exhibit A.3).

( ) Standard errors appear in parentheses. Because results are rounded to the nearest whole number, some totals may appear inconsistent.

Lower Quarter TIMSS International Benchmark – Example Item 16

An Item That Students Reaching the Lower Quarter International Benchmark Are Likely to Answer Correctly*

8th Grade Mathematics

Content Area: Data Representation, Analysis and Probability

Description: Reads a thermometer and locates the reading in a table.

This table shows temperatures at various times on four days.

TEMPERATURE					
	6 a.m.	9 a.m.	Noon	3 p.m.	6 p.m.
Monday	15°	17°	24°	21°	16°
Tuesday	20°	16°	15°	10°	9°
Wednesday	8°	14°	16°	19°	15°
Thursday	8°	11°	19°	26°	20°

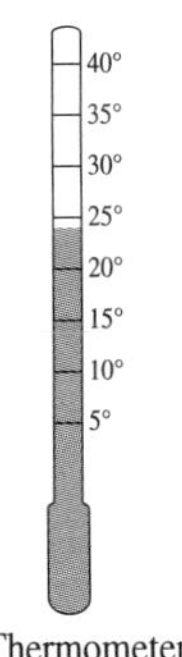

Thermometer

On which day and at what time was the temperature shown in the table the same as that shown on the thermometer.

A. Monday, Noon

B. Tuesday, 6 a.m.

C. Wednesday, 3 p.m.

D. Thursday, 3 p.m.

	Overall Percent Correct	
Naperville Sch. Dist. #203, IL	99 (1.0)	▲
Japan	96 (0.8)	▲
Singapore	95 (0.9)	▲
Belgium (Flemish) †	95 (1.5)	▲
First in the World Consort., IL	95 (2.7)	▲
Academy School Dist. #20, CO	92 (2.1)	▲
Korea, Rep. of	92 (0.9)	▲
England †	92 (2.2)	▲
Chinese Taipei	91 (1.2)	▲
Czech Republic	91 (1.9)	▲
Illinois	91 (1.8)	▲
Project SMART Consortium, OH	91 (3.7)	●
Indiana †	91 (1.9)	▲
SW Math/Sci. Collaborative, PA	91 (1.8)	▲
Hong Kong, SAR †	90 (1.5)	▲
Netherlands †	90 (2.6)	▲
Massachusetts	90 (2.0)	▲
Canada	89 (2.6)	▲
United States	89 (1.2)	▲
Fremont/Lincoln/WestSide PS, NE	89 (2.2)	▲
Montgomery County, MD [2]	89 (3.2)	●
North Carolina	89 (2.2)	▲
Idaho	89 (2.6)	▲
Oregon	88 (1.9)	▲
Michigan Invitational Group, MI	88 (3.3)	●
Texas	88 (2.3)	▲
Guilford County, NC [2]	88 (4.1)	●
Michigan	88 (2.7)	▲
Pennsylvania	87 (3.6)	●
Connecticut	87 (3.6)	●
Missouri	87 (1.9)	▲
Maryland	87 (1.8)	▲
Delaware Science Coalition, DE	87 (3.2)	●
South Carolina	87 (2.1)	▲
Chicago Public Schools, IL	86 (3.5)	●
Russian Federation	85 (2.6)	●
Italy	81 (2.0)	●
Jersey City Public Schools, NJ	81 (2.1)	●
Miami-Dade County PS, FL	76 (5.2)	●
Rochester City Sch. Dist., NY	73 (4.7)	●
International Avg. (All Countries)	79 (0.3)	

▲ Participant average significantly higher than international average

● No statistically significant difference between participant average and international average

▼ Participant average significantly lower than international average

Significance tests adjusted for multiple comparisons

SOURCE: IEA Third International Mathematics and Science Study (TIMSS), 1998-1999.

* This item was answered correctly by a majority of students reaching this benchmark.

States in *italics* did not fully satisfy guidelines for sample participation rates (see Appendix A for details).

† Met guidelines for sample participation rates only after replacement schools were included (see Exhibit A.6).

2 National Defined Population covers less than 90 percent of National Desired Population (see Exhibit A.3).

( ) Standard errors appear in parentheses. Because results are rounded to the nearest whole number, some totals may appear inconsistent.

What Issues Emerge from the Benchmark Descriptions?

The benchmark descriptions and example items strongly suggest a gradation in achievement, from the top-performing students' ability to generalize and solve non-routine or contextualized problems to the lower-performing students being able primarily to use routine, mainly numeric procedures. The fact that even at the Median Benchmark students demonstrate only limited achievement in problem solving beyond straightforward one-step problems may suggest a need to reconsider the role, or priority, of problem solving in mathematics curricula.

The choices teachers make determine, to a large extent, what students learn. According to the NCTM's "The Teaching Principle," in effective teaching worthwhile mathematical problems are used to introduce important ideas and engage students' thinking. The TIMSS 1999 Benchmarking results show that higher achievement is related to the emphasis that teachers place on reasoning and problem-solving activities (see Chapter 6, Exhibit 6.11). This finding is consistent with the video study component of TIMSS conducted in 1995. Analyses of videotapes of mathematics classes revealed that in the typical mathematics lesson in Japan students worked on developing solution procedures to report to the class that were often expected to be original constructions. In contrast, in the typical U.S. lesson students essentially practiced procedures that had been demonstrated by the teacher.

In looking across the item-level results, it is also important to note the variation in performance across the topics covered. On the 16 items presented in this chapter, there was a substantial range in performance for many Benchmarking participants. For example, students in the Benchmarking entities performed relatively well on the items requiring rounding (Exhibits 2.13 and 2.17), and students in Texas did very well on the subtraction questions (Exhibits 2.18 and 2.19). Conversely, students in the Benchmarking entities had particular difficulty with measurement items containing figures (Exhibits 2.2 and 2.9). In some cases, differences of this sort will result from intended differences in emphasis in state or district curricula. It is likely, however, that variation in results may be unintended, and the findings will provide important information about strengths and weaknesses in intended or implemented curricula. For example, Maryland, the Michigan Invitational Group, Chicago, Rochester, and Miami-Dade may not have anticipated performing below the international average on a relatively straightforward word problem involving proportional reasoning (Exhibit 2.8). At the very least, an in-depth examination of the TIMSS 1999 results may reveal aspects of curricula that merit further investigation.

CHAPTER 3 Average Achievement in the Mathematics Content Areas

Chapter 3 presents results by the major content areas in mathematics to provide information about the possible effects of curricular variation on average achievement. Average performance is provided for five content areas: fractions and number sense; measurement; data representation, analysis, and probability; geometry; and algebra.

As delineated by the curriculum of the countries around the world and in the Benchmarking entities, mathematics contains a range of content areas (see Chapter 5 on curriculum). For example, almost all TIMSS 1999 countries and Benchmarking participants reported some elements of arithmetic as well as algebra and geometry in the eighth-grade mathematics curriculum. Since these content areas can differ in complexity, enter the curriculum at different times, receive varying degrees of emphasis, or even be taught as separate courses, Chapter 3 presents results by the major content areas in mathematics. For each Benchmarking entity, average achievement is shown for each content area and compared with the international average for that content area, and average achievement in the content areas is profiled in relation to overall mathematics achievement. Results are also provided by gender. These different perspectives are provided to identify the relative strengths and weaknesses of students in the different mathematics content areas as well as the possible effects of curricular variation on average achievement.

The TIMSS 1999 mathematics test for the eighth grade was designed to enable reporting by five content areas in accordance with the TIMSS mathematics framework. These areas, with their main topics, are:

- Fractions and number sense

  Includes whole numbers, fractions and decimals, integers, exponents, estimation and approximation, proportionality

- Measurement

  Includes standard and non-standard units, common measures, perimeter, area, volume, estimation of measures

- Data representation, analysis, and probability

  Includes representing and interpreting tables, charts, and graphs; range, mean; informal likelihood, simple numerical probability

- Geometry

  Includes points, lines, planes, angles, visualization, triangles, polygons, circles, transformations, symmetry, congruence, similarity, constructions

- Algebra

  Includes number patterns, representation of numerical situations, solving simple linear equations, operations with expressions, representations of relations and functions.

How Does Achievement Differ Across Mathematics Content Areas?

Exhibit 3.1 presents average achievement in each of the five mathematics content areas for the Benchmarking states, districts, and consortia. The Benchmarking jurisdictions as well as selected reference countries are displayed in decreasing order of achievement for each content area, and symbols indicate whether performance is statistically significantly above or below the international average for all of the countries that participated in TIMSS 1999. To allow comparison of the relative performance of each country in each content area, the international average for each content area was scaled to be 487, the same as the overall international average.

The six countries scoring highest in the overall mathematics assessment – Singapore, Korea, Chinese Taipei, Hong Kong, Japan, and Belgium (Flemish) – were also the highest-scoring countries (though not always in the same rank order) in each content area. Correspondingly, the Naperville School District and the First in World Consortium were the highest-scoring Benchmarking entities, performing significantly above the international average, and generally about the same as Belgium (Flemish), in each area.

In contrast to the consistent performance across content areas displayed by the highest-performing entities, performance varied substantially for some middle-performing entities, including the United States. The United States performed significantly above the international average in fractions and number sense; data representation, analysis, and probability; and algebra. In contrast, however, it performed similarly to the international average in measurement and geometry. The same pattern occurred in several of the Benchmarking jurisdictions, including the Project SMART Consortium, Texas, Indiana, Michigan, the Southwest Pennsylvania Math and Science Collaborative, Massachusetts, Oregon, and Guilford County. Montgomery County, the Michigan Invitational Group, and the Academy School District performed above the international average in measurement as well as in the three areas in which the U.S. did relatively well, but like the U.S. performed only at the international average in geometry. Although students in Pennsylvania and Illinois performed above the international average in fractions and number sense as well as in algebra, they performed similarly to the international average in the other three areas.

Exhibits B.1 through B.5 in Appendix B compare average achievement among individual entities for each of the content areas. The exhibits show whether or not the differences in average achievement between pairs of participating entities are statistically significant.

Exhibit 3.1 Average Achievement in Mathematics Content Areas

TIMSS 1999 Benchmarking
Boston College

8th Grade Mathematics

Fractions and Number Sense
Average Scale Score

(61 items)

Participant	Significance	Average Scale Score
Singapore	▲	608 (5.6)
Hong Kong, SAR †	▲	579 (4.5)
Chinese Taipei	▲	576 (4.2)
Korea, Rep. of	▲	570 (2.7)
Japan	▲	570 (2.6)
Naperville Sch. Dist. #203, IL	▲	569 (3.9)
First in the World Consort., IL	▲	561 (4.9)
Belgium (Flemish) †	▲	557 (3.1)
Netherlands †	▲	545 (7.1)
Montgomery County, MD ²	▲	540 (5.1)
Michigan Invitational Group, MI	▲	535 (5.1)
Academy School Dist. #20, CO	▲	534 (2.8)
Canada	▲	533 (2.5)
Project SMART Consortium, OH	▲	527 (7.9)
Texas	▲	527 (8.9)
Indiana †	▲	526 (7.6)
Michigan	▲	525 (7.2)
SW Math/Sci. Collaborative, PA	▲	524 (6.6)
Connecticut	▲	522 (7.9)
Massachusetts	▲	521 (5.9)
Oregon	▲	521 (6.2)
Pennsylvania	▲	517 (5.3)
Illinois	▲	516 (6.2)
Russian Federation	▲	513 (6.4)
Guilford County, NC ²	▲	513 (7.3)
United States	▲	509 (4.2)
South Carolina	○	509 (7.0)
Czech Republic	▲	507 (4.8)
Idaho	○	505 (6.9)
Maryland	○	501 (5.9)
Fremont/Lincoln/WestSide PS, NE	○	498 (6.4)
England †	○	497 (3.8)
North Carolina	○	497 (7.0)
Missouri	○	497 (4.8)
Delaware Science Coalition, DE	○	487 (8.3)
Jersey City Public Schools, NJ	○	483 (7.3)
Chicago Public Schools, IL	○	474 (6.1)
Italy	▽	471 (5.0)
Rochester City Sch. Dist., NY	▽	458 (5.7)
Miami-Dade County PS, FL	▽	434 (9.0)

Scale: 200 500 800

International Avg. (All Countries) 487 (0.7)

Measurement
Average Scale Score

(24 items)

Participant	Significance	Average Scale Score
Singapore	▲	599 (6.3)
Korea, Rep. of	▲	571 (2.8)
Hong Kong, SAR †	▲	567 (5.8)
Chinese Taipei	▲	566 (3.4)
Japan	▲	558 (2.4)
Belgium (Flemish) †	▲	549 (4.0)
Naperville Sch. Dist. #203, IL	▲	549 (3.4)
Netherlands †	▲	538 (5.8)
First in the World Consort., IL	▲	535 (5.8)
Czech Republic	▲	535 (5.0)
Russian Federation	▲	527 (6.0)
Canada	▲	521 (2.4)
Michigan Invitational Group, MI	▲	516 (5.8)
Montgomery County, MD ²	▲	516 (4.3)
England †	▲	507 (3.8)
Academy School Dist. #20, CO	▲	507 (3.5)
Italy	○	501 (5.0)
Oregon	○	500 (6.3)
Project SMART Consortium, OH	○	498 (7.8)
SW Math/Sci. Collaborative, PA	○	495 (7.0)
Michigan	○	494 (7.4)
Connecticut	○	493 (8.3)
Massachusetts	○	491 (7.0)
Illinois	○	491 (6.3)
Pennsylvania	○	489 (6.0)
Indiana †	○	489 (6.8)
Texas	○	489 (9.1)
Guilford County, NC ²	○	487 (7.1)
United States	○	482 (3.9)
Idaho	○	482 (8.1)
Maryland	○	482 (5.9)
South Carolina	○	475 (7.1)
Missouri	○	474 (6.3)
Fremont/Lincoln/WestSide PS, NE	○	474 (8.7)
North Carolina	○	472 (7.5)
Delaware Science Coalition, DE	▽	459 (8.7)
Jersey City Public Schools, NJ	▽	450 (9.1)
Chicago Public Schools, IL	▽	439 (8.1)
Rochester City Sch. Dist., NY	▽	417 (6.2)
Miami-Dade County PS, FL	▽	407 (8.9)

Scale: 200 500 800

International Avg. (All Countries) 487 (0.7)

SOURCE: IEA Third International Mathematics and Science Study (TIMSS), 1998-1999.

▲ Participant average significantly higher than international average

○ Participant average not significantly different from international average

▽ Participant average significantly lower than international average

Significance tests adjusted for multiple comparisons

States in *italics* did not fully satisfy guidelines for sample participation rates (see Appendix A for details).

† Met guidelines for sample participation rates only after replacement schools were included (see Exhibit A.6).

2 National Defined Population covers less than 90% of National Desired Population (see Exhibit A.3).

( ) Standard errors appear in parentheses. Because results are rounded to the nearest whole number, some totals may appear inconsistent.

Data Representation, Analysis, and Probability
Average Scale Score

(21 items)

Participant	Significance	Average Scale Score
Korea, Rep. of	▲	576 (4.2)
Singapore	▲	562 (6.2)
Chinese Taipei	▲	559 (5.1)
Naperville Sch. Dist. #203, IL	▲	559 (4.9)
First in the World Consort., IL	▲	558 (7.3)
Japan	▲	555 (2.3)
Hong Kong, SAR †	▲	547 (5.4)
Belgium (Flemish) †	▲	544 (3.8)
Montgomery County, MD [2]	▲	541 (4.8)
Michigan Invitational Group, MI	▲	538 (6.9)
Netherlands †	▲	538 (7.9)
Project SMART Consortium, OH	▲	534 (8.6)
Academy School Dist. #20, CO	▲	527 (4.1)
Texas	▲	527 (10.2)
Massachusetts	▲	521 (6.3)
Canada	▲	521 (4.5)
Guilford County, NC [2]	▲	520 (10.1)
SW Math/Sci. Collaborative, PA	▲	518 (6.5)
Indiana †	▲	518 (6.3)
Michigan	▲	517 (6.8)
Connecticut	○	516 (9.9)
Oregon	▲	516 (7.0)
Czech Republic	▲	513 (5.9)
Pennsylvania	○	510 (8.6)
Illinois	○	510 (7.1)
South Carolina	○	507 (7.5)
England †	○	506 (8.0)
United States	▲	506 (5.2)
Maryland	○	504 (6.4)
North Carolina	○	502 (5.8)
Russian Federation	○	501 (4.8)
Idaho	○	501 (7.2)
Missouri	○	500 (5.0)
Fremont/Lincoln/WestSide PS, NE	○	496 (10.8)
Delaware Science Coalition, DE	○	493 (9.7)
Jersey City Public Schools, NJ	○	488 (9.6)
Italy	○	484 (4.5)
Chicago Public Schools, IL	○	472 (7.2)
Rochester City Sch. Dist., NY	▽	465 (6.2)
Miami-Dade County PS, FL	▽	445 (9.0)

Scale: 200 – 500 – 800

International Avg. (All Countries): 487 (0.7)

Geometry
Average Scale Score

(21 items)

Participant	Significance	Average Scale Score
Japan	▲	575 (5.1)
Korea, Rep. of	▲	573 (3.9)
Singapore	▲	560 (6.7)
Chinese Taipei	▲	557 (5.8)
Hong Kong, SAR †	▲	556 (4.9)
Belgium (Flemish) †	▲	535 (4.1)
Naperville Sch. Dist. #203, IL	▲	528 (4.2)
Russian Federation	▲	522 (6.0)
First in the World Consort., IL	▲	519 (8.6)
Netherlands †	▲	515 (5.5)
Czech Republic	▲	513 (5.5)
Canada	▲	507 (4.7)
Montgomery County, MD [2]	○	501 (4.5)
Academy School Dist. #20, CO	○	499 (5.0)
Michigan Invitational Group, MI	○	495 (8.3)
Guilford County, NC [2]	○	491 (7.5)
Texas	○	486 (7.9)
Michigan	○	486 (8.0)
Oregon	○	486 (6.8)
Illinois	○	483 (6.8)
SW Math/Sci. Collaborative, PA	○	482 (8.9)
Italy	○	482 (5.6)
Project SMART Consortium, OH	○	477 (8.1)
Massachusetts	○	477 (6.1)
South Carolina	○	476 (7.8)
Indiana †	○	476 (7.6)
North Carolina	○	475 (5.6)
United States	○	473 (4.4)
Pennsylvania	○	473 (4.7)
England †	▽	471 (4.2)
Connecticut	○	470 (7.7)
Fremont/Lincoln/WestSide PS, NE	▽	467 (5.6)
Maryland	▽	466 (6.0)
Missouri	▽	466 (5.6)
Idaho	▽	465 (6.5)
Jersey City Public Schools, NJ	▽	458 (7.6)
Delaware Science Coalition, DE	▽	457 (6.2)
Chicago Public Schools, IL	▽	457 (6.4)
Rochester City Sch. Dist., NY	▽	433 (6.3)
Miami-Dade County PS, FL	▽	423 (7.8)

Scale: 200 – 500 – 800

International Avg. (All Countries): 487 (0.7)

▲ Participant average significantly higher than international average

○ Participant average not significantly different from international average

▽ Participant average significantly lower than international average

Significance tests adjusted for multiple comparisons

SOURCE: IEA Third International Mathematics and Science Study (TIMSS), 1998-1999.

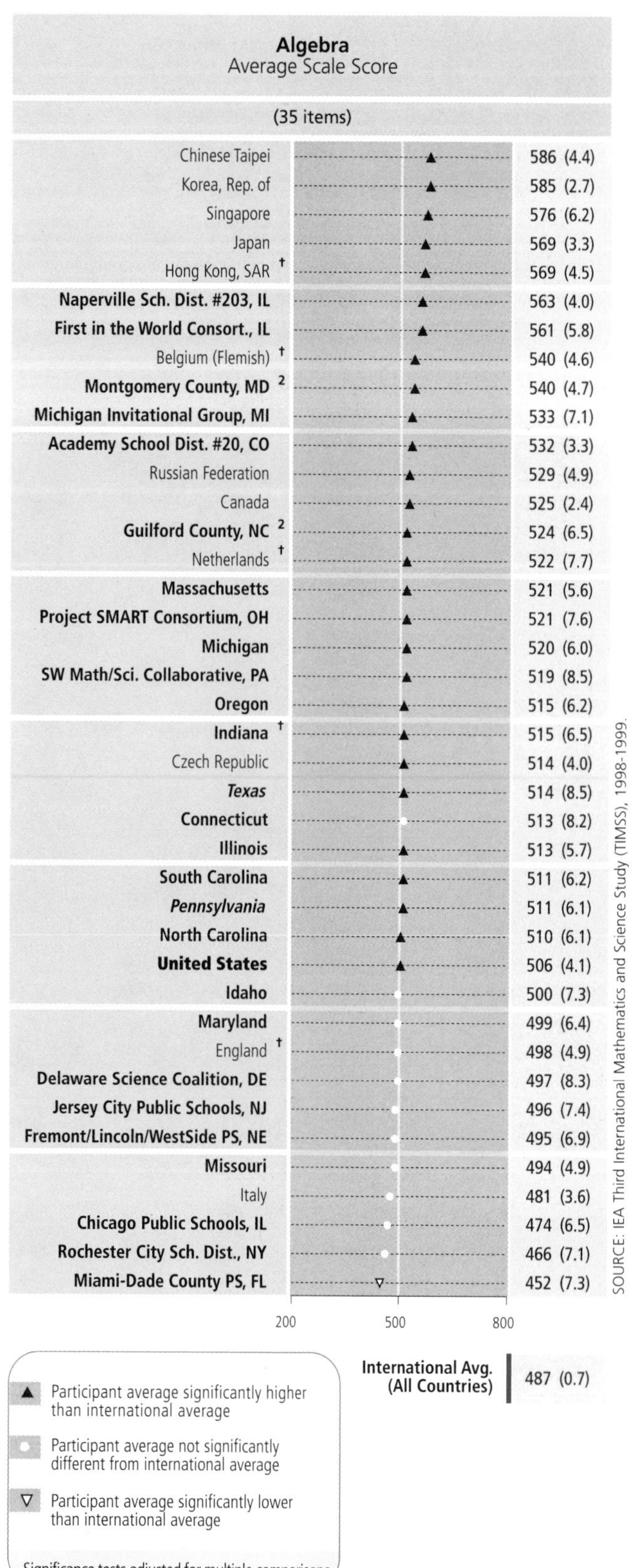

States in *italics* did not fully satisfy guidelines for sample participation rates (see Appendix A for details).

† Met guidelines for sample participation rates only after replacement schools were included (see Exhibit A.6).

2 National Defined Population covers less than 90% of National Desired Population (see Exhibit A.3).

( ) Standard errors appear in parentheses. Because results are rounded to the nearest whole number, some totals may appear inconsistent.

In Which Content Areas Are Students Relatively Strong or Weak?

For purposes of comparison, Exhibit 3.2 profiles the relative performance in mathematics content areas within the comparison countries, while Exhibit 3.3 provides the corresponding information for the Benchmarking states and Exhibit 3.4 for the districts and consortia. These exhibits display the difference between average performance in each content area and average mathematics performance overall, highlighting any variation. The profiles reveal that as in the participating countries, students in many of the Benchmarking jurisdictions performed relatively better or worse in several content areas than they did overall. For example, students in all the Benchmarking entities generally followed the U.S. pattern of performing better than they did overall in fractions and number sense; data representation, analysis, and probability; and algebra, but less well in measurement and geometry.

In particular, a number of jurisdictions had relatively worse geometry performance, including Connecticut, Idaho, Indiana, Maryland, Massachusetts, Missouri, and Pennsylvania among the states. Districts and consortia with such results were the Academy School District, the Delaware Science Coalition, First in the World, the Fremont/Lincoln/Westside Public Schools, the Michigan Invitational Group, Montgomery County, Naperville, and Project SMART. Students' relatively low achievement in geometry is most likely related to less coverage of geometry topics in mathematics classrooms (see Chapter 5).

Among other notable findings, students in North and South Carolina did relatively well in algebra compared with their overall performance, and those in the Rochester City School District had particular difficulty in the area of measurement. Differences in relative performance may be related to one or more of a number of factors, such as emphases in intended curricula or widely used textbooks, strengths or weaknesses in curriculum implementation, and the grade level at which topics are introduced. For the Benchmarking entities, the patterns of relative strengths and weaknesses profiled in Exhibits 3.3 and 3.4 are sometimes reflected in strengths and weaknesses relative to other countries and the United States (shown in Exhibit 3.1).

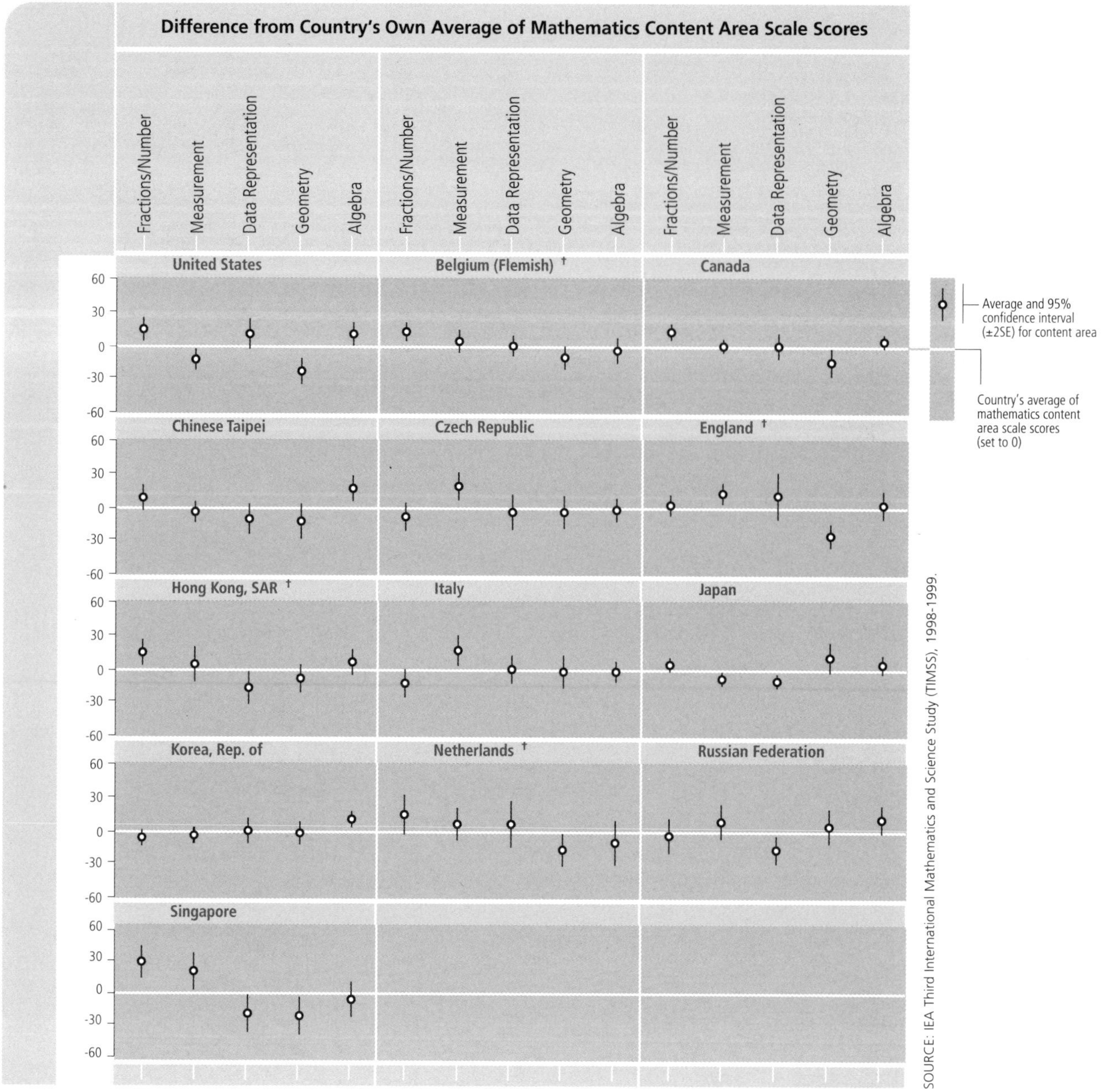

Exhibit 3.2 Countries' Profiles of Relative Performance in Mathematics Content Areas

8th Grade Mathematics

Difference from Country's Own Average of Mathematics Content Area Scale Scores

Fractions/Number | Measurement | Data Representation | Geometry | Algebra

Countries shown: United States; Belgium (Flemish) †; Canada; Chinese Taipei; Czech Republic; England †; Hong Kong, SAR †; Italy; Japan; Korea, Rep. of; Netherlands †; Russian Federation; Singapore

Scale: 60, 30, 0, -30, -60

Average and 95% confidence interval (±2SE) for content area

Country's average of mathematics content area scale scores (set to 0)

SOURCE: IEA Third International Mathematics and Science Study (TIMSS), 1998-1999.

† Met guidelines for sample participation rates only after replacement schools were included (see Exhibit A.6).

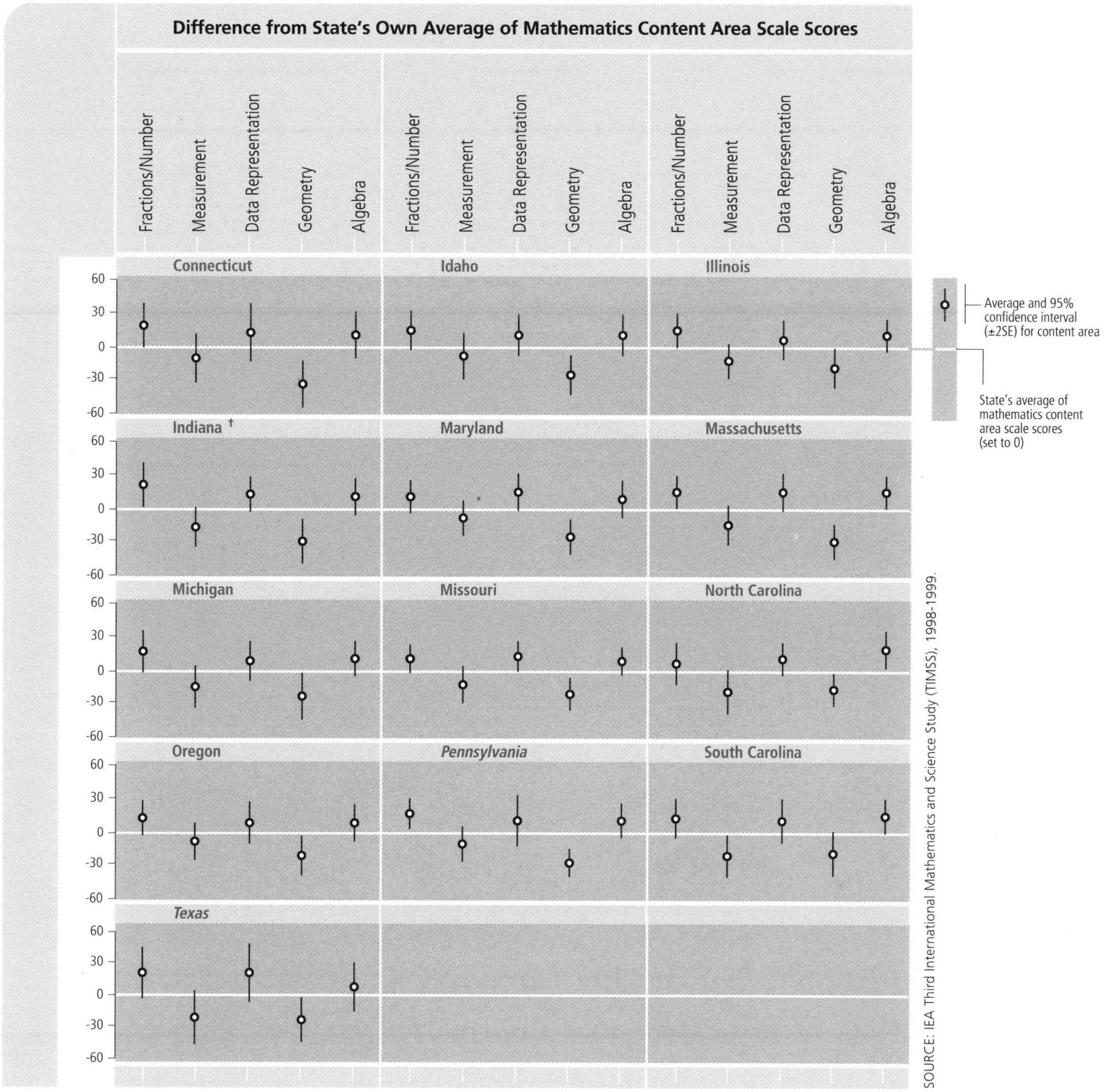

Difference from State's Own Average of Mathematics Content Area Scale Scores

Fractions/Number | Measurement | Data Representation | Geometry | Algebra

Connecticut | Idaho | Illinois
Indiana † | Maryland | Massachusetts
Michigan | Missouri | North Carolina
Oregon | *Pennsylvania* | South Carolina
Texas

60 | 30 | 0 | -30 | -60

Average and 95% confidence interval (±2SE) for content area

State's average of mathematics content area scale scores (set to 0)

SOURCE: IEA Third International Mathematics and Science Study (TIMSS), 1998-1999.

States in *italics* did not fully satisfy guidelines for sample participation rates (see Appendix A for details).

† Met guidelines for sample participation rates only after replacement schools were included (see Exhibit A.6).

Districts' and Consortia's Profiles of Relative Performance in Mathematics Content Areas

8th Grade Mathematics

Difference from District's Own Average of Mathematics Content Area Scale Scores

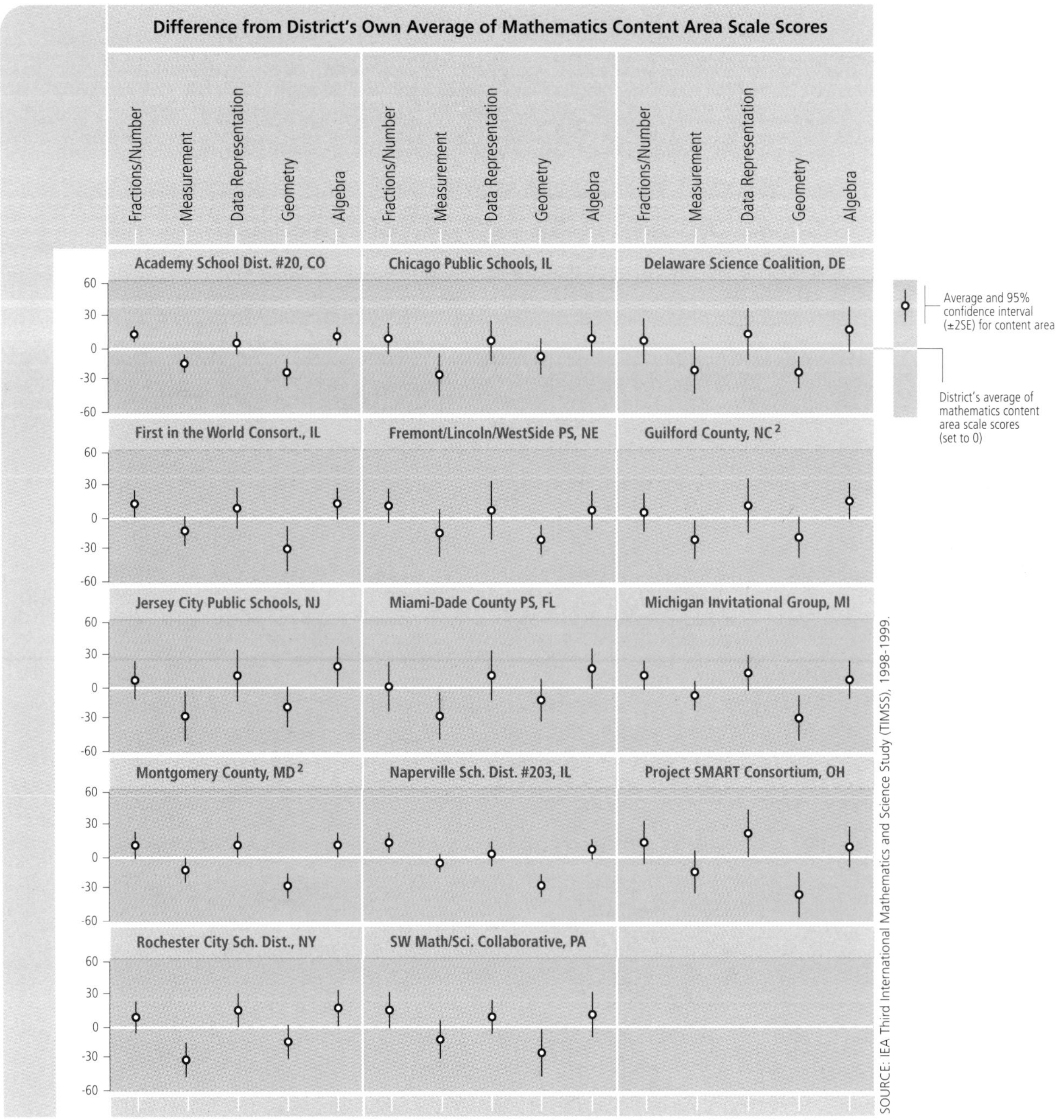

SOURCE: IEA Third International Mathematics and Science Study (TIMSS), 1998-1999.

2 National Defined Population covers less than 90 percent of National Desired Population (see Exhibit A.3).

What Are the Gender Differences in Achievement for the Content Areas?

Exhibit 3.5 displays average achievement in mathematics content areas by gender for the Benchmarking entities as well as the comparison countries. The most striking feature of the exhibit is the very small number of statistically significant differences. There were no significant gender differences in average achievement in any Benchmarking jurisdiction, except that boys had higher average achievement than girls in fractions and number sense in Pennsylvania – for the Southwest Pennsylvania Math and Science Collaborative and for the state as a whole. Even though the United States had higher average achievement for boys than for girls in measurement, there were no significant differences in the Benchmarking entities.

An important stage of item selection for the TIMSS 1999 assessment was the examination of item statistics to detect items that differentiated between groups, including girls and boys, at the country level. Such items were scrutinized and retained when there was no apparent source of gender bias. It is therefore likely that the absence of significant gender differences in the averages for girls and boys in a country is due partly to a balance between items on which one or the other gender tends to perform better. It is also reasonable to assume that where significant differences do occur, they result from gender differences in one or more of the factors in student backgrounds and schooling that have consistently been found to affect achievement in mathematics.

In spite of there being few statistically significant differences in the average achievement of girls and boys in the content areas, it is interesting to look at the patterns of the differences. Consistent with the differences in the international averages, there was a strong tendency across the Benchmarking entities for boys to have higher average achievement than girls in fractions and number sense, measurement, and geometry. The results were more mixed in data representation, analysis, and probability and in algebra.

Exhibit 3.5 Average Achievement in Mathematics Content Areas by Gender

	Average Scale Scores for Mathematics Content Areas					
	Fractions and Number Sense		Measurement		Data Representation, Analysis, and Probability	
	Girls	Boys	Girls	Boys	Girls	Boys
Countries						
United States	505 (4.5)	514 (5.0)	475 (4.0)	489 (4.9) ▲	503 (7.0)	508 (6.3)
Belgium (Flemish) †	555 (6.0)	558 (7.7)	550 (6.5)	547 (8.2)	549 (6.7)	539 (8.8)
Canada	530 (2.4)	536 (3.4)	519 (4.6)	523 (4.4)	520 (5.2)	522 (6.6)
Chinese Taipei	574 (4.9)	579 (5.2)	563 (3.3)	569 (5.2)	557 (5.5)	561 (7.9)
Czech Republic	498 (5.7)	517 (6.1)	525 (6.1)	545 (6.6)	502 (7.0)	524 (6.9)
England †	487 (6.0)	507 (5.4)	500 (6.4)	515 (5.4)	498 (6.8)	513 (10.9)
Hong Kong, SAR †	579 (4.5)	578 (6.1)	567 (5.7)	567 (7.3)	546 (5.3)	548 (7.4)
Italy	463 (6.7)	479 (4.8)	494 (5.7)	508 (5.6)	483 (7.3)	484 (6.2)
Japan	563 (3.4)	576 (4.0)	556 (3.5)	559 (3.0)	552 (5.5)	559 (3.8)
Korea, Rep. of	566 (4.3)	573 (3.3)	567 (3.8)	575 (3.2)	574 (6.2)	579 (5.4)
Netherlands †	540 (7.9)	551 (7.5)	535 (7.5)	540 (6.2)	534 (10.3)	541 (8.3)
Russian Federation	510 (6.2)	516 (7.1)	524 (7.0)	529 (6.1)	502 (7.0)	501 (9.4)
Singapore	607 (6.2)	609 (6.8)	597 (7.3)	601 (9.0)	563 (6.8)	561 (8.8)
States						
Connecticut	514 (7.7)	530 (9.2)	484 (9.0)	503 (8.9)	512 (10.4)	520 (10.6)
Idaho	505 (6.9)	506 (8.5)	479 (8.7)	485 (8.4)	503 (8.3)	499 (8.8)
Illinois	511 (7.0)	522 (6.6)	489 (8.3)	494 (6.4)	506 (8.1)	514 (8.2)
Indiana †	518 (7.9)	534 (8.1)	481 (7.8)	497 (8.4)	514 (8.2)	522 (6.5)
Maryland	496 (6.4)	507 (6.4)	477 (6.7)	487 (5.8)	502 (7.7)	507 (8.5)
Massachusetts	516 (6.5)	526 (5.8)	486 (7.6)	497 (7.2)	520 (7.5)	523 (6.8)
Michigan	518 (7.2)	532 (7.7)	488 (7.7)	501 (8.5)	512 (7.9)	523 (7.3)
Missouri	494 (5.1)	500 (6.2)	470 (7.6)	478 (6.8)	499 (5.7)	500 (6.8)
North Carolina	492 (8.7)	502 (6.8)	471 (8.2)	473 (9.1)	504 (7.5)	499 (7.6)
Oregon	518 (6.9)	524 (7.0)	497 (7.3)	503 (7.4)	516 (8.4)	516 (11.5)
Pennsylvania	510 (5.8)	524 (5.6) ▲	482 (6.0)	497 (7.4)	508 (9.0)	513 (11.7)
South Carolina	506 (8.0)	512 (6.7)	473 (7.8)	477 (7.4)	508 (8.0)	506 (10.6)
Texas	526 (8.4)	527 (10.3)	482 (8.7)	495 (10.9)	530 (9.7)	524 (12.2)
Districts and Consortia						
Academy School Dist. #20, CO	530 (3.7)	539 (4.2)	500 (4.0)	514 (5.4)	530 (4.2)	524 (5.9)
Chicago Public Schools, IL	472 (6.3)	477 (6.5)	436 (9.8)	443 (7.3)	465 (9.5)	479 (7.9)
Delaware Science Coalition, DE	480 (8.9)	494 (10.7)	452 (10.1)	466 (10.8)	491 (10.2)	495 (13.0)
First in the World Consort., IL	556 (5.5)	566 (6.3)	530 (6.7)	540 (7.8)	548 (10.3)	568 (7.4)
Fremont/Lincoln/WestSide PS, NE	492 (7.1)	504 (8.0)	469 (9.0)	478 (10.2)	492 (12.5)	499 (11.7)
Guilford County, NC 2	507 (7.8)	519 (8.2)	479 (9.4)	496 (7.6)	512 (10.8)	530 (11.2)
Jersey City Public Schools, NJ	479 (9.4)	486 (7.8)	446 (10.8)	454 (10.4)	487 (9.8)	489 (12.0)
Miami-Dade County PS, FL	432 (9.9)	437 (9.7)	405 (8.3)	410 (11.4)	444 (9.9)	446 (10.7)
Michigan Invitational Group, MI	538 (5.3)	533 (5.5)	512 (7.6)	520 (8.3)	547 (10.0)	530 (6.3)
Montgomery County, MD 2	534 (6.5)	546 (6.3)	514 (5.9)	518 (6.9)	543 (6.9)	540 (5.6)
Naperville Sch. Dist. #203, IL	564 (4.9)	575 (4.1)	546 (5.0)	551 (4.5)	555 (7.7)	562 (9.3)
Project SMART Consortium, OH	524 (8.8)	530 (8.3)	496 (8.6)	499 (8.7)	539 (10.0)	529 (9.8)
Rochester City Sch. Dist., NY	451 (8.2)	465 (5.9)	405 (8.2)	431 (7.7)	465 (8.0)	464 (11.4)
SW Math/Sci. Collaborative, PA	517 (6.4)	531 (7.5) ▲	487 (6.9)	502 (9.0)	513 (7.8)	524 (7.8)
International Avg. (All Countries)	484 (0.9)	491 (0.9) ▲	483 (1.0)	491 (1.0) ▲	486 (1.1)	489 (1.1)

SOURCE: IEA Third International Mathematics and Science Study (TIMSS), 1998-1999.

▲ Significantly higher than other gender

Significance tests adjusted for multiple comparisons

States in *italics* did not fully satisfy guidelines for sample participation rates (see Appendix A for details).

† Met guidelines for sample participation rates only after replacement schools were included (see Exhibit A.6).

2 National Defined Population covers less than 90 percent of National Desired Population (see Exhibit A.3).

( ) Standard errors appear in parentheses. Because results are rounded to the nearest whole number, some totals may appear inconsistent.

	Average Scale Scores for Mathematics Content Areas			
	Geometry		Algebra	
	Girls	Boys	Girls	Boys
Countries				
United States	469 (5.5)	477 (5.1)	507 (4.3)	504 (4.6)
Belgium (Flemish) †	538 (6.9)	531 (9.1)	545 (6.8)	535 (8.8)
Canada	511 (6.5)	503 (4.9)	526 (3.7)	524 (5.2)
Chinese Taipei	555 (7.1)	560 (6.8)	585 (4.5)	588 (6.1)
Czech Republic	506 (7.6)	520 (4.9)	513 (3.9)	516 (6.7)
England †	467 (4.8)	474 (6.7)	493 (6.0)	502 (5.1)
Hong Kong, SAR †	558 (6.1)	554 (6.4)	570 (4.8)	568 (5.6)
Italy	476 (8.6)	489 (5.1)	481 (5.4)	481 (4.0)
Japan	572 (5.8)	578 (5.8)	568 (4.2)	571 (3.6)
Korea, Rep. of	569 (7.3)	578 (4.8)	585 (3.7)	585 (3.9)
Netherlands †	516 (7.0)	515 (5.2)	522 (9.3)	522 (7.4)
Russian Federation	518 (7.2)	526 (7.4)	533 (5.7)	524 (6.3)
Singapore	556 (9.2)	565 (6.5)	578 (6.7)	574 (7.9)
States				
Connecticut	465 (10.5)	475 (8.8)	510 (8.4)	516 (8.9)
Idaho	462 (9.3)	468 (7.1)	504 (7.0)	496 (8.5)
Illinois	479 (8.5)	487 (9.5)	514 (7.6)	511 (5.2)
Indiana †	471 (8.9)	481 (8.0)	516 (6.7)	514 (7.0)
Maryland	462 (5.6)	471 (7.8)	499 (7.4)	500 (7.4)
Massachusetts	475 (6.0)	478 (7.1)	522 (6.0)	521 (6.2)
Michigan	480 (7.0)	493 (10.8)	517 (6.6)	523 (6.6)
Missouri	464 (7.3)	468 (8.8)	495 (6.0)	493 (5.4)
North Carolina	473 (8.3)	478 (7.3)	512 (6.4)	507 (6.8)
Oregon	485 (8.9)	487 (6.9)	522 (6.4)	509 (7.1)
Pennsylvania	466 (5.9)	479 (5.5)	512 (7.2)	510 (7.1)
South Carolina	474 (9.8)	479 (9.1)	514 (6.4)	508 (7.0)
Texas	484 (7.3)	489 (9.8)	514 (9.0)	514 (9.0)
Districts and Consortia				
Academy School Dist. #20, CO	495 (6.5)	504 (5.5)	534 (3.6)	531 (5.4)
Chicago Public Schools, IL	457 (5.8)	456 (9.4)	475 (6.9)	473 (7.7)
Delaware Science Coalition, DE	456 (6.3)	458 (9.1)	495 (8.5)	499 (10.0)
First in the World Consort., IL	519 (7.2)	518 (12.5)	561 (7.6)	560 (6.3)
Fremont/Lincoln/WestSide PS, NE	461 (6.6)	473 (8.0)	496 (8.2)	495 (8.7)
Guilford County, NC [2]	487 (8.2)	495 (9.1)	522 (7.5)	527 (6.5)
Jersey City Public Schools, NJ	454 (10.3)	462 (7.2)	498 (8.0)	494 (7.8)
Miami-Dade County PS, FL	420 (8.8)	425 (9.3)	457 (8.1)	448 (7.7)
Michigan Invitational Group, MI	500 (8.9)	489 (10.2)	540 (6.6)	525 (8.8)
Montgomery County, MD [2]	500 (8.0)	502 (5.1)	542 (5.3)	537 (6.5)
Naperville Sch. Dist. #203, IL	522 (7.3)	534 (7.4)	561 (3.7)	565 (5.4)
Project SMART Consortium, OH	470 (9.7)	484 (10.9)	524 (7.0)	518 (9.3)
Rochester City Sch. Dist., NY	427 (10.7)	438 (9.5)	466 (8.5)	467 (8.0)
SW Math/Sci. Collaborative, PA	476 (9.3)	489 (9.9)	515 (8.9)	523 (8.7)
International Avg. (All Countries)	485 (1.2)	489 (1.1)	489 (0.9)	485 (0.9)

SOURCE: IEA Third International Mathematics and Science Study (TIMSS), 1998-1999.

▲ Significantly higher than other gender

Significance tests adjusted for multiple comparisons

CHAPTER

4 Students' Backgrounds and Attitudes Towards Mathematics

There is abundant evidence that student achievement is related to home background factors, and to students' activities and attitudes. To help interpret the achievement results, Chapter 4 provides detailed information about students' home backgrounds, how they spend their time out of school, their self-concept in mathematics, and their attitudes towards mathematics.

4

To provide an educational context for interpreting the achievement results of the Benchmarking participants, TIMSS collected detailed information from students about their home backgrounds, how they spend their time, and their attitudes towards mathematics. This chapter presents eighth-grade students' responses to a subset of these questions. One set addresses home resources and support for academic achievement. Another examines how much out-of-school time students spend on their schoolwork. A third addresses students' self-concept in mathematics and their feelings towards mathematics.

In an effort to summarize this information concisely and focus attention on educationally relevant support and practice, TIMSS sometimes has combined information from individual questions to form an index that was more global and reliable than the component questions (e.g., home educational resources). According to their responses, students were placed in a "high," "medium," or "low" category. Cutoff points were established so that the high level of an index corresponds to conditions or activities generally associated with good educational practice and high academic achievement. For each index, the percentages of students in each category are presented in relation to their mathematics achievement. The data from the component questions and more detail about some areas are provided in the reference section of this report (see reference section R1).

What Educational Resources Do Students Have in Their Homes?

There is no shortage of evidence that students from homes with extensive educational resources have higher achievement in mathematics and other subjects than those from less advantaged backgrounds. TIMSS in 1995 showed that this was true of students from homes with large numbers of books, with a range of educational study aids, or with parents with university-level education.[1] The TIMSS 1999 international report presented combined student responses to these three variables in an index of home educational resources (HER) that was clearly related to achievement in mathematics.[2]

Exhibit 4.1 summarizes the home educational resources index in a two-page display. The index is described on the first page. Students at the high level of this index reported coming from homes with more than 100 books, with all three study aids (a computer, a study desk or table for the student's own use, and a dictionary), and where at least one

1 Beaton, A.E., Mullis, I.V.S., Martin, M.O., Gonzalez, E.J., Kelly, D.L., and Smith, T.A. (1996), *Mathematics Achievement in the Middle School Years: IEA's Third International Mathematics and Science Study*, Chestnut Hill, MA: Boston College.

2 Mullis, I.V.S., Martin, M.O., Gonzalez, E.J., Gregory, K.D., Garden, R.A., O'Connor, K.M., Chrostowski, S.J., and Smith, T.A. (2000), *TIMSS 1999 International Mathematics Report: Findings from IEA's Repeat of the Third International Mathematics and Science Study at the Eighth Grade*, Chestnut Hill, MA: Boston College.

parent finished university. Students at the low level had 25 or fewer books in the home, not all three study aids, and parents that had not completed secondary education. The remaining students were assigned to the medium level.

The first page of the display also presents the percentage of students at each level of the index for each Benchmarking participant and for selected reference countries, together with the average mathematics achievement for those students. Standard errors are also shown. Entities are ordered by the percentage of students at the high index level. The international average across all TIMSS 1999 countries is shown at the bottom. The second page of the display graphically shows the percentage of students at the high index level for each entity. There was a substantial difference in the average mathematics achievement of students at the index levels in every entity for which data were available. This is reflected in the international average for the TIMSS 1999 countries, where the achievement difference between students at the high level (559) and the low level (431) amounted to 128 score points.

Relative to other countries, the United States had a large percentage of students at the high level of the home educational resources index (22 percent). Of the TIMSS 1999 countries included in Exhibit 4.1, only Canada had a comparable percentage of students at the high level (27 percent). The relatively high standing of the United States on this index was reflected in the results for the Benchmarking jurisdictions, most of which had larger percentages of students in the high category of home educational resources than did most of the comparison countries.

The Benchmarking participants with the greatest percentages of students at the high level included the Naperville School District (56 percent), the First in the World Consortium (45 percent), the Academy School District (44 percent), and Montgomery County (39 percent). Together with the Michigan Invitational Group (29 percent), these were also among the top-performing jurisdictions in mathematics. The four urban Benchmarking school districts that had the lowest student achievement in mathematics – the Jersey City Public Schools, the Chicago Public Schools, the Rochester City School District, and the Miami-Dade County Public Schools – also had the lowest percentages of students at the high level of the home educational resources index (only 7 to 10 percent).

Since the association between home educational resources and mathematics achievement is well documented in TIMSS and in extensive educational research, low average student achievement in the less wealthy areas most likely reflects the low level of educational resources in students' homes. These effects can be found even when children begin school. For example, kindergartners whose mothers have higher levels of education are more likely to be able pass through four levels of mathematics proficiency that involve such tasks as reading numerals, counting, and sequencing numbers. Similarly, first-time kindergartners whose families have not received or are not receiving welfare services are more likely than kindergartners from families receiving welfare to pass through the mathematics proficiency levels.[3]

However, since there is far from a one-to-one correspondence between high performance and home resources, clearly other influences are also at work. For example, Chinese Taipei had about the same percentage of students (eight percent) at the high index level as Rochester, Chicago, Jersey City, and Miami-Dade, but the average mathematics achievement of its students at that level was considerably higher. In fact, the international average for all 38 TIMSS 1999 countries was just nine percent. There is also evidence that financial resources alone will not result in high academic achievement. According to OECD analyses for 1994, U.S. schools ranked third highest among 22 countries in per-student expenditures on primary schools and third highest among 23 countries on secondary schools.[4]

Exhibits R1.1 through R1.3 in the reference section present more detailed information on the student responses that were combined in the home educational resources index. Exhibit R1.1 shows the percentage of eighth-grade students in each of the Benchmarking jurisdictions and comparison countries who had a dictionary, study desk or table, or computer, and shows that students reporting having all three had higher average mathematics achievement than those without all three.

Exhibit R1.2 shows for each entity the percentage of students at each of five ranges of numbers of books in the home in relation to average mathematics achievement. In most jurisdictions, the more books students reported in the home, the higher their mathematics achievement.

The percentages of students in each of five categories of parents' educational level are shown in Exhibit R1.3, together with their average mathematics achievement. Although countries did their best to use educational categories that were comparable across all countries, the range of educational provision made this difficult. About half of the participating countries had to modify the response options presented to students in the questionnaire in order to conform to their national education system. Exhibit R1.4 provides details of how these modifications were aligned with the categories of parents' education

text continue on page 114

3 West, J., Denton, K., and Germino-Hausken, E. (2000), *America's Kindergartners: Findings from the Early Childhood Longitudinal Study, Kindergarten Class of 1998-99*, NCES 2000-070, Washington, DC: National Center for Education Statistics.

4 *Education at a Glance: OECD Indicators* (1997), Paris, France: Organization for Economic Cooperation and Development. The OECD adjusted the expenditure estimates for the purchasing power of each country's currency.

Index of Home Educational Resources

Index based on students' responses to three questions about home educational resources: number of books in the home; educational aids in the home (computer, study desk/table for own use, dictionary); parents' education (see reference exhibits R1.1-R1.3). High level indicates more than 100 books in the home; all three educational aids; and either parent's highest level of education is finished university. Low level indicates 25 or fewer books in the home; not all three educational aids; and both parents' highest level of education is some secondary or less or is not known. Medium level includes all other possible combinations of responses. See reference exhibit R1.4 for national definitions of educational levels; response categories were defined by each country to conform to their own educational system and may not be strictly comparable across countries.

	High HER		Medium HER		Low HER	
	Percent of Students	Average Achievement	Percent of Students	Average Achievement	Percent of Students	Average Achievement
Naperville Sch. Dist. #203, IL	56 (1.3)	583 (3.5)	43 (1.3)	553 (3.3)	0 (0.2)	~ ~
First in the World Consort., IL	45 (2.5)	580 (7.2)	53 (2.5)	546 (6.1)	2 (0.3)	~ ~
Academy School Dist. #20, CO	44 (1.6)	550 (3.1)	55 (1.6)	513 (2.6)	1 (0.3)	~ ~
Montgomery County, MD	39 (2.5)	578 (5.8)	59 (2.4)	515 (3.9)	2 (0.8)	~ ~
Michigan Invitational Group, MI	29 (2.6)	557 (8.5)	70 (2.6)	523 (5.8)	1 (0.3)	~ ~
Connecticut	29 (2.8)	554 (9.4)	68 (2.5)	499 (8.0)	3 (0.8)	426 (10.2)
Oregon	28 (2.6)	556 (5.9)	68 (2.6)	502 (5.5)	3 (0.6)	421 (15.4)
Canada	27 (1.0)	552 (4.1)	71 (1.0)	525 (2.2)	2 (0.2)	~ ~
Michigan	27 (2.9)	557 (7.8)	71 (2.7)	505 (6.3)	2 (0.5)	~ ~
Guilford County, NC	26 (2.0)	558 (9.2)	72 (1.7)	499 (7.6)	3 (0.4)	451 (16.0)
Maryland	26 (2.0)	544 (6.4)	71 (1.8)	481 (5.9)	3 (0.5)	415 (13.2)
Massachusetts	25 (2.1)	555 (6.6)	72 (1.8)	502 (5.8)	3 (0.6)	449 (14.0)
SW Math/Sci. Collaborative, PA	25 (2.8)	560 (9.5)	72 (2.9)	505 (6.8)	3 (0.8)	441 (16.2)
Fremont/Lincoln/WestSide PS, NE	24 (1.7)	528 (11.1)	72 (1.7)	477 (8.7)	3 (0.4)	424 (15.7)
Indiana	23 (2.6)	553 (7.9)	74 (2.4)	506 (6.3)	3 (0.5)	442 (9.2)
Pennsylvania	22 (2.7)	549 (9.7)	75 (2.6)	498 (4.8)	2 (0.4)	~ ~
Delaware Science Coalition, DE	22 (2.6)	538 (10.2)	75 (2.4)	466 (7.1)	3 (0.9)	406 (16.2)
United States	22 (1.5)	555 (5.1)	73 (1.4)	492 (3.1)	4 (0.5)	427 (6.4)
Illinois	22 (2.7)	562 (6.5)	74 (2.6)	498 (6.0)	4 (0.7)	438 (7.6)
Project SMART Consortium, OH	22 (2.3)	557 (11.0)	76 (2.1)	513 (6.5)	2 (0.5)	~ ~
Texas	21 (2.8)	581 (6.6)	70 (2.1)	512 (8.0)	9 (1.6)	432 (11.4)
Idaho	21 (1.8)	532 (6.7)	74 (1.6)	492 (6.5)	5 (1.1)	403 (13.2)
Missouri	17 (1.4)	527 (8.5)	79 (1.4)	485 (5.0)	4 (0.5)	434 (7.9)
South Carolina	17 (1.6)	560 (8.4)	79 (1.6)	493 (7.3)	4 (0.6)	439 (7.1)
North Carolina	16 (1.9)	546 (9.4)	81 (1.6)	489 (5.9)	4 (0.6)	422 (10.8)
Korea, Rep. of	14 (0.8)	637 (2.8)	80 (0.8)	583 (1.9)	5 (0.3)	513 (5.0)
Czech Republic	13 (0.8)	560 (6.8)	83 (0.8)	517 (3.9)	4 (0.5)	460 (11.3)
Chicago Public Schools, IL	10 (2.4)	489 (12.0)	81 (1.8)	463 (5.3)	9 (1.4)	432 (9.4)
Miami-Dade County PS, FL	10 (2.2)	505 (16.5)	80 (2.3)	419 (8.4)	11 (1.4)	367 (12.8)
Netherlands	9 (1.1)	575 (10.4)	89 (1.1)	538 (7.1)	2 (0.8)	~ ~
Russian Federation	9 (0.8)	560 (8.3)	86 (0.7)	527 (5.9)	6 (0.5)	474 (12.6)
Rochester City Sch. Dist., NY	8 (1.5)	497 (18.8)	82 (1.4)	445 (5.5)	10 (0.9)	416 (7.9)
Belgium (Flemish)	8 (0.7)	599 (6.5)	86 (1.3)	559 (3.9)	6 (1.3)	490 (11.7)
Chinese Taipei	8 (0.7)	666 (7.2)	84 (0.7)	586 (3.6)	8 (0.6)	502 (6.6)
Jersey City Public Schools, NJ	7 (1.2)	514 (18.6)	82 (1.3)	477 (8.5)	11 (1.0)	440 (8.9)
Italy	6 (0.6)	528 (7.3)	81 (0.8)	484 (3.7)	14 (0.8)	434 (6.4)
Singapore	5 (0.7)	663 (10.0)	87 (0.6)	605 (6.0)	8 (0.7)	552 (7.3)
Hong Kong, SAR	3 (0.3)	612 (8.8)	78 (0.8)	586 (4.2)	19 (0.9)	566 (5.2)
England	– –	– –	– –	– –	– –	– –
Japan	– –	– –	– –	– –	– –	– –
International Avg. (All Countries)	9 (0.1)	559 (2.3)	72 (0.2)	487 (0.8)	19 (0.2)	431 (1.2)

SOURCE: IEA Third International Mathematics and Science Study (TIMSS), 1998-1999.

States in *italics* did not fully satisfy guidelines for sample participation rates (see Appendix A for details).

( ) Standard errors appear in parentheses. Because results are rounded to the nearest whole number, some totals may appear inconsistent.

A dash (–) indicates data are not available. A tilde (~) indicates insufficient data to report achievement.

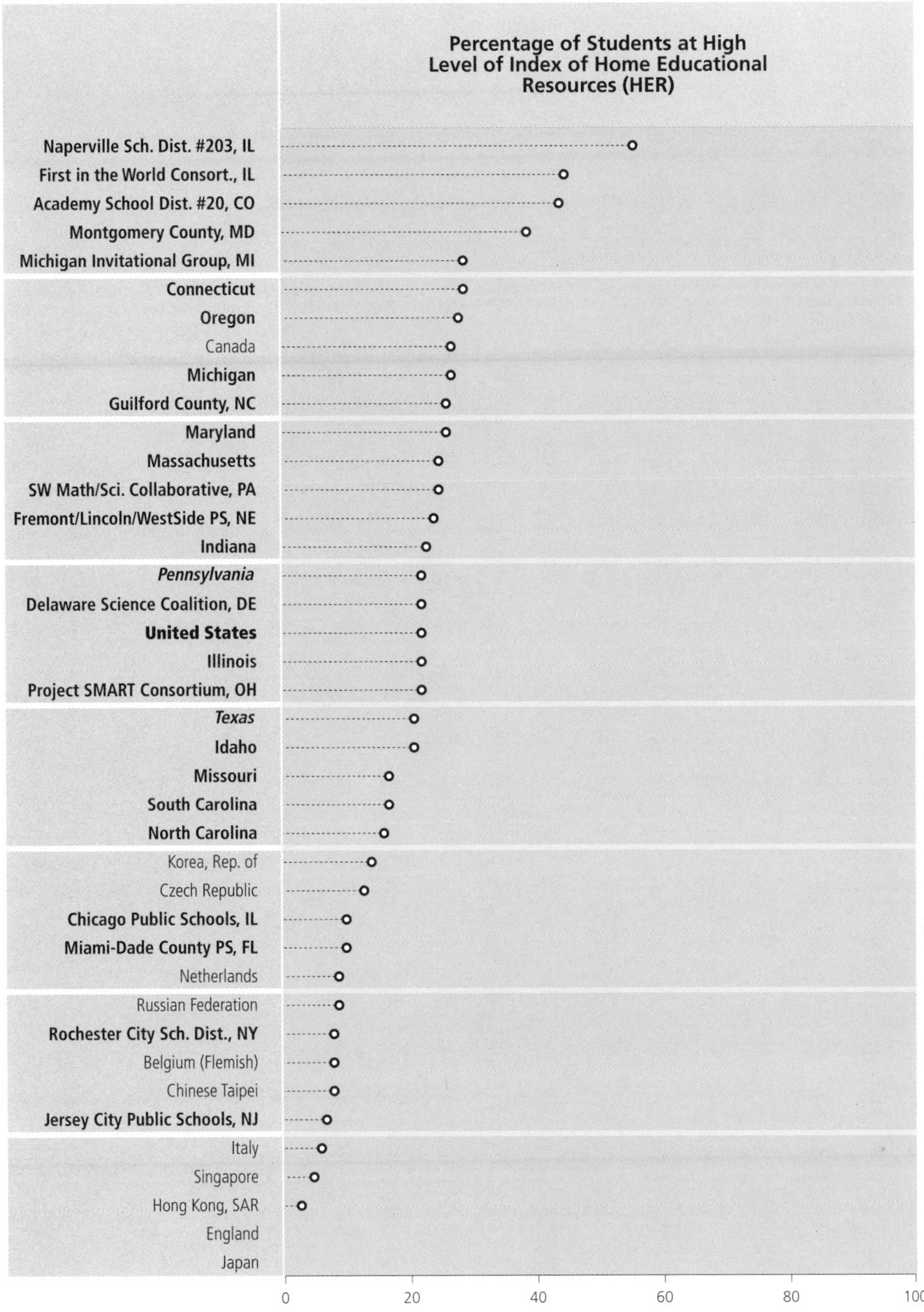

Percentage of Students at High Level of Index of Home Educational Resources (HER)

Naperville Sch. Dist. #203, IL
First in the World Consort., IL
Academy School Dist. #20, CO
Montgomery County, MD
Michigan Invitational Group, MI
Connecticut
Oregon
Canada
Michigan
Guilford County, NC
Maryland
Massachusetts
SW Math/Sci. Collaborative, PA
Fremont/Lincoln/WestSide PS, NE
Indiana
Pennsylvania
Delaware Science Coalition, DE
United States
Illinois
Project SMART Consortium, OH
Texas
Idaho
Missouri
South Carolina
North Carolina
Korea, Rep. of
Czech Republic
Chicago Public Schools, IL
Miami-Dade County PS, FL
Netherlands
Russian Federation
Rochester City Sch. Dist., NY
Belgium (Flemish)
Chinese Taipei
Jersey City Public Schools, NJ
Italy
Singapore
Hong Kong, SAR
England
Japan

0 20 40 60 80 100

SOURCE: IEA Third International Mathematics and Science Study (TIMSS), 1998-1999.

text continued from page 111

used in this report. Despite the different educational approaches, structures, and organizations across the TIMSS 1999 countries, it is clear that parents' education is positively related to students' mathematics achievement. The pattern across countries was that eighth-grade students whose parents had more education were also those who had higher achievement in mathematics. The same was true for nearly all Benchmarking jurisdictions.

As information technology and the Internet become more and more important as an educational resource, those who do not have access to this technology will be increasingly at a disadvantage. To provide information about this "digital divide," Exhibit 4.2 presents the percentage of students in each entity that reported having a computer at home, together with their average mathematics achievement. Compared with some of the reference countries as well as the international average (45 percent), students in the Benchmarking jurisdictions reported relatively high levels of computer ownership; more than 70 percent of students in each state reported having a computer at home. In the wealthier districts and consortia such as the Academy School District, the First in the World Consortium, Montgomery County, and the Naperville School District, more than 90 percent of students so reported. Even in the less advantaged public school districts, more than half the students reported having a computer at home. In almost every entity, students with a computer at home had higher average mathematics achievement than those without.

Students who speak a language (or languages) in the home that is different from the language spoken in school sometimes benefit from being multilingual. However, when they are still developing proficiency in the language of instruction they can be at a disadvantage in learning situations. Exhibit 4.3 contains students' reports of how frequently they speak the language of the TIMSS test at home in relation to their average mathematics achievement. Students from homes where the language of the test is always or almost always spoken had higher average achievement than those who spoke it less frequently. In all of the Benchmarking states except Massachusetts and Texas, 90 percent or more of the students reported always or almost always speaking the language of the test at home. The percentage of students speaking the language of the test at home was lower in a number of school districts, however, particularly the public school systems in Chicago, Jersey City, and Miami-Dade.

Exhibit 4.4 presents students' reports of their race/ethnicity. Across the United States as a whole, 63 percent reported that they were white, 15 percent black, 12 percent Hispanic, five percent Asian or Pacific Islander, one percent American Indian or Alaskan Native, and four percent other.

There was a pronounced relationship between race/ethnicity and mathematics achievement, with Asian/Pacific Islander students having the highest average achievement, followed by white, Hispanic, and black students. This pattern was found for many of the Benchmarking participants. Because minority students are often concentrated in urban schools, the resource disparities between urban and non-urban schools summarized in the introduction to this report are particularly troubling in light of the persistent achievement gaps between many minority and non-minority students.

Among Benchmarking states, Maryland, North Carolina, and South Carolina had more than 30 percent black students, and Texas more than 30 percent Hispanic. Racial composition varied even more among the Benchmarking districts and consortia. Predominantly white jurisdictions included the Academy School District, the Fremont/Lincoln/Westside Public Schools, the Michigan Invitational Group, Naperville, and the Southwest Pennsylvania Math and Science Collaborative, with more than 80 percent white students. Ethnically more diverse jurisdictions included Chicago (47 percent black, 37 percent Hispanic), Jersey City (35 percent black, 35 percent Hispanic, 16 percent Asian/Pacific Islander), Miami-Dade (31 percent black, 55 percent Hispanic), Montgomery County (16 percent black, 12 percent Hispanic, 15 percent Asian/Pacific Islander), and Rochester (56 percent black, 16 percent Hispanic).

By the end of the eighth grade, students in most countries can say what their expectations are for further education. Although one-quarter or more of the students in some countries did not know, Exhibit 4.5 shows that, on average across countries, more than half the students reported that they expected to finish university (a four-year degree program or equivalent). The United States was among the countries that had the highest percentage, with almost 80 percent expecting to finish university. In almost every country, also, there was a positive association between educational expectations and mathematics achievement. Among Benchmarking participants, the percentage of students expecting to finish university was also high, even in areas with low student achievement, as more than 70 percent of students in all Benchmarking entities reported that they expected to finish university.

Exhibits R1.5 to R1.7 in the reference section present eighth-grade students' reports about how they, their mothers, and their friends feel about the importance of doing well in various academic and non-academic activities. On average across the TIMSS 1999 countries, more than 90 percent of students reported that they and their mothers agreed that it was important to do well in mathematics, science, and language. Somewhat fewer reported that their friends agreed (77 to 86 percent). As might be anticipated, slightly more students reported that they and their friends felt it was important to have fun (92 percent) than reported that their mothers found this important (85 percent). More moderate agreement was reported for the importance of doing well in sports (from 81 to 87 percent). In general, the reports of students in the Benchmarking jurisdictions resembled those in the United States overall. It is noteworthy, however, that students in the U.S. and in many Benchmarking jurisdictions were less likely than their counterparts internationally, on average, to report that their friends think it is important to do well in mathematics, science, and language, and were more likely to report that they, their mothers, and their friends think it is important to have fun.

Students were also asked why they needed to do well in mathematics (see Exhibit R1.8). In most entities, getting into their desired secondary school or university was a stronger motivating factor than was pleasing their parents or getting their desired job.

8th Grade Mathematics

	Have Computer at Home		Do Not Have Computer at Home	
	Percent of Students	Average Achievement	Percent of Students	Average Achievement
Countries				
United States	80 (1.2)	515 (3.8)	20 (1.2)	459 (4.7)
Belgium (Flemish)	86 (1.0)	564 (3.5)	14 (1.0)	523 (7.4)
Canada	85 (0.8)	536 (2.5)	15 (0.8)	505 (4.5)
Chinese Taipei	63 (1.0)	605 (3.9)	37 (1.0)	552 (4.4)
Czech Republic	47 (1.2)	536 (4.8)	53 (1.2)	506 (4.7)
England	85 (0.8)	503 (4.1)	15 (0.8)	466 (6.2)
Hong Kong, SAR	72 (1.3)	589 (4.0)	28 (1.3)	566 (5.8)
Italy	63 (1.0)	488 (4.1)	37 (1.0)	465 (4.2)
Japan	52 (0.9)	592 (2.3)	48 (0.9)	566 (2.3)
Korea, Rep. of	67 (0.9)	600 (1.8)	33 (0.9)	561 (3.0)
Netherlands	96 (1.0)	542 (7.3)	4 (1.0)	513 (11.1)
Russian Federation	22 (1.2)	531 (6.5)	78 (1.2)	525 (6.4)
Singapore	80 (1.3)	614 (6.1)	20 (1.3)	567 (7.3)
States				
Connecticut	88 (1.7)	521 (8.4)	12 (1.7)	449 (9.3)
Idaho	82 (2.1)	505 (6.6)	18 (2.1)	452 (9.2)
Illinois	80 (2.1)	521 (6.7)	20 (2.1)	464 (6.4)
Indiana	81 (1.5)	523 (7.2)	19 (1.5)	479 (7.2)
Maryland	86 (1.4)	504 (5.9)	14 (1.4)	442 (7.4)
Massachusetts	87 (1.6)	520 (5.7)	13 (1.6)	469 (8.0)
Michigan	85 (1.7)	526 (6.6)	15 (1.7)	468 (9.9)
Missouri	76 (1.8)	501 (5.2)	24 (1.8)	456 (6.7)
North Carolina	74 (1.8)	507 (7.2)	26 (1.8)	461 (6.3)
Oregon	86 (1.7)	524 (5.4)	14 (1.7)	457 (7.0)
Pennsylvania	83 (2.0)	516 (6.0)	17 (2.0)	466 (7.4)
South Carolina	75 (2.2)	514 (7.2)	25 (2.2)	465 (8.3)
Texas	73 (3.3)	540 (7.5)	27 (3.3)	464 (9.3)
Districts and Consortia				
Academy School Dist. #20, CO	96 (0.5)	531 (1.9)	4 (0.5)	484 (10.8)
Chicago Public Schools, IL	61 (1.7)	471 (7.0)	39 (1.7)	450 (5.9)
Delaware Science Coalition, DE	82 (1.6)	489 (9.3)	18 (1.6)	438 (9.2)
First in the World Consort., IL	96 (0.6)	563 (5.7)	4 (0.6)	476 (14.5)
Fremont/Lincoln/WestSide PS, NE	81 (1.6)	500 (8.8)	19 (1.6)	435 (12.1)
Guilford County, NC	81 (1.6)	524 (7.5)	19 (1.6)	469 (9.6)
Jersey City Public Schools, NJ	58 (2.3)	488 (11.8)	42 (2.3)	459 (5.4)
Miami-Dade County PS, FL	66 (2.8)	438 (10.7)	34 (2.8)	391 (8.0)
Michigan Invitational Group, MI	89 (1.6)	538 (5.6)	11 (1.6)	486 (10.4)
Montgomery County, MD	91 (1.4)	546 (3.8)	9 (1.4)	458 (7.1)
Naperville Sch. Dist. #203, IL	98 (0.4)	570 (2.8)	2 (0.4)	~ ~
Project SMART Consortium, OH	83 (1.2)	527 (8.3)	17 (1.2)	489 (6.6)
Rochester City Sch. Dist., NY	61 (2.3)	451 (8.0)	39 (2.3)	440 (6.6)
SW Math/Sci. Collaborative, PA	82 (1.9)	528 (6.6)	18 (1.9)	468 (10.5)
International Avg. (All Countries)	45 (0.2)	509 (1.1)	55 (0.2)	470 (0.8)

SOURCE: IEA Third International Mathematics and Science Study (TIMSS), 1998-1999.

Background data provided by students.

States in *italics* did not fully satisfy guidelines for sample participation rates (see Appendix A for details).

( ) Standard errors appear in parentheses. Because results are rounded to the nearest whole number, some totals may appear inconsistent.

A tilde (~) indicates insufficient data to report achievement.

Exhibit 4.3 Frequency with Which Students Speak Language of the Test at Home

8th Grade Mathematics

	Always or Almost Always		Sometimes		Never	
	Percent of Students	Average Achievement	Percent of Students	Average Achievement	Percent of Students	Average Achievement
Countries						
United States	90 (1.0)	509 (3.8)	9 (1.0)	456 (8.2)	1 (0.1)	~ ~
Belgium (Flemish)	86 (1.3)	566 (3.2)	8 (0.7)	531 (8.0)	6 (0.9)	522 (13.5)
Canada	91 (0.6)	532 (2.5)	8 (0.5)	523 (6.6)	2 (0.2)	~ ~
Chinese Taipei	67 (1.4)	606 (3.9)	31 (1.3)	545 (5.3)	2 (0.2)	~ ~
Czech Republic	98 (0.5)	523 (4.0)	1 (0.3)	~ ~	1 (0.2)	~ ~
England	95 (0.9)	500 (4.2)	5 (0.8)	471 (12.1)	0 (0.1)	~ ~
Hong Kong, SAR r	80 (2.4)	571 (4.5)	17 (1.9)	600 (8.5)	3 (0.5)	609 (12.2)
Italy	77 (1.1)	493 (3.5)	20 (1.0)	434 (5.6)	4 (0.5)	442 (11.8)
Japan	97 (0.3)	581 (1.8)	3 (0.3)	532 (11.5)	0 (0.1)	~ ~
Korea, Rep. of	96 (0.3)	589 (2.0)	4 (0.3)	545 (4.9)	0 (0.0)	~ ~
Netherlands	86 (2.4)	544 (7.8)	8 (1.2)	529 (9.0)	6 (1.8)	531 (13.7)
Russian Federation	94 (2.3)	527 (5.9)	5 (2.3)	527 (36.9)	1 (0.2)	~ ~
Singapore	27 (1.8)	629 (7.1)	63 (1.6)	595 (6.4)	10 (0.5)	601 (8.2)
States						
Connecticut	90 (1.4)	517 (8.8)	8 (1.4)	472 (13.4)	2 (0.3)	~ ~
Idaho	92 (1.4)	501 (6.7)	7 (1.3)	430 (13.3)	1 (0.3)	~ ~
Illinois	91 (1.3)	515 (6.6)	8 (1.2)	471 (10.1)	1 (0.2)	~ ~
Indiana	96 (0.6)	518 (7.1)	3 (0.5)	477 (15.8)	1 (0.3)	~ ~
Maryland	91 (0.8)	497 (5.9)	8 (0.7)	493 (10.2)	1 (0.3)	~ ~
Massachusetts	88 (1.6)	518 (5.7)	10 (1.4)	493 (11.7)	2 (0.3)	~ ~
Michigan	96 (0.6)	520 (7.2)	3 (0.4)	484 (13.2)	1 (0.2)	~ ~
Missouri	95 (0.6)	494 (5.5)	4 (0.5)	453 (11.5)	1 (0.2)	~ ~
North Carolina	96 (0.5)	498 (7.0)	3 (0.4)	471 (13.2)	1 (0.2)	~ ~
Oregon	92 (1.1)	520 (5.9)	7 (0.9)	456 (12.0)	1 (0.4)	~ ~
Pennsylvania	95 (1.1)	510 (6.3)	5 (0.9)	472 (13.7)	1 (0.3)	~ ~
South Carolina	97 (0.4)	504 (7.7)	2 (0.4)	~ ~	0 (0.2)	~ ~
Texas	82 (2.9)	532 (8.4)	17 (2.8)	464 (10.6)	1 (0.4)	~ ~
Districts and Consortia						
Academy School Dist. #20, CO	93 (0.8)	531 (2.0)	6 (0.7)	507 (12.5)	1 (0.3)	~ ~
Chicago Public Schools, IL	77 (4.7)	464 (6.5)	21 (4.6)	461 (8.8)	2 (0.7)	~ ~
Delaware Science Coalition, DE	91 (0.9)	485 (9.0)	6 (0.9)	454 (13.7)	3 (0.5)	434 (24.6)
First in the World Consort., IL	85 (1.3)	564 (5.9)	14 (1.3)	531 (7.8)	1 (0.3)	~ ~
Fremont/Lincoln/WestSide PS, NE	92 (1.1)	493 (8.9)	7 (0.9)	447 (10.0)	1 (0.3)	~ ~
Guilford County, NC	95 (0.7)	516 (7.3)	4 (0.7)	500 (16.4)	1 (0.5)	~ ~
Jersey City Public Schools, NJ	74 (1.5)	474 (9.3)	26 (1.4)	485 (9.1)	1 (0.3)	~ ~
Miami-Dade County PS, FL	59 (4.1)	428 (9.2)	36 (3.6)	420 (11.4)	5 (0.8)	394 (17.6)
Michigan Invitational Group, MI	96 (0.6)	535 (6.1)	3 (0.5)	509 (22.7)	1 (0.3)	~ ~
Montgomery County, MD	83 (1.9)	544 (4.0)	15 (2.0)	512 (10.4)	2 (0.6)	~ ~
Naperville Sch. Dist. #203, IL	93 (0.5)	570 (2.9)	6 (0.6)	573 (7.6)	1 (0.2)	~ ~
Project SMART Consortium, OH	95 (0.9)	523 (7.7)	4 (0.7)	485 (11.4)	1 (0.3)	~ ~
Rochester City Sch. Dist., NY	86 (1.3)	450 (6.7)	13 (1.1)	437 (8.2)	2 (0.6)	~ ~
SW Math/Sci. Collaborative, PA	98 (0.4)	518 (7.2)	1 (0.3)	~ ~	1 (0.2)	~ ~
International Avg. (All Countries)	79 (0.3)	493 (0.8)	17 (0.2)	466 (2.3)	5 (0.1)	455 (4.1)

SOURCE: IEA Third International Mathematics and Science Study (TIMSS), 1998-1999.

Background data provided by students.

States in *italics* did not fully satisfy guidelines for sample participation rates (see Appendix A for details).

( ) Standard errors appear in parentheses. Because results are rounded to the nearest whole number, some totals may appear inconsistent.

A tilde (~) indicates insufficient data to report achievement.

An "r" indicates a 70-84% student response rate.

	White		Black		Hispanic	
	Percent of Students	Average Achievement	Percent of Students	Average Achievement	Percent of Students	Average Achievement
States						
Connecticut	74 (4.5)	533 (6.8)	10 (3.0)	432 (12.5)	9 (2.2)	451 (13.5)
Idaho	83 (2.0)	506 (6.5)	1 (0.3)	~ ~	10 (1.7)	432 (8.9)
Illinois	65 (3.4)	533 (5.3)	17 (2.9)	449 (7.8)	12 (2.3)	462 (10.3)
Indiana	83 (2.3)	525 (7.2)	10 (2.2)	438 (6.5)	3 (0.6)	493 (11.9)
Maryland	55 (4.2)	521 (4.7)	30 (3.9)	438 (7.0)	4 (0.6)	487 (12.8)
Massachusetts	74 (3.4)	524 (5.1)	7 (1.6)	464 (20.4)	8 (1.4)	464 (11.0)
Michigan	82 (3.4)	532 (5.9)	10 (3.4)	418 (9.5)	3 (0.6)	481 (15.6)
Missouri	78 (3.2)	505 (4.9)	15 (3.1)	426 (12.3)	2 (0.4)	~ ~
North Carolina	62 (3.5)	521 (6.7)	31 (3.2)	447 (7.9)	3 (0.5)	474 (14.1)
Oregon	80 (1.9)	523 (5.4)	1 (0.5)	~ ~	8 (1.1)	452 (13.6)
Pennsylvania	78 (4.5)	519 (5.6)	12 (3.7)	446 (16.8)	3 (1.3)	476 (7.1)
South Carolina	63 (4.0)	533 (6.0)	32 (4.0)	446 (7.0)	1 (0.4)	~ ~
Texas	47 (5.2)	562 (5.0)	13 (2.5)	464 (16.7)	32 (4.7)	476 (8.6)
Districts and Consortia						
Academy School Dist. #20, CO	82 (1.0)	535 (2.4)	3 (0.5)	484 (15.7)	7 (0.6)	496 (8.7)
Chicago Public Schools, IL	11 (3.2)	499 (12.5)	47 (10.6)	447 (8.4)	37 (8.9)	468 (10.0)
Delaware Science Coalition, DE	63 (2.3)	501 (9.3)	24 (2.0)	435 (6.2)	5 (0.7)	465 (12.4)
First in the World Consort., IL	74 (1.8)	564 (5.6)	1 (0.3)	~ ~	7 (0.8)	478 (5.0)
Fremont/Lincoln/WestSide PS, NE	83 (1.6)	498 (8.1)	3 (0.8)	437 (28.6)	4 (0.7)	404 (14.6)
Guilford County, NC	57 (2.1)	544 (6.8)	35 (2.3)	463 (8.6)	2 (0.5)	~ ~
Jersey City Public Schools, NJ	7 (0.9)	513 (14.7)	35 (1.7)	442 (7.7)	35 (1.1)	474 (6.4)
Miami-Dade County PS, FL	7 (2.5)	501 (24.8)	31 (5.6)	381 (11.5)	55 (6.8)	438 (8.5)
Michigan Invitational Group, MI	88 (1.2)	534 (6.0)	4 (1.0)	473 (14.5)	1 (0.5)	~ ~
Montgomery County, MD	50 (2.7)	564 (6.2)	16 (1.3)	482 (9.3)	12 (1.8)	480 (13.0)
Naperville Sch. Dist. #203, IL	82 (1.0)	569 (2.6)	1 (0.4)	~ ~	2 (0.5)	~ ~
Project SMART Consortium, OH	79 (1.9)	530 (8.4)	10 (1.5)	476 (5.5)	4 (0.7)	475 (12.5)
Rochester City Sch. Dist., NY	16 (2.2)	504 (12.0)	56 (2.6)	428 (6.1)	16 (1.7)	443 (6.5)
SW Math/Sci. Collaborative, PA	87 (2.9)	526 (6.9)	10 (2.6)	440 (11.9)	1 (0.3)	~ ~
United States	63 (2.4)	525 (4.6)	15 (1.9)	444 (5.5)	12 (1.6)	457 (6.4)

SOURCE: IEA Third International Mathematics and Science Study (TIMSS), 1998-1999.

Background data provided by students.

States in *italics* did not fully satisfy guidelines for sample participation rates (see Appendix A for details).

( ) Standard errors appear in parentheses. Because results are rounded to the nearest whole number, some totals may appear inconsistent.

A tilde (~) indicates insufficient data to report achievement.

	Asian/ Pacific Islander		American Indian/ Alaskan Native		Other	
	Percent of Students	Average Achievement	Percent of Students	Average Achievement	Percent of Students	Average Achievement
States						
Connecticut	2 (0.4)	~ ~	0 (0.2)	~ ~	4 (0.6)	481 (13.8)
Idaho	2 (0.5)	~ ~	2 (0.5)	~ ~	2 (0.3)	~ ~
Illinois	4 (0.9)	544 (11.9)	0 (0.2)	~ ~	2 (0.4)	~ ~
Indiana	2 (0.4)	~ ~	1 (0.3)	~ ~	2 (0.4)	~ ~
Maryland	5 (0.6)	551 (7.0)	1 (0.2)	~ ~	5 (0.6)	511 (12.5)
Massachusetts	5 (0.8)	559 (19.8)	1 (0.2)	~ ~	5 (0.8)	490 (13.4)
Michigan	2 (0.3)	~ ~	1 (0.2)	~ ~	3 (0.3)	490 (14.1)
Missouri	1 (0.3)	~ ~	1 (0.4)	~ ~	3 (0.4)	450 (15.3)
North Carolina	1 (0.3)	~ ~	1 (0.4)	~ ~	2 (0.4)	~ ~
Oregon	4 (0.7)	531 (10.0)	3 (0.5)	482 (11.7)	4 (0.5)	517 (10.0)
Pennsylvania	3 (1.4)	526 (17.1)	1 (0.2)	~ ~	3 (0.5)	512 (12.1)
South Carolina	1 (0.2)	~ ~	1 (0.2)	~ ~	2 (0.3)	~ ~
Texas	4 (1.4)	569 (24.1)	1 (0.1)	~ ~	3 (0.4)	515 (16.7)
Districts and Consortia						
Academy School Dist. #20, CO	4 (0.6)	527 (10.7)	1 (0.3)	~ ~	4 (0.5)	511 (12.1)
Chicago Public Schools, IL	2 (1.0)	~ ~	1 (0.2)	~ ~	2 (0.5)	~ ~
Delaware Science Coalition, DE	2 (0.6)	~ ~	1 (0.2)	~ ~	5 (0.9)	475 (13.6)
First in the World Consort., IL	15 (1.7)	591 (11.4)	1 (0.4)	~ ~	2 (0.8)	~ ~
Fremont/Lincoln/WestSide PS, NE	3 (0.5)	476 (17.6)	2 (0.4)	~ ~	5 (0.9)	475 (19.3)
Guilford County, NC	4 (0.4)	529 (14.2)	1 (0.2)	~ ~	2 (0.5)	~ ~
Jersey City Public Schools, NJ	16 (1.7)	533 (16.2)	0 (0.2)	~ ~	7 (0.8)	504 (16.5)
Miami-Dade County PS, FL	2 (0.6)	~ ~	1 (0.1)	~ ~	5 (1.1)	426 (24.1)
Michigan Invitational Group, MI	3 (0.5)	580 (16.4)	0 (0.2)	~ ~	3 (0.3)	533 (19.2)
Montgomery County, MD	15 (1.4)	564 (6.7)	1 (0.2)	~ ~	6 (0.8)	535 (14.3)
Naperville Sch. Dist. #203, IL	12 (0.8)	599 (5.9)	0 (0.1)	~ ~	3 (0.5)	549 (8.6)
Project SMART Consortium, OH	3 (0.5)	550 (23.1)	1 (0.2)	~ ~	3 (0.7)	519 (15.8)
Rochester City Sch. Dist., NY	3 (0.5)	500 (22.4)	2 (0.5)	~ ~	7 (1.0)	465 (13.3)
SW Math/Sci. Collaborative, PA	1 (0.4)	~ ~	0 (0.1)	~ ~	2 (0.4)	~ ~
United States	5 (1.3)	539 (10.7)	1 (0.2)	~ ~	4 (0.3)	496 (9.5)

SOURCE: IEA Third International Mathematics and Science Study (TIMSS), 1998-1999.

Exhibit 4.5 Students' Expectations for Finishing School*

8th Grade Mathematics

	Finish University[1]		Some Vocational/ Technical Education or University Only[2]		Finish Secondary School Only[3]		Some Secondary School Only		Don't Know	
	Percent of Students	Average Achievement	Percent of Students	Average Achievement	Percent of Students	Average Achievement	Percent of Students	Average Achievement	Percent of Students	Average Achievement
Countries										
United States	78 (1.2)	516 (3.8)	9 (0.6)	466 (5.1)	5 (0.4)	426 (6.2)	1 (0.1)	~ ~	7 (0.5)	474 (5.9)
Belgium (Flemish)	26 (1.1)	605 (6.4)	30 (0.9)	563 (3.8)	16 (0.9)	509 (4.5)	0 (0.0)	~ ~	29 (1.0)	544 (2.9)
Canada	76 (0.9)	539 (2.6)	13 (0.6)	522 (4.7)	4 (0.3)	482 (7.7)	1 (0.1)	~ ~	7 (0.6)	497 (6.0)
Chinese Taipei	62 (1.4)	624 (3.7)	24 (1.0)	527 (3.0)	2 (0.3)	~ ~	0 (0.1)	~ ~	11 (0.6)	534 (7.2)
Czech Republic	38 (1.8)	564 (4.1)	5 (0.6)	542 (7.1)	39 (1.5)	496 (3.3)	8 (1.0)	452 (7.1)	10 (0.8)	493 (7.6)
England	– –	– –	– –	– –	– –	– –	– –	– –	– –	– –
Hong Kong, SAR	63 (1.7)	601 (3.8)	20 (0.9)	562 (4.9)	10 (0.8)	529 (7.7)	1 (0.2)	~ ~	6 (0.4)	562 (6.8)
Italy	33 (1.3)	517 (4.1)	19 (0.9)	487 (4.4)	31 (1.1)	463 (4.0)	7 (0.6)	396 (10.4)	9 (0.7)	461 (8.7)
Japan	38 (0.9)	614 (2.7)	18 (0.6)	564 (2.6)	18 (0.7)	532 (3.0)	1 (0.1)	~ ~	25 (0.7)	572 (3.1)
Korea, Rep. of	77 (0.7)	605 (1.9)	8 (0.4)	521 (4.2)	4 (0.3)	500 (6.3)	0 (0.1)	~ ~	11 (0.5)	551 (4.3)
Netherlands	22 (2.8)	582 (9.6)	30 (1.8)	549 (5.7)	29 (2.6)	507 (9.0)	1 (0.2)	~ ~	18 (0.9)	533 (8.1)
Russian Federation	61 (1.5)	547 (5.4)	19 (1.0)	505 (6.1)	7 (0.5)	481 (10.4)	2 (0.5)	~ ~	11 (0.7)	496 (7.8)
Singapore	57 (2.1)	625 (6.1)	26 (1.6)	576 (5.5)	2 (0.3)	~ ~	0 (0.0)	~ ~	15 (0.7)	587 (8.2)
States										
Connecticut	80 (1.6)	524 (9.5)	8 (1.0)	468 (10.8)	4 (0.5)	441 (8.8)	1 (0.2)	~ ~	7 (0.8)	483 (8.9)
Idaho	72 (2.0)	511 (6.3)	11 (0.9)	480 (8.5)	7 (0.9)	425 (8.9)	1 (0.2)	~ ~	9 (0.9)	458 (10.9)
Illinois	81 (1.2)	521 (7.1)	9 (0.8)	465 (7.6)	4 (0.7)	443 (9.3)	0 (0.1)	~ ~	6 (0.6)	487 (9.1)
Indiana	79 (1.6)	527 (6.6)	9 (0.9)	471 (8.1)	4 (0.6)	449 (13.1)	1 (0.2)	~ ~	7 (0.7)	486 (13.3)
Maryland	80 (1.2)	506 (6.6)	9 (0.7)	456 (8.4)	4 (0.5)	415 (9.6)	1 (0.2)	~ ~	6 (0.6)	481 (7.4)
Massachusetts	78 (1.5)	526 (5.9)	10 (0.6)	477 (8.3)	5 (0.7)	429 (11.3)	1 (0.1)	~ ~	6 (0.7)	493 (7.7)
Michigan	83 (1.1)	527 (7.4)	7 (0.7)	473 (9.3)	3 (0.4)	454 (11.0)	1 (0.1)	~ ~	6 (0.5)	483 (14.0)
Missouri	72 (1.5)	504 (5.8)	12 (0.9)	468 (6.5)	8 (0.8)	426 (8.2)	1 (0.2)	~ ~	7 (0.6)	468 (7.6)
North Carolina	79 (1.5)	508 (7.4)	9 (0.7)	455 (6.5)	6 (0.7)	432 (8.8)	1 (0.1)	~ ~	4 (0.4)	461 (10.4)
Oregon	76 (1.9)	529 (5.9)	10 (0.9)	485 (9.1)	5 (0.8)	439 (7.8)	1 (0.2)	~ ~	9 (0.9)	472 (10.5)
Pennsylvania	77 (1.4)	518 (6.8)	9 (0.7)	478 (8.6)	5 (0.6)	448 (10.2)	1 (0.1)	~ ~	7 (0.6)	481 (9.6)
South Carolina	80 (1.3)	519 (8.1)	9 (0.8)	437 (7.8)	6 (0.6)	415 (8.6)	0 (0.1)	~ ~	5 (0.5)	458 (9.8)
Texas	80 (2.0)	534 (7.9)	7 (0.8)	459 (10.8)	6 (1.3)	427 (16.4)	1 (0.3)	~ ~	6 (0.7)	492 (15.7)
Districts and Consortia										
Academy School Dist. #20, CO	83 (1.1)	537 (2.1)	5 (0.6)	482 (11.4)	3 (0.4)	463 (12.5)	1 (0.3)	~ ~	8 (0.9)	512 (8.5)
Chicago Public Schools, IL	74 (1.8)	474 (6.7)	11 (0.8)	434 (10.1)	8 (1.2)	414 (8.4)	1 (0.3)	~ ~	6 (0.9)	456 (14.7)
Delaware Science Coalition, DE	74 (2.2)	498 (9.0)	11 (0.8)	444 (8.6)	7 (1.1)	417 (12.5)	1 (0.4)	~ ~	7 (1.0)	431 (8.9)
First in the World Consort., IL	92 (1.1)	564 (5.4)	3 (0.8)	494 (12.1)	1 (0.5)	~ ~	0 (0.2)	~ ~	4 (0.8)	540 (19.3)
Fremont/Lincoln/WestSide PS, NE	74 (2.3)	506 (8.8)	7 (1.1)	442 (19.1)	5 (1.3)	404 (9.7)	1 (0.2)	~ ~	12 (1.4)	458 (12.4)
Guilford County, NC	89 (1.5)	521 (7.4)	5 (0.9)	460 (13.4)	3 (0.8)	419 (15.2)	0 (0.3)	~ ~	3 (0.6)	481 (16.3)
Jersey City Public Schools, NJ	80 (1.6)	485 (9.7)	8 (0.9)	443 (10.4)	6 (0.8)	442 (13.4)	0 (0.0)	~ ~	6 (0.8)	439 (17.1)
Miami-Dade County PS, FL	76 (2.4)	440 (8.8)	10 (1.3)	372 (11.6)	6 (0.7)	361 (13.1)	1 (0.2)	~ ~	7 (1.0)	365 (18.8)
Michigan Invitational Group, MI	80 (2.1)	543 (5.2)	9 (1.6)	503 (9.0)	5 (0.7)	459 (11.0)	1 (0.3)	~ ~	5 (0.8)	495 (16.8)
Montgomery County, MD	85 (1.0)	547 (4.1)	6 (0.9)	472 (12.7)	2 (0.3)	~ ~	1 (0.3)	~ ~	7 (0.6)	521 (9.5)
Naperville Sch. Dist. #203, IL	94 (0.8)	572 (2.8)	3 (0.5)	532 (10.8)	1 (0.3)	~ ~	0 (0.1)	~ ~	3 (0.5)	519 (17.4)
Project SMART Consortium, OH	81 (2.1)	533 (7.9)	8 (1.1)	468 (9.1)	4 (0.8)	469 (11.2)	1 (0.3)	~ ~	7 (0.8)	479 (9.3)
Rochester City Sch. Dist., NY	76 (1.6)	455 (6.5)	9 (1.1)	421 (14.6)	7 (0.9)	392 (16.1)	1 (0.3)	~ ~	8 (1.0)	436 (13.0)
SW Math/Sci. Collaborative, PA	80 (2.1)	528 (6.6)	8 (0.8)	476 (10.0)	5 (0.5)	450 (12.5)	0 (0.1)	~ ~	7 (1.2)	478 (12.4)
International Avg. (All Countries)	52 (0.3)	517 (0.8)	17 (0.1)	469 (1.0)	15 (0.2)	442 (1.0)	3 (0.1)	390 (3.1)	14 (0.1)	462 (1.1)

SOURCE: IEA Third International Mathematics and Science Study (TIMSS), 1998-1999.

Background data provided by students.

* Response categories were defined by each country to conform to their own educational system and may not be strictly comparable across countries. See Reference Exhibit R1.4 for country definitions of educational levels.

1 In most countries, finish university is defined as completion of at least a 4-year degree program at a university or an equivalent institute of higher education. For the United States, includes community college, college, or university.

2 In some countries, may include higher post-secondary education levels.

3 In most countries, finish secondary school corresponds to completion of an upper-secondary track terminating after 11 to 13 years of schooling (ISCED level 3 vocational, apprenticeship or academic tracks).

States in *italics* did not fully satisfy guidelines for sample participation rates (see Appendix A for details).

How Much of Their Out-of-School Time Do Students Spend on Homework During the School Week?

One of the main ways for students to consolidate and extend classroom learning is to spend time out of school studying or doing homework. Well-chosen homework assignments can reinforce classroom learning, and by providing a challenge can encourage students to extend their understanding of the subject matter. Homework also allows students who are having trouble keeping up with their classmates to review material taught in class.

To summarize the amount of time typically devoted to homework in each country and Benchmarking jurisdiction, TIMSS constructed an index of out-of-school study time (OST) that assigns students to a high, medium, or low level based on the amount of time they reported studying mathematics, science, and other subjects. Students at the high level reported spending more than three hours each day out of school studying all subjects combined. Students at the medium level reported spending more than one hour but not more than three, while those at the low level reported one hour or less per day.

Exhibit 4.6 shows the percentages of students at each level of this index, and their average mathematics achievement, for Benchmarking participants and comparison countries. On average across all the TIMSS 1999 countries, 38 percent of eighth-grade students were at the high level of the out-of-school study time index, and a further 48 percent were at the medium level. Only 14 percent, on average, were at the low level, with just one hour of homework or less each day. The United States was one of the countries with relatively little emphasis on homework, with just 22 percent of students at the high level and 23 percent at the low level. Among Benchmarking participants, the jurisdictions that reported the greatest amount of out-of-school study time included the Jersey City and Chicago Public Schools, and the Academy School District, which each had more than one-third of their students at the high level of the index.

On average internationally, and in many of the Benchmarking entities, students at the low index level had lower average mathematics achievement than their classmates who reported more out-of-school study time. However, spending a lot of time studying was not necessarily associated with higher achievement. In many of the Benchmarking entities, students at the medium level of the study index had average achievement that was as high as or higher than that of students at the high level. This pattern suggests that, compared with their higher-achieving counterparts, the lower-performing students may do less homework, either because they simply do not do it or because their teachers do not assign it, or more homework, perhaps in an effort to keep up academically.

More detailed information on the amount of time students reported spending on mathematics homework is presented in Exhibit 4.7. The results reveal that while students on average across all the TIMSS 1999 countries spent 1.1 hours per day doing mathematics homework, students in most of the Benchmarking jurisdictions and the United States spent somewhat less. The exhibit also shows the percentages of students that reported spending one hour or more, less than one hour, and no time at all studying mathematics or doing mathematics homework on a normal school day, together with their average mathematics achievement. On average across all countries, 40 percent of students reported spending one hour or more per day doing mathematics homework. None of the Benchmarking states reported this much homework, but three school districts did – the Academy School District (41 percent), the Chicago Public Schools (48 percent), and the Jersey City Public Schools (44 percent). The next highest levels of mathematics homework were reported in Illinois, North Carolina, Guilford County, the Miami-Dade County Public Schools, and Montgomery County, where 30 percent or more of students reported spending one hour or more. At least 20 percent of the students in Missouri, Texas, and the Project SMART Consortium reported spending no time at all doing mathematics homework on a normal school day.

Further detail on the student data that underlie the out-of-school study time index appears in Exhibit R1.9 in the reference section. In comparison with the approximately one hour each day spent on mathematics homework, the TIMSS 1999 countries on average reported 2.8 hours of homework in total. None of the Benchmarking jurisdictions reached this level, the highest being 2.7 hours in Chicago and Jersey City, and the lowest 1.8 hours in Texas, the Fremont/Lincoln/Westside Public Schools, and Project SMART. To provide a fuller picture of how students spend their out-of-school time on a school day, Exhibit R1.10, also in the reference section, gives students' reports on how they spend their daily leisure time. The two most popular activities internationally were watching television or videos and playing or talking with friends (each about two hours per day). Among Benchmarking participants, students generally reported spending a little more time on these activities and on sports, and less time reading for enjoyment. For example, in the four jurisdictions with the lowest average mathematics achievement – the public school systems of Jersey City, Chicago, Rochester, and Miami-Dade – students reported watching television or videos for about three to three and one-half hours (as well as playing computer games for about one hour).

Exhibits 4.6-4.7

Index of Out-of-School Study Time

Index based on students' responses to three questions about out-of-school study time: time spent after school studying mathematics or doing mathematics homework; time spent after school studying science or doing science homework; time spent after school studying or doing homework in school subjects other than mathematics and science (see reference exhibit R1.9). Number of hours based on: no time = 0, less than 1 hour = 0.5, 1-2 hours = 1.5, 3-5 hours = 4, more than 5 hours = 7. High level indicates more than three hours studying all subjects combined. Medium level indicates more than one hour to three hours studying all subjects combined. Low level indicates one hour or less studying all subjects combined.

	High OST		Medium OST		Low OST	
	Percent of Students	Average Achievement	Percent of Students	Average Achievement	Percent of Students	Average Achievement
Singapore	59 (1.2)	608 (5.8)	35 (0.9)	609 (7.4)	7 (0.6)	559 (10.2)
Italy	58 (1.3)	489 (4.1)	36 (1.2)	487 (4.6)	6 (0.6)	405 (9.1)
Russian Federation	48 (1.3)	540 (4.7)	46 (1.2)	532 (7.0)	6 (0.6)	479 (9.3)
Belgium (Flemish)	41 (1.3)	554 (3.3)	52 (1.1)	571 (3.8)	7 (1.0)	517 (16.4)
Jersey City Public Schools, NJ	37 (2.4)	489 (10.5)	47 (1.8)	479 (8.9)	16 (1.7)	452 (8.7)
Chicago Public Schools, IL	37 (2.1)	469 (7.2)	51 (1.6)	468 (6.1)	12 (1.2)	451 (11.6)
Academy School Dist. #20, CO	34 (1.3)	538 (3.2)	55 (1.4)	533 (3.0)	11 (0.9)	501 (6.4)
Montgomery County, MD	28 (1.4)	551 (8.5)	57 (2.3)	547 (4.3)	15 (1.5)	496 (6.7)
First in the World Consort., IL	27 (2.4)	551 (7.6)	61 (2.2)	566 (6.5)	12 (1.1)	549 (11.7)
Guilford County, NC	26 (1.6)	507 (7.4)	62 (1.9)	522 (8.7)	12 (1.0)	498 (14.3)
Naperville Sch. Dist. #203, IL	25 (1.4)	568 (5.2)	63 (1.7)	574 (3.4)	12 (0.9)	560 (7.9)
Miami-Dade County PS, FL	25 (1.5)	429 (12.7)	51 (1.3)	436 (9.6)	24 (2.4)	405 (8.0)
Massachusetts	25 (1.7)	515 (6.8)	62 (1.6)	526 (5.9)	13 (1.2)	469 (8.2)
Illinois	25 (1.6)	505 (8.7)	58 (1.2)	518 (7.1)	17 (1.4)	501 (6.1)
Canada	24 (0.8)	516 (3.5)	59 (1.0)	540 (2.8)	18 (0.8)	528 (4.1)
Connecticut	24 (1.1)	506 (9.8)	62 (1.7)	528 (8.9)	15 (1.5)	474 (7.9)
North Carolina	23 (1.2)	490 (7.9)	57 (1.3)	510 (7.1)	19 (1.6)	469 (8.0)
Rochester City Sch. Dist., NY	23 (1.8)	450 (8.9)	56 (2.3)	458 (6.9)	21 (2.2)	422 (9.2)
Chinese Taipei	23 (1.0)	625 (4.5)	42 (0.8)	602 (3.9)	35 (1.3)	542 (4.4)
United States	22 (0.8)	508 (4.8)	56 (0.9)	517 (4.1)	23 (1.3)	477 (3.9)
South Carolina	21 (1.3)	488 (9.3)	57 (1.1)	518 (7.6)	22 (1.4)	490 (8.7)
Michigan	20 (1.1)	516 (8.3)	59 (1.0)	527 (7.1)	20 (1.3)	499 (8.7)
Maryland	20 (1.0)	501 (8.2)	60 (1.3)	506 (5.6)	20 (1.3)	466 (7.6)
Oregon	19 (1.1)	524 (8.1)	55 (1.5)	526 (5.6)	25 (1.7)	491 (5.5)
Netherlands	19 (1.4)	521 (11.5)	74 (1.3)	548 (6.5)	7 (1.0)	529 (12.8)
Missouri	18 (1.5)	485 (7.0)	54 (1.5)	499 (6.0)	28 (1.6)	480 (6.3)
Texas	18 (1.4)	527 (12.0)	49 (2.2)	532 (7.4)	33 (2.6)	506 (10.6)
Delaware Science Coalition, DE	18 (1.0)	474 (10.9)	58 (2.1)	500 (9.3)	24 (1.9)	450 (7.9)
Pennsylvania	17 (1.9)	496 (8.4)	59 (2.0)	521 (5.1)	24 (1.9)	490 (8.1)
Indiana	17 (1.3)	510 (8.3)	58 (1.5)	526 (7.1)	25 (2.0)	500 (8.4)
Idaho	17 (1.3)	490 (8.6)	55 (1.9)	509 (6.4)	28 (2.1)	479 (9.6)
Project SMART Consortium, OH	17 (1.0)	515 (9.2)	58 (1.2)	532 (7.8)	26 (1.6)	503 (9.0)
Japan	17 (0.9)	586 (2.9)	49 (0.9)	587 (2.1)	35 (1.3)	564 (3.1)
Michigan Invitational Group, MI	17 (1.1)	535 (11.9)	63 (1.8)	539 (4.1)	20 (1.9)	512 (10.0)
Hong Kong, SAR	16 (0.8)	600 (5.3)	42 (0.9)	595 (3.9)	42 (1.4)	564 (5.0)
Czech Republic	16 (1.1)	500 (5.7)	62 (1.4)	527 (4.7)	22 (1.3)	519 (6.5)
Fremont/Lincoln/WestSide PS, NE	16 (1.8)	480 (10.1)	54 (1.6)	510 (7.7)	30 (2.2)	464 (11.4)
Korea, Rep. of	16 (0.7)	612 (4.3)	43 (0.7)	601 (2.5)	41 (1.0)	565 (2.5)
SW Math/Sci. Collaborative, PA	15 (1.1)	506 (6.9)	61 (1.6)	528 (7.0)	24 (1.9)	499 (10.7)
England	– –	– –	– –	– –	– –	– –
International Avg. (All Countries)	38 (0.2)	492 (0.9)	48 (0.2)	497 (0.8)	14 (0.1)	463 (1.6)

SOURCE: IEA Third International Mathematics and Science Study (TIMSS), 1998-1999.

States in *italics* did not fully satisfy guidelines for sample participation rates (see Appendix A for details). A dash (–) indicates data are not available.

( ) Standard errors appear in parentheses. Because results are rounded to the nearest whole number, some totals may appear inconsistent.

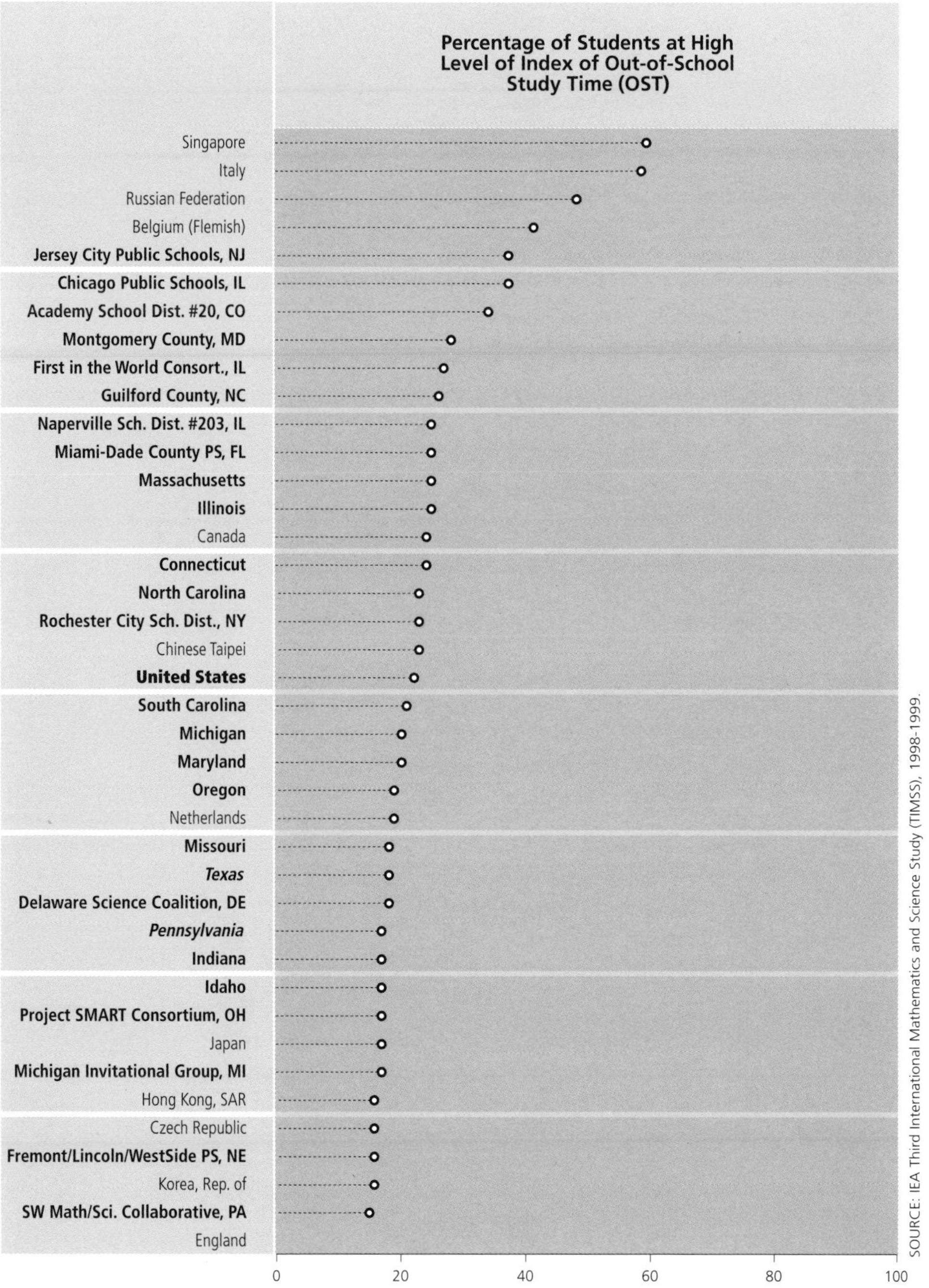

Percentage of Students at High Level of Index of Out-of-School Study Time (OST)

Singapore
Italy
Russian Federation
Belgium (Flemish)
Jersey City Public Schools, NJ
Chicago Public Schools, IL
Academy School Dist. #20, CO
Montgomery County, MD
First in the World Consort., IL
Guilford County, NC
Naperville Sch. Dist. #203, IL
Miami-Dade County PS, FL
Massachusetts
Illinois
Canada
Connecticut
North Carolina
Rochester City Sch. Dist., NY
Chinese Taipei
United States
South Carolina
Michigan
Maryland
Oregon
Netherlands
Missouri
Texas
Delaware Science Coalition, DE
Pennsylvania
Indiana
Idaho
Project SMART Consortium, OH
Japan
Michigan Invitational Group, MI
Hong Kong, SAR
Czech Republic
Fremont/Lincoln/WestSide PS, NE
Korea, Rep. of
SW Math/Sci. Collaborative, PA
England

0 20 40 60 80 100

SOURCE: IEA Third International Mathematics and Science Study (TIMSS), 1998-1999.

Exhibit 4.7 Total Amount of Out-of-School Time Students Spend Studying Mathematics or Doing Mathematics Homework on a Normal School Day

8th Grade Mathematics

	One Hour or More		Less Than One Hour		No Time		Average Hours[1]
	Percent of Students	Average Achievement	Percent of Students	Average Achievement	Percent of Students	Average Achievement	
Countries							
United States	27 (1.1)	505 (4.5)	58 (0.7)	514 (4.0)	15 (1.1)	466 (4.8)	0.8 (0.02)
Belgium (Flemish)	47 (1.2)	550 (3.1)	50 (1.0)	573 (3.7)	3 (0.8)	476 (21.2)	1.1 (0.03)
Canada	28 (1.0)	510 (3.3)	61 (1.0)	542 (2.8)	11 (0.8)	527 (5.2)	0.8 (0.02)
Chinese Taipei	25 (1.0)	627 (4.7)	44 (0.8)	604 (3.5)	31 (1.3)	529 (4.8)	0.7 (0.02)
Czech Republic	20 (1.1)	493 (5.2)	68 (1.3)	528 (4.6)	12 (1.0)	525 (9.2)	0.7 (0.02)
England	– –	– –	– –	– –	– –	– –	– –
Hong Kong, SAR	24 (1.1)	600 (4.8)	51 (0.9)	591 (3.9)	25 (1.2)	552 (6.1)	0.7 (0.02)
Italy	57 (1.3)	482 (4.0)	39 (1.2)	488 (4.5)	5 (0.5)	400 (9.5)	1.3 (0.03)
Japan	20 (0.9)	585 (2.5)	54 (0.9)	586 (2.0)	26 (1.2)	558 (3.8)	0.6 (0.01)
Korea, Rep. of	21 (0.9)	610 (4.1)	45 (0.7)	598 (2.0)	34 (1.0)	560 (2.6)	0.6 (0.02)
Netherlands	14 (1.5)	507 (12.2)	78 (1.3)	546 (6.7)	8 (1.1)	559 (14.0)	0.6 (0.02)
Russian Federation	45 (1.5)	530 (5.2)	49 (1.3)	537 (6.7)	6 (0.5)	483 (10.0)	1.1 (0.03)
Singapore	61 (1.1)	604 (5.7)	34 (1.0)	612 (7.6)	5 (0.5)	562 (10.7)	1.3 (0.02)
States							
Connecticut	27 (1.1)	504 (9.6)	61 (1.5)	526 (9.1)	12 (1.2)	468 (9.2)	0.8 (0.02)
Idaho	25 (1.6)	494 (9.4)	56 (2.0)	508 (6.3)	19 (1.8)	464 (9.9)	0.7 (0.02)
Illinois	32 (1.5)	501 (9.9)	56 (1.2)	520 (6.7)	12 (1.0)	487 (5.5)	0.8 (0.02)
Indiana	24 (1.8)	512 (8.9)	58 (1.5)	526 (6.8)	17 (1.7)	485 (8.4)	0.7 (0.03)
Maryland	25 (1.1)	496 (7.8)	61 (1.6)	503 (6.2)	14 (1.3)	460 (8.6)	0.8 (0.02)
Massachusetts	27 (1.4)	507 (6.2)	62 (1.4)	525 (5.7)	10 (1.0)	466 (9.8)	0.8 (0.02)
Michigan	26 (1.6)	521 (8.2)	60 (1.2)	525 (7.5)	13 (1.4)	478 (7.9)	0.8 (0.03)
Missouri	23 (2.1)	489 (9.0)	55 (1.9)	500 (5.5)	22 (1.4)	468 (5.8)	0.7 (0.03)
North Carolina	30 (1.6)	494 (8.8)	59 (1.3)	506 (6.6)	11 (1.0)	449 (8.8)	0.8 (0.02)
Oregon	26 (1.5)	526 (7.2)	59 (1.2)	520 (5.6)	15 (1.1)	480 (6.9)	0.8 (0.02)
Pennsylvania	21 (1.9)	500 (10.6)	64 (1.4)	518 (5.5)	16 (1.5)	479 (7.2)	0.7 (0.03)
South Carolina	28 (1.3)	495 (8.8)	58 (1.0)	517 (7.6)	14 (1.2)	463 (8.8)	0.8 (0.02)
Texas	27 (2.0)	534 (10.3)	51 (1.5)	530 (8.0)	22 (2.3)	486 (11.3)	0.8 (0.04)
Districts and Consortia							
Academy School Dist. #20, CO	41 (1.6)	536 (3.5)	50 (1.4)	533 (3.2)	9 (0.7)	483 (7.0)	1.0 (0.03)
Chicago Public Schools, IL	48 (2.5)	460 (6.0)	44 (1.7)	472 (6.7)	8 (1.5)	439 (10.9)	1.2 (0.06)
Delaware Science Coalition, DE	21 (1.0)	473 (10.4)	61 (2.1)	497 (9.0)	17 (1.7)	437 (11.6)	0.7 (0.03)
First in the World Consort., IL	29 (1.5)	553 (6.9)	65 (1.7)	566 (6.9)	6 (1.1)	526 (17.6)	0.8 (0.02)
Fremont/Lincoln/WestSide PS, NE	20 (2.7)	474 (10.0)	61 (3.2)	508 (8.2)	19 (1.7)	444 (8.8)	0.7 (0.05)
Guilford County, NC	35 (1.4)	508 (7.0)	58 (1.7)	523 (9.0)	7 (0.9)	475 (13.4)	0.9 (0.03)
Jersey City Public Schools, NJ	44 (2.0)	475 (10.6)	46 (1.8)	485 (8.5)	10 (1.5)	450 (8.2)	1.1 (0.05)
Miami-Dade County PS, FL	32 (1.2)	417 (12.3)	51 (1.6)	436 (9.7)	17 (2.0)	400 (6.7)	0.9 (0.03)
Michigan Invitational Group, MI	25 (2.0)	537 (10.8)	60 (1.5)	539 (4.5)	15 (1.6)	501 (9.9)	0.7 (0.03)
Montgomery County, MD	35 (2.4)	544 (7.6)	56 (2.3)	545 (4.0)	9 (1.1)	474 (7.6)	0.9 (0.04)
Naperville Sch. Dist. #203, IL	28 (1.4)	562 (5.4)	66 (1.4)	576 (3.7)	6 (0.8)	536 (12.7)	0.8 (0.02)
Project SMART Consortium, OH	21 (1.0)	517 (8.7)	59 (1.4)	531 (8.1)	20 (1.5)	494 (8.2)	0.6 (0.02)
Rochester City Sch. Dist., NY	29 (2.2)	441 (9.3)	56 (2.1)	459 (7.0)	15 (1.9)	415 (8.1)	0.8 (0.05)
SW Math/Sci. Collaborative, PA	20 (2.0)	516 (7.5)	67 (1.6)	524 (7.5)	13 (1.3)	484 (11.4)	0.7 (0.03)
International Avg. (All Countries)	40 (0.2)	486 (0.9)	50 (0.2)	495 (0.8)	10 (0.1)	455 (1.7)	1.1 (0.00)

SOURCE: IEA Third International Mathematics and Science Study (TIMSS), 1998-1999.

Background data provided by students.

1 Average hours based on: No time=0; less than 1 hour=.5; 1-2 hours=1.5; 3-5 hours=4; more than 5 hours=7.

States in *italics* did not fully satisfy guidelines for sample participation rates (see Appendix A for details).

( ) Standard errors appear in parentheses. Because results are rounded to the nearest whole number, some totals may appear inconsistent.

A dash (–) indicates data are not available.

How Do Students Perceive Their Ability in Mathematics?

To investigate how students think of their abilities in mathematics, TIMSS created an index of students' self-concept in mathematics (SCM). It is based on student's responses to five statements about their mathematics ability:

- I would like mathematics much more if it were not so difficult
- Although I do my best, mathematics is more difficult for me than for many of my classmates
- Nobody can be good in every subject, and I am just not talented in mathematics
- Sometimes when I do not understand a new topic in mathematics initially, I know that I will never really understand it
- Mathematics is not one of my strengths.

Students who disagreed or strongly disagreed with all five statements were assigned to the high level of the index, while students who agreed or strongly agreed with all five were assigned to the low level. The medium level includes all other combinations of responses. (As an example of one of the components of the index, Exhibit R1.11 in the reference section shows the percentages of agreement for the statement "mathematics is not one of my strengths.")

The percentages of eighth-grade students at each index level, and their average mathematics achievement, are presented in Exhibit 4.8. Across participating countries, the United States was among those with the greatest percentages of students at the high level of the self-concept index: 31 percent compared with 18 percent on average across all countries. Several of the Benchmarking participants had even greater percentages at the high level, notably the Naperville School District and the First in the World Consortium, with 40 percent or more of students at this level.

Although there was a clear positive association between self-concept and mathematics achievement within every country and within every Benchmarking jurisdiction, the relationship across entitiess was more complex. Several countries with high average mathematics achievement, including Singapore, Hong Kong, Chinese Taipei, Korea, and Japan, had relatively low percentages of students (15 percent or less) in the high self-concept category. Since all of these are Asian Pacific countries, they may share cultural traditions that encourage a modest self-concept.

text continued on page 132

Exhibit 4.8 Index of Students' Self-Concept in Mathematics (SCM)

TIMSS 1999
Benchmarking
Boston College

8th Grade Mathematics

Index of Students' Self-Concept in Mathematics

Index based on students' responses to five statements about their mathematics ability: 1) I would like mathematics much more if it were not so difficult; 2) although I do my best, mathematics is more difficult for me than for many of my classmates; 3) nobody can be good in every subject, and I am just not talented in mathematics; 4) sometimes, when I do not understand a new topic in mathematics initially, I know that I will never really understand it; 5) mathematics is not one of my strengths. High level indicates student disagrees or strongly disagrees with all five statements. Low level indicates student agrees or strongly agrees with all five statements. Medium level includes all other possible combinations of responses.

	High SCM		Medium SCM		Low SCM	
	Percent of Students	Average Achievement	Percent of Students	Average Achievement	Percent of Students	Average Achievement
Russian Federation	45 (1.5)	568 (4.7)	44 (1.1)	510 (6.5)	11 (0.8)	470 (10.9)
Naperville Sch. Dist. #203, IL	44 (1.4)	597 (3.9)	49 (1.7)	554 (3.1)	7 (0.8)	507 (7.6)
First in the World Consort., IL	40 (2.5)	590 (6.9)	55 (3.1)	545 (6.1)	5 (1.1)	481 (9.0)
SW Math/Sci. Collaborative, PA	36 (1.9)	553 (7.8)	56 (1.6)	504 (7.6)	8 (0.7)	447 (11.7)
Chicago Public Schools, IL	36 (2.8)	505 (6.7)	56 (2.7)	445 (6.0)	8 (1.2)	404 (9.1)
North Carolina	36 (1.7)	533 (7.5)	54 (1.4)	484 (6.7)	10 (0.8)	430 (8.7)
Michigan	36 (1.6)	554 (7.4)	53 (1.7)	508 (6.7)	11 (0.8)	452 (6.4)
Oregon	35 (1.6)	552 (5.8)	55 (1.3)	505 (5.6)	9 (0.9)	444 (7.5)
Illinois	35 (1.8)	549 (6.9)	56 (1.5)	495 (7.0)	9 (0.9)	448 (7.5)
Connecticut	35 (2.0)	547 (10.0)	56 (1.8)	502 (8.5)	9 (1.0)	448 (9.5)
Canada	35 (1.0)	573 (2.9)	56 (1.0)	517 (2.4)	9 (0.5)	459 (6.1)
Fremont/Lincoln/WestSide PS, NE	34 (2.1)	539 (9.7)	51 (1.7)	479 (9.1)	14 (1.5)	406 (8.6)
Project SMART Consortium, OH	34 (2.2)	562 (8.4)	56 (2.0)	509 (7.1)	10 (1.2)	448 (7.5)
Academy School Dist. #20, CO	34 (1.4)	560 (3.4)	58 (1.5)	521 (2.8)	8 (0.8)	460 (8.5)
Pennsylvania	34 (1.7)	543 (8.3)	56 (1.3)	499 (5.5)	10 (0.9)	443 (6.3)
Montgomery County, MD	33 (1.7)	572 (6.1)	58 (1.5)	529 (3.4)	9 (1.1)	473 (9.8)
Michigan Invitational Group, MI	33 (2.3)	568 (6.1)	55 (2.2)	527 (4.7)	12 (1.0)	465 (13.0)
Guilford County, NC	33 (2.7)	535 (7.7)	60 (2.7)	508 (8.0)	8 (1.1)	469 (13.5)
Massachusetts	33 (1.9)	553 (6.4)	58 (1.5)	503 (5.5)	10 (1.0)	446 (8.1)
Indiana	32 (1.9)	557 (6.9)	57 (1.5)	504 (6.5)	12 (1.1)	457 (9.6)
Rochester City Sch. Dist., NY	31 (1.6)	486 (6.6)	54 (1.6)	440 (8.0)	15 (1.2)	402 (8.1)
Maryland	31 (1.4)	535 (5.7)	58 (1.0)	487 (6.2)	11 (0.9)	432 (7.6)
Delaware Science Coalition, DE	31 (1.5)	528 (9.7)	57 (1.8)	472 (7.9)	12 (1.1)	418 (12.6)
United States	31 (1.0)	551 (4.6)	58 (0.8)	493 (3.9)	11 (0.6)	435 (5.6)
Idaho	31 (1.9)	534 (7.6)	58 (1.5)	488 (6.4)	11 (0.9)	429 (9.7)
Jersey City Public Schools, NJ	30 (2.8)	531 (8.9)	60 (2.4)	459 (6.8)	9 (1.2)	413 (9.3)
England	30 (1.3)	543 (5.0)	61 (1.2)	487 (3.9)	9 (0.6)	430 (6.5)
Texas	29 (1.5)	565 (9.0)	60 (1.3)	513 (9.1)	11 (1.1)	447 (10.6)
South Carolina	28 (1.8)	548 (6.9)	61 (1.4)	492 (7.7)	11 (0.9)	441 (8.0)
Missouri	27 (1.6)	527 (7.0)	60 (1.6)	484 (5.1)	12 (0.8)	441 (8.6)
Netherland	27 (2.0)	578 (7.0)	65 (1.8)	532 (7.7)	8 (0.9)	490 (9.8)
Belgium (Flemish)	25 (0.8)	600 (5.4)	62 (0.8)	554 (3.3)	13 (1.1)	506 (7.8)
Italy	24 (0.9)	539 (3.8)	63 (0.9)	474 (3.8)	13 (0.8)	412 (5.4)
Miami-Dade County PS, FL	23 (2.2)	478 (13.0)	60 (1.8)	420 (9.2)	17 (2.1)	364 (8.2)
Czech Republic	19 (1.2)	585 (5.7)	66 (1.0)	515 (4.0)	15 (1.0)	461 (5.5)
Singapore	15 (1.0)	656 (8.8)	74 (0.8)	603 (5.7)	11 (0.7)	547 (7.1)
Hong Kong, SAR	14 (0.7)	624 (4.6)	71 (0.8)	585 (3.8)	14 (0.8)	531 (6.3)
Chinese Taipei	11 (0.5)	660 (6.0)	75 (0.7)	591 (3.9)	14 (0.7)	506 (4.2)
Korea, Rep. of	10 (0.5)	646 (4.0)	85 (0.5)	585 (1.8)	5 (0.3)	515 (5.7)
Japan	6 (0.4)	634 (6.2)	82 (0.5)	581 (1.8)	12 (0.5)	536 (3.8)
International Avg. (All Countries)	18 (0.2)	547 (1.1)	67 (0.2)	486 (0.7)	15 (0.1)	436 (0.9)

SOURCE: IEA Third International Mathematics and Science Study (TIMSS), 1998-1999.

States in *italics* did not fully satisfy guidelines for sample participation rates (see Appendix A for details).

( ) Standard errors appear in parentheses. Because results are rounded to the nearest whole number, some totals may appear inconsistent.

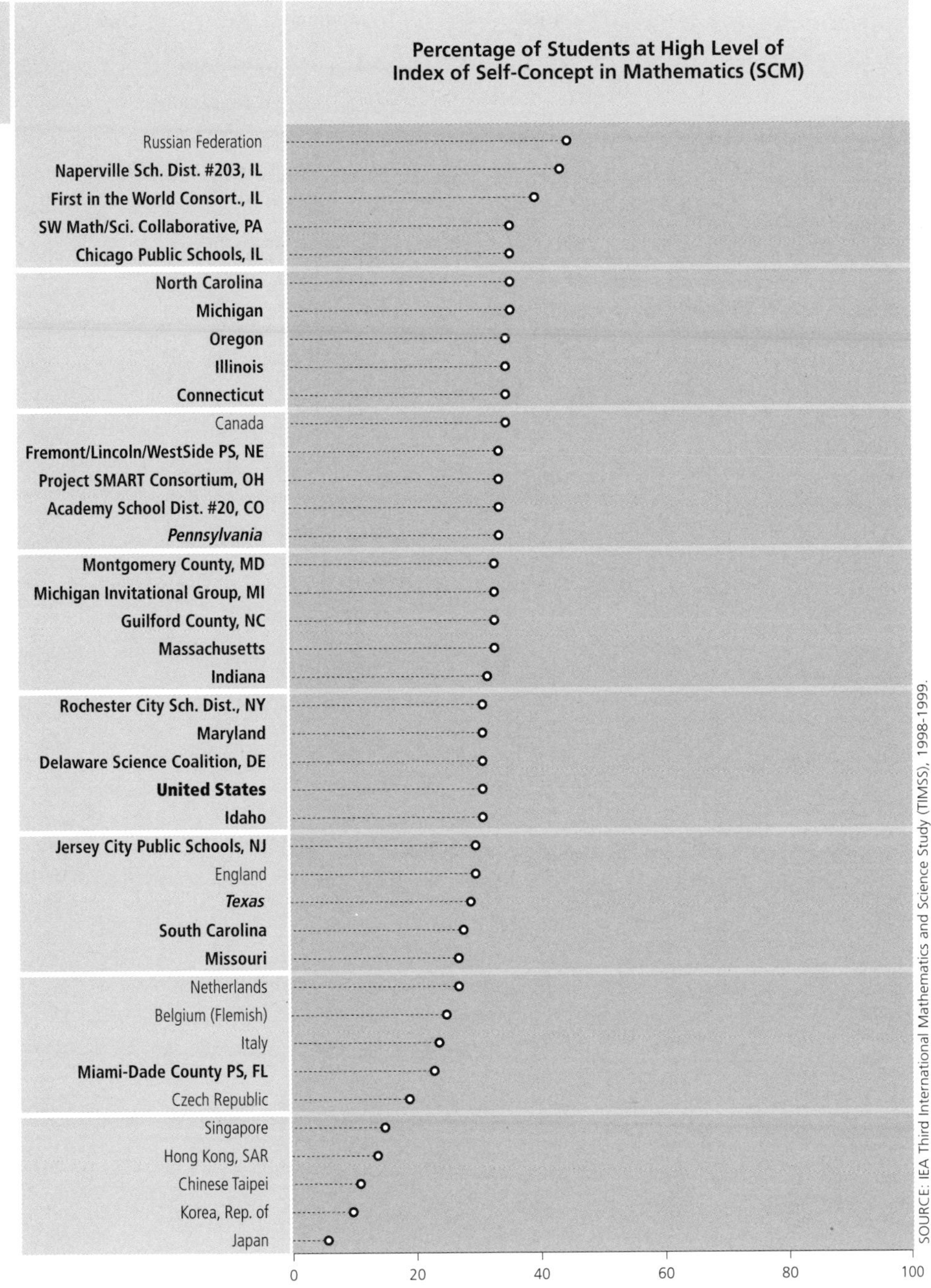
Percentage of Students at High Level of Index of Self-Concept in Mathematics (SCM)
Russian Federation
Naperville Sch. Dist. #203, IL
First in the World Consort., IL
SW Math/Sci. Collaborative, PA
Chicago Public Schools, IL
North Carolina
Michigan
Oregon
Illinois
Connecticut
Canada
Fremont/Lincoln/WestSide PS, NE
Project SMART Consortium, OH
Academy School Dist. #20, CO
Pennsylvania
Montgomery County, MD
Michigan Invitational Group, MI
Guilford County, NC
Massachusetts
Indiana
Rochester City Sch. Dist., NY
Maryland
Delaware Science Coalition, DE
United States
Idaho
Jersey City Public Schools, NJ
England
Texas
South Carolina
Missouri
Netherlands
Belgium (Flemish)
Italy
Miami-Dade County PS, FL
Czech Republic
Singapore
Hong Kong, SAR
Chinese Taipei
Korea, Rep. of
Japan
0
20
40
60
80
100
SOURCE: IEA Third International Mathematics and Science Study (TIMSS), 1998-1999.

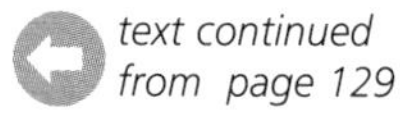
text continued from page 129

Exhibit 4.9 presents the percentages of girls and boys in the Benchmarking entities and in the comparison countries at the high, medium, and low levels of the mathematics self-concept index. Despite the gender differences in the United States as a whole, there were few significant differences among Benchmarking participants. There were greater percentages of boys at the high index level in Connecticut, Indiana, Massachusetts, and the Delaware Science Coalition. Indiana, Massachusetts, Michigan, the Academy School District, and the Delaware Science Coalition had greater percentages of girls at the medium level, and Montgomery County had a greater percentage of girls at the low level.

Exhibit 4.9 Index of Students' Self-Concept in Mathematics (SCM) by Gender

8th Grade Mathematics

	High SCM Percent of Students		**Medium** SCM Percent of Students		**Low** SCM Percent of Students	
	Girls	Boys	Girls	Boys	Girls	Boys
Countries						
United States	28 (1.3)	34 (1.2) ▲	61 (1.2) ▲	54 (1.0)	11 (0.7)	11 (0.7)
Belgium (Flemish)	24 (1.3)	26 (1.2)	61 (1.5)	63 (1.2)	16 (1.4) ▲	11 (1.1)
Canada	31 (1.4)	39 (1.1) ▲	59 (1.6) ▲	52 (1.0)	9 (0.7)	9 (0.5)
Chinese Taipei	7 (0.5)	14 (0.8) ▲	79 (0.8) ▲	72 (1.0)	14 (0.8)	14 (0.9)
Czech Republic	16 (1.3)	22 (1.5) ▲	69 (1.3)	63 (1.3)	15 (1.0)	15 (1.5)
England	24 (1.5)	36 (1.8) ▲	65 (1.5) ▲	57 (1.7)	11 (1.0)	7 (0.7)
Hong Kong, SAR	11 (0.9)	18 (0.9) ▲	74 (1.2) ▲	69 (1.0)	15 (1.1)	14 (1.1)
Italy	22 (1.1)	25 (1.3)	64 (1.3)	63 (1.3)	14 (1.0)	13 (1.0)
Japan	3 (0.4)	8 (0.7) ▲	80 (0.9)	83 (0.9)	17 (0.8) ▲	8 (0.5)
Korea, Rep. of	7 (0.6)	12 (0.7) ▲	87 (0.6) ▲	84 (0.7)	6 (0.4)	4 (0.4)
Netherlands	21 (2.1)	33 (2.6) ▲	69 (1.8)	61 (2.7)	10 (1.2)	6 (1.0)
Russian Federation	48 (1.8)	42 (1.8)	42 (1.5)	45 (1.4)	10 (0.9)	13 (1.0)
Singapore	13 (0.9)	17 (1.4)	77 (0.9) ▲	72 (1.0)	11 (0.8)	12 (0.9)
States						
Connecticut	31 (2.2)	40 (2.5) ▲	59 (2.2)	52 (2.1)	10 (1.1)	8 (1.1)
Idaho	29 (2.2)	33 (2.2)	61 (2.0)	56 (1.9)	10 (1.2)	11 (1.0)
Illinois	32 (2.3)	38 (2.1)	59 (1.9)	52 (1.8)	8 (1.0)	9 (1.2)
Indiana	27 (1.8)	36 (2.5) ▲	61 (1.7) ▲	52 (2.2)	11 (1.3)	12 (1.0)
Maryland	30 (1.8)	33 (1.8)	59 (1.3)	57 (1.7)	11 (0.9)	10 (1.1)
Massachusetts	28 (2.3)	37 (1.9) ▲	62 (1.9) ▲	53 (1.8)	9 (1.2)	10 (1.2)
Michigan	33 (1.6)	39 (2.3)	56 (1.8) ▲	49 (2.1)	11 (1.0)	11 (1.1)
Missouri	27 (1.4)	28 (2.3)	61 (1.6)	60 (2.2)	12 (1.2)	12 (1.0)
North Carolina	37 (1.9)	35 (2.0)	54 (1.6)	54 (2.0)	8 (0.9)	11 (1.3)
Oregon	33 (2.3)	38 (1.8)	59 (2.2)	52 (2.0)	8 (1.3)	11 (1.2)
Pennsylvania	31 (2.3)	37 (2.0)	59 (1.8)	53 (1.5)	11 (1.3)	10 (1.2)
South Carolina	24 (2.1)	32 (2.5)	64 (1.6)	57 (2.3)	12 (1.1)	10 (1.3)
Texas	26 (2.3)	32 (1.7)	64 (1.9)	57 (1.9)	10 (1.4)	12 (1.4)
Districts and Consortia						
Academy School Dist. #20, CO	30 (1.9)	38 (1.9)	63 (1.8) ▲	53 (1.9)	7 (1.2)	9 (1.2)
Chicago Public Schools, IL	35 (2.9)	37 (3.1)	56 (2.6)	56 (3.3)	9 (1.4)	7 (1.2)
Delaware Science Coalition, DE	26 (2.1)	37 (1.8) ▲	63 (2.1) ▲	51 (2.3)	11 (1.2)	12 (1.9)
First in the World Consort., IL	38 (2.7)	41 (4.0)	56 (2.8)	55 (4.4)	6 (1.4)	4 (0.9)
Fremont/Lincoln/WestSide PS, NE	31 (3.9)	38 (1.4)	54 (2.9)	49 (2.5)	16 (2.2)	13 (1.9)
Guilford County, NC	31 (3.0)	35 (3.3)	62 (3.0)	57 (3.3)	7 (1.1)	8 (1.6)
Jersey City Public Schools, NJ	27 (2.2)	34 (4.1)	62 (2.4)	58 (3.8)	11 (1.5)	8 (1.6)
Miami-Dade County PS, FL	23 (2.6)	24 (2.9)	60 (2.3)	59 (2.0)	17 (2.0)	17 (3.0)
Michigan Invitational Group, MI	30 (3.8)	37 (3.3)	60 (3.2)	49 (2.9)	10 (1.5)	14 (1.8)
Montgomery County, MD	33 (2.7)	34 (1.6)	56 (2.1)	59 (1.8)	11 (1.3) ▲	7 (1.2)
Naperville Sch. Dist. #203, IL	41 (1.9)	47 (2.2)	51 (2.3)	46 (2.1)	8 (1.2)	6 (0.9)
Project SMART Consortium, OH	31 (2.8)	38 (2.1)	58 (2.7)	53 (2.4)	11 (1.5)	9 (1.3)
Rochester City Sch. Dist., NY	31 (1.6)	32 (2.9)	56 (2.2)	51 (3.0)	13 (2.1)	16 (1.7)
SW Math/Sci. Collaborative, PA	32 (2.4)	41 (2.6)	59 (2.0)	52 (2.8)	9 (1.1)	7 (0.7)
International Avg. (All Countries)	17 (0.2)	20 (0.2) ▲	68 (0.2) ▲	66 (0.2)	16 (0.2) ▲	15 (0.2)

SOURCE: IEA Third International Mathematics and Science Study (TIMSS), 1998-1999.

▲ Significantly higher than other gender

Significance tests adjusted for multiple comparisons

Background data provided by students.

States in *italics* did not fully satisfy guidelines for sample participation rates (see Appendix A for details).

( ) Standard errors appear in parentheses. Because results are rounded to the nearest whole number, some totals may appear inconsistent.

What Are Students' Attitudes Towards Mathematics?

Generating positive attitudes towards mathematics among students is an important goal of mathematics education in many jurisdictions. To gain some understanding of eighth-graders' views about the utility of mathematics and their enjoyment of it as a school subject, TIMSS created an index of positive attitudes towards mathematics (PATM). Students were asked to state their agreement with the following five statements:

- I like mathematics
- I enjoy learning mathematics
- Mathematics is boring[5]
- Mathematics is important to everyone's life
- I would like a job that involved using mathematics.

For each statement, students responded on a four-point scale indicating whether their feelings about mathematics were strongly positive, positive, negative, or strongly negative. The responses were averaged, with students being placed in the high category if their average indicated a positive or strongly positive attitude. Students with a negative or strongly negative attitude on average were placed in the low category. The students between these extremes were placed in the medium category. The results are presented in Exhibit 4.10. (Additional information on students' liking mathematics, one of the components of the index, is provided in Exhibit R1.12 in the reference section.)

Internationally, eighth graders generally had positive attitudes towards mathematics, with 37 percent on average across all TIMSS 1999 countries in the high category and a further 52 percent in the medium category. Only 11 percent of students were in the low category. The percentage for the United States did not vary much from the international average for the high category, but was greater in the low category (16 percent). Benchmarking jurisdictions with large percentages of students at the high level included Jersey City, Chicago, and North Carolina (44 percent or more). Jurisdictions with students having somewhat less favorable attitudes included Massachusetts, Oregon, and the Academy School District, where 28 to 29 percent of the students were at the high level. The reference countries with the least positive attitudes were Japan and Korea (9 percent in the high category). Since these are countries with high average mathematics achievement, it may be that the students follow a demanding mathematics curriculum that leads to high achievement but little enthu-

5 The response categories for this statement were reversed in constructing the index.

siasm for the subject matter. However, there was a clear positive association between attitudes towards mathematics and mathematics achievement on average across all the TIMSS 1999 countries and in many of the Benchmarking entities.

Exhibit 4.11 shows the percentages of girls and boys in each of the comparison countries and Benchmarking jurisdictions at each level of the index of positive attitudes towards mathematics. Although the United States, like many of the other countries, had significantly different percentages of girls and boys at the index levels, there were essentially no significant differences among the Benchmarking participants. The only significant difference was in Massachusetts, with a greater percentage of girls at the medium level.

Index of Students' Positive Attitudes Towards Mathematics

Index based on students' responses to five statements about mathematics: 1) I like mathematics; 2) I enjoy learning mathematics; 3) mathematics is boring (reversed scale); 4) mathematics is important to everyone's life; 5) I would like a job that involved using mathematics. Average is computed across the five items based on a 4-point scale: 1 = strongly negative; 2 = negative; 3 = positive; 4 = strongly positive. High level indicates average is greater than 3. Medium level indicates average is greater than 2 and less than or equal to 3. Low level indicates average is less than or equal to 2.

	High PATM		Medium PATM		Low PATM	
	Percent of Students	Average Achievement	Percent of Students	Average Achievement	Percent of Students	Average Achievement
Jersey City Public Schools, NJ	51 (2.4)	499 (8.4)	41 (1.8)	462 (9.5)	8 (1.2)	409 (9.3)
Chicago Public Schools, IL	47 (3.0)	478 (7.9)	45 (2.6)	453 (6.5)	8 (1.7)	437 (10.6)
Singapore	45 (1.0)	620 (6.4)	48 (0.9)	595 (6.7)	7 (0.5)	568 (9.1)
North Carolina	44 (1.4)	509 (7.7)	46 (1.2)	489 (7.5)	9 (0.8)	466 (8.0)
England	41 (1.3)	506 (5.4)	51 (1.2)	495 (4.5)	8 (0.5)	478 (8.1)
Guilford County, NC	40 (2.0)	513 (10.0)	49 (1.6)	516 (8.2)	10 (0.9)	510 (10.8)
Rochester City Sch. Dist., NY r	39 (2.1)	467 (7.0)	49 (1.9)	449 (7.9)	12 (1.5)	414 (11.7)
Illinois	39 (1.5)	526 (8.7)	50 (1.3)	503 (6.3)	12 (0.7)	484 (8.2)
Miami-Dade County PS, FL	38 (2.6)	440 (10.6)	48 (2.3)	413 (10.1)	14 (1.7)	414 (9.0)
South Carolina	38 (1.4)	510 (8.7)	49 (0.9)	501 (8.2)	13 (1.1)	490 (7.0)
Texas	37 (1.4)	537 (10.7)	50 (1.1)	513 (9.2)	13 (1.0)	504 (11.5)
Russian Federation	36 (1.3)	555 (5.3)	58 (1.2)	518 (6.3)	5 (0.4)	496 (8.3)
SW Math/Sci. Collaborative, PA	36 (1.7)	536 (7.6)	49 (1.5)	509 (7.9)	14 (1.3)	496 (10.9)
Naperville Sch. Dist. #203, IL	36 (1.4)	595 (3.7)	50 (1.8)	562 (3.4)	14 (1.3)	530 (6.7)
Fremont/Lincoln/WestSide PS, NE	36 (1.5)	513 (12.0)	51 (1.8)	488 (9.0)	14 (1.0)	437 (8.4)
Italy	35 (1.2)	512 (4.2)	51 (1.1)	469 (4.3)	14 (0.8)	449 (5.1)
Canada	35 (0.9)	552 (3.4)	51 (1.0)	526 (2.7)	14 (0.7)	500 (4.6)
Maryland	35 (1.5)	514 (5.9)	50 (1.0)	490 (6.4)	15 (1.2)	480 (7.6)
United States	35 (1.1)	522 (4.5)	49 (0.7)	500 (3.9)	16 (0.7)	481 (4.7)
Indiana	35 (2.1)	537 (7.0)	49 (1.7)	508 (7.5)	16 (0.9)	495 (10.5)
Delaware Science Coalition, DE	35 (1.8)	506 (9.9)	50 (1.5)	476 (9.3)	16 (1.5)	466 (10.7)
Pennsylvania	34 (1.6)	527 (8.6)	51 (1.0)	503 (5.8)	15 (1.5)	480 (9.5)
Connecticut	34 (1.7)	528 (11.6)	51 (1.3)	509 (8.4)	15 (1.4)	497 (9.1)
Michigan	33 (1.5)	538 (9.3)	51 (1.3)	516 (6.8)	16 (1.1)	486 (5.6)
Michigan Invitational Group, MI	33 (1.2)	560 (7.1)	50 (1.3)	528 (6.0)	17 (1.9)	497 (10.9)
First in the World Consort., IL	33 (1.4)	576 (6.5)	52 (1.8)	559 (6.5)	15 (1.6)	525 (10.5)
Project SMART Consortium, OH	31 (2.1)	552 (8.1)	52 (1.8)	517 (7.1)	17 (1.4)	482 (9.9)
Idaho	31 (2.1)	518 (7.5)	51 (1.5)	492 (7.5)	18 (1.5)	468 (8.0)
Missouri	31 (1.9)	508 (7.3)	52 (1.2)	488 (4.8)	17 (1.2)	468 (7.5)
Montgomery County, MD	30 (1.8)	553 (5.8)	53 (1.7)	535 (3.6)	18 (1.6)	523 (6.9)
Massachusetts	29 (1.8)	534 (7.0)	52 (1.1)	511 (6.0)	19 (1.4)	491 (7.4)
Oregon	29 (1.8)	536 (6.9)	55 (1.6)	513 (6.7)	15 (1.6)	489 (7.7)
Hong Kong, SAR	28 (0.9)	613 (4.1)	61 (0.8)	578 (4.1)	11 (0.6)	533 (4.8)
Academy School Dist. #20, CO	28 (1.2)	549 (4.1)	52 (1.4)	527 (2.9)	21 (1.4)	509 (5.4)
Belgium (Flemish)	25 (0.9)	598 (4.7)	53 (0.9)	555 (3.5)	22 (1.1)	523 (4.5)
Chinese Taipei	23 (0.8)	643 (5.1)	59 (0.8)	582 (4.1)	18 (0.7)	529 (5.4)
Czech Republic	19 (1.2)	559 (6.2)	63 (1.2)	515 (4.9)	18 (1.0)	500 (5.8)
Netherland	17 (1.4)	555 (11.7)	63 (1.0)	543 (7.1)	20 (1.4)	522 (8.4)
Japan	9 (0.5)	619 (5.4)	61 (0.7)	585 (2.0)	29 (0.9)	554 (2.9)
Korea, Rep. of	9 (0.4)	647 (4.2)	65 (0.8)	591 (2.1)	26 (0.8)	560 (2.6)
International Avg. (All Countries)	37 (0.2)	512 (0.9)	52 (0.2)	481 (0.8)	11 (0.1)	473 (1.2)

SOURCE: IEA Third International Mathematics and Science Study (TIMSS), 1998-1999.

States in *italics* did not fully satisfy guidelines for sample participation rates (see Appendix A for details). An "r" indicates a 70-84% student response rate.

( ) Standard errors appear in parentheses. Because results are rounded to the nearest whole number, some totals may appear inconsistent.

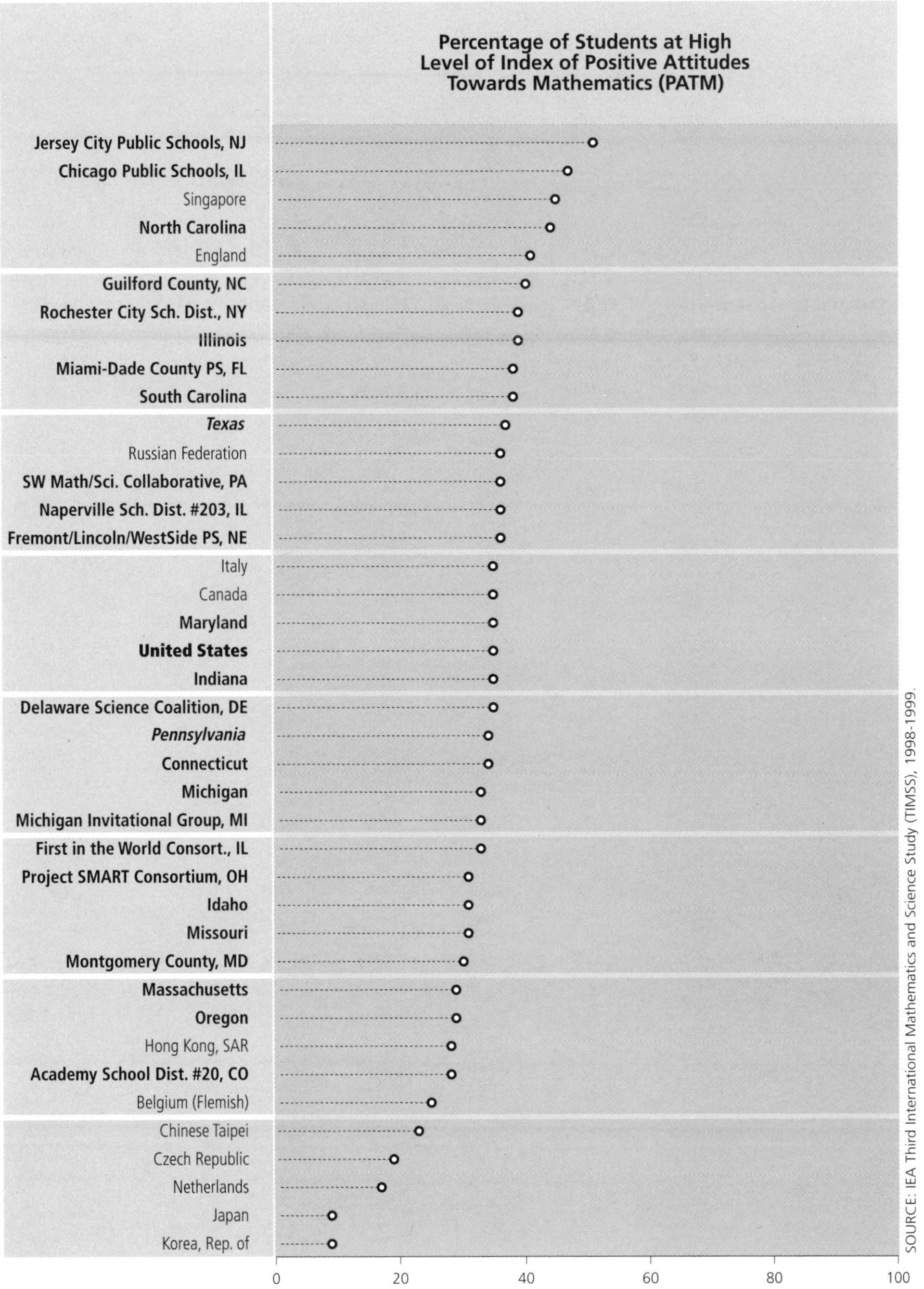

Percentage of Students at High Level of Index of Positive Attitudes Towards Mathematics (PATM)

Jersey City Public Schools, NJ
Chicago Public Schools, IL
Singapore
North Carolina
England
Guilford County, NC
Rochester City Sch. Dist., NY
Illinois
Miami-Dade County PS, FL
South Carolina
Texas
Russian Federation
SW Math/Sci. Collaborative, PA
Naperville Sch. Dist. #203, IL
Fremont/Lincoln/WestSide PS, NE
Italy
Canada
Maryland
United States
Indiana
Delaware Science Coalition, DE
Pennsylvania
Connecticut
Michigan
Michigan Invitational Group, MI
First in the World Consort., IL
Project SMART Consortium, OH
Idaho
Missouri
Montgomery County, MD
Massachusetts
Oregon
Hong Kong, SAR
Academy School Dist. #20, CO
Belgium (Flemish)
Chinese Taipei
Czech Republic
Netherlands
Japan
Korea, Rep. of

0 20 40 60 80 100

SOURCE: IEA Third International Mathematics and Science Study (TIMSS), 1998-1999.

	High PATM Percent of Students		Medium PATM Percent of Students		Low PATM Percent of Students	
	Girls	Boys	Girls	Boys	Girls	Boys
Countries						
United States	32 (1.3)	37 (1.2) ▲	52 (1.1) ▲	46 (0.9)	16 (0.7)	16 (1.1)
Belgium (Flemish)	24 (1.4)	26 (1.7)	53 (1.8)	53 (1.4)	23 (1.6)	21 (1.3)
Canada	31 (1.1)	38 (1.2) ▲	53 (1.4) ▲	48 (1.1)	15 (0.9)	13 (0.9)
Chinese Taipei	18 (0.9)	27 (1.1) ▲	61 (1.0)	58 (1.0)	21 (0.9) ▲	15 (0.8)
Czech Republic	16 (1.5)	22 (1.7)	64 (1.7)	61 (1.4)	20 (1.4)	17 (1.3)
England	35 (1.7)	48 (1.7) ▲	55 (1.5) ▲	47 (1.5)	10 (0.8) ▲	6 (0.7)
Hong Kong, SAR	22 (1.1)	34 (1.2) ▲	65 (1.0) ▲	57 (1.1)	13 (0.8) ▲	8 (0.6)
Italy	33 (1.6)	38 (1.4)	52 (1.5)	49 (1.4)	15 (1.0)	13 (1.0)
Japan	6 (0.5)	13 (0.7) ▲	59 (1.0)	64 (1.0) ▲	36 (1.2) ▲	23 (0.9)
Korea, Rep. of	8 (0.6)	10 (0.6)	64 (1.2)	66 (1.0)	28 (1.3)	25 (0.9)
Netherlands	12 (1.5)	23 (1.8) ▲	62 (1.4)	63 (1.9)	26 (1.9) ▲	14 (1.4)
Russian Federation	37 (1.6)	36 (1.6)	58 (1.5)	59 (1.4)	5 (0.5)	5 (0.6)
Singapore	41 (1.4)	48 (1.4) ▲	52 (1.1) ▲	45 (1.3)	7 (0.7)	7 (0.7)
States						
Connecticut	33 (2.1)	34 (2.0)	52 (1.7)	49 (1.9)	15 (1.8)	16 (1.5)
Idaho	30 (2.1)	32 (3.0)	55 (2.1)	47 (2.1)	15 (1.7)	21 (1.9)
Illinois	38 (2.3)	39 (1.3)	51 (2.0)	48 (1.4)	10 (0.9)	13 (1.0)
Indiana	31 (2.2)	38 (2.5)	53 (1.8)	45 (2.4)	16 (1.2)	17 (1.3)
Maryland	32 (2.0)	38 (1.5)	52 (1.5)	48 (1.1)	16 (1.4)	15 (1.4)
Massachusetts	26 (2.0)	32 (2.2)	56 (1.6) ▲	48 (1.6)	18 (1.7)	19 (1.8)
Michigan	30 (2.0)	36 (2.0)	54 (1.9)	48 (1.8)	16 (1.3)	16 (1.5)
Missouri	32 (2.4)	30 (1.8)	53 (1.9)	50 (1.7)	15 (1.2)	20 (1.8)
North Carolina	44 (2.2)	44 (1.9)	48 (2.1)	45 (1.6)	8 (0.9)	11 (1.1)
Oregon	26 (2.5)	32 (2.1)	57 (1.6)	54 (2.5)	17 (2.0)	13 (1.5)
Pennsylvania	32 (2.0)	37 (1.8)	52 (1.5)	49 (2.1)	15 (1.5)	14 (1.7)
South Carolina	35 (2.1)	40 (2.2)	52 (2.2)	47 (1.8)	13 (1.5)	13 (1.2)
Texas	35 (2.3)	38 (1.2)	53 (2.0)	48 (1.3)	11 (1.4)	15 (1.0)
Districts and Consortia						
Academy School Dist. #20, CO	24 (2.1)	32 (1.7)	54 (2.2)	50 (2.1)	22 (2.1)	19 (1.7)
Chicago Public Schools, IL	46 (3.1)	48 (3.7)	46 (3.0)	45 (2.9)	8 (1.5)	8 (2.1)
Delaware Science Coalition, DE	31 (2.3)	39 (2.1)	52 (1.8)	47 (2.4)	17 (2.1)	14 (1.8)
First in the World Consort., IL	32 (3.0)	33 (3.1)	54 (3.0)	50 (2.1)	15 (2.1)	16 (2.4)
Fremont/Lincoln/WestSide PS, NE	32 (2.1)	39 (3.1)	53 (2.1)	48 (2.7)	14 (2.5)	13 (1.2)
Guilford County, NC	40 (2.6)	41 (2.1)	51 (2.4)	47 (2.1)	9 (0.9)	12 (1.6)
Jersey City Public Schools, NJ	50 (2.7)	52 (3.5)	42 (2.5)	40 (2.8)	8 (1.2)	8 (2.1)
Miami-Dade County PS, FL	37 (3.0)	40 (2.9)	49 (3.0)	47 (2.3)	14 (2.5)	14 (1.5)
Michigan Invitational Group, MI	33 (2.5)	33 (2.1)	51 (2.3)	49 (2.2)	16 (3.4)	19 (2.1)
Montgomery County, MD	29 (2.1)	30 (2.0)	52 (2.0)	54 (2.1)	19 (2.0)	16 (1.7)
Naperville Sch. Dist. #203, IL	34 (2.1)	38 (2.0)	53 (2.4)	48 (2.2)	14 (1.7)	14 (1.7)
Project SMART Consortium, OH	28 (2.6)	34 (2.3)	53 (2.8)	50 (1.9)	19 (1.7)	16 (1.8)
Rochester City Sch. Dist., NY r	35 (2.9)	44 (2.4)	53 (2.7)	45 (2.2)	13 (1.7)	11 (2.2)
SW Math/Sci. Collaborative, PA	35 (2.3)	38 (2.5)	51 (2.3)	48 (2.2)	14 (1.3)	15 (2.0)
International Avg. (All Countries)	35 (0.2)	39 (0.2) ▲	53 (0.2) ▲	51 (0.2)	12 (0.2) ▲	10 (0.1)

SOURCE: IEA Third International Mathematics and Science Study (TIMSS), 1998-1999.

▲ Significantly higher than other gender

Significance tests adjusted for multiple comparisons

Background data provided by students.

States in *italics* did not fully satisfy guidelines for sample participation rates (see Appendix A for details).

( ) Standard errors appear in parentheses. Because results are rounded to the nearest whole number, some totals may appear inconsistent.

An "r" indicates a 70-84% student response rate.

CHAPTER 5

The Mathematics Curriculum

The first part of Chapter 5 presents information about the curricular goals in the TIMSS 1999 countries and Benchmarking states, districts, and consortia. The ways in which the curriculum is supported and monitored within each entity, and the relationship between the curriculum and system-wide testing, are examined. The second part of the chapter contains teachers' reports about the mathematics topics actually studied in their classrooms.

5

In comparing achievement across systems, it is important to consider differences in students' curricular experiences and how they may affect the mathematics they have studied. At the most fundamental level, students' opportunity to learn the content, skills, and processes tested in the TIMSS 1999 assessment depends to a great extent on the curricular goals and intentions inherent in each system's policies for mathematics education. Just as important as what students are expected to learn, however, is what their teachers choose to teach them, which ultimately determines the mathematics students are taught.

Teachers' instructional programs are usually guided by an "official curriculum" that describes the mathematics education that should be provided. The official curriculum can be communicated by documents or statements of various sorts (often called guides, guidelines, standards, or frameworks) prepared by the education ministry or by national or regional education departments. These documents, together with supporting material such as instructional guides or mandated textbooks, are referred to as the *intended curriculum.*

To collect information about the intended mathematics curriculum at the eighth grade, the coordinators in each participating country and Benchmarking jurisdiction responsible for implementing the study completed questionnaires and participated in interviews. Information was gathered about factors related to supporting and monitoring the implementation of the official curriculum, including instructional materials, audits, and assessments aligned with the curriculum.

In many cases, teachers need to interpret and modify the intended curriculum according to their perceptions of the needs and abilities of their classes, and this evolves into the *implemented curriculum.* Research has shown that, even in highly regulated education systems, this is not identical to the intended curriculum. Furthermore, what is actually implemented is often inconsistent across an education system. Studies, including the Second International Mathematics Study, suggest that the implemented curriculum in the United States varies considerably from classroom to classroom – calling for more research into not only what is intended to be taught but what content is covered.[1] To collect data about the implemented curriculum, the mathematics teachers of the students tested in TIMSS 1999 completed questionnaires about whether students had been taught the various mathematics topics covered in the test.

1 Mayer, D.P., Mullens, J.E., and Moore, M.T. (2000), *Monitoring School Quality: An Indicators Report,* NCES 2001-030, Washington, DC: National Center for Education Statistics.

Does Decision Making About the Intended Curriculum Take Place at the National, State, or Local Level?

Depending on the education system, students' learning goals are set at different levels of authority. Some systems are highly centralized, with the ministry of education (or highest authority in the system) being exclusively responsible for the major decisions governing the direction of education. In others, such decisions are made regionally or locally. Each approach has its strengths and weaknesses. Centralized decision making can add coherence and uniformity in curriculum coverage, but may constrain a school or teacher's flexibility in tailoring instruction to the needs of students.

Exhibit 5.1 presents information for each TIMSS 1999 country about the highest level of authority responsible for making curricular decisions and gives the curriculum's current status. The data reveal that 35 of the 38 countries reported that the specifications for students' curricular goals were developed as national curricula. Australia determined curricula at the state level, with local input; the United States did so at both the state and local (district and school) levels, with variability across states; and Canada did so at the provincial level.

In recent decades, it has become common for intended curricula to be updated regularly. At the time of the TIMSS 1999 testing, the official mathematics curricula in 29 countries had been in place for less than a decade, and more than half of them were in revision. Of the eight countries with a mathematics curriculum of more than 10 years' standing, five were being revised. In Australia, Canada, and the United States, curriculum change is made at the state, provincial, or local level, and some mathematics curricula were in revision at the time of testing. The curricula in these three countries were relatively recent, having been developed within the 10 years preceding the study.

The development and implementation of academic content standards and subject-specific curriculum frameworks has been a central focus of educational change in the United States at both the state and local level. There has been concerted effort across the United States in writing and revising academic standards that has very much included attention to mathematics. Much of this effort has been based on work done at the national level during this period to develop standards aimed at increasing the mathematics competencies of all students. Since 1989, when the National Council of Teachers of Mathematics (NCTM) published *Curriculum and Education Standards for School Mathematics*, the mathematics education community has had the benefit of a unified set of goals for mathematics teaching and learning. The NCTM standards have been a springboard for state and local efforts to focus and improve mathematics education.[2] All states except Iowa (which as a matter of policy publishes no state standards) now have content or curriculum standards in mathematics, and many educational jurisdictions have worked successfully to improve their initial standards in clarity and content.[3]

In all 13 states that participated in TIMSS 1999 Benchmarking, curriculum frameworks or content standards in mathematics were published between 1995 and 2000 (see Exhibit 5.2). Four states detailed the standards for every grade including the eighth grade, seven states detailed them by a cluster or pair of grades that included the eighth grade, and two states reported the eighth grade as a benchmark grade at which certain standards should be met. Most states provided standards documents to guide districts and schools in developing their own curriculum, while some states, such as North Carolina, developed a statewide curriculum for all schools to use.

Exhibit 5.3 presents information about the curriculum of participating districts and consortia. Of the eight districts that participated, one reported that it uses the statewide curriculum in all schools (Guilford County); five had a district-wide curriculum that supported the state-developed frameworks or standards (the Jersey City Public Schools, the Miami-Dade County Public Schools, Montgomery County, the Naperville School District, and the Rochester City School District); and two had a curriculum developed at the school level (the Academy School District and the Chicago Public Schools), with Chicago also offering an optional structured curriculum district-wide. Each participating consortium indicated that all or most of its districts developed their own curriculum at the district level.

2 Kelly, D.L., Mullis, I.V.S., and Martin, M.O. (2000), *Profiles of Student Achievement in Mathematics at the TIMSS International Benchmarks: U.S. Performance and Standards in an International Context*, Chestnut Hill, MA: Boston College.

3 Raimi, R.A. (2000), "The State of State Standards in Mathematics" in C.E. Finn and M.J. Petrilli (eds.), *The State of State Standards*, Washington, DC: Thomas B. Fordham Foundation; Glidden, H. (1999), *Making Standards Matter 1999*, Washington, DC: American Federation of Teachers.

	National or Regional Curriculum	Year Curriculum Introduced	Status of Curriculum
United States [1]	Regional & Local	1994-1999	As of 1999, 49 of 50 states completed standards
Australia	Regional & Local	1995-1998	In revision (2 states); not being revised (3 states); no curriculum statement (3 states)
Belgium (Flemish)	National	1997	As introduced
Bulgaria	National	1997	As introduced
Canada	Regional	1997-1998 (most provinces)	As introduced
Chile	National	1980	In revision
Chinese Taipei	National	1997	In revision
Cyprus	National	1987	In revision
Czech Republic	National	1996	In revision
England	National	1995	In revision, same structure with minor revisions (to be implemented 2000/01)
Finland	National	1994	As introduced
Hong Kong, SAR	National	1987	In revision
Hungary	National	1986	In revision
Indonesia	National	1994	In revision
Iran, Islamic Rep.	National	1985	As introduced
Israel	National	1990	As introduced
Italy	National	1979	As introduced
Japan	National	1993	As introduced
Jordan	National	1993-1994	In revision
Korea, Rep. of	National	1995	As introduced
Latvia (LSS)	National	1992	In revision
Lithuania	National	1997	In revision
Macedonia, Rep. of	National	1979 (adaptations in 1995)	As introduced
Malaysia	National	1990	In revision
Moldova	National	1991	In revision
Morocco	National	1991	In revision
Netherlands	National	1993	As introduced
New Zealand	National	1993	As introduced
Philippines	National	1998	In revision
Romania	National	1993	In revision
Russian Federation	National	1997	In revision
Singapore	National	1993	In revision
Slovak Republic	National	–	–
Slovenia	National	1983	In revision
South Africa	National	1996	In revision
Thailand	National	1990	In revision
Tunisia	National	1997	As introduced
Turkey	National	1991	In revision

SOURCE: IEA Third International Mathematics and Science Study (TIMSS), 1998-1999.

Background data provided by National Research Coordinators.

1 United States: The NCTM standards were developed in 1989 and revised for 2000. As of 1999, most states had developed content standards. Currently, many states are in the process of updating and revising their standards.

A dash (–) indicates data are not available.

	Curriculum Framework/Content Standards and Year[1]	Grades/Clusters Detailed in Framework/Standards
Connecticut	Connecticut's K-12 Mathematics Curriculum Framework (1998)	Grade clusters: K-4, 5-8, 9-12
Idaho	Skills-Based Scope and Sequence Guides K-6 (1996); Achievement Standards K-8 (In draft); Achievement Standards 9-12 (1999)	Every grade: K-6; Grade clusters: 7-8, 9-12
Illinois	Illinois Learning Standards for Mathematics (1997)	Grade clusters: Early Elementary, Late Elementary, Middle/Junior High School, Early High School, Late High School
Indiana	Indiana Mathematics Proficiency Guide (1997); revised Indiana Academic Standards for Mathematics (2000)	Every grade K-8, individual courses in high school
Maryland	Learning Outcomes (1990); Content Standards (2000)	Grade clusters: K-3, 4-5, 6-8, 9-12
Massachusetts	Massachusetts Mathematics Curriculum Frameworks (1996; revised 2000)	Grade clusters: pK-4, 5-8, 9-10, 11-12; revised pairs: pK-K, 1-2, 3-4, 5-6, 7-8, 9-10, 11-12
Michigan	Michigan Curriculum Frameworks (1995); Michigan Essential Goals and Objectives for Mathematics Education (1985)	Grade clusters: Elementary, Middle, High School
Missouri	Frameworks for Curriculum Development in Mathematics (1996)	Grade clusters: K-4, 5-8, 9-12
North Carolina	North Carolina Standard Course of Study (1998)	Every grade: K-8, individual courses in high school
Oregon	Oregon Mathematics Content Standards (1996, 1998)	Benchmark grades: 3, 5, 8, 10, 12
Pennsylvania	Academic Standards (1999)	Benchmark grades: 3, 5, 8, 11
South Carolina	South Carolina Curriculum Standards (1998)	Every grade: K-8, individual courses in high school
Texas	Texas Essential Knowledge and Skills (1998)	Every grade: K-8, individual courses in high school

SOURCE: IEA Third International Mathematics and Science Study (TIMSS), 1998-1999.

Background data provided by coordinators from participating jurisdictions.

1 Indicates year(s) in which curriculum frameworks/content standards were instituted.

	Level of Curriculum Development
Academy School Dist. #20, CO	Curriculum is developed at the school level. Curriculum is currently in revision to reflect state standards.
Chicago Public Schools, IL	Curriculum is developed at the school level. The district writes standards statements which are aligned with state standards; schools translate these into a curriculum. The district also offers an optional structured curriculum.
Delaware Science Coalition, DE	Curriculum is created at the district-level based on the state content standards.
First in the World Consort., IL	Most districts within the Consortium have district-wide objectives and/or a curriculum based on state standards.
Fremont/Lincoln/ WestSide PS, NE	Each district has locally-developed standards and a curriculum based on the state standards.
Guilford County, NC	The district uses state-developed curriculum, the North Carolina Standard Course of Study.
Jersey City Public Schools, NJ	The mathematics curriculum (pK-12) is developed by the district and is aligned with the New Jersey Core Curriculum Content Standards.
Miami-Dade County PS, FL	The district has developed a mathematics curriculum, Competency-Based Curriculum (CBC), which is correlated to the Florida Sunshine State Standards for Mathematics. Most recently, the state has developed Grade Level Expectations (GLEs) that further define what a student should know and be able to do at specific grade levels. The district is currently making revisions to the CBC to reflect the GLEs.
Michigan Invitational Group, MI	Most districts have district-wide curriculum guides aligned to the state standards.
Montgomery County, MD	The district develops curriculum based on state standards.
Naperville Sch. Dist. #203, IL	The district develops curriculum based on state standards. District level mathematics curriculum is being revised for 2000-01.
Project SMART Consortium, OH	Each district in the SMART Consortium has a separate curriculum. In 2001, SMART will be adopting a mathematics curriculum for project schools.
Rochester City Sch. Dist., NY	New York State has developed a core curriculum for all grade levels. The Rochester City School District has written aligned curricula for pre-K through grade 8. The curricula for grades 9-12 are currently under revision.
SW Math/Sci. Collaborative, PA	Each district in the consortium has a separate curriculum. District-level curriculum is not necessarily based on the state standards.

SOURCE: IEA Third International Mathematics and Science Study (TIMSS), 1998-1999.

Background data provided by coordinators from participating jurisdictions.

How Do Education Systems Support and Monitor Curriculum Implementation?

During the past decade, content-driven systemic school reform has emerged as a promising model for school improvement.[4] That is, curriculum frameworks establishing what students should know and be able to do provide a coherent direction for improving the quality of instruction. Teacher preparation, instructional materials, and other aspects of the system are then aligned to reflect the content of the frameworks in an integrated way to reinforce and sustain high-quality teaching and learning in schools and classrooms.

Education systems use different ways to achieve this desired connection between the intended and the implemented curriculum. The methods used by the TIMSS 1999 countries to monitor curriculum implementation are shown in Exhibit 5.4, and by states, districts, and consortia in Exhibits 5.5 through 5.7. For example, teachers can be trained in the content and pedagogical approaches specified in the curriculum guides. Another way to help ensure alignment is to develop instructional materials, including textbooks, instructional guides, and ministry notes, that are tailored to the curriculum. Systems can also monitor implementation of the intended curriculum by means of school inspection or audit.

Of the methods for supporting and monitoring curriculum implementation shown in Exhibit 5.4, 10 countries reported using all six, and a further 14 countries used five. Support for the national/regional mathematics curriculum as part of pre-service education was reported by 26 of the 38 countries. Nearly all countries (34) used in-service teacher education, and most countries (31) used mandated or recommended textbooks. Ministry notes and directives were used in 30 countries, as was a system of school inspection or audit.

States, districts, and consortia provided data on policies related to textbook selection, pedagogical guides, and accreditation. As shown in Exhibit 5.5, seven of the Benchmarking states reported that they do not select textbooks for use at the local level. The other six states issue a list of books from which districts can choose. Almost all districts and consortia reported that their state does not select textbooks, while three reported state involvement in textbook selection. Ten jurisdictions indicated that textbooks were chosen or recommended at the district level, and four that selection occurs at the school level or, in the consortia, at the school and district level depending on the district.

4 O'Day, J.A. and Smith, M.S. (1993), "Systemic Reform and Educational Opportunity" in S.H. Fuhrman (ed.), *Designing Coherent Education Policy: Improving the System*, San Francisco, CA: Jossey-Bass, Inc.

As shown in Exhibit 5.6, nine of the 13 Benchmarking states developed materials that included pedagogical guidance for instruction and implementation of the curriculum frameworks and standards. Twelve districts and consortia had at least state- or district-level guides to support curriculum implementation.

As shown in Exhibit 5.7, six of the participating states had accreditation systems, four of which included student performance on the state assessment in their accreditation review (Indiana, Michigan, Missouri, and Oregon). Two states without accreditation systems, Illinois and Texas, made periodic site visits to evaluate schools. Only one consortium, the Michigan Invitational Group, reported having an accreditation system at the state level. The Academy School District in Colorado reported that the state was in the process of implementing a system for 2001.

Exhibit 5.4 Countries' Use of Methods to Support or Monitor Implementation of the Curriculum*

	Pre-Service Teacher Education	In-Service Teacher Education	Mandated or Recommended Textbook(s)	Instructional or Pedagogical Guide	Ministry Notes and Directives	System of School Inspection or Audit
United States [1]	+	+	+	+	+	+
Australia [2]	●	●		●	●	
Belgium (Flemish)	●	●		●	●	●
Bulgaria	●	●	●		●	●
Canada [3]	●	●	●	●	●	
Chile			●		●	
Chinese Taipei	●	●	●	●		●
Cyprus		●	●		●	●
Czech Republic	●		●		●	●
England	●	●				●
Finland	●	●	●	●		
Hong Kong, SAR	●	●	●	●	●	●
Hungary	●	●	●	●	●	
Indonesia		●	●	●	●	●
Iran, Islamic Rep.	●	●	●	●	●	●
Israel	●	●	●		●	
Italy		●		●	●	●
Japan		●	●	●	●	●
Jordan		●	●	●	●	●
Korea, Rep. of	●	●	●	●	●	●
Latvia (LSS)	●	●	●	●	●	●
Lithuania		●	●		●	
Macedonia, Rep. Of	●	●	●	●		●
Malaysia	●	●	●	●	●	●
Moldova		●	●		●	●
Morocco	●	●	●	●	●	●
Netherlands	●	●		●	●	●
New Zealand	●	●				●
Philippines		●	●	●	●	●
Romania	●	●	●	●	●	●
Russian Federation	●	●	●	●	●	●
Singapore	●	●	●	●	●	●
Slovak Republic	●		●		●	●
Slovenia	●	●	●	●		●
South Africa	●	●	●	●		●
Thailand	●	●	●	●	●	●
Tunisia		●	●	●	●	●
Turkey		●	●		●	●

SOURCE: IEA Third International Mathematics and Science Study (TIMSS), 1998-1999.

● Country reported that method is used to support or monitor the implementation of the national/regional curriculum at grade 8

+ Not applicable nationally

Background data provided by National Research Coordinators.

* Other than system-wide assessments and public examinations described in Exhibits 5.8 and 5.9, respectively.

1 United States: Methods are implemented by individual states and vary from state to state. As of 1998, 13 states have policies on textbook/materials selection; 8 states have policies recommending textbook/materials.

2 Australia: Results shown are for the majority of states/territories.

3 Canada: Results shown are for the majority of provinces.

States', Districts' and Consortia's Use of Textbooks and Instructional Materials to Support Implementation of the Curriculum

8th Grade Mathematics

States	Policy on Textbooks and Instructional Materials
Connecticut	The state does not select textbooks.
Idaho	The state approves a list of textbooks and materials from which districts/schools must choose. The textbooks selection criteria include alignment with Idaho Skills-Based Scope and Sequence Guide, which specifies skills that all students should know at different levels. Schools are required to select all their basic instructional materials from the Idaho Adoption Guide produced by the adoption committee. Schools not choosing from this adoption list can lose accreditation points.
Illinois	The state does not select textbooks.
Indiana	The state selects a list of textbooks from which districts/schools can choose; however, waivers are granted. The state texts are not necessarily based on the state standards. The state intends to align textbooks selections with Indiana's new Academic Standards (2000).
Maryland	The state does not select textbooks.
Massachusetts	The state does not select textbooks.
Michigan	The state does not select textbooks.
Missouri	The state does not select textbooks.
North Carolina	The state selects a list of textbooks and materials based on the curriculum from which districts can choose.
Oregon	The state selects a list of textbooks and materials from which districts can choose. Districts may submit a waiver for an independent adoption to select textbooks and instructional materials of their own choice. These district-level adoptions must meet the state selection criteria.
Pennsylvania	The state does not select textbooks.
South Carolina	The state selects a list of textbooks and materials from which districts can choose. The state funds the instructional materials that are selected from the state approved list.
Texas	The State Textbook Review Committee selects textbooks and instructional materials to support the state curriculum framework. Districts choose textbooks and/or instructional materials using local criteria. The state funds the purchase of textbooks and/or instructional materials that are on the selected list. Districts may waiver, at own expense, from selected textbooks or instructional materials.

SOURCE: IEA Third International Mathematics and Science Study (TIMSS), 1998-1999.

Background data provided by coordinators from participating jurisdictions.

Districts and Consortia	Policy on Textbooks and Instructional Materials
Academy School Dist. #20, CO	STATE: The state does not select textbooks. LOCAL: Schools can select materials based on guidelines with acceptance by the Board of Education.
Chicago Public Schools, IL	STATE: The state does not select textbooks. LOCAL: Schools in districts choose instructional materials.
Delaware Science Coalition, DE	STATE: The state does not select textbooks. LOCAL: Textbook selection may be made at the school or district level. Due to the influence of two NSF-funded Teacher Enhancement Grants in Delaware, by Fall 2000 every school district in the state will be using an NSF-funded standards-based mathematics curriculum with some students.
First in the World Consort., IL	STATE: The state does not select textbooks. LOCAL: Texts and materials selected and recommended at the district level. The FIW Consortium is reviewing materials to recommend as well.
Fremont/Lincoln/ WestSide PS, NE	STATE: The state does not select textbooks. LOCAL: Districts select textbooks/textbook series and schools select supplemental materials.
Guilford County, NC	STATE: The state selects a list of textbooks and materials based on the state content standards from which districts can choose. LOCAL: One textbook used throughout county. A system-wide committee reviews the state selected list and one textbook per grade level is selected to be used system-wide.
Jersey City Public Schools, NJ	STATE: The state does not select textbooks. LOCAL: A committee is formed at the district level to facilitate the selection of mathematics textbooks and materials. There is a "standard operating procedure" for the formulation of the committee so as to include all constituent groups. All selected textbooks and materials are aligned with the district's grade-level mathematics curricula, the NJ Core Curriculum Content Standards in mathematics, and the national standards in mathematics.
Miami-Dade County PS, FL	STATE: The state recommends the textbooks and instructional materials. LOCAL: The district selection committee narrows the selection to two or three textbooks. The schools pick one of the selected textbooks. The new legislation makes waivers for using non-adopted texts more difficult, but schools are allotted some money to spend on non-state adopted materials with review at the district level.
Michigan Invitational Group, MI	STATE: The state does not select textbooks. LOCAL: Textbook selection is made at the school/district level.
Montgomery County, MD	STATE: The state does not select textbooks. LOCAL: The district recommends a few textbooks.
Naperville Sch. Dist. #203, IL	STATE: The state does not select textbooks. LOCAL: District uses criteria based on the learning outcomes to select instructional materials. No one textbook selected.
Project SMART Consortium, OH	STATE: The state does not select textbooks, but approves a liberal textbook list from which districts can choose. LOCAL: The districts select instructional materials that are closely aligned to the curriculum.
Rochester City Sch. Dist., NY	STATE: The state does not select textbooks. LOCAL: The district chooses one text series for all schools to use.
SW Math/Sci. Collaborative, PA	STATE: The state does not select textbooks. LOCAL: Each district selects a textbook. The Collaborative encourages consideration of exemplary NSF-developed materials.

SOURCE: IEA Third International Mathematics and Science Study (TIMSS), 1998-1999.

States	Pedagogical Guides
Connecticut	The "Guide to K-12 Program Development in Mathematics" (1999) provides a curriculum framework with content standards and performance standards as well as "illustrative lessons" for each content standard at each grade band. In addition, the state provides curriculum handbooks with objectives, sample lessons, sample test items, and teacher resources. Prototype assessments with high-quality student responses are also distributed.
Idaho	Pedagogical guides are not available at the state level.
Illinois	Performance descriptors have been completed in draft form. Classroom assessment tasks and student-work examplars will be available Summer 2001.
Indiana	The "Indiana Mathematics Proficiency Guide" (1997) contains grade specific standards with ideas for activities including examples that clarify the skills, and ways to incorporate communication, reasoning, problem solving, connections, and technology into the mathematics classroom. New Curriculum Frameworks are being written to support Indiana's new Academic Standards (2000).
Maryland	The guide "Better Mathematics: Building Effective Teaching Through Educational Research" focuses on appropriate teaching methods.
Massachusetts	The curriculum frameworks provide teaching activities for each learning standard.
Michigan	Toolkits are designed to support the implementation of the curriculum frameworks including kits on planning subject area instructional units, curriculum integration, designing classroom assessments, and connecting with the learner. The "Mathematics Teaching and Learning Sample Activities" was developed specifically to assist in teaching the mathematics frameworks.
Missouri	The Curriculum Frameworks provide appropriate teaching activities by discipline with examples of how "Show-Me Standards" may be taught and assessed.
North Carolina	The development of a curriculum enhancement guide is in process.
Oregon	"Teaching and Learning to Standards" supports the Oregon content standards and provides best practices, example lessons, teaching strategies, tools and on-line resources.
Pennsylvania	Pedagogical guides are not available at the state level.
South Carolina	The "South Carolina Standards Implementation Guide" includes information on standards-based education in the State, standards-based assessment practices, samples of standards-based instructional modules, tips and tools for educators (vignettes, content briefs, etc.), glossary of terms, and a list of websites.
Texas	The Educator's Guides include objectives for mathematics (grade 3 - high school algebra). The Supplement to the Educator's Guide includes additional information on teaching the objectives and sample problems. Study Guides are provided to students performing below the standard on state assessments. These Study Guides, for use by students, parents, and teachers, include sample problems and activities.

SOURCE: IEA Third International Mathematics and Science Study (TIMSS), 1998-1999.

Background data provided by coordinators from participating jurisdictions.

	Pedagogical Guides
Districts and Consortia	
Academy School Dist. #20, CO	No specific "how-to" instructional manuals are provided. The district provides all schools with best-practice examples from NCTM. The state has provided districts with grade-appropriate sample assessments, released items, and samples of scored student work which the district has expanded upon.
Chicago Public Schools, IL	An optional structured curriculum provides daily lesson plans at all grade levels. For high schools, test blueprints of the "Chicago Academic Standards Exam" (CASE) are provided to teachers for instructional purposes.
Delaware Science Coalition, DE	The "Delaware Curriculum Framework" (1995) contains several classroom activities and a vignette for each standard at each grade band. A "Teacher's Desk Reference" has been published that provides indicators at each grade level serving as a reference for district curriculum committees in developing local curriculum and as a reference for teachers in planning lessons and units of study.
First in the World Consort., IL	Each district in the consortium develops mathematics guides to support their own curriculum (teacher guides, manipulations, peer coaching, etc.).
Fremont/Lincoln/ WestSide PS, NE	Two of the districts have curriculum guides in mathematics with instructional activities. The third district uses commercially-developed materials.
Guilford County, NC	There is a state-written book, Strategies for Instruction, detailing best practices, lessons, assessments, and teaching methods based on the North Carolina Course of Study.
Jersey City Public Schools, NJ	The "New Jersey Framework for Teaching" in Mathematics, published in May 1996, discusses essential components of a quality K-12 mathematics program. The framework is not a curriculum, but a comprehensive digest of activities, curriculum connections, and instructional strategies related to the NJ Core Curriculum Content Standards in Mathematics. In addition to the state standards and the state frameworks, the district's curriculum guides provide content guidelines based on grade-level competencies. In the district curriculum materials, manipulatives, resources, and learning activities are provided at each grade level.
Miami-Dade County PS, FL	The Florida Department of Education released the "Curriculum Planning Tool" (CPT) which includes a bank of activities linked to the strands and standards. It also maintains a website with information of Grade Level Expectations and other guidelines for instruction. The state also produced the Florida Curriculum Frameworks for Mathematics and a "Mathematics Best Practices" CD-ROM. All guides and curriculum materials developed at the district level are aligned with the Sunshine State Standards. Some of the district level guides are: "The Competency-Based Curriculum" (1992, revised 1999), Supplement to the "Competency-Based Curriculum" (1999), "Here Comes the Sunshine State Standards" (1998), "Awesome Activities for Achieving Success on the Sunshine State Standards K-8" (1999), "Focus on Algebra I through a Sunshine State Standards Lens" (1999), and "Summer School Curriculum K-5" (1996, 2000), "Summer 2000 Balanced Assessment for Middle School", and numerous packages of materials produced for individual workshops.
Michigan Invitational Group, MI	Toolkits are designed to support the implementation of the curriculum frameworks including kits on planning subject area instructional units, curriculum integration, designing classroom assessments, and connecting with the learner. The "Mathematics Teaching and Learning Sample Activities" was developed specifically to assist in teaching the mathematics frameworks.
Montgomery County, MD	"Better Mathematics" produced at state level and "State of the Art Instruction that Ensures Classroom Success for Every Student: A Handbook for Educators" produced at the local level, both address pedagogy. Local curriculum documents are written for each mathematics course which include: goals, objectives, lessons, and strategies. The curriculum document exists for K-12.
Naperville Sch. Dist. #203, IL	District-level guide connects outcomes to resources and provides general teaching strategies and guidance for using manipulatives.
Project SMART Consortium, OH	There are not pedagogical guides at the state level. As soon as the state "Draft Content Mathematics Standards" are approved by the Ohio State Board of Education (early 2001) plans are underway to provide pedagogical guides to locals. Ohio is a local-control state, therefore, many locals have developed various types of mathematics guides.
Rochester City Sch. Dist., NY	New York State provides core curriculum guides based on the standards at all grades levels. The district has developed mathematics curriculum guides and pacing charts that align NYS standards with instruction for students in grades pK-8. Guides for grades 9-10 are being developed.
SW Math/Sci. Collaborative, PA	In 2000, the Collaborative and the local intermediate unit convened teachers from 30 districts to develop a grade-by-grade conceptual framework linked to lessons from exemplary materials.

SOURCE: IEA Third International Mathematics and Science Study (TIMSS), 1998-1999.

States', Districts' and Consortia's Use of Accreditation to Support Implementation of the Curriculum

8th Grade Mathematics

States	Use of Accreditation
Connecticut	No accreditation system.
Idaho	Accreditation requires that curriculum developed at the local level be aligned with state standards. Schools must establish educational standards for all grade levels and develop high school exiting standards for graduation; these standards must be aligned with exiting standards established by the State Board of Education. It also requires that schools participate in state testing and adhere to textbook adoption policies.
Illinois	There are periodic quality-assurance site visits to schools.
Indiana	The accreditation system requires K-8 schools to self-report alignment of curriculum with state standards (proficiencies); grade 9-12 schools submit a master schedule and course descriptions to verify compliance with state standards. Performance on the ISTEP+ is also considered in accreditation. Technical assistance is available to schools that do not meet the accreditation standards.
Maryland	No accreditation system.
Massachusetts	No accreditation system.
Michigan	State-level accreditation is based in part on student performance on state assessments. The system is being revised to include successful achievement as well as continuous improvement.
Missouri	The Missouri School Improvement Program, designed to accredit districts, assesses districts progress on the Show-Me Standards as measured by the Missouri Assessment Program. There are "success teams" that help districts improve student achievement in all subject areas.
North Carolina	No accreditation system.
Oregon	All schools are state accredited through a system of "standard" assurances, Consolidated District and School Improvement Plans, Annual Performance Reports, and Schools Reviews. State accreditation is based on the Oregon Performance Accountability System (OPAS), that assesses school mathematics performance. Any school falling in the low or unacceptable category receives targeted assistance including alignment with standards, instructional improvement and professional development.
Pennsylvania	No accreditation system.
South Carolina	The accreditation system is in revision. Schools must meet a battery of standards in the current accreditation system, but student academic performance is not included. The new accreditation system will include student academic performance and will go into effect in 2001.
Texas	Although not considered an accreditation system, the state's accountability system includes a variety of on-site evaluations designed to provide feedback for improvement.

SOURCE: IEA Third International Mathematics and Science Study (TIMSS), 1998-1999.

Background data provided by coordinators from participating jurisdictions.

	Use of Accreditation
Districts and Consortia	
Academy School Dist. #20, CO	The state will be implementing an accreditation system beginning in Fall 2001 based primarily on the success and/or progress on the standards-referenced state assessment (CSAP).
Chicago Public Schools, IL	No accreditation system.
Delaware Science Coalition, DE	No accreditation system.
First in the World Consort., IL	No accreditation system.
Fremont/Lincoln/ WestSide PS, NE	No accreditation system.
Guilford County, NC	No accreditation system.
Jersey City Public Schools, NJ	No accreditation system.
Miami-Dade County PS, FL	No accreditation system.
Michigan Invitational Group, MI	State-level accreditation is based in part on student performance on state assessments. The system is being revised to include successful achievement as well as continuous improvement.
Montgomery County, MD	No accreditation system.
Naperville Sch. Dist. #203, IL	No accreditation system.
Project SMART Consortium, OH	No accreditation system.
Rochester City Sch. Dist., NY	No accreditation system.
SW Math/Sci. Collaborative, PA	No accreditation system.

SOURCE: IEA Third International Mathematics and Science Study (TIMSS), 1998-1999.

What TIMSS 1999 Countries Have Assessments And Exams in Mathematics?

Assessments and exams that are aligned with the intended curriculum provide a means for evaluating system- and student-level achievement. System-wide assessments are designed primarily to inform policy makers about matters such as national standards of achievement of the intended curriculum objectives, strengths and weaknesses in the curriculum or how it is being implemented, and whether educational achievement is improving or deteriorating. The primary purpose of national public examinations, while providing information of interest to national and regional policy makers, is to provide information for making decisions about individual students.

Exhibit 5.8 shows that about two-thirds of the participating countries had national assessments in mathematics, with half of those assessing all students and half sampling students. Most countries tested two or three grades, with Hong Kong (nine grades) and Korea (seven grades) testing the most grades. Generally, the purpose of system-wide assessments was to provide feedback to government policy makers and the public, although some countries provided feedback to individual schools. For example, in Singapore the 20 schools found to provide the greatest value-added measures received monetary rewards, as did teachers of the top 25 percent of classes in Chile.

Using public examinations as a way to select students for university or academic tracks in secondary school can be an important motivating factor for student achievement (see Exhibit 5.9). Thirty-seven countries reported having public examinations or awards, at one or more grades, that included testing achievement in mathematics. Most countries held their examinations in the final year of schooling for certification and selection to higher education (often, university education). In about one-third of the countries, public examinations were also used for selection or course assignment (tracking) within secondary schools.

	System-Wide Assessments[1]	Grades		Purpose/Consequences
		Entire Grade Level	Sample from Grade Level	
United States	Yes		4, 8, 12	National and state-level feedback
Australia [2]	Yes	3, 5 (all states) 7 (four states)		System-level, school-level, and individual student-level feedback
Belgium (Flemish)	No			
Bulgaria	Yes		4, 8	System-level feedback, administered only in 1998
Canada [3]	Yes	3, 6, 9 (5 provinces); 5, 8, 11 (1 province); 4, 7, 10 (1 province); 12 (1 province)	Ages 13 and 16 nationally (most provinces)	System- and school-level feedback
Chile	Yes	4, 8, 10		System-level, school-level, class-level feedback; top 25% of teachers are given monetary rewards; usually one grade level assessed each year
Chinese Taipei	No			
Cyprus	No			
Czech Republic	No			
England	Yes	1, 5, 8		School-level feedback; course selection and placement for grade 9
Finland	Yes		4, 6, 9	System-level feedback
Hong Kong, SAR	Yes		1 - 9	System-level feedback
Hungary	Yes		4, 6, 8, 10, 12	System-level, school-level, and individual-level feedback
Indonesia	Yes		various grades	System-level feedback, assessments given irregularly at different primary grades
Iran, Islamic Rep.	No			
Israel	Yes		4, 8	System-level feedback
Italy	Yes		6, 8, 10, 13	System-level feedback; first administered in 1999 with a grade 4 assessment instituted in 2000
Japan	Yes		5, 6, 7, 8, 9	System-level feedback
Jordan	Yes		4, 5, 8, 10	System-level feedback; monitoring reform impact; curricular revisions
Korea, Rep. Of	Yes	4, 5, 6, 7, 8, 10, 11		System-level feedback
Latvia (LSS)	No			
Lithuania	No			
Macedonia, Rep. Of	Yes		4, 5, 6, 7, 8	System-level feedback and research purposes (projects and curriculum development)
Malaysia	Yes	6, 9, 11, 13		System- and school-level feedback; "good schools" publicized
Moldova	No			
Morocco	Yes	6, 9, 10, 11, 12		System- and school-level feedback
Netherlands	Yes	10, 11, 12	6	System-level feedback
New Zealand	Yes		3, 7	System-level feedback
Philippines	Yes	6, 10		System- and school-level feedback (the assessment was sample-based up until 1999)
Romania	No			
Russian Federation	Yes		various grades	Irregularly for research purposes
Singapore	Yes	6,10,12		System- and school-level feedback; selection into courses, certification and entry to university
Slovak Republic	No			
Slovenia	No			Assessments administered in grades 1-8 from 1991-1996
South Africa	No			
Thailand	Yes	6, 9, 12		System-level feedback
Tunisia	Yes	4, 6, 9, 13		System- and school-level feedback; may lead to redistribution of teachers in the regions; assessments at grades 4 and 6 developed regionally
Turkey	Yes		5, 8, 11	System- and school-level feedback

SOURCE: IEA Third International Mathematics and Science Study (TIMSS), 1998-1999.

Background data provided by National Research Coordinators.

1 Public examinations are also used for system-wide assessment purposes in these countries: Malaysia, Morocco, Netherlands, Philippines, Singapore, Tunisia, and Turkey.

2 Australia: System-wide assessments are administered in 3 of 8 states/territories.

3 Canada: System-wide assessments are administered in 5 of 10 provinces.

	Public Exams/ Awards	Grade(s)	Purpose/Consequences
United States [1]	Yes	varies	Primarily feedback to system and schools; in 8 states grade promotion is dependent on results; in 18 states graduation is dependent on results of grade 12 exams
Australia	Yes	12	Certification and selection for tertiary education
Belgium (Flemish)	No		
Bulgaria	Yes	7/8, 12	Candidates for profile schools (grade 7 or 8); certification and entrance to university -- not taken by all students (grade 12)
Canada [2]	Yes	3,6,8 (1province); 10, 11(1 province); 12 (4 provinces)	Feedback to system and schools; certification (grade 12)
Chile	Yes	12	Entry to university
Chinese Taipei	Yes	9, 12	Entry to secondary school (grade 9); entry to university (grade 12)
Cyprus	Yes	12	Certification and entry to university (grade 12); a certification exam occurs on a local level for grade 9
Czech Republic	Yes	13	Certification (mathematics can be chosen as one of four subjects for leaving examination)
England	Yes	10, 12	Certification (grade 10), certification and entry to university (grade 12); feedback to system and schools
Finland	Yes	12	Certification and selection for tertiary education
Hong Kong, SAR	Yes	6, 11, 13	School placement (grade 6); certification and placement for 12th grade (grade 11); placement in tertiary institutions (grade 13)
Hungary	Yes	12	Certification and entry to university
Indonesia	Yes	6, 9, 12	Leaving exam and selection for junior secondary school (grade 6); selection for senior secondary school (grade 9); leaving exam (grade 12); system-level feedback, in some cases school- and classroom-level feedback
Iran, Islamic Rep.	Yes	11, 12	Certification (grade 11); entry to tertiary education (grade 12); in addition, provincial exams are administered at grade 8
Israel	Yes	11 or 12	Entry to higher education
Italy	Yes	13	Certification and entry to university
Japan	Yes	9, 12	Entry to prefectural and municipal upper secondary schools (grade 9); entry to national, prefectural and municipal universities (grade 12)
Jordan	Yes	12	Certification and entry to tertiary education
Korea, Rep. of	Yes	12	College entrance exam for selection of students
Latvia (LSS)	Yes	9, 12	Certification
Lithuania	Yes	9, 12	Graduation from Basic and Upper Secondary schools
Macedonia, Rep. Of	Yes	12	Certification and entry to university; the exam constitutes 40% of the required points for entry to university with the remaining points based on university entry exams
Malaysia	Yes	6, 9, 11, 13	Feedback to system and schools; achievement test (grade 6); entry to course tracks (grade 9); certification and end of secondary (grade 11); certification and entry to university (grade 13)
Moldova	Yes	9, 11/12	Certification and selection for high school (grade 9); graduation (grade 11 or 12 depending on school)
Morocco	Yes	6, 9, 10, 11, 12	Remedial test for retention purposes (grade 6); certification, selection to secondary and selection to courses (grade 9); certification and entry to tertiary (grade 12); feedback to system and schools
Netherlands	Yes	10, 11, 12	End-of-track examinations; exams recommended at grades 6 and 8
New Zealand	Yes	10, 12	Certification and course selection (grade 10); entry to tertiary education (grade 12); feedback to system and schools; informal between-school comparisons
Philippines	Yes	6, 10	Feedback to system and schools
Romania	Yes	8, 12	Certification (grade 8); certification (grade 12; mathematics can be chosen as one of 7 subjects)
Russian Federation	Yes	9, 11	Certification
Singapore	Yes	6, 10, 12	Selection into courses; certification and entry to university; feedback to system and schools
Slovak Republic	Yes	12	Certification (mathematics can be chosen as one of four subjects for leaving exam)
Slovenia	Yes	8, 12	Entry to secondary school (grade 8); certification and entry to tertiary education (grade 12)
South Africa	Yes	12	Certification and selection for tertiary education
Thailand	Yes	12	Entry to university
Tunisia	Yes	6, 9, 13	Regional exam for promotion (grade 6); feedback to system and schools, selection for schools and courses, and promotion (grade 9); certification and entry to university (grade 13)
Turkey	Yes	8, 11	Placement in specialized schools for some students (grade 8); entry to university (grade 11)

SOURCE: IEA Third International Mathematics and Science Study (TIMSS), 1998-1999.

Background data provided by National Research Coordinators.

1 United States: As of 1997-1998, public examinations are administered in 47 of 50 states at grades 7-8 or 9-12.

2 Canada: Public examinations are administered in 5 of 10 provinces.

What Benchmarking Jurisdictions Have Assessments in Mathematics?

Across the United States, many states are conducting assessments based on their own content standards and are assessing whether students in their schools are meeting these standards for academic achievement. Forty-three states have some type of criterion-referenced mathematics assessment aligned to state standards.[5]

All 13 Benchmarking states had developed or were developing state-level mathematics assessments aligned with their state curriculum frameworks or content standards. As summarized in Exhibits 5.10 and 5.11, most of them reported recently revising or developing their criterion-referenced assessment to align with their current eighth-grade framework/standards. Assessments in Connecticut, Idaho, Indiana, Maryland, Massachusetts, North Carolina, and Texas were reported to be in revision, and those in Illinois, Michigan, and South Carolina to be in development. In addition to these criterion-referenced assessments, seven states (Idaho, Illinois, Indiana, Maryland, Missouri, North Carolina, and South Carolina) reported using norm-referenced mathematics tests to assess student mathematics achievement statewide.

All the Benchmarking states except Pennsylvania have participated in recent state mathematics assessments as part of the National Assessment of Educational Progress (NAEP). Ten of the 13 states participated in both 1996 and 2000, and Idaho and Oregon in one of the years.

As shown in Exhibit 5.12, six of the Benchmarking states use or plan to use performance on a mathematics assessment as a requirement for graduation from high school. In Indiana and Texas, the exit exam was based on the state mathematics standards. In Maryland, North Carolina, and South Carolina, they were basic skills competency tests not based on state standards, but these states were in the process of changing to standards-based exit exams. Massachusetts was planning to institute a standards-based exit exam beginning with the class of 2003.

Benchmarking states reported a range of other consequences of their mathematics assessments for students, apart from their use as a graduation requirement. For example, Connecticut, Oregon, and Pennsylvania reported that they affix a certificate or seal to students' diplomas to show that they have met the performance goal on the state high school mathematics assessment; Illinois, North Carolina, Oregon, and South Carolina reported a policy of using assessment results to assist in making promotion decisions; Texas was phasing in a promotion policy; and Connecticut

5 Orlofsky, G.F. and Olson, L. (2001), "The State of the States" in *Quality Counts 2001, A Better Balance: Standards, Tests, and the Tools to Succeed*, Education Week 20(17).

was encouraging its districts to reevaluate their social promotion policies. As an incentive, students meeting the standards in Michigan and Missouri could receive state funds to support their academic careers through scholarship money and funds for advanced course work, respectively. No consequences for students based on test results were reported in Idaho, Maryland, and Massachusetts, and no additional consequences beyond that of the high school exit exam for students in Indiana.

Benchmarking states also reported a range of consequences at the district or school level. Connecticut, Massachusetts, Michigan, and North Carolina reported that additional funding was made available to low-performing schools and districts to support remediation. In Indiana, Oregon, and South Carolina, districts were required to provide remediation to students with low scores on the state assessments. States had the right to take over schools or districts in Maryland, Massachusetts, Michigan, and Pennsylvania. While consequences of assessments for schools or districts usually involved remediation activities or sanctions, Connecticut, Indiana, and Maryland provided monetary rewards to districts and/or schools that showed improvement.

As shown in Exhibit 5.13, almost all the Benchmarking districts and consortia (13 of 14) participated in the mathematics assessments administered by their state. The Fremont/Lincoln/Westside Public Schools of Nebraska was the only district or consortium that reported having no state-administered assessments. Most districts and consortia also conducted district-wide assessments at the local level. Four districts reported using local standards-based assessments: Jersey City, Miami-Dade, Montgomery County, and Naperville. The Chicago Public Schools and the First in the World Consortium reported that they are developing district-wide mathematics assessments. Some districts in the Project SMART Consortium also administered district-developed assessments. Eight districts and consortia reported that norm-referenced tests were used for student assessment at the district level. Guilford County was the only district or consortium that reported having no assessments beyond those administered by the state.

Exhibit 5.10 States' Mathematics Assessments

8th Grade Mathematics

	State-Developed Criterion-Referenced Mathematics Assessment[1]	Other Mathematics Assessments	Participated in NAEP	
			1996	2000
Connecticut	Connecticut Mastery Test (CMT): In revision - Grades 4, 6, 8 Connecticut Academic Performance Test (CAPT): In revision - Grade 10	None	Yes	Yes
Idaho	Direct Mathematics Assessment (DMA): In revision - Grades 4, 8 (2001-02)	ITBS: Grades 3-8 TAP: Grades 9-11	No	Yes
Illinois	Illinois Goal Assessment Program (IGAP): Grades 3, 6, 8, 10 (1988-99) Illinois Standard Achievement Test (ISAT): Grades 3, 5, 8 (2000) Prairie State Achievement Examination (PSAE): Grade 11 (2001)	ISAT is also reported as a norm-referenced assessment: Grades 3, 5, 8, 10	Yes[2]	Yes
Indiana	Indiana Statewide Testing for Educational Progress-Plus (ISTEP+): In revision - Grades 3, 6, 8, 10	ISTEP+ includes a norm-referenced component: Grades 3, 6, 8, 10	Yes	Yes
Maryland	Maryland School Performance Assessment Program (MSPAP): In revision - Grades 3, 5, 8	CTBS/5: Grades 2, 4, 6 Maryland Functional Tests: Grades 9, 11	Yes	Yes
Massachusetts	Massachusetts Comprehensive Assessment System (MCAS): Grades 4, 8, 10 (Revised 2000)	None	Yes	Yes
Michigan	Michigan Educational Assessment Program (MEAP): Grades 4, 7, 11. In revision/development - Grades 4, 8, 11.	None	Yes	Yes
Missouri	Missouri Assessment Program (MAP): Grades 4, 8, 10	MAP includes the Terra Nova	Yes	Yes
North Carolina	North Carolina Testing Program: In revision - end-of-grade exams in Grades 3-8, North Carolina Competency Test, end-of-course exams in high school North Carolina High School Comprehensive - Grade 10 In development - Grade 11	ITBS: Grades 4 and 8	Yes	Yes
Oregon	Oregon State-wide Assessment System: Grades 3, 5, 8, 10	None	Yes	No
Pennsylvania	Pennsylvania System of School Assessment (PSSA): Grades 5, 8, 11	None	No	No
South Carolina	Basic Skills Assessment Program (1981-1999) Palmetto Achievement Challenge Test (PACT): Grades 3-8 (2000) In development - Grade 10 (2002-03)	MAT7: Grades 4, 5, 7, 9, 11 (1995-1999) Terra Nova: Grades 3, 6, 9 (1999) Terra Nova: Grades 5, 8, 11 (2000) Terra Nova: Grades 4, 7, 10 (2001) Terra Nova: Grades 3, 6, 9 (2002)	Yes	Yes
Texas	Texas Assessment of Academic Skills (TAAS): Grades 3-8, 10, end-of-course tests in high school (Revised 2000)	None	Yes	Yes

SOURCE: IEA Third International Mathematics and Science Study (TIMSS), 1998-1999.

Background data provided by coordinators from participating jurisdictions.

1 Specifically developed to be aligned with the curriculum framework/content standards indicated in Exhibit 5.2.

2 Illinois participated in NAEP in 1996 but results were not reported due to low participation rates.

	Status of State-Developed Mathematics Assessment
Connecticut	The Connecticut Mastery Test (CMT) was developed to be aligned with Connecticut's 1981 Guide to Curriculum Development in Mathematics. The Connecticut Academic Performance Test (CAPT), first administered in 1995, was developed to be aligned with the 1987 Common Core of Learning. The assessments are being revised for the 2000-01 school year based on Connecticut's 1998 K-12 Mathematics Curriculum Framework. The CMT is administered in the fall and the CAPT is administered in the spring.
Idaho	The Idaho Direct Mathematics Assessment (DMA) is administered at grades 4 and 8. This formative, performance assessment was aligned with state standards for the 2001 and 2002 assessments. The Grade 11 assessment was field tested in December 2000 and will be administered in 2002.
Illinois	Illinois Standard Achievement Test (ISAT) administered at grades 3, 5, 8, replaced the Illinois Goal Assessment Program (IGAP) which was administered from 1988-1999 at grades 3, 6, 8, and 10. Beginning in 2001, the state will give new high school tests, the Prairie State Achievement Examination (PSAE), based on the 1997 Illinois Learning Standards for Mathematics.
Indiana	The Indiana Statewide Testing for Educational Progress (ISTEP+) is a state developed assessment system designed to assess the standards detailed in the 1997 Proficiency Guide. The assessments are administered at grades 3, 6, 8, and 10. Voluntary state assessments of high school courses (Core 40 assessments) are available. All assessments are being revised for 2002 based on Indiana's Academic Standards (2000).
Maryland	The Maryland School Performance Assessment Program (MSPAP) assesses students at grades 3, 5, and 8. Currently, the MSPAP is based on the 1990 Learning Outcomes. By 2003, the MSPAP will be revised to assess the 2000 standards. The High School Assessment, in development, is proposed as an end-of-course test which will be part of the graduation requirement.
Massachusetts	Massachusetts Comprehensive Assessment System (MCAS) was first administered in 1998 to grades 4, 8, and 10. Grade 6 will be included from 2001. The Mathematics MCAS was developed to assess the 1996 Curriculum Frameworks which are currently in revision. The Mathematics revision was released in November 2000.
Michigan	The Michigan Educational Assessment Program (MEAP) currently administers assessments at grades 4, 7, and 11. The tests at grades 4 and 7 are based upon the Michigan Essential Goals and Objectives for Mathematics Education (1985). These tests are being revised to assess the 1995 Michigan Curriculum Frameworks and will be administered at grades 4 and 8 starting in 2001/2002. The Grade 11 test was first administered in 1996 and revised in 1998 based on the 1995 High School Proficiency Test Framework and will be revised for the 2002 administration to assess the 1995 curriculum framework.
Missouri	The Missouri Assessment Program (MAP) has been developed for mathematics in grades 4, 8, and 10. Each test includes multiple-choice, short constructed-response, and performance-event items. The test consist of three sessions. The first two sessions include items designed to assess the Show-Me Standards (1996) which are directly related to the curriculum frameworks. Items that match the Show-Me Standards from the norm-referenced Terra Nova are administered in the third session.
North Carolina	The North Carolina Testing Program includes the end-of-grade exams, first administered in 1994, at grades 3-8, and the end-of-course exams (Algebra, Geometry, Algebra II) in high school. These tests are currently based on the 1989 Standard Course of Study. The new tests will be revised to assess the 1998 curriculum by 2000-01. The North Carolina High School Competency Test is administered at grade 10 to measure student growth from grade 8 to grade 10. Students who do not score at the proficient level on the grade 8 end-of-grade exam are required to pass the North Carolina Competency Test in order to graduate from high school. The North Carolina Competency Test will be replaced by an 11th grade exit exam, developed to assess the high school standards through the eleventh grade.
Oregon	The Oregon Statewide Assessment System includes a knowledge and skills state test at grades 3, 5, 8, and 10; a performance state test at grades 4, 5, 6, 8, and 10; and local Classroom Work Samples at grades 3, 5, 8, and 10. All assessments are based on the content standards. As of 1999-2000, the mathematics knowledge and skills tests are achievement level tests: Levels A, B, and C. Students are administered one of the three versions of the test based on their ability level.
Pennsylvania	The Pennsylvania System of School Assessment (PSSA) is administered at grades 5, 8, and 11 and were revised for the 1999 administration to assess the 1999 standards.
South Carolina	The Palmetto Achievement Challenge Test (PACT) is administered at grades 3-8 and is based on the standards. PACT replaces the Basic Skills Assessment Program (BSAP) administered from 1981-1999 at grades 3, 6, and 8. Currently, the basic skills exit exam is given at 10th grade. As of 2002-03, the PACT High School exit exam, based on the 10th grade standards, will be required for graduation.
Texas	The Texas Assessment of Academic Skills (TAAS) were revised to more specifically assess the current standards for the 2000 administration. TAAS is administered in grades 3-8 and the TAAS end-of-course tests are administered in high school. The 10th grade standards-based exit-level exam is based on the 8th, 9th, and 10th grade standards.

SOURCE: IEA Third International Mathematics and Science Study (TIMSS), 1998-1999.

Background data provided by coordinators from participating jurisdictions.

	Assessment	Graduation Requirement	Other Consequences
Connecticut	Connecticut Mastery Test (CMT); Connecticut Academic Performance Test (CAPT)	No	STUDENT: Students meeting the state performance goal on the 10th grade CAPT assessment receive a certificate of mastery. This certificate is affixed to students' official transcripts. Students who do not meet the state goal may retake the test in grades 11 and 12. Results are reported publicly (e.g., newspapers) but there are no direct consequences. DISTRICT/SCHOOL: Based on test results, districts are encouraged to reevaluate their social promotion policy and curriculum. The State Board of Education developed a list of schools in need of improvement based on student performance and performance trends on the CMT. Targeted assistance for these schools is being discussed. Currently, districts with low-performance on the CMT receive additional funding to support remediation. Monetary awards are given to districts that increase the percent of students meeting the state goals on the CMT.
Idaho	Direct Mathematics Assessment	No	STUDENT: No consequences for students. DISTRICT/SCHOOL: Schools are expected to address student performance issues in their accreditation school improvement plans.
Illinois	Illinois Standards Achievement Tests (ISAT); Prairie State Achievement Examination (PSAE)	No	STUDENT: Test results may be used, in conjunction with other data, to make decisions about student's promotion/retention, summer school requirements, and remediation. DISTRICT/SCHOOL: Test results are considered at both the district and school levels as part of the state accountability system. Schools receive a measure of improvement based on the percentage of students in each performance level on the ISAT.
Indiana	Indiana Statewide Testing for Educational Progress-Plus (ISTEP+)	Students must pass the grade 10 test that is based on the 9th grade standards to graduate. As of 2000, students who fail parts of the exam but meet other criteria may still be allowed to graduate.	STUDENT: No additional consequences for the student. DISTRICT/SCHOOL: The state gives monetary rewards to schools that evidence improvement. Districts are required to provide remediation to low-performing students.
Maryland	Maryland School Performance Assessment Program (MSPAP); High School Assessment (HSA)	The HSA is being phased in and will be required for graduation with the class of 2007. Currently, the Maryland Functional Tests are required for graduation.	STUDENT: There are no student-level consequences based on the MSPAP since each student is given only a portion of the assessment. DISTRICT/SCHOOL: The MSPAP is a school accountability assessment. Part of schools' performance rating is based on MSPAP assessment scores. Schools that improve significantly over a two-year period receive monetary rewards. Schools are required to develop school improvement plans for areas in which standards were not met. The State Board of Education has the right to reconstitute schools based on low MSPAP test scores and lack of improvement. Thus far, three schools in Maryland have been reconstituted.
Massachusetts	Massachusetts Comprehensive Assessment System (MCAS)	Beginning with the class of 2003, students must pass the 10th grade assessments in English Language Arts and Mathematics to graduate.	STUDENT: No additional consequences for the student. DISTRICT/SCHOOL: Results are being used as a high-stakes accountability measure to evaluate performance and improvement for schools and districts. Schools will be rated based on performance and progress. Recognized schools may be eligible for an Exemplary Schools Program. Low performance and inadequate progress may result in the removal of principals and/or state-takeover of districts. Targeted resources and funding will be provided to low-performing schools and districts.
Michigan	Michigan Educational Assessment Program (MEAP)	No	STUDENT: Students meeting the standards on the 11th grade assessments qualify for college scholarship money. In the future, students that meet the standards on the 8th grade assessments will qualify for scholarship money, as well. DISTRICTS/SCHOOL: Low-performing schools receive additional teacher training and resources. Low-performance and inadequate progress may result in state-takeover of school districts.
Missouri	Missouri Assessment Program (MAP)	No	STUDENT: Students scoring at the lowest performance level must retake a shortened version of the exam the following year. Students performing at proficient or above on the 10th grade test receive state funds for college-level courses or Advanced Placement Exams. DISTRICT/SCHOOL: Test results will be a part of district-level accreditation.

SOURCE: IEA Third International Mathematics and Science Study (TIMSS), 1998-1999.

Background data provided by coordinators from participating jurisdictions.

	Assessment	Graduation Requirement	Other Consequences
North Carolina	North Carolina Testing Program	Beginning with the class of 2003, students will have to pass a new 11th grade exit exam which will replace the current 8th and 10th grade competency tests. Currently, all students must pass the high school competency test to graduate.	STUDENT: North Carolina requires districts to consider student performance on the state assessments when making promotion decisions. Students are given several chances to perform to expectations on these exams. North Carolina is implementing a new promotion policy based on performance on the assessments. Beginning in 2000-01, 5th graders must perform at Level III for promotion. In 2001-02, 3rd and 8th graders must perform at Level III for promotion. Beginning in 2000-01, the Algebra I End-of-Course test will comprise 25 percent of each student's final grade for Algebra I. DISTRICT/SCHOOL: Schools are rated based on student performance and improvement. Monetary awards are given to schools that meet or exceed their goals. The state funds intervention at schools that have been low-performing. In addition, state-appointed assistant teams support low-performing schools in meeting the standards. North Carolina schools' accountability status is based on assessment results. Beginning in 2001, districts will not promote students not performing at grade level and intervention for these students will be required.
Oregon	Oregon State-wide Assessment System	No	STUDENT: Students who meet the performance standard on the state-level and local standards-based assessments receive Certificates of Initial Mastery in each area in which the standard is met. Students who do not meet the performance standard have an opportunity to take the test again. Low-performing students receive additional support and individual instruction to help them meet the standards. These students can change schools if instruction at one school is not meeting their needs. Districts may use the results of the tests to determine student promotion. DISTRICT/SCHOOL: Test results are part of the accountability system. Districts must meet set goals for the assessments to avoid possible sanctions.
Pennsylvania	Pennsylvania System of School Assessment (PSSA)	No	STUDENT: As of 2003, students who achieve a score of proficient or above on the 11th grade assessment will receive a seal on their diploma indicating their achievement. DISTRICT/SCHOOL: Beginning in 2001-02, Pennsylvania will require districts to provide extra academic assistance to students who are not meeting the 3rd and 5th grade mathematics standards. The recently passed Empowerment Act makes provisions for the state to take over districts in part due to low mathematics scores.
South Carolina	Palmetto Achievement Challenge Tests (PACT)	Beginning with the class of 2003, students will have to pass a standards-based exam to graduate. Currently, passing a basic skills exam is required.	STUDENT: The promotion policy considers students' performance on the state assessments. DISTRICT/SCHOOL: Schools are rated based on student performance and improvement. Accreditation of schools will take into account student performance on the state assessments. Districts are required to provide remediation to low-performing students.
Texas	Texas Assessment of Academic Skills (TAAS)	Students must pass the 10th-grade standards-based exit-level exam or the end-of-course exams.	STUDENT: Students may retake the high school exit-level exams, if necessary. A new promotion policy based on the assessments is being phased in for 5th and 8th grade students starting with students who enter kindergarten in 1999. DISTRICT/SCHOOL: School rating takes into account results on state assessments. Districts are required to offer remediation to low-performing students.

SOURCE: IEA Third International Mathematics and Science Study (TIMSS), 1998-1999.

	Mathematics Assessments	
	State	**Local**
Academy School Dist. #20, CO	Colorado Student Assessment Program (CSAP) includes a mathematics assessment at grade 5 starting in Spring 1999, at grade 8 in 2000, and at grade 10 in 2001. As of 2001, districts will be evaluated based on their achievement or progress on the state assessment. Intervention teams will be provided to those districts in need.	In addition to the CSAP, students take Terra Nova at grade 4; ITBS at grades 3, 5, and 7; and ITED at grade 10. District-developed performance assessment units are optional.
Chicago Public Schools, IL	Starting in 2000, the Illinois Standard Achievement Test (ISAT), administered at grades 3, 5, and 8, replaced the Illinois Goal Assessment Program (IGAP) which was administered from 1988-1999 at 3, 6, 8, and 10. The ISAT is reported as a criterion-referenced and norm-referenced assessment. Beginning in 2001, the state will give new high school tests, the Prairie State Achievement Examination (PSAE), based on the 1997 Illinois Learning Standards for Mathematics. ISAT results may be used, in conjunction with other data, to make decisions about student's promotion/retention, summer school requirements, and remediation. Test results are considered at the district and school level as part of the state accountability system. Schools receive a measure of improvement based on the percentage of students in each performance level on the ISAT.	Chicago Academic Standards Exam was developed to assess the district framework and is being piloted in 1999-2000. Chicago also uses ITBS (3-8) and TAP (9-11). Students who have low scores on the ITBS in grades 3, 6, and 8 have to attend summer school prior to promotion. Chicago schools not meeting minimum school-wide levels on local assessments are put into a system of intervention, remediation, and/or probation. Schools with these designations receive additional supervision, support, and guidance. The state uses a similar process with the state assessments. For schools below level in both assessments, the state and district combine efforts.
Delaware Science Coalition, DE	The Delaware Student Testing Program (DSTP) is administered at grades 3, 5, 8, and 10. Accountability legislation has been passed for districts, schools, teachers, and students that is tied to assessment results. Following the 2001 administration, schools and districts will be evaluated based on improvement and sustained achievement on test scores. Policies to set acceptable improvement levels are under development. Following the 2002 administration, a portion of a teacher's annual appraisal will be tied to assessment results.	There are no district-wide assessments based on the standards. Legislation calls for standards-based testing at all grades beginning in 2001. These tests have not been identified/developed yet.
First in the World Consort., IL	Starting in 2000, the Illinois Standard Achievement Test (ISAT), administered at grades 3, 5, and 8, replaced the Illinois Goal Assessment Program (IGAP) which was administered from 1988-1999 at 3, 6, 8, and 10. The ISAT is reported as a criterion-referenced and norm-referenced assessment. Beginning in 2001, the state will give new high school tests, the Prairie State Achievement Examination (PSAE), based on the 1997 Illinois Learning Standards for Mathematics. ISAT results may be used, in conjunction with other data, to make decisions about student's promotion/retention, summer school requirements, and remediation. Test results are considered at the district and school level as part of the state accountability system. Schools receive a measure of improvement based on the percentage of students in each performance level on the ISAT.	The consortium also participated in TIMSS in 1996 and is developing assessments for districts' use.
Fremont/Lincoln/ WestSide PS, NE	There are no assessments at the state level. Assessing students is local responsibility.	Fremont administers the ITBS (grades 3-9, 11), Lincoln administers the MAT (grades 2-9, 11), and Westside administers SAT-9 (grades 3, 5, 7), Explore (grade 8), and PLAN (grade 10).
Guilford County, NC	The North Carolina Testing Program includes the end-of-grade tests (1994) administered at grades 3-8 and the end-of-course exams given in high school. These tests are currently in line with the 1989 Course of Study. The new test will be revised to assess the 1998 curriculum by 2000-01. The 8th and 10th grade competency test will be replaced by an 11th grade standards-based exit exam. State end-of-course exams are used to rate individual schools. State assistance teams may be sent to low-performing schools.	There are no additional district-wide assessments.
Jersey City Public Schools, NJ	Starting in May 1999, the New Jersey Elementary School Proficiency Assessment (ESPA) was administered at grade 4. The ESPA contains a mathematics component. Similarly, beginning in March 1999, the NJ Grade Eight Proficiency Assessment (GEPA) was administered at grade 8. This test replaced the Early Warning Test which had been previously administered to the eighth graders. Both the ESPA and the GEPA are tests of excellence and measure student performance in relation to the NJ Core Content Curriculum Standards in Mathematics. The High School Proficiency Assessment (HSPA) is presently in development at the state level and will be used beginning in the spring 2001 for first time juniors (class of 2002) as the mandated test for graduation. Presently, the High School Proficiency Test (HSPT), which contains a mathematics component, has been administered statewide since the early 1990s as the mandated test for graduation.	In addition to the state assessments, at the elementary level, the district has developed district-wide mid-terms in mathematics in grades 3-8. These exams have been administered since 1999. The district exams are designed to measure student progress and are aligned to the district curriculum and to the ESPA and GEPA in format and content. At the high school level, mid-terms and final exams are given in the areas of Algebra I and II, Geometry, Pre-Calculus, Calculus I, and AP Calculus (general level and honors courses).

SOURCE: IEA Third International Mathematics and Science Study (TIMSS), 1998-1999.

Background data provided by coordinators from participating jurisdictions.

	Mathematics Assessments	
	State	**Local**
Miami-Dade County PS, FL	Florida's Comprehensive Assessment Test (FCAT) is administered at grades 5, 8, and 10. The FCAT is a multiple-choice and performance-based assessment that includes both criterion-referenced and norm-referenced components. An extension of the FCAT will be in place in 2000 to assess grades 3, 4, 6, 7, and 9 using multiple-choice items. The 11th grade minimum skills graduation test that is not aligned with the standards is being phased out. Schools are graded on student performance on the FCAT in mathematics, reading and writing. Several levels of support are provided to the schools that are not performing well on the state assessment. Instructional supervisors, educational specialists, and other professionals assist with efforts to employ intervention strategies to support curriculum implementation of the Florida Sunshine State Standards.	The SAT-9 NRT Mathematics is administered to students in grade 2. The EXPLORE, which has mathematics and science assessments, is administered to all grade 8 students. District-developed standards-based assessments are used to monitor student progress in mathematics at grades 5, 8, and 10.
Michigan Invitational Group, MI	The Michigan Educational Assessment Program (MEAP) assesses students at grades 4, 7, and 11. The grade 4 and 7 tests are based upon the Michigan Essential Goals and Objectives for Mathematics Education (1998). These tests are in revision and will be administered at Grades 4 and 8 starting in 2001/02. The Grade 11 test, first administered in 1996 and revised in 1998, is based on the 1995 High School Proficiency Test Framework.	Districts administer norm-referenced tests.
Montgomery County, MD	The Maryland School Performance Assessment Program (MSPAP) assesses students at grades 3, 5, and 8. Currently, the MSPAP is based on the 1990 Learning Outcomes. By 2003, the MSPAP will be revised to assess the 2000 standards.	A criterion-referenced assessment has been developed to assess the curriculum for grades 3-8.
Naperville Sch. Dist. #203, IL	Starting in 2000, the Illinois Standard Achievement Test (ISAT), administered at grades 3, 5, and 8 replaced the Illinois Goal Assessment Program (IGAP) which was administered from 1988-1999 at 3, 6, 8, and 10. The ISAT is reported as a criterion-referenced and norm-referenced assessment. Beginning in 2001, the state will give new high school tests, the Prairie State Achievement Examination (PSAE), based on the 1997 Illinois Learning Standards for Mathematics. ISAT results may be used, in conjunction with other data, to make decisions about student's promotion/retention, summer school requirements, and remediation. Test results are considered at the district and school level as part of the state accountability system. Schools receive a measure of improvement based on the percentage of students in each performance level on the ISAT.	Criterion-referenced and performance-based assessments developed to assess the curriculum are administered at all grades.
Project SMART Consortium, OH	There are state assessments developed to assess the standards at grades 4, 6, 9, and 12. Students must pass the 9th grade mathematics test to graduate from high school. These tests are based on 8th grade standards. The class of 2005 will be the first required to pass the new 10th grade tests based on the 10th-grade standards to graduate. Mathematics performance on state assessments is tied to the local district report card accountability system. If seventy-five percent of students do not perform at the state pass rate, the district must put an intervention system in place.	District assessments are given at grades 1-3, 5, and 7 to assess student progress. Tests are both commercial achievement tests and district-developed assessments.
Rochester City Sch. Dist., NY	Beginning in 1999, New York assessed student performance using state-developed tests based on the standards. New York is phasing out the high school competency exams administered to students in grades 9-12. All the students in the class of 2003 will be required to take the New York State Regents Examinations. Students in grade 4 take a NYS Elementary Mathematics Assessment. Students in grade 8 take a NYS Intermediate Mathematics Assessment. New York State has developed a school accountability system that will be phased in by 2003. School districts must provide academic intervention services to students scoring below the state designated performance level on state assessments or to students at risk of not achieving the state learning standards.	The Stanford 9 is administered to students (grades 1-7) not assessed by state programs.
SW Math/Sci. Collaborative, PA	The Pennsylvania System of School Assessment (PSSA) is administered at grades 5, 8, and 11. Beginning in 2001-02, districts will be required to provide extra academic assistance to students who are not meeting the 3rd and 5th grade mathematics standards. The recently passed Empowerment Act makes provisions for the state to take over districts in part due to low mathematics scores.	Each district has its own assessment system in addition to the state assessments. Many of these are standardized tests like the IOWA or Stanford. Some districts use New Standards-Reference Exams.

SOURCE: IEA Third International Mathematics and Science Study (TIMSS), 1998-1999.

How Do Education Systems Deal with Individual Differences?

The challenge of maximizing opportunity to learn for students with widely differing abilities and interests is met differently in different education systems. Exhibit 5.14 summarizes questionnaire and interview data on how selected comparison countries, as well as states, districts, and consortia, organized their curricula to deal with this issue.

Some participants indicated using more than one method of dealing with individual differences among students, and in these cases the category describing the main method was reported. In the United States, and in Canada, Chinese Taipei, Hong Kong, and Korea among the comparison countries, the same curriculum was intended for all students, but it was recommended that teachers adapt the level and scope of their teaching to the abilities and interests of their students. In the Czech Republic and England, the mathematics curriculum was taught at different levels to different groups, four in the Czech Republic and nine in England – so many because in England the levels are defined in terms of progressively more complex performance to be demonstrated. Another approach to differentiated provision was followed in Belgium (Flemish), the Netherlands, the Russian Federation, and Singapore, which assign different curricula to students of different levels of ability and interest. Two of the comparison countries, Italy and Japan, reported that their official mathematics curricula did not address the issue of differentiating instruction for eighth-grade students with different abilities or interests.

All of the Benchmarking states and most of the districts and consortia generally resembled the United States in that they provided the same curriculum for all, but expected teachers to adapt the level and scope of their teaching to their students' needs. The First in the World Consortium and Miami-Dade provided the same curriculum to all, but at different levels for different groups, while Naperville provided a different curriculum to students of different abilities.

Schools' reports on how they organize to accommodate students with different abilities or interests are shown in Exhibit R2.1 in the reference section. Compared with the international average, substantial percentages of students in many Benchmarking jurisdictions were in schools reporting that different classes study different content, including the states, districts and consortia reporting that their frameworks or standards were developed for all students with teachers adapting to students' needs.

	Curriculum Addresses Differentiation	Approaches to Addressing Students with Different Abilities or Interests at Grade 8			
		Same Curriculum for All Students, and Teachers Adapt to Students' Needs	Same Curriculum with Different Levels for Different Groups	Different Curricula for Different Groups	Number of Curriculum Levels
Countries					
United States [1]	Yes	Yes	No	No	1
Belgium (Flemish)	Yes	No	No	Yes	2
Canada	Yes	Yes	No	No	1
Chinese Taipei	Yes	Yes	No	No	1
Czech Republic	Yes	No	Yes	No	4
England [2]	Yes	No	Yes	No	9
Hong Kong, SAR	Yes	Yes	No	No	1
Italy	No				
Japan	No				
Korea, Rep. Of	Yes	Yes	No	No	1
Netherlands	Yes	No	No	Yes	4
Russian Federation	Yes	No	No	Yes	2
Singapore	Yes	No	No	Yes	3
States					
Connecticut	Yes	Yes	No	No	1
Idaho	Yes	Yes	No	No	1
Illinois	Yes	Yes	No	No	1
Indiana	Yes	Yes	No	No	1
Maryland	Yes	Yes	No	No	1
Massachusetts	Yes	Yes	No	No	1
Michigan	Yes	Yes	No	No	1
Missouri	Yes	Yes	No	No	1
North Carolina	Yes	Yes	No	No	1
Oregon	Yes	Yes	No	No	1
Pennsylvania	Yes	Yes	No	No	1
South Carolina	Yes	Yes	No	No	1
Texas	Yes	Yes	No	No	1
Districts and Consortia					
Academy School Dist. #20, CO	Yes	Yes	No	No	1
Chicago Public Schools, IL	Yes	Yes	No	No	1
Delaware Science Coalition, DE	Yes	Yes	No	No	1
First in the World Consort., IL	Yes	No	Yes	No	3
Fremont/Lincoln/WestSide PS, NE	Yes	Yes	No	No	1
Guilford County, NC	Yes	Yes	No	No	1
Jersey City Public Schools, NJ	Yes	Yes	No	No	1
Miami-Dade County PS, FL	Yes	No	Yes	No	2
Michigan Invitational Group, MI	Yes	Yes	No	No	1
Montgomery County, MD	Yes	Yes	No	No	1
Naperville Sch. Dist. #203, IL	Yes	No	No	Yes	2
Project SMART Consortium, OH	Yes	Yes	No	No	1
Rochester City Sch. Dist., NY	Yes	Yes	No	No	1
SW Math/Sci. Collaborative, PA [3]	–	–	–	–	–

SOURCE: IEA Third International Mathematics and Science Study (TIMSS), 1998-1999.

Background data provided by coordinators from participating jurisdictions.

1 United States: Most state standards are designed for all students.

2 England: While there is one "programme of study" for grades 6-8,the document identifies nine performance-levels describing the types and range of performance that pupils working at a particular level should demonstrate.

3 SW Math/Sci. Collaborative: Covering a workforce region of 118 autonomous districts, the Collaborative cannot provide a representative response for these questions.

A dash (–) indicates data are not available.

What Are the Major Characteristics of the Intended Curriculum?

Exhibit 5.15 indicates the relative emphasis given to various aspects of mathematics instruction in the intended curriculum. As might be anticipated for students at this point in their schooling, major emphasis in the comparison countries was most commonly placed on understanding mathematical concepts and mastering basic skills. Assessing student learning was also given major emphasis in most countries. "Real-life" applications of mathematics were stressed in the curriculum of most countries. In the Netherlands, for example, this approach was reported to be emphasized even more heavily than either understanding mathematics concepts or mastering basic skills. Communicating mathematically, an aspect of teaching and learning that has received increasing attention in recent years, was given major or moderate emphasis in the curriculum of most of the comparison countries. Adopting a multicultural approach, working on mathematics projects, solving non-routine problems, deriving formal proofs, and integrating mathematics with other school subjects all received less emphasis.

In general, curricular emphasis among the Benchmarking participants was very similar to that in the United States as a whole. A majority of the Benchmarking entities placed major emphasis in their curricula on mastering basic skills, understanding mathematics concepts, real-life applications of mathematics, communicating mathematically, and assessing student learning. With only one exception, all the other entities place moderate emphasis in each of these areas.

It is possible that in some entities some of the approaches and processes reported as being given minor or no emphasis in the intended curriculum may receive more emphasis in the implemented curriculum. Conversely, it is also possible that some of the approaches and processes reported as being given major or moderate emphasis in the intended curriculum may receive less emphasis in the implemented curriculum.

Exhibit 5.15 Emphasis on Approaches and Processes

8th Grade Mathematics

	Mastering Basic Skills	Understanding Mathematics Concepts	Real-life Applications of Mathematics	Communicating Mathematically	Solving Non-Routine Problems	Deriving Formal Proofs	Working on Mathematics Projects	Integration of Mathematics with Other School Subjects	Thematic Approach	Multicultural Approach	Assessing Student Learning
Countries											
United States	●	●	◉	◉	●	○	◉	◉	◉	◉	●
Belgium (Flemish)	●	●	◉	●	◉	○	○	◉	○	○	●
Canada [1]	◉	●	●	◉	◉	○	●	●	●	○	●
Chinese Taipei	●	●	●	◉	◉	○	○	◉	◉	○	●
Czech Republic	●	●	◉	●	◉	◉	○	◉	●	○	●
England	●	●	●	●	◉	○	○	○	○	◉	●
Hong Kong, SAR	●	●	●	●	◉	◉	○	○	○	○	●
Italy	○	●	●	◉	◉	○	○	●	●	●	◉
Japan	●	●	◉	◉	●	●	◉	○	○	○	●
Korea, Rep. Of	●	●	●	○	◉	●	◉	◉	◉	○	◉
Netherlands	◉	◉	●	●	◉	○	●	●	●	◉	◉
Russian Federation	●	●	◉	●	◉	●	○	◉	●	○	●
Singapore	●	●	●	●	●	◉	○	○	○	○	●
States											
Connecticut	●	●	●	●	●	◉	◉	●	◉	●	●
Idaho	●	●	●	●	◉	○	◉	◉	◉	○	○
Illinois	●	●	●	●	●	○	●	●	●	●	●
Indiana	●	●	●	●	◉	○	○	◉	○	○	●
Maryland	◉	●	●	●	●	○	◉	●	◉	◉	●
Massachusetts	●	◉	●	◉	◉	○	◉	◉	○	○	◉
Michigan	●	●	●	●	●	◉	◉	◉	○	○	●
Missouri	●	●	●	●	●	◉	◉	◉	◉	◉	●
North Carolina	◉	●	●	●	◉	◉	◉	●	●	◉	●
Oregon	●	●	◉	◉	◉	○	○	○	○	○	◉
Pennsylvania	●	●	●	◉	◉	◉	◉	◉	◉	◉	●
South Carolina	◉	●	●	●	◉	○	◉	●	◉	○	●
Texas	●	●	●	●	●	◉	◉	◉	◉	●	●
Districts and Consortia											
Academy School Dist. #20, CO	–	–	–	–	–	–	–	–	–	–	–
Chicago Public Schools, IL	●	●	●	●	◉	○	◉	◉	◉	◉	●
Delaware Science Coalition, DE	●	◉	◉	◉	◉	○	◉	◉	○	○	●
First in the World Consort., IL	◉	●	◉	◉	●	◉	◉	◉	◉	○	●
Fremont/Lincoln/WestSide PS, NE	●	●	●	●	●	◉	●	◉	○	○	●
Guilford County, NC	●	●	●	●	◉	◉	◉	◉	◉	◉	●
Jersey City Public Schools, NJ	●	●	●	●	◉	○	●	●	●	◉	●
Miami-Dade County PS, FL	◉	●	●	●	●	◉	◉	◉	○	◉	●
Michigan Invitational Group, MI	●	●	●	●	●	○	●	◉	◉	○	●
Montgomery County, MD	●	●	◉	◉	○	◉	◉	◉	◉	○	●
Naperville Sch. Dist. #203, IL	●	●	◉	◉	●	○	◉	◉	○	○	●
Project SMART Consortium, OH	◉	●	◉	◉	◉	○	○	○	○	○	◉
Rochester City Sch. Dist., NY	●	●	●	●	●	◉	◉	◉	◉	◉	●
SW Math/Sci. Collaborative, PA [2]	–	–	–	–	–	–	–	–	–	–	–

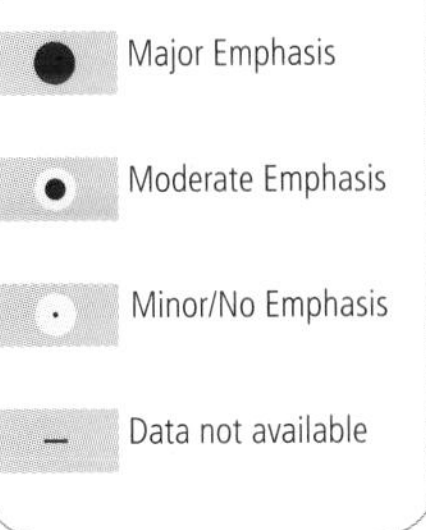

SOURCE: IEA Third International Mathematics and Science Study (TIMSS), 1998-1999.

Background data provided by coordinators from participating jurisdictions.

1 Canada: Results shown are for the majority of provinces.

2 SW Math/Sci. Collaborative: Covering a workforce region of 118 autonomous districts, the Collaborative cannot provide a representative response for these questions.

What Mathematics Content Do Teachers Emphasize at the Eighth Grade?

Teachers of the mathematics classes tested were asked what subject matter they emphasized most in their classes (e.g., geometry, algebra, various combinations of content, etc.). Their responses, presented in Exhibit 5.16, reveal that most eighth-grade students around the world are being taught mathematics with an integration of content areas. Internationally on average, more than half the students were taught a combination of mathematics topics (i.e., combined algebra, geometry, number, etc.), and almost 20 percent were in classes emphasizing algebra and geometry combined.

Just as in TIMSS 1995,[6] the mathematics curriculum in the U.S. at the eighth grade does not appear to be as advanced as in other countries. About one-third of the U.S. eighth-grade students were in mathematics classes where the emphasis was on the combination of algebra, geometry, number, etc., but more than one-quarter were in classes emphasizing mainly number. None of the reference countries except Canada had a comparable proportion of students in classes emphasizing mainly number, and across all the TIMSS 1999 countries a mere 14 percent of students were in such classes.

Even when U.S. eighth graders were being taught algebra, it was usually as a single emphasis. More than one-quarter of the students were in classes emphasizing only algebra, compared with six percent in classes with a combined algebra and geometry emphasis. This is almost a reverse of the international pattern of 20 percent in algebra and geometry combined compared with eight percent in algebra only.

The Benchmarking states generally resembled the United States overall in the percentages of students in classes emphasizing various mathematics subject matter. Relative emphasis on mathematics subject matter varied more across the districts and consortia. Similar to the United States overall, most Benchmarking jurisdictions had much higher percentages of students whose teachers reported emphasizing mainly number at the eighth grade than did those in the top-performing comparison countries. These data suggest that many students in the U.S. continue to be taught number concepts at the eighth grade while their peers in other countries study topics in geometry and algebra, as discussed below. This is supported by previous TIMSS studies that showed that U.S. eighth-grade students who were not in Algebra 1 courses (approximately 75 to 80 percent of students) continued to receive instruction in arithmetic,

6 Peak, L. (1996), *Pursuing Excellence: A Study of U.S. Eighth-Grade Mathematics and Science Teaching, Learning, Curriculum, and Achievement in International Context*, NCES 97-198, Washington, DC: National Center for Education Statistics.

estimation, and "measurement – units" compared with their peers internationally who have completed these topics and received more focused instruction on integers, rational numbers, "exponents, roots and radicals," and on geometry, algebra, and proportionality topics.[7]

In the Benchmarking states, the percentages of students in classes emphasizing mainly number is striking, and ranged from 20 percent in Indiana and Massachusetts to 39 percent in Idaho and Illinois. In Chicago and the Fremont/Lincoln/Westside Public Schools, 47 and 40 percent of students, respectively, had teachers who reported emphasizing mainly number at the eighth grade, while only four percent had teachers who did so in high-performing Naperville. Less than 10 percent of students were in mainly number classes in only six of the Benchmarking jurisdictions: the First in the World Consortium, Guilford County, Jersey City, the Michigan Invitational Group, Naperville, and Rochester.

There was even more variation among districts and consortia in the percentage of students in classes emphasizing algebra, ranging from two to five percent in Chicago, Jersey City, and Rochester to 91 percent in Naperville. Districts and consortia with more than one-third of their students in classes emphasizing algebra were the Academy School District, First in the World, Guilford County, Miami-Dade, the Michigan Invitational Group, Montgomery County, Naperville, and the Southwest Pennsylvania Math and Science Collaborative. Nearly all Benchmarking jurisdictions had no more than three percent of their students in classes emphasizing geometry. Only the Academy School District and the First in the World Consortium had appreciable percentages of students in such classes (14 and 18 percent, respectively).

7 Schmidt, W.H., McKnight, C.C., and Raizen, S.A. (1997), *A Splintered Vision: An Investigation of U.S. Science and Mathematics Education*, Dordrecht, the Netherlands: Kluwer Academic Publishers.

Exhibit 5.16 Subject Matter Emphasized Most in Mathematics Class

TIMSS 1999 Benchmarking
Boston College

8th Grade Mathematics

	Percentage of Students Whose Teachers Report the Subject Matter Emphasized Most in Their Grade 8 Mathematics Class					
	Mainly Number	Combined Algebra, Geometry, Number, etc.	Combined Algebra and Geometry	Algebra	Geometry	Other
Countries						
United States	28 (3.0)	32 (3.4)	6 (1.6)	27 (2.7)	1 (0.8)	6 (1.4)
Belgium (Flemish)	10 (3.3)	65 (3.6)	17 (2.3)	3 (1.2)	2 (1.3)	3 (2.3)
Canada	r 26 (3.0)	r 53 (2.8)	r 6 (1.6)	r 6 (1.4)	r 1 (0.0)	r 9 (1.9)
Chinese Taipei	2 (1.1)	57 (4.2)	24 (3.6)	4 (1.7)	9 (2.6)	4 (1.6)
Czech Republic	0 (0.2)	76 (3.9)	19 (3.9)	4 (1.2)	0 (0.0)	0 (0.0)
England	s 0 (0.0)	s 100 (0.0)	s 0 (0.0)	s 0 (0.0)	s 0 (0.0)	s 0 (0.0)
Hong Kong, SAR	7 (2.4)	60 (4.8)	11 (2.8)	13 (3.3)	4 (1.8)	5 (2.1)
Italy	2 (1.0)	67 (3.8)	22 (3.3)	5 (1.8)	4 (1.4)	1 (0.0)
Japan	7 (2.0)	30 (4.1)	35 (4.0)	16 (3.1)	9 (2.5)	4 (1.6)
Korea, Rep. of	6 (1.9)	51 (4.0)	20 (3.1)	20 (3.4)	2 (1.1)	2 (0.9)
Netherlands	4 (3.2)	77 (4.6)	13 (2.9)	2 (1.1)	1 (0.8)	3 (1.6)
Russian Federation	0 (0.0)	0 (0.0)	100 (0.0)	0 (0.0)	0 (0.0)	0 (0.0)
Singapore	8 (2.3)	46 (4.5)	12 (2.9)	29 (3.7)	0 (0.0)	5 (1.7)
States						
Connecticut	r 22 (4.1)	r 29 (4.7)	r 3 (2.1)	r 35 (6.9)	r 3 (2.6)	r 9 (4.0)
Idaho	r 39 (7.0)	r 23 (5.9)	r 4 (2.4)	r 30 (5.5)	r 0 (0.0)	r 4 (2.9)
Illinois	39 (5.2)	27 (4.5)	7 (2.3)	23 (4.7)	1 (0.6)	2 (1.5)
Indiana	20 (4.5)	31 (5.9)	6 (3.0)	40 (7.4)	0 (0.0)	3 (2.3)
Maryland	r 26 (5.5)	r 28 (5.7)	r 7 (2.7)	r 37 (5.1)	r 1 (0.7)	r 2 (1.1)
Massachusetts	20 (3.3)	22 (4.9)	2 (1.7)	44 (5.2)	0 (0.2)	12 (3.2)
Michigan	23 (3.9)	23 (4.9)	6 (1.4)	43 (4.2)	1 (1.1)	4 (2.2)
Missouri	31 (5.6)	29 (5.8)	3 (1.9)	27 (4.7)	2 (1.5)	8 (2.4)
North Carolina	26 (4.6)	43 (6.3)	4 (2.2)	24 (3.3)	1 (0.1)	3 (1.6)
Oregon	30 (4.9)	30 (5.9)	2 (1.4)	29 (5.0)	3 (1.5)	6 (2.0)
Pennsylvania	23 (5.7)	27 (6.5)	6 (2.1)	39 (5.0)	1 (0.5)	5 (1.7)
South Carolina	28 (5.6)	19 (4.7)	8 (3.2)	38 (5.4)	2 (1.5)	6 (2.7)
Texas	r 33 (5.8)	r 26 (6.0)	r 0 (0.0)	r 35 (6.1)	r 1 (0.1)	r 5 (2.4)
Districts and Consortia						
Academy School Dist. #20, CO	18 (0.3)	17 (0.3)	0 (0.0)	49 (0.4)	14 (0.3)	3 (0.1)
Chicago Public Schools, IL	47 (10.6)	38 (8.6)	13 (6.4)	2 (1.7)	0 (0.0)	0 (0.0)
Delaware Science Coalition, DE	r 22 (6.3)	r 38 (6.4)	r 0 (0.0)	r 25 (5.9)	r 0 (0.0)	r 15 (5.4)
First in the World Consort., IL	9 (3.9)	32 (4.3)	5 (3.5)	35 (8.5)	18 (8.0)	0 (0.0)
Fremont/Lincoln/WestSide PS, NE	40 (9.1)	15 (8.4)	5 (5.1)	22 (6.7)	0 (0.0)	18 (3.3)
Guilford County, NC	9 (4.2)	36 (7.7)	4 (1.0)	44 (7.9)	0 (0.0)	6 (3.5)
Jersey City Public Schools, NJ	r 8 (3.4)	r 73 (6.1)	r 8 (1.8)	r 5 (3.4)	r 0 (0.0)	r 6 (2.8)
Miami-Dade County PS, FL	s 18 (5.9)	s 33 (8.6)	s 4 (3.4)	s 40 (9.6)	s 0 (0.0)	s 6 (4.4)
Michigan Invitational Group, MI	9 (2.8)	35 (8.5)	4 (0.2)	50 (8.5)	0 (0.0)	2 (0.1)
Montgomery County, MD	s 30 (4.7)	s 15 (3.3)	s 3 (1.7)	s 48 (5.9)	s 1 (0.4)	s 4 (2.9)
Naperville Sch. Dist. #203, IL	4 (2.0)	1 (0.0)	5 (0.4)	91 (2.1)	0 (0.0)	0 (0.0)
Project SMART Consortium, OH	34 (7.5)	24 (4.8)	1 (1.2)	31 (8.0)	2 (2.2)	7 (4.0)
Rochester City Sch. Dist., NY	7 (2.1)	70 (4.5)	18 (3.4)	5 (2.0)	0 (0.0)	0 (0.0)
SW Math/Sci. Collaborative, PA	20 (5.8)	24 (5.7)	11 (4.5)	36 (5.7)	3 (2.1)	6 (2.2)
International Avg. (All Countries)	14 (0.4)	55 (0.6)	19 (0.5)	8 (0.4)	3 (0.2)	2 (0.2)

SOURCE: IEA Third International Mathematics and Science Study (TIMSS), 1998-1999.

Background data provided by teachers.

States in *italics* did not fully satisfy guidelines for sample participation rates (see Appendix A for details).

( ) Standard errors appear in parentheses. Because results are rounded to the nearest whole number, some totals may appear inconsistent.

An "r" indicates teacher response data available for 70-84% of students. An "s" indicates teacher response data available for 50-69% of students.

Are There Policies on Using Calculators?

Official policies on calculator use are summarized in Exhibit 5.17. In general, the curricula in the comparison countries included policies on using calculators, either without restriction (three countries) or with some restrictions (seven countries). Several countries commented that calculators were not permitted in the lower grades or that their use in these grades was limited. Across the United States as a whole, policy varied from state to state, and this was reflected among the Benchmarking states, with four states, Idaho, Indiana, Massachusetts, and North Carolina, reporting calculator use under restricted circumstances and the other nine reporting unrestricted use.

Exhibit 5.17 Policy on Calculator Usage*

	Curriculum Contains Recommendations About Use of Calculators	Type of Policy	Comments
Countries			
United States	Yes	Varies from state to state	
Belgium (Flemish)	Yes	Restricted Use	Calculators are permitted on a limited basis so that students can master the basic skills of computation and mental calculation. Calculator usage increases and is compulsory after grade 9.
Canada	Yes	Unrestricted, 2 provinces, Restricted, 8 provinces	In general, calculator use is encouraged, except in lower grades in some provinces.
Chinese Taipei	Yes	Restricted Use	Calculators are not allowed on entrance exams so teachers limit their use in the classroom.
Czech Republic	Yes	Restricted Use	Computational skills are practiced without calculators.
England	Yes	Restricted Use	Calculator use increases as students progress through school. The emphasis is on pupils having a range of skills: calculator, pencil and paper, and mental computation. Graphic calculators are required at higher levels.
Hong Kong, SAR	Yes	Unrestricted Use	Calculators may be used for exploration only from grades 1 to 6. No restrictions are set on the use of calculators for students from grade 7 onwards.
Italy	No		
Japan	Yes	Unrestricted Use	Calculators are not permitted until grade 5.
Korea, Rep. of	Yes	Restricted Use	Currently, calculators are not used in class. However, the new curriculum, to be implemented in 2000/1, recommends the wide use of calculators.
Netherlands	Yes	Unrestricted Use	Calculators are compulsory at national exam level. In grades 11-12 the graphic calculator is compulsory for mathematics students.
Russian Federation	Yes	Restricted Use	There is some use of calculators in elementary school. Recommended use of calculators on a level with oral and written calculations in secondary school. Students are not allowed to use calculators on public exams in grades 9 and 11.
Singapore	Yes	Restricted Use	In primary school, students are not allowed to use calculators in mathematics. In secondary school, the use of calculators is allowed from grade 7, though the use is restricted.
States			
Connecticut	Yes	Unrestricted Use	Calculator use is not permitted on the grade 4 test. It is permitted on two of the three testing sessions for the grade 6 and 8 tests and on all parts of the grade 10 test. It is recommended that students use the type of calculator with which they are most familiar.
Idaho	Yes	Restricted Use	Calculators should be used when appropriate with greater use after grade 4.
Illinois	Yes	Unrestricted Use	Calculators are expected to be used as a tool while supporting computation and estimation skills. Calculators are allowed on the grade 8 assessment.
Indiana	Yes	Restricted Use	In the early grades it is used as a means to explore number patterns and to solve problems. In the later grades, calculators are to be used as a tool for exploring higher order concepts.
Maryland	Yes	Unrestricted Use	Calculators are used as a tools in mathematics. Local systems and teachers decide when they are appropriate to use.
Massachusetts	Yes	Restricted Use	Elementary students should learn basic arithmetic operations independent of calculator use; middle and secondary students may use graphing calculators to enhance, rather than replace, their understanding and skills. Calculators are allowed on specified portions of grades 8 and 10 assessments.
Michigan	Yes	Unrestricted Use	Calculators are used as a tool in mathematics. Local systems and teachers decide when they are appropriate to use. Unrestricted use of calculators is allowed on the state assessment.
Missouri	Yes	Unrestricted Use	Calculators are not allowed on grade 4 assessment but are allowed at later grades.
North Carolina	Yes	Restricted Use	The curriculum does not contain an explicit policy on classroom use of calculators. In the classroom, calculator use changes as the mathematical processes become more advanced. Early learners use a 4-function calculator and later progress to a scientific calculator. Older students use graphing calculators. The emphasis is on the use of the appropriate calculator for each grade level. Policy does dictate calculator usage on statewide assessments. For the end-of-grade tests, 4-function calculators are not permitted on the computation part of the test, but are allowed on the application part. Graphing calculators are used in Algebra I; the most advanced calculator allowed in Algebra II and Geometry is a symbolic manipulation calculator.
Oregon	Yes	Unrestricted Use	Standards call for students to be proficient both with and without calculators. The state requires the selection and use of appropriate methods and tools for computing with numbers including mental calculations, paper and pencil, calculators, and computers. Restriction is left to the discretion of the district.
Pennsylvania	Yes	Unrestricted Use	The standards document includes the use of technology in the classroom which specifically includes calculator usage. Restrictions are at the discretion of the districts and schools. The state test does include calculator usage. Some questions on the tests must be answered without the use of calculators in order to assess students' computational skills.
South Carolina	Yes	Unrestricted Use	Calculator usage is advocated by the standards at all levels. However, the testing program does not include calculators.
Texas	Yes	Unrestricted Use	The standards documents indicate the use of calculators in 7th and 8th grade mathematics. The standards also require computation without the use of calculators. Calculators are not permitted on state assessments with the exception of the high school Algebra end-of-course test.

SOURCE: IEA Third International Mathematics and Science Study (TIMSS), 1998-1999.

Background data provided by coordinators from participating jurisdictions.

* The use of calculators on TIMSS was not allowed in 1995 or in 1999.

1 Michigan Invitational Group: The consortium cannot provide a representative response for these questions.

2 SW Math/Sci. Collaborative: Covering a workforce region of 118 autonomous districts, the Collaborative cannot provide a representative response for these questions.

A dash (–) indicates data are not available.

	Curriculum Contains Recommendations About Use of Calculators	Type of Policy	Comments
Districts and Consortia			
Academy School Dist. #20, CO	No	–	In practice, calculator usage increases in middle school and high school.
Chicago Public Schools, IL	Yes	Restricted Use	In early grades, calculators are used to explore different aspects of number sense. As students progress through school, the calculator is used to perform complicated computations.
Delaware Science Coalition, DE	Yes	Unrestricted Use	The standards require the appropriate selection of methods of calculation including mental math, paper and pencil, calculators, and computers. The use of grade-level appropriate calculators is also recommended. In K-5, a basic 4-function calculator or one using an algebraic operating system is used. In middle and high school, a scientific or graphing calculator is used.
First in the World Consort., IL	Yes	Unrestricted/Restricted Use	Restriction varies across districts. Most districts that prescribe to the Everyday Mathematics program use calculators at primary grades to develop number sense with patterns and estimation. Calculator usage for computational purposes is not allowed until the middle grades. Graphing calculators are generally introduced in the accelerated grades 6-8 pre-algebra/algebra courses.
Fremont/Lincoln/WestSide PS, NE	Yes	Restricted Use	Calculators are used as problem-solving instruments but not used for regular computational instruction and practice.
Guilford County, NC	Yes	Restricted Use	Calculators are used on 70% of the end-of-grade tests in grades 3-8.
Jersey City Public Schools, NJ	Yes	Restricted Use	At the elementary level, the district encourages all students from grades 3-8 to utilize the calculator as a resource tool in the classroom as well as to use the calculator on certain parts of the Fourth Grade Elementary School Proficiency Assessment (ESPA) and on all of the Grade Eight Proficiency Assessment (GEPA). At the high school level, the district encourages all students from grades 9-12 to utilize the calculator as a resource tool in the classroom as well as to use the calculator on the HSPT 11.
Miami-Dade County PS, FL	Yes	Unrestricted Use	Basic 4-function calculators are mainly used at the elementary level. Scientific and graphing calculators are used more frequently at the senior high school level.
Michigan Invitational Group, MI [1]	–	–	–
Montgomery County, MD	Yes	Unrestricted Use	–
Naperville Sch. Dist. #203, IL	Yes	Restricted Use	Calculators are used across all grade levels. Restrictions vary depending on the instructional purpose and the critical mathematics objective. The sophistication of the calculator increases with the grade level to the use of graphing calculators for all 8th grade students.
Project SMART Consortium, OH	Yes	Unrestricted Use	Calculators of various types are used in classrooms. Students will use scientific calculators during the grade 10 state assessment.
Rochester City Sch. Dist., NY	Yes	Restricted Use	Calculators are mandated for NYS Regents' examinations (grades 9-12), and the NYS Intermediate Mathematics Examination (grade 8). Calculators are at the discretion of the building for standardized and district-developed assessments.
SW Math/Sci. Collaborative, PA [2]	–	–	–

SOURCE: IEA Third International Mathematics and Science Study (TIMSS), 1998-1999.

What Mathematics Topics Are Included in the Intended Curriculum?

In the course of their meetings on planning and implementation of TIMSS 1999, the National Research Coordinators developed a list of mathematics topics that they agreed covered most of the content in the intended mathematics curriculum in their respective countries. These topics, presented in Exhibit 5.18, built on the topics covered in the TIMSS 1995 mathematics test and included in the teacher questionnaire. They represent all topics likely to have been included in the curricula of the 38 participating countries up to and including eighth grade. From the following choices, the coordinators from the participating entities indicated the percentages of students in their own countries or jurisdictions expected to have been taught each topic up to and including eighth grade:

- All or almost all students (at least 90 percent)
- About half of the students
- Only the more able students (top track – about 25 percent)
- Only the most advanced students (10 percent or less).

Exhibit 5.19 summarizes the data according to the percentage of topics intended to be taught to all or almost all students (at least 90 percent) in each entity, across the entire list of topics and for each content area. Information on specific topics in the intended curricula for each content area is presented in Exhibits R2.2 through R2.6 in the reference section of this report.

Internationally on average, curricular guidelines up to and including eighth grade called for nearly all students to have been taught three-fourths of the topics overall. The greatest percentage of topics intended to be taught to 90 percent or more of the students was in fractions and number sense (86 percent, on average across countries) and in measurement (83 percent). About two-thirds of the topics in geometry (67 percent) and algebra (68 percent), internationally on average, were expected to have been taught to nearly all students. Four of the comparison countries, Italy, Japan, Korea, and Singapore, reported that at least 10 of the 11 algebra topics (91 percent ore more) were intended to be taught to at least 90 percent of the students.

In the United States overall, 93 percent of the mathematics topics – compared with the international average of 75 percent – were intended to be taught to 90 percent or more of the students. This relatively high level of coverage resulted from the inclusion of 100 percent of the topics in fractions and number sense, measurement, and data representation, analysis, and probability, and more than 80 percent of the topics in geometry and algebra. These results are supported by research based on TIMSS data from 1995 that shows that the U.S. is one of a number of countries whose mathematics curricula cover many topics each year and are comparatively more diverse than the curricula of many countries whose curricula are more focused.[8]

Benchmarking participants generally resembled the United States in topic coverage in the intended curriculum, although there were differences, particularly among the districts and consortia. With Connecticut the sole exception, all Benchmarking jurisdictions reported that at least 88 percent of the fractions and number sense topics were included in the curriculum for almost all students. Data representation, analysis, and probability was included in the curriculum for almost all students in almost all Benchmarking jurisdictions, but the coverage of geometry and algebra was much more variable. Among states the percentage of geometry topics intended for almost all students ranged from 54 percent in Idaho to 100 percent in Pennsylvania, and among districts and consortia from 46 percent in Chicago to 85 percent in First in the World, Jersey City, Miami-Dade, Montgomery County, and Naperville. Among states the percentage of algebra topics included ranged from 55 percent in Massachusetts and Missouri to 100 percent in Illinois and Pennsylvania, and among districts and consortia from just nine percent in Chicago to 91 percent in the Delaware Science Coalition, First in the World, and Miami-Dade.

It should be noted that some countries reported having different curricula or different levels of curriculum for different groups of students, as detailed in Exhibit 5.14. Not surprisingly, then, these countries often reported that about half, only the more able (25 percent), or the top 10 percent of students were expected to have been taught substantial percentages of the topics, in particular those in geometry and algebra. The two comparison countries with the lowest percentages of topics overall intended to be taught to nearly all students have differentiated curricula (England and the Netherlands). It should also be noted that if content within a topic area required different responses, coordinators from participating entities chose the response that best represented the entire topic area and noted the discrepancy (see Exhibit A.8 in the appendix for details).

8 Schmidt, W.H., McKnight, C.C., Valverde, G.A., Houang, R.T., and Wiley, D.E. (1997), *Many Visions, Many Aims Volume 1: A Cross-National Investigation of Curricular Intentions in School Mathematics*, Dordrecht, the Netherlands: Kluwer Academic Publishers.

Mathematics Topics Included in the TIMSS Questionnaires

8th Grade Mathematics

Fractions and Number Sense

- ■ Whole numbers - including place values, factorization and operations (+, -, x, ÷)
- ■ Understanding and representing common fractions
- ■ Computations with common fractions
- ■ Understanding and representing decimal fractions
- ■ Computations with decimal fractions
- ■ Relationships between common and decimal fractions, ordering of fractions
- ■ Rounding whole numbers and decimal fractions
- ■ Estimating the results of computations
- ■ Number lines
- ◆ Whole number powers of integers
- ■ Computations with percentages and problems involving percentages
- ■ Simple computations with negative numbers
- ■ Square roots (of perfect squares less than 144), small integer exponents
- ◆ Prime factors, highest common factor, lowest common multiple, rules for divisibility
- ◆ Sets, subsets, union, intersection, Venn diagrams
- ◆ Rate problems
- ■ Concepts of ratio and proportion; ratio and proportion problems

Measurement

- ■ Units of measurement; standard metric units
- ■ Reading measurement instruments
- ■ Estimates of measurement; accuracy of measurement
- ◆ Conversions of units between measurement systems
- ■ Perimeter and area of simple shapes – triangles, rectangles and circles
- ■ Perimeter and area of combined shapes
- ■ Volume of rectangular solids – i.e., Volume = length x width x height
- ◆ Volume of other solids (e.g., pyramids, cylinders, cones, spheres)
- ◆ Computing with measurements (+, -, x, ÷)
- ■ Scales applied to maps and models

Data Representation, Analysis, and Probability

- ◆ Collecting and graphing data from a survey
- ■ Representation and interpretation of data in graphs, charts, and tables
- ■ Arithmetic mean
- ◆ Median and mode
- ■ Simple probabilities – understanding and calculations

■ Topics included in the curriculum and teacher questionnaires (intended and implemented curriculum).

◆ Topics also included in the curriculum questionnaire (intended curriculum).

SOURCE: IEA Third International Mathematics and Science Study (TIMSS), 1998-1999.

Geometry

- ■ Cartesian coordinates of points in a plane
- ■ Coordinates of points on a given straight line
- ■ Simple two dimensional geometry – angles on a straight line, parallel lines, triangles and quadrilaterals
- ■ Congruence and similarity
- ◆ Angles – (acute, right, supplementary, etc.)
- ◆ Pythagorean theorem (without proof)
- ■ Symmetry and transformations (reflection and rotation)
- ■ Visualization of three-dimensional shapes
- ◆ Geometric constructions with straight-edge and compass
- ◆ Regular polygons and their properties – names (e.g., hexagon and octagon), sum of angles, etc.
- ◆ Proofs (formal deductive demonstrations of geometric relationships)
- ◆ Sine, cosine, and tangent in right-angle triangles
- ◆ Nets of solids

Algebra

- ■ Number patterns and simple relations
- ◆ Writing expressions for general terms in number pattern sequence
- ◆ Translating from verbal descriptions to symbolic expressions
- ■ Simple algebraic expressions
- ◆ Evaluating simple algebraic expressions by substitution of given value of variables
- ■ Representing situations algebraically; formulas
- ■ Solving simple equations
- ■ Solving simple inequalities
- ◆ Solving simultaneous equations in two variables
- ◆ Interpreting linear relations
- ◆ Using the graph of a relationship to interpolate/extrapolate

SOURCE: IEA Third International Mathematics and Science Study (TIMSS), 1998-1999.

■ Topics included in the curriculum and teacher questionnaires (intended and implemented curriculum).

◆ Topics also included in the curriculum questionnaire (intended curriculum).

Exhibit 5.19 Mathematics Topics in the Intended Curriculum for At Least 90% of Students, Up to and Including Eighth Grade

TIMSS 1999 Benchmarking
Boston College

8th Grade Mathematics

	Percentage of Topics Intended to Be Taught to All or Almost All (at least 90%) Students					
	Overall	Fractions and Number Sense	Measurement	Data Representation, Analysis, and Probability	Geometry	Algebra
Countries						
United States	93	100	100	100	85	82
Belgium (Flemish)	80	100	90	80	62	64
Canada	82	94	90	100	77	55
Chinese Taipei	59	82	50	40	46	55
Czech Republic	77	94	90	80	69	45
England	25	29	30	40	23	9
Hong Kong, SAR	79	94	80	40	77	73
Italy	91	100	80	80	92	91
Japan	89	82	100	80	85	100
Korea, Rep. of	80	82	100	80	54	91
Netherlands	46	53	40	60	54	27
Russian Federation	75	88	60	100	62	73
Singapore	89	94	100	80	77	91
States						
Connecticut	73	76	80	100	62	64
Idaho	73	88	70	100	54	64
Illinois	89	94	80	100	77	100
Indiana	84	88	90	100	69	82
Maryland	86	100	90	100	77	64
Massachusetts	82	94	100	100	69	55
Michigan	80	94	60	100	77	73
Missouri	80	100	90	100	62	55
North Carolina	88	100	90	100	77	73
Oregon	82	88	90	100	69	73
Pennsylvania	100	100	100	100	100	100
South Carolina	89	88	90	100	92	82
Texas	89	94	100	100	85	73
Districts and Consortia						
Academy School Dist. #20, CO [1]	–	–	–	–	–	–
Chicago Public Schools, IL	61	88	90	60	46	9
Delaware Science Coalition, DE	86	88	90	100	69	91
First in the World Consort., IL	95	100	100	100	85	91
Fremont/Lincoln/WestSide PS, NE	91	100	100	100	77	82
Guilford County, NC	89	100	90	100	77	82
Jersey City Public Schools, NJ	89	94	100	100	85	73
Miami-Dade County PS, FL	93	94	100	100	85	91
Michigan Invitational Group, MI	84	94	70	100	77	82
Montgomery County, MD	84	88	80	100	85	73
Naperville Sch. Dist. #203, IL	91	100	90	100	85	82
Project SMART Consortium, OH	84	94	80	100	69	82
Rochester City Sch. Dist., NY	91	100	100	100	77	82
SW Math/Sci. Collaborative, PA [2]	–	–	–	–	–	–
International Avg. (All Countries)	75	86	83	60	67	68

SOURCE: IEA Third International Mathematics and Science Study (TIMSS), 1998-1999.

Background data provided by coordinators from participating jurisdictions according to the official curriculum. Coordinators indicated the percentage of students who should have been taught each of the topics listed in Exhibit 5.18. The response categories were: all or almost all of the students (at least 90%); about half of the students; only the more able students (top track - about 25%); only the most advanced students (10% or less); not included in curriculum through grade 8. (See Reference Exhibits R2.2-R2.6 for detail by topic.)

1 Academy School Dist. #20: As a district that has site-based curriculum development, the district cannot provide a representative response for these questions.

2 SW Math/Sci. Collaborative: Covering a workforce region of 118 autonomous districts, the Collaborative cannot provide a representative response for these questions.

A dash (–) indicates data are not available.

Have Students Been Taught the Topics Tested by TIMSS?

In interpreting the achievement results, it is important to consider how extensively the topics tested are taught in the participating entities. As shown in Exhibits 5.20 through 5.24, the five major mathematics content areas assessed in TIMSS 1999 were represented by 34 topic areas. For each area, teachers indicated whether their students had been taught the topics before this year (i.e., the eighth grade), one to five periods this year, more than five periods this year; whether the topics had not yet been taught; or whether the teacher did not know. Exhibits 5.20 through 5.24 show the percentages of students in each entity reported to have been taught each topic before or during the year of testing.

According to their teachers, nearly all students in all the comparison countries had been taught the topics in fractions and number sense, as shown in Exhibit 5.20. The international average for each topic exceeded 90 percent of students, with the exception of "square roots (of perfect squares less than 144), small integer exponents" and "concepts of ratio and proportions; ratio and proportion problems," with averages of 83 and 87 percent, respectively. Teachers in the United States overall as well as in the Benchmarking jurisdictions reported similar percentages, with 90 percent or more of the students in each jurisdiction being taught each topic with the exception of the two topics relating to square roots and ratio/proportion.

However, Exhibit R2.7 in the reference section indicates that internationally many students had instruction in these topics before the eighth grade, while students in several Benchmarking jurisdictions were taught them during that grade. For example, high-performing Chinese Taipei reported that 90 percent of its students were taught more than 80 percent of the fractions and number sense topics before the eighth grade and not again during the eighth grade. Only eight percent of U.S. students were taught more than 80 percent of these topics before the eighth grade only. Similarly, all but one of the Benchmarking jurisdictions had less than one-fifth of their students taught more than 80 percent of fraction and number sense topics before the eighth grade only. In the U.S. overall and across the Benchmarking jurisdictions, a larger proportion of students were taught, or were continued to be taught, fractions and number sense topics at the eighth grade than were students internationally. This echoes the findings of the TIMSS 1995 curricula analysis that showed that states in the U.S. intended to cover far more than the average number of mathematics topics commonly covered internationally, and that topics in the U.S. were often added as students progressed through school at the same rate as in other countries but without dropping other topics that had been taught previously.[9]

Instructional coverage was high for the measurement topics presented in Exhibit 5.21. At least 87 percent of students, on average internationally, were taught six of the seven topics. The topic with the lowest coverage was "scales applied to maps and models," with an international average of 77 percent. Two topics, "units of measurement; standards metric units" and "perimeter and area of simple shapes – triangles, rectangles, and circles," were taught to 96 percent of students on average internationally. The United States as a whole and most of the Benchmarking jurisdictions reported percentages above the international average for a majority of the topics. While teachers in Jersey City reported that all students were taught all measurement topics, teachers in the Fremont/Lincoln/Westside Public Schools reported percentages of students below the international averages for six of the seven measurement topics.

As indicated by Exhibit R2.8 in the reference section, measurement topics received less emphasis in the eighth grade than did fractions and number sense topics (see Exhibit R2.7). As with fractions and number sense, substantial percentages of students internationally had studied the measurement topics before the eighth grade, whereas among the Benchmarking jurisdictions, greater percentages began or continued to study them during the eighth grade. Montgomery County was the only jurisdiction reporting a greater percentage of students than internationally (22 percent, on average) who were taught more than 80 percent of the measurement topics before the eighth grade and not again during the eighth grade.

Corresponding to the reports for the intended curricula, teachers reported lower average percentages internationally across the data representation, analysis, and probability topics, shown in Exhibit 5.22. Teachers were asked about three topics in this content area, including "representation and interpretation of data in graphs, charts, and tables" and "arithmetic mean." While the international average for students who were taught these two topics was 75 and 70 percent, respectively, all Benchmarking jurisdictions and the United States overall reported that at least 88 percent of their students were taught each of these topics. The international average percentage of students taught the other topic in this content area, "simple probabilities – understanding and calculations," was

9 Schmidt, W.H., McKnight, C.C., and Raizen, S.A. (1997), *A Splintered Vision: An Investigation of U.S. Science and Mathematics Education*, Dordrecht, the Netherlands: Kluwer Academic Publishers.

43 percent. Coverage of this topic varied widely, from just three or four percent in Japan and Chinese Taipei to 99 percent in Korea. The Benchmarking jurisdictions generally resembled the United States overall, where 79 percent were taught this topic.

For students in most countries, the data representation, analysis, and probability topics received moderate attention in the eighth grade, with few students having been taught them only in earlier grades, and one-third having not yet been taught half or more of the topics by the end of the eighth grade (see Exhibit R2.9). In comparison, however, relatively greater percentages of students in the United States and in the Benchmarking entities were reported to have been taught these topics during the eighth grade. In the U.S. overall, 79 percent of students were taught more than half the topics during the eighth grade. All Benchmarking jurisdictions had a much greater percentage of students than internationally (39 percent, on average) who were taught more than half the topics during the eighth grade, ranging from 60 percent in Rochester to 99 percent in Chicago.

Teachers reported a range of instructional coverage across topics in geometry, presented in Exhibit 5.23. "Simple two dimensional geometry – angles on a straight line, parallel lines, triangles and quadrilaterals" was reported to have been taught internationally on average to 95 percent of the students, and "visualization of three-dimensional shapes" to only 57 percent. The topics showing the greatest variation across countries were "symmetry and transformations" and "visualization of three-dimensional shapes." For example, the percentage of students taught "symmetry and transformations" ranged from less than 30 percent in Chinese Taipei to 98 percent in Japan. The other four geometry topics were taught to more than 90 percent of the students in high-performing Japan, Korea, and Singapore. The United States was similar to the international averages in coverage of the geometry topics, as were most of the Benchmarking participants, although they did show variation, particularly the districts and consortia. For example, in Jersey City, Montgomery County, and Naperville, 90 percent of more of the students were taught each of the geometry topics. However, in the Academy School District, Miami-Dade, and Rochester, less than 50 percent of the students were taught "symmetry and transformations" and "visualization of three-dimensional shapes," the two topics that had the lowest coverage both internationally and in the U.S.

As shown in Exhibit R2.10 in the reference section, only small percentages of students had completed instruction in the geometry topics before the eighth grade, and relatively large percentages had not yet been introduced to many geometry topics by the end of the eighth grade. According to the teachers in the United States, 25 percent of the students had not been taught half or more of the geometry topics by the end of eighth grade, close to the international average of 22 percent. This was exceeded only by Chinese Taipei (33 percent) among the comparison countries. In the Czech Republic, Italy, Japan, Korea, and Singapore, less than only 10 percent of the students had not yet been taught half or more of these topics. One-quarter or more of the students in six Benchmarking states and four districts and consortia had not been taught half or more of the geometry topics by the end of the eighth grade, with the greatest percentage in the Academy School District (49 percent).

Teachers across countries reported that most students had been taught the algebra topics, as shown in Exhibit 5.24. More than 85 percent of students internationally, in the U.S. overall, and in all the Benchmarking entities were taught each of these topics, with the exception of "solving simple inequalities," which had an international average of 66 percent. The percentages of students taught the algebra topics in the United States and in the Benchmarking entities generally exceeded the international averages. In North Carolina, the Academy School District, Jersey City, Montgomery County, and Naperville, 90 percent or more of the students were taught each of the algebra topics.

For many jurisdictions, teachers reported presenting algebra topics during the eighth grade for substantial percentages of students (see Exhibit R2.11). Teachers in all Benchmarking jurisdictions except Rochester reported that at least half the students were taught more than half the topics for more than five periods during the eighth grade. Similarly, teachers in all Benchmarking jurisdictions reported that less than 10 percent of the students had been taught half or more of the topics before the eighth grade only. In contrast, 85 percent of the students in Chinese Taipei and 35 percent in Japan were taught the topics before the eighth grade.

Exhibits 5.20-5.24

Exhibit 5.20 Percentages of Students Taught Fractions and Number Sense Topics*

8th Grade Mathematics

	Whole numbers - including place values, factorization and operations (+, −, x, ÷)	Understanding and representing common fractions	Computations with common fractions	Understanding and representing decimal fractions	Computations with decimal fractions	Relationships between common and decimal fractions, ordering of fractions	Rounding whole numbers and decimal fractions
Countries							
United States	100 (0.2)	100 (0.0)	100 (0.0)	98 (0.8)	98 (0.8)	98 (0.8)	99 (0.7)
Belgium (Flemish)	95 (3.1)	99 (1.2)	97 (2.4)	88 (2.9)	83 (2.2)	89 (4.1)	90 (3.5)
Canada	r 99 (0.6)	r 100 (0.3)	r 100 (0.3)	r 99 (0.5)	r 98 (0.8)	r 99 (0.4)	r 100 (0.3)
Chinese Taipei	100 (0.0)	100 (0.3)	100 (0.3)	100 (0.3)	99 (0.7)	100 (0.3)	98 (1.1)
Czech Republic	100 (0.0)	100 (0.0)	100 (0.0)	100 (0.0)	100 (0.0)	100 (0.0)	100 (0.0)
England	s 100 (0.1)	s 99 (0.5)	s 93 (2.0)	s 97 (0.9)	s 95 (1.1)	s 94 (1.1)	s 97 (0.9)
Hong Kong, SAR	98 (1.1)	99 (0.8)	99 (0.8)	99 (0.8)	100 (0.0)	99 (0.8)	100 (0.4)
Italy	100 (0.0)	100 (0.0)	100 (0.0)	100 (0.5)	100 (0.0)	100 (0.0)	100 (0.4)
Japan	99 (1.0)	98 (1.4)	100 (0.0)	98 (1.4)	100 (0.0)	99 (1.0)	92 (2.7)
Korea, Rep. of	92 (2.1)	96 (1.5)	96 (1.6)	97 (1.4)	96 (1.6)	96 (1.7)	94 (2.0)
Netherlands	r 74 (5.8)	100 (0.3)	100 (0.3)	r 96 (3.2)	r 96 (3.3)	r 96 (3.3)	100 (0.0)
Russian Federation	– –	– –	– –	– –	– –	– –	– –
Singapore	100 (0.0)	100 (0.0)	100 (0.0)	99 (0.9)	100 (0.0)	100 (0.0)	100 (0.0)
States							
Connecticut	r 99 (0.9)	r 100 (0.0)	r 100 (0.0)	r 100 (0.0)	r 99 (0.9)	r 100 (0.0)	r 100 (0.0)
Idaho	r 100 (0.3)	r 100 (0.0)	r 100 (0.0)	r 97 (2.1)	r 97 (2.2)	r 98 (1.5)	r 100 (0.3)
Illinois	100 (0.0)	100 (0.0)	100 (0.0)	99 (1.1)	97 (2.2)	99 (1.1)	99 (1.1)
Indiana	100 (0.0)	100 (0.0)	100 (0.0)	100 (0.0)	99 (1.4)	98 (1.5)	100 (0.0)
Maryland	r 100 (0.0)	r 100 (0.0)	r 100 (0.0)	r 98 (1.8)	r 98 (1.8)	r 98 (1.7)	r 98 (1.7)
Massachusetts	r 100 (0.0)	100 (0.0)	100 (0.0)	r 99 (1.1)	r 98 (1.5)	r 99 (1.2)	r 99 (1.1)
Michigan	100 (0.0)	100 (0.0)	100 (0.0)	99 (1.0)	99 (1.0)	100 (0.3)	100 (0.3)
Missouri	100 (0.0)	100 (0.0)	100 (0.0)	100 (0.0)	99 (0.9)	100 (0.2)	100 (0.0)
North Carolina	100 (0.0)	100 (0.0)	100 (0.0)	99 (1.1)	98 (1.5)	99 (1.1)	100 (0.0)
Oregon	100 (0.0)	100 (0.0)	100 (0.0)	98 (1.3)	98 (1.4)	100 (0.2)	100 (0.2)
Pennsylvania	100 (0.0)	100 (0.0)	100 (0.0)	100 (0.4)	99 (1.0)	r 100 (0.1)	95 (4.8)
South Carolina	100 (0.0)	100 (0.0)	100 (0.0)	100 (0.0)	99 (0.9)	99 (0.9)	100 (0.0)
Texas	100 (0.0)	100 (0.0)	100 (0.0)	98 (1.3)	98 (1.3)	99 (1.3)	99 (1.2)
Districts and Consortia							
Academy School Dist. #20, CO	100 (0.0)	100 (0.0)	100 (0.0)	98 (0.2)	98 (0.2)	100 (0.0)	100 (0.0)
Chicago Public Schools, IL	100 (0.0)	100 (0.0)	100 (0.0)	100 (0.0)	100 (0.0)	100 (0.0)	100 (0.0)
Delaware Science Coalition, DE	r 100 (0.0)	r 100 (0.4)	r 100 (0.4)	r 99 (0.5)	r 99 (0.5)	r 99 (0.5)	r 100 (0.4)
First in the World Consort., IL	r 100 (0.0)	r 100 (0.0)	r 100 (0.0)	r 100 (0.0)	r 100 (0.0)	r 100 (0.0)	r 100 (0.0)
Fremont/Lincoln/WestSide PS, NE	100 (0.0)	100 (0.0)	100 (0.0)	100 (0.0)	100 (0.0)	100 (0.0)	100 (0.0)
Guilford County, NC	100 (0.0)	100 (0.0)	100 (0.0)	100 (0.0)	98 (2.0)	100 (0.0)	100 (0.0)
Jersey City Public Schools, NJ	r 100 (0.0)	100 (0.0)	100 (0.0)	100 (0.0)	100 (0.0)	100 (0.0)	100 (0.0)
Miami-Dade County PS, FL	s 100 (0.0)	s 99 (0.5)	s 99 (0.6)	s 97 (2.5)	s 97 (2.5)	s 96 (2.7)	s 97 (2.2)
Michigan Invitational Group, MI	100 (0.0)	100 (0.0)	100 (0.0)	100 (0.0)	100 (0.0)	100 (0.0)	100 (0.0)
Montgomery County, MD	s 100 (0.0)	s 100 (0.0)	s 100 (0.0)	s 100 (0.0)	s 100 (0.0)	s 100 (0.0)	s 100 (0.0)
Naperville Sch. Dist. #203, IL	100 (0.0)	100 (0.0)	100 (0.0)	100 (0.0)	100 (0.0)	100 (0.0)	100 (0.0)
Project SMART Consortium, OH	100 (0.0)	100 (0.2)	100 (0.2)	100 (0.2)	100 (0.2)	100 (0.2)	100 (0.0)
Rochester City Sch. Dist., NY	100 (0.0)	100 (0.0)	100 (0.0)	100 (0.0)	100 (0.0)	100 (0.0)	100 (0.0)
SW Math/Sci. Collaborative, PA	100 (0.0)	100 (0.0)	98 (0.4)	98 (1.9)	97 (2.0)	99 (0.6)	100 (0.0)
International Avg. (All Countries)	98 (0.3)	99 (0.2)	98 (0.2)	98 (0.2)	98 (0.2)	98 (0.2)	95 (0.3)

SOURCE: IEA Third International Mathematics and Science Study (TIMSS), 1998-1999.

Background data provided by teachers.

* Taught before or during this school year.

States in *italics* did not fully satisfy guidelines for sample participation rates (see Appendix A for details).

( ) Standard errors appear in parentheses. Because results are rounded to the nearest whole number, some totals may appear inconsistent.

A dash (–) indicates data are not available.

An "r" indicates teacher response data available for 70-84% of students. An "s" indicates teacher response data available for 50-69% of students.

	Estimating the results of computations	Number lines	Computations with percentages and problems involving percentages	Simple computations with negative numbers	Square roots (of perfect squares less than 144), small integer exponents	Concepts of ratio and proportions; ratio and proportion problems
Countries						
United States	100 (0.2)	99 (0.5)	96 (1.4)	97 (1.1)	82 (3.7)	93 (1.8)
Belgium (Flemish)	r 94 (2.0)	96 (2.5)	93 (2.1)	89 (2.6)	80 (2.2)	70 (2.8)
Canada	r 100 (0.3)	r 100 (0.1)	r 98 (0.8)	r 97 (1.6)	r 96 (1.2)	r 95 (1.3)
Chinese Taipei	95 (2.0)	99 (0.8)	94 (1.9)	100 (0.3)	96 (1.6)	90 (2.6)
Czech Republic	100 (0.0)	100 (0.0)	100 (0.0)	100 (0.0)	100 (0.0)	100 (0.2)
England	s 96 (1.7)	s 97 (1.3)	s 96 (1.3)	s 96 (1.3)	s 87 (2.0)	s 79 (2.7)
Hong Kong, SAR	r 94 (2.2)	92 (2.6)	95 (1.9)	99 (0.8)	98 (1.2)	91 (2.5)
Italy	94 (2.0)	99 (0.8)	96 (1.6)	98 (1.1)	100 (0.0)	99 (0.8)
Japan	r 89 (3.3)	100 (0.0)	100 (0.0)	100 (0.0)	14 (3.0)	97 (1.6)
Korea, Rep. of	89 (2.5)	98 (1.2)	92 (2.0)	95 (1.8)	64 (4.1)	90 (2.3)
Netherlands	r 99 (1.0)	99 (0.9)	98 (1.2)	98 (1.4)	92 (3.1)	r 80 (5.8)
Russian Federation	– –	– –	– –	– –	– –	– –
Singapore	100 (0.4)	100 (0.0)	100 (0.0)	100 (0.0)	100 (0.0)	100 (0.0)
States						
Connecticut	r 100 (0.0)	r 99 (1.2)	r 99 (1.4)	r 91 (3.3)	r 84 (5.1)	r 93 (3.5)
Idaho	r 99 (0.9)	r 96 (2.2)	r 94 (2.5)	r 92 (3.5)	r 80 (3.9)	r 89 (3.5)
Illinois	100 (0.0)	99 (0.1)	96 (2.0)	97 (1.6)	82 (5.1)	97 (1.8)
Indiana	100 (0.0)	99 (1.0)	94 (2.8)	95 (1.6)	76 (6.6)	95 (2.2)
Maryland	r 100 (0.0)	r 100 (0.0)	r 98 (1.1)	r 93 (3.2)	r 73 (5.1)	r 97 (1.6)
Massachusetts	100 (0.0)	r 99 (0.2)	r 97 (1.9)	97 (1.8)	r 74 (4.9)	r 89 (3.3)
Michigan	100 (0.0)	100 (0.0)	97 (2.1)	99 (0.7)	r 80 (3.6)	92 (3.8)
Missouri	100 (0.0)	100 (0.4)	93 (3.8)	98 (2.0)	77 (6.1)	93 (4.0)
North Carolina	100 (0.0)	100 (0.0)	100 (0.0)	99 (1.1)	92 (2.9)	98 (1.7)
Oregon	100 (0.0)	100 (0.0)	91 (4.4)	98 (1.1)	81 (5.5)	89 (3.9)
Pennsylvania	100 (0.0)	100 (0.0)	94 (2.2)	98 (0.9)	89 (2.6)	92 (2.3)
South Carolina	100 (0.0)	99 (0.9)	97 (2.0)	98 (1.3)	97 (1.6)	96 (2.3)
Texas	100 (0.2)	100 (0.1)	98 (1.3)	98 (1.3)	97 (1.6)	97 (2.0)
Districts and Consortia						
Academy School Dist. #20, CO	100 (0.0)	100 (0.0)	96 (0.2)	100 (0.0)	92 (0.2)	93 (0.3)
Chicago Public Schools, IL	100 (0.0)	100 (0.0)	93 (4.8)	96 (3.6)	90 (6.0)	98 (2.0)
Delaware Science Coalition, DE	r 97 (2.4)	r 98 (2.0)	r 92 (0.9)	r 99 (0.7)	r 77 (6.6)	r 87 (5.2)
First in the World Consort., IL	r 100 (0.0)	r 100 (0.0)	r 98 (2.4)	r 100 (0.0)	r 94 (2.9)	r 97 (2.7)
Fremont/Lincoln/WestSide PS, NE	100 (0.0)	100 (0.0)	95 (5.1)	100 (0.0)	93 (1.8)	r 86 (8.0)
Guilford County, NC	100 (0.0)	100 (0.0)	100 (0.0)	100 (0.0)	97 (2.3)	98 (2.4)
Jersey City Public Schools, NJ	r 100 (0.0)	100 (0.0)	100 (0.0)	100 (0.0)	r 100 (0.0)	100 (0.0)
Miami-Dade County PS, FL	s 100 (0.0)	s 100 (0.0)	s 94 (3.2)	s 94 (3.6)	s 81 (6.2)	s 91 (4.7)
Michigan Invitational Group, MI	100 (0.0)	100 (0.0)	100 (0.0)	98 (2.2)	83 (6.9)	97 (2.0)
Montgomery County, MD	s 100 (0.0)	s 100 (0.0)	s 100 (0.1)	s 100 (0.0)	s 100 (0.2)	s 98 (2.0)
Naperville Sch. Dist. #203, IL	100 (0.0)	100 (0.0)	100 (0.0)	100 (0.0)	96 (1.8)	100 (0.0)
Project SMART Consortium, OH	100 (0.0)	94 (4.3)	98 (1.4)	91 (5.1)	83 (4.0)	97 (2.6)
Rochester City Sch. Dist., NY	100 (0.0)	100 (0.0)	92 (3.1)	98 (1.7)	49 (3.9)	90 (4.1)
SW Math/Sci. Collaborative, PA	100 (0.0)	100 (0.0)	90 (5.4)	91 (3.5)	86 (4.8)	82 (6.4)
International Avg. (All Countries)	93 (0.4)	92 (0.3)	95 (0.3)	97 (0.2)	83 (0.4)	87 (0.4)

SOURCE: IEA Third International Mathematics and Science Study (TIMSS), 1998-1999.

Exhibit 5.21 Percentages of Students Taught Measurement Topics*

8th Grade Mathematics

	Units of measurement, standard metric units	Reading measurement instruments	Estimates of measurement, accuracy of measurement	Perimeter and area of simple shapes – triangles, rectangles, and circles	Perimeter and area of combined shapes	Volume of rectangular solids – i.e., volume= length × width × height	Scales applied to maps and models
Countries							
United States	96 (1.0)	r 92 (1.7)	r 91 (1.2)	95 (1.4)	90 (1.6)	83 (2.0)	r 84 (2.5)
Belgium (Flemish)	95 (1.8)	r 83 (3.8)	r 85 (4.1)	98 (1.2)	r 85 (3.9)	89 (3.5)	88 (2.2)
Canada	r 99 (0.5)	r 97 (1.2)	r 97 (1.0)	r 97 (0.9)	r 96 (1.3)	r 68 (2.7)	r 92 (2.1)
Chinese Taipei	96 (1.7)	95 (2.0)	90 (2.7)	100 (0.3)	92 (2.3)	99 (0.7)	74 (3.8)
Czech Republic	100 (0.2)	r 99 (0.6)	97 (1.2)	100 (0.0)	90 (3.2)	100 (0.0)	98 (1.2)
England	s 98 (0.9)	s 96 (1.3)	s 86 (2.8)	s 98 (1.0)	s 96 (1.1)	s 93 (1.4)	s 76 (2.6)
Hong Kong, SAR	98 (1.2)	96 (1.9)	92 (2.5)	100 (0.0)	99 (0.8)	98 (1.5)	91 (2.7)
Italy	100 (0.0)	96 (1.6)	90 (2.3)	99 (0.8)	96 (1.3)	95 (1.4)	91 (2.2)
Japan	90 (2.5)	r 84 (3.3)	r 66 (4.2)	99 (0.7)	78 (3.3)	98 (1.4)	84 (3.1)
Korea, Rep. of	85 (2.7)	84 (2.7)	93 (2.1)	98 (1.2)	95 (1.8)	98 (1.0)	73 (3.4)
Netherlands	r 93 (4.7)	s 54 (8.4)	r 78 (6.3)	98 (1.2)	84 (4.9)	89 (4.9)	88 (5.3)
Russian Federation	– –	– –	– –	– –	– –	– –	– –
Singapore	100 (0.0)	r 98 (1.2)	98 (1.3)	100 (0.0)	100 (0.0)	100 (0.0)	96 (1.6)
States							
Connecticut	r 95 (2.6)	s 95 (2.5)	s 91 (3.6)	r 95 (2.3)	r 90 (3.5)	r 81 (4.3)	s 88 (4.0)
Idaho	r 91 (4.5)	s 93 (4.4)	s 90 (4.4)	r 93 (2.3)	r 79 (5.1)	r 73 (6.3)	r 79 (5.7)
Illinois	98 (1.6)	99 (0.5)	r 98 (1.7)	98 (2.0)	93 (2.9)	91 (3.3)	90 (3.5)
Indiana	99 (1.1)	98 (1.2)	92 (3.9)	95 (2.3)	85 (4.9)	83 (4.6)	82 (5.0)
Maryland	r 97 (1.8)	r 93 (3.2)	r 97 (1.8)	r 97 (2.0)	r 94 (2.8)	r 83 (4.9)	r 91 (3.6)
Massachusetts	r 95 (2.2)	r 94 (2.7)	r 89 (4.3)	r 93 (3.3)	r 89 (3.5)	r 75 (6.3)	r 82 (5.5)
Michigan	98 (1.3)	r 98 (1.3)	r 97 (1.4)	96 (1.8)	r 91 (2.7)	91 (3.0)	r 89 (4.4)
Missouri	94 (4.4)	96 (3.0)	r 89 (4.6)	99 (0.9)	84 (4.6)	76 (4.5)	89 (4.4)
North Carolina	95 (1.8)	r 92 (2.4)	r 86 (3.2)	98 (1.4)	91 (3.0)	90 (3.9)	r 92 (3.1)
Oregon	100 (0.4)	96 (1.1)	96 (1.8)	97 (1.5)	90 (3.6)	82 (3.6)	88 (3.8)
Pennsylvania	96 (2.1)	86 (6.4)	r 94 (2.9)	97 (1.4)	89 (3.2)	74 (7.0)	92 (2.3)
South Carolina	98 (1.6)	100 (0.0)	98 (1.6)	98 (1.6)	93 (2.7)	92 (2.8)	97 (1.8)
Texas	98 (0.2)	98 (1.6)	97 (1.9)	100 (0.0)	95 (2.4)	96 (2.2)	94 (3.4)
Districts and Consortia							
Academy School Dist. #20, CO	92 (0.2)	87 (0.3)	79 (0.3)	97 (0.2)	89 (0.3)	76 (0.3)	85 (0.3)
Chicago Public Schools, IL	97 (2.7)	97 (2.7)	100 (0.0)	100 (0.0)	97 (2.6)	91 (5.3)	82 (8.8)
Delaware Science Coalition, DE	r 96 (2.9)	r 95 (3.5)	r 91 (3.1)	r 99 (0.7)	r 89 (2.7)	r 93 (3.3)	r 77 (5.7)
First in the World Consort., IL	r 96 (3.5)	r 96 (3.6)	r 91 (5.2)	r 95 (2.9)	r 94 (2.9)	r 92 (3.8)	r 98 (1.7)
Fremont/Lincoln/WestSide PS, NE	r 89 (6.7)	r 87 (6.7)	r 75 (6.7)	r 100 (0.0)	r 71 (6.4)	r 82 (3.5)	r 76 (9.5)
Guilford County, NC	96 (2.9)	93 (3.9)	93 (3.8)	92 (3.9)	91 (3.6)	91 (3.5)	86 (3.8)
Jersey City Public Schools, NJ	r 100 (0.0)	r 100 (0.0)	r 100 (0.0)	100 (0.0)	100 (0.0)	100 (0.0)	r 100 (0.0)
Miami-Dade County PS, FL	s 92 (4.4)	s 83 (6.9)	s 82 (7.4)	s 99 (1.5)	s 95 (3.4)	s 89 (5.2)	s 68 (8.4)
Michigan Invitational Group, MI	95 (1.7)	100 (0.0)	100 (0.0)	98 (1.7)	98 (1.7)	96 (2.8)	91 (4.1)
Montgomery County, MD	s 98 (0.5)	s 99 (0.5)	s 96 (0.9)	s 100 (0.0)	s 99 (0.6)	s 94 (2.6)	s 97 (2.9)
Naperville Sch. Dist. #203, IL	100 (0.0)	100 (0.0)	100 (0.0)	100 (0.0)	100 (0.0)	100 (0.0)	93 (1.1)
Project SMART Consortium, OH	97 (0.9)	95 (2.1)	95 (1.6)	97 (3.4)	96 (2.9)	94 (4.3)	97 (2.7)
Rochester City Sch. Dist., NY	100 (0.0)	93 (2.0)	r 92 (2.2)	89 (1.5)	71 (3.2)	66 (4.4)	88 (3.7)
SW Math/Sci. Collaborative, PA	94 (4.4)	92 (5.0)	92 (4.2)	95 (2.7)	86 (4.1)	76 (6.4)	79 (6.9)
International Avg. (All Countries)	96 (0.3)	89 (0.5)	87 (0.5)	96 (0.3)	89 (0.5)	87 (0.5)	77 (0.6)

SOURCE: IEA Third International Mathematics and Science Study (TIMSS), 1998-1999.

Background data provided by teachers.

* Taught before or during this school year.

States in *italics* did not fully satisfy guidelines for sample participation rates (see Appendix A for details).

( ) Standard errors appear in parentheses. Because results are rounded to the nearest whole number, some totals may appear inconsistent.

A dash (–) indicates data are not available.

An "r" indicates teacher response data available for 70-84% of students. An "s" indicates teacher response data available for 50-69% of students.

		Representation and interpretation of data in graphs, charts, and tables		Arithmetic mean		Simple probabilities – understanding and calculations
Countries						
United States		96 (1.2)		93 (1.6)		79 (2.3)
Belgium (Flemish)		86 (4.1)		93 (2.1)	r	24 (3.0)
Canada	r	91 (2.4)	r	81 (2.7)	r	72 (3.3)
Chinese Taipei		11 (2.3)		12 (2.7)		4 (1.6)
Czech Republic		49 (5.6)		88 (3.4)		7 (2.8)
England	s	99 (0.4)	s	93 (2.3)	s	90 (2.4)
Hong Kong, SAR		65 (4.5)		30 (4.1)		10 (2.8)
Italy		84 (3.0)		62 (3.6)		49 (3.8)
Japan		43 (4.7)		38 (4.5)		3 (1.4)
Korea, Rep. of		95 (1.7)		78 (3.2)		99 (0.6)
Netherlands		87 (4.7)		77 (5.7)	r	46 (6.5)
Russian Federation		– –		– –		– –
Singapore		97 (1.7)		88 (3.2)	s	17 (4.2)
States						
Connecticut	r	100 (0.0)	s	98 (1.6)	s	81 (5.0)
Idaho	r	93 (2.8)	r	90 (3.7)	r	77 (5.6)
Illinois		99 (1.4)		98 (1.2)		82 (4.3)
Indiana		93 (6.4)		94 (3.7)		83 (4.9)
Maryland	r	100 (0.4)	r	97 (0.8)	r	82 (3.8)
Massachusetts	r	95 (2.0)	r	95 (2.1)	r	84 (2.8)
Michigan		98 (1.7)	r	97 (1.6)	r	87 (2.9)
Missouri		99 (1.4)		96 (2.8)		76 (5.0)
North Carolina		91 (3.4)		90 (4.0)		69 (6.8)
Oregon		98 (1.2)		96 (1.4)		92 (2.7)
Pennsylvania		92 (2.4)		92 (2.9)		79 (6.0)
South Carolina		100 (0.0)		97 (2.1)		97 (1.8)
Texas		97 (1.9)	r	98 (1.3)		100 (0.3)
Districts and Consortia						
Academy School Dist. #20, CO		98 (0.2)		96 (0.2)		83 (0.3)
Chicago Public Schools, IL		100 (0.0)		95 (3.6)		94 (3.4)
Delaware Science Coalition, DE	r	95 (3.6)	r	94 (3.2)	r	89 (4.7)
First in the World Consort., IL	r	100 (0.0)	r	100 (0.0)	r	73 (7.4)
Fremont/Lincoln/WestSide PS, NE		97 (0.2)		88 (2.9)		79 (7.5)
Guilford County, NC		97 (2.2)		88 (3.2)		87 (3.7)
Jersey City Public Schools, NJ		100 (0.0)		100 (0.0)		100 (0.0)
Miami-Dade County PS, FL	s	95 (3.8)	s	94 (4.1)	s	80 (7.5)
Michigan Invitational Group, MI		98 (2.2)		93 (6.3)		94 (4.3)
Montgomery County, MD	s	96 (3.2)	s	96 (2.6)	s	92 (3.3)
Naperville Sch. Dist. #203, IL		100 (0.0)		100 (0.0)		100 (0.0)
Project SMART Consortium, OH		95 (3.1)		97 (2.6)		89 (3.9)
Rochester City Sch. Dist., NY		91 (1.8)		88 (3.2)		85 (1.8)
SW Math/Sci. Collaborative, PA		99 (1.0)		89 (4.8)	r	86 (5.3)
International Avg. (All Countries)		75 (0.6)		70 (0.6)		43 (0.6)

SOURCE: IEA Third International Mathematics and Science Study (TIMSS), 1998-1999.

Background data provided by teachers.

* Taught before or during this school year.

States in *italics* did not fully satisfy guidelines for sample participation rates (see Appendix A for details).

( ) Standard errors appear in parentheses. Because results are rounded to the nearest whole number, some totals may appear inconsistent.

A dash (–) indicates data are not available.

An "r" indicates teacher response data available for 70-84% of students. An "s" indicates teacher response data available for 50-69% of students.

Percentages of Students Taught Geometry Topics*

8th Grade Mathematics

	Cartesian coordinates of points in a plane	Coordinates of points on a given straight line	Simple two dimensional geometry – angles on a straight line, parallel lines, triangles and quadrilaterals	Congruence and similarity	Symmetry and transformations (reflection and rotation)	Visualization of three-dimensional shapes
Countries						
United States	r 83 (2.4)	82 (2.5)	89 (2.0)	r 80 (2.6)	r 62 (2.9)	r 61 (2.7)
Belgium (Flemish)	78 (3.0)	r 54 (3.9)	91 (4.1)	79 (2.5)	87 (2.9)	57 (4.0)
Canada	r 81 (2.5)	r 84 (2.6)	r 94 (1.8)	r 84 (2.7)	r 78 (2.4)	r 63 (3.2)
Chinese Taipei	100 (0.0)	99 (0.9)	78 (3.5)	60 (4.3)	29 (3.7)	42 (4.1)
Czech Republic	94 (2.6)	88 (4.9)	100 (0.0)	86 (3.7)	98 (1.1)	73 (5.2)
England	s 94 (1.3)	s 79 (3.1)	s 95 (1.6)	s 54 (4.1)	s 88 (2.6)	s 75 (3.0)
Hong Kong, SAR	98 (1.3)	95 (1.9)	97 (1.6)	89 (2.8)	r 31 (4.6)	r 29 (4.7)
Italy	93 (1.9)	79 (3.0)	98 (1.2)	91 (2.0)	65 (3.8)	89 (2.4)
Japan	100 (0.0)	99 (1.0)	97 (1.4)	98 (1.2)	98 (1.3)	82 (2.9)
Korea, Rep. of	98 (1.1)	99 (0.7)	99 (0.7)	99 (0.7)	71 (3.7)	52 (4.2)
Netherlands	r 97 (1.5)	r 97 (1.5)	98 (1.1)	49 (5.8)	78 (5.3)	r 60 (6.2)
Russian Federation	– –	– –	– –	– –	– –	– –
Singapore	91 (2.6)	93 (2.4)	96 (1.8)	96 (1.9)	84 (3.4)	r 72 (4.4)
States						
Connecticut	s 71 (6.1)	r 82 (5.6)	r 85 (4.0)	r 67 (6.6)	r 60 (6.0)	s 56 (6.9)
Idaho	r 64 (5.2)	r 71 (5.3)	r 81 (6.1)	r 71 (5.1)	s 57 (5.4)	s 50 (7.8)
Illinois	89 (3.7)	87 (4.1)	96 (2.4)	88 (4.2)	70 (5.5)	r 80 (5.2)
Indiana	77 (5.6)	82 (3.4)	85 (4.9)	75 (6.1)	r 54 (6.6)	r 54 (6.8)
Maryland	r 83 (3.8)	r 76 (4.0)	r 80 (5.3)	r 68 (5.7)	r 59 (6.7)	r 51 (5.3)
Massachusetts	r 88 (3.7)	r 77 (5.0)	r 84 (4.7)	r 63 (5.6)	r 59 (6.1)	r 57 (7.7)
Michigan	r 86 (3.3)	r 92 (3.0)	r 96 (1.7)	r 88 (3.4)	r 78 (5.2)	r 77 (5.0)
Missouri	83 (3.8)	75 (4.8)	91 (4.5)	84 (4.3)	61 (5.4)	54 (6.6)
North Carolina	94 (2.5)	92 (2.9)	93 (2.5)	90 (3.0)	77 (4.5)	74 (5.6)
Oregon	85 (5.2)	86 (4.6)	92 (2.4)	87 (4.2)	75 (5.5)	r 61 (6.6)
Pennsylvania	78 (5.8)	76 (6.2)	94 (1.7)	82 (5.0)	r 57 (7.5)	r 58 (9.0)
South Carolina	90 (3.4)	93 (2.0)	93 (2.6)	90 (4.0)	82 (4.3)	r 82 (5.1)
Texas	96 (2.1)	91 (4.1)	96 (2.1)	98 (1.7)	97 (1.7)	87 (4.6)
Districts and Consortia						
Academy School Dist. #20, CO	87 (0.3)	82 (0.3)	64 (0.4)	70 (0.4)	47 (0.4)	41 (0.4)
Chicago Public Schools, IL	86 (6.7)	89 (5.6)	96 (3.4)	95 (4.8)	70 (9.0)	78 (6.7)
Delaware Science Coalition, DE	r 84 (4.7)	r 83 (4.8)	r 87 (4.5)	r 79 (5.7)	r 72 (6.4)	r 61 (7.4)
First in the World Consort., IL	99 (1.5)	99 (1.5)	96 (2.8)	93 (3.2)	70 (3.7)	75 (4.8)
Fremont/Lincoln/WestSide PS, NE	97 (2.8)	r 94 (4.3)	r 97 (3.1)	r 74 (9.7)	r 45 (8.7)	r 56 (8.8)
Guilford County, NC	88 (4.7)	92 (3.1)	90 (4.0)	92 (3.1)	r 83 (5.9)	r 89 (4.8)
Jersey City Public Schools, NJ	r 95 (0.4)	97 (0.3)	97 (0.3)	r 98 (1.6)	100 (0.0)	94 (2.8)
Miami-Dade County PS, FL	s 66 (9.6)	s 74 (10.9)	s 87 (5.8)	s 68 (10.0)	s 24 (8.8)	s 32 (10.9)
Michigan Invitational Group, MI	87 (5.2)	93 (2.2)	98 (0.1)	87 (6.9)	72 (7.6)	72 (6.2)
Montgomery County, MD	s 94 (3.9)	s 97 (2.7)	s 100 (0.0)	s 100 (0.0)	s 97 (0.8)	s 92 (3.9)
Naperville Sch. Dist. #203, IL	100 (0.0)	100 (0.0)	97 (2.6)	97 (2.6)	93 (2.8)	90 (2.8)
Project SMART Consortium, OH	72 (5.1)	84 (6.0)	96 (2.9)	89 (4.3)	65 (5.8)	77 (5.9)
Rochester City Sch. Dist., NY	98 (1.7)	r 78 (2.7)	98 (1.7)	r 67 (5.3)	24 (5.7)	r 44 (4.6)
SW Math/Sci. Collaborative, PA	79 (6.3)	78 (7.6)	80 (5.6)	82 (5.1)	r 57 (7.3)	r 70 (6.6)
International Avg. (All Countries)	85 (0.4)	84 (0.5)	95 (0.3)	72 (0.6)	63 (0.6)	57 (0.7)

SOURCE: IEA Third International Mathematics and Science Study (TIMSS), 1998-1999.

Background data provided by teachers.

* Taught before or during this school year.

States in *italics* did not fully satisfy guidelines for sample participation rates (see Appendix A for details).

( ) Standard errors appear in parentheses. Because results are rounded to the nearest whole number, some totals may appear inconsistent.

A dash (–) indicates data are not available.

An "r" indicates teacher response data available for 70-84% of students. An "s" indicates teacher response data available for 50-69% of students.

Exhibit 5.24 Percentages of Students Taught Algebra Topics*

	Number patterns and simple relations	Simple algebraic expressions	Representing situations algebraically; formulas	Solving simple equations	Solving simple inequalities
Countries					
United States	97 (1.1)	98 (0.9)	96 (1.1)	98 (0.6)	83 (2.3)
Belgium (Flemish)	r 86 (2.9)	84 (1.9)	84 (3.1)	85 (2.8)	r 9 (2.1)
Canada	r 98 (1.0)	r 98 (0.8)	r 92 (2.1)	r 94 (2.3)	r 50 (3.2)
Chinese Taipei	92 (2.5)	99 (0.8)	99 (0.8)	98 (1.2)	43 (4.2)
Czech Republic	r 99 (1.2)	100 (0.0)	97 (1.9)	96 (2.0)	32 (5.2)
England	s 98 (0.6)	s 96 (1.1)	s 89 (1.8)	s 93 (1.5)	s 39 (3.7)
Hong Kong, SAR	r 87 (3.0)	100 (0.0)	100 (0.0)	100 (0.0)	27 (4.0)
Italy	98 (1.2)	100 (0.4)	95 (1.7)	95 (1.7)	27 (2.9)
Japan	r 94 (2.4)	100 (0.0)	98 (1.2)	100 (0.0)	99 (0.7)
Korea, Rep. of	95 (1.3)	99 (0.7)	96 (1.6)	99 (0.7)	99 (1.0)
Netherlands	87 (4.9)	r 86 (4.9)	81 (6.0)	76 (5.3)	r 39 (6.4)
Russian Federation	– –	– –	– –	– –	– –
Singapore	98 (1.4)	100 (0.0)	100 (0.0)	100 (0.0)	93 (2.3)
States					
Connecticut	r 92 (2.8)	r 95 (2.5)	r 95 (2.4)	r 95 (2.5)	r 79 (4.3)
Idaho	r 88 (5.3)	r 88 (5.3)	r 86 (5.5)	r 93 (3.5)	r 73 (7.0)
Illinois	100 (0.4)	99 (0.1)	95 (2.1)	100 (0.0)	86 (3.3)
Indiana	95 (2.6)	96 (1.7)	92 (2.6)	95 (2.0)	73 (7.2)
Maryland	r 91 (3.0)	r 95 (2.5)	r 91 (3.4)	r 95 (2.6)	r 73 (4.2)
Massachusetts	99 (1.1)	99 (0.7)	r 92 (3.0)	r 94 (3.0)	r 78 (4.2)
Michigan	99 (0.7)	99 (1.0)	98 (1.1)	96 (1.5)	r 84 (3.9)
Missouri	100 (0.2)	99 (1.0)	99 (0.4)	94 (3.3)	80 (4.3)
North Carolina	99 (1.1)	100 (0.0)	99 (1.4)	100 (0.0)	r 90 (3.2)
Oregon	99 (0.8)	100 (0.3)	93 (2.4)	99 (0.3)	84 (4.2)
Pennsylvania	97 (1.5)	98 (1.5)	97 (1.5)	99 (1.0)	86 (2.0)
South Carolina	97 (1.7)	97 (1.7)	96 (2.5)	97 (1.7)	88 (4.3)
Texas	100 (0.3)	96 (2.3)	98 (1.4)	98 (1.5)	87 (4.0)
Districts and Consortia					
Academy School Dist. #20, CO	100 (0.0)	98 (0.1)	96 (0.2)	100 (0.0)	94 (0.1)
Chicago Public Schools, IL	98 (2.4)	100 (0.0)	97 (2.7)	100 (0.0)	87 (5.8)
Delaware Science Coalition, DE	r 95 (3.4)	r 95 (3.4)	r 91 (4.8)	r 95 (3.3)	r 74 (6.6)
First in the World Consort., IL	r 100 (0.0)	r 100 (0.0)	r 100 (0.0)	r 100 (0.0)	r 87 (4.1)
Fremont/Lincoln/WestSide PS, NE	100 (0.0)	95 (0.2)	97 (3.0)	100 (0.0)	69 (10.3)
Guilford County, NC	100 (0.0)	100 (0.0)	98 (1.5)	100 (0.0)	89 (5.0)
Jersey City Public Schools, NJ	r 100 (0.0)	r 100 (0.0)	r 100 (0.0)	r 100 (0.0)	r 96 (2.3)
Miami-Dade County PS, FL	s 93 (4.4)	s 93 (4.1)	s 95 (3.3)	s 90 (6.9)	s 78 (10.1)
Michigan Invitational Group, MI	98 (1.5)	98 (2.2)	98 (2.2)	100 (0.0)	75 (7.5)
Montgomery County, MD	s 96 (3.2)	s 94 (3.3)	s 95 (3.3)	s 95 (3.2)	s 92 (4.3)
Naperville Sch. Dist. #203, IL	100 (0.0)	100 (0.0)	100 (0.0)	100 (0.0)	94 (3.6)
Project SMART Consortium, OH	97 (3.4)	92 (4.8)	97 (2.8)	94 (4.4)	79 (7.4)
Rochester City Sch. Dist., NY	100 (0.0)	93 (2.0)	93 (2.0)	100 (0.0)	63 (4.8)
SW Math/Sci. Collaborative, PA	100 (0.2)	99 (1.0)	97 (2.1)	98 (1.7)	75 (5.1)
International Avg. (All Countries)	88 (0.5)	94 (0.4)	90 (0.4)	94 (0.3)	66 (0.5)

SOURCE: IEA Third International Mathematics and Science Study (TIMSS), 1998-1999.

Background data provided by teachers.

* Taught before or during this school year.

States in *italics* did not fully satisfy guidelines for sample participation rates (see Appendix A for details).

( ) Standard errors appear in parentheses. Because results are rounded to the nearest whole number, some totals may appear inconsistent.

A dash (–) indicates data are not available.

An "r" indicates teacher response data available for 70-84% of students. An "s" indicates teacher response data available for 50-69% of students.

What Can Be Learned About the Mathematics Curriculum?

In contrast to the United States, most countries around the world have well-established, centrally-mandated national curricula. Recently, however, states and districts in the U.S. have been making great strides in establishing content standards and curriculum frameworks to guide curriculum implementation in schools. Furthermore, many education systems in the U.S. have begun to assess whether the intended curriculum in mathematics is being attained or learned by their students.

Although effort has been made to develop rigorous curriculum standards, the intended mathematics curriculum in the United States overall and in many Benchmarking jurisdictions does not seem as advanced or focused as that in other countries. Students in the U.S. are generally taught more topics with less depth, with each often spread over the course of more grades, than are their peers in other nations.[10] This lack of focus has been cited as a potential explanation for the relatively poor academic performance of U.S. students compared with those in other nations.[11] Thoroughly examining the Benchmarking jurisdictions' results in an international context can provide insights into what students are expected to learn in mathematics, what is taught in classrooms, and what policies and practices provide the best match between the intended and the implemented curriculum to improve student achievement.

10 Schmidt, W.H., McKnight, C.C., and Raizen, S.A. (1997), *A Splintered Vision: An Investigation of U.S. Science and Mathematics Education*, Dordrecht, the Netherlands: Kluwer Academic Publishers.

11 Mayer, D.P., Mullens, J.E., and Moore, M.T. (2000), *Monitoring School Quality: An Indicators Report*, NCES 2001-030, Washington, DC: National Center for Education Statistics.

CHAPTER 6 Teachers and Instruction

Chapter 6 presents information about mathematics teachers and instruction. Teachers' reports are given on their educational background, teaching preparation, and instructional practices. Information is also provided about how teachers spend their time related to teaching tasks, the materials used in instruction, the activities students do in class, the use of calculators and computers in mathematics lessons, the role of homework, and the reliance on different types of assessment.

Teachers and the instructional approaches they use determine the mathematics students learn. They structure the content and pace of lessons, introducing new material, selecting various instructional activities, and monitoring students' developing understanding of the concepts studied. Teachers may help students use technology and tools to investigate mathematical ideas, analyze students' work for misconceptions, and promote positive attitudes towards mathematics. They may also assign homework and conduct formal and informal assessments to evaluate achievement. Chapter 6 presents mathematics teachers' reports on some of these issues.

Because the sampling for the teacher questionnaires was based on participating students, teachers' responses do not necessarily represent all eighth-grade mathematics teachers in each participating entity. Rather, they represent teachers of the representative samples of students assessed. It is important to note that when information from the teacher questionnaire is reported, the student is always the unit of analysis. That is, the data shown are the percentages of *students* whose teachers reported on various characteristics or instructional strategies. Using the student as the unit of analysis makes it possible to describe the mathematics instruction received by representative samples of students. Although this perspective may differ from that obtained by simply collecting information from teachers, it is consistent with the TIMSS goals of examining the educational contexts and performance of students.

The teachers who completed the questionnaires were the mathematics teachers of the students who took the TIMSS 1999 test. The general sampling procedure was to sample a mathematics class from each participating school, administer the test to those students, and ask their teacher to complete the questionnaire. Thus, the information about instruction is tied directly to the students tested. Sometimes, however, teachers did not complete the questionnaire assigned to them, so most entities had some percentage of students for whom no teacher questionnaire information is available. The exhibits in this chapter have special notations on this point. For a TIMSS 1999 participating entity (country, state, district, or consortium) where teacher responses are available for 70 to 84 percent of the students, an "r" is included next to the data. Where teacher responses are available for 50 to 69 percent of students, an "s" is included; where they are available for less than 50 percent, an "x" replaces the data.

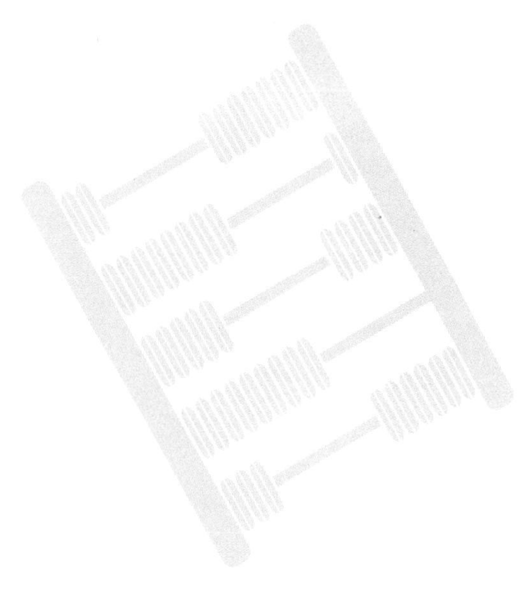

What Preparation Do Teachers Have for Teaching Mathematics?

This section presents information about background characteristics of mathematics teachers, including age and gender, major area of study, and certification. Teachers' confidence in teaching various mathematics topics is also discussed.

As shown by the international average at the bottom of Exhibit 6.1, the majority of the eighth-grade students internationally were taught mathematics by teachers in their 30s and 40s. If there were a steady replenishing of the teaching force, one might expect approximately equivalent percentages of students taught by teachers in their 20s, 30s, 40s, and 50s. Very few countries, however, had a comparatively younger teaching force. Internationally on average, only 16 percent of students were taught by teachers younger than age 30. Although 21 percent of students internationally were taught by teachers age 50 or older, the teaching force was relatively older in a number of countries.

Most Benchmarking participants did not differ substantially from the international profile. However, the Academy School District and the Jersey City Public Schools had no students with teachers in their 20s and had larger percentages of students with teachers in their 40s and 50s than internationally. Similarly, the Chicago Public Schools, the Miami-Dade County Public Schools, the Project SMART Consortium, and the Southwest Pennsylvania Math and Science Collaborative had more than 65 percent of their students taught by teachers 40 years or older compared with 54 percent internationally. On the other hand, the teachers in the Fremont/Lincoln/Westside Public Schools were younger than the international average – 67 percent of the students had teachers under age 40 compared with 46 percent internationally.

Internationally on average, 60 percent of eighth-grade students were taught mathematics by females and 40 percent by males, and similar percentages were found in a number of countries. None of the TIMSS 1999 Benchmarking states differed from the international profile of having more students taught by female mathematics teachers than males. In South Carolina, in particular, 85 percent of the students were taught mathematics by female teachers. Among the Benchmarking districts and consortia, the First in the World Consortium, the Fremont/Lincoln/Westside Public Schools, Guilford County, and Montgomery County had more than three-fourths of their students taught by female mathematics teachers. In comparison, the Michigan Invitational Group, the Naperville School District, and the Southwest Pennsylvania Math and Science Collaborative had more male than female teachers.

Exhibit 6.2 presents teachers' reports about their major areas of study during their post-secondary teacher preparation programs. Teachers' undergraduate and graduate studies give some indication of their preparation to teach mathematics. Also, research shows that higher achievement in mathematics is associated with teachers having a bachelor's and/or master's degree in mathematics.[1] According to their teachers, however, U.S. eighth-grade students were less likely than those in other countries to be taught mathematics by teachers with a major area of study in mathematics.

On average internationally, 71 percent of students were taught by teachers who had mathematics as a major area of study. (Note that teachers can have dual majors, or different majors at the undergraduate and graduate level.) This compares with 41 percent for the United States, a figure not too different from that for many Benchmarking participants, although there was a range of 16 percent in Jersey City to 73 percent in First in the World and Naperville. Suffice it to say that in the United States and most Benchmarking entities, a smaller percentage of students than the international average was taught by mathematics teachers with a major in mathematics. Canada and Italy were the only nations that reported lower percentages than the United States.

Internationally on average, 31 percent of the students were taught by teachers with mathematics education as a major area of study. In comparison, more than half of the students were taught by teachers with this major in the states of Illinois, Michigan, and Pennsylvania, as well as in the districts and consortia of Chicago, First in the World, the Fremont/Lincoln/Westside Public Schools, Guilford County, Project SMART, Rochester, and the Southwest Pennsylvania Math and Science Collaborative.

Internationally on average, 32 percent of the students were taught by teachers with education as a major area of study. Significantly more students in the United States (54 percent) had mathematics teachers with an education major than did students internationally. In general across the Benchmarking participants, about twice as many teachers reported an education major as did internationally. It is clear that teachers in the United States have less "in field" mathematics preparation than their counterparts around the world.

To gauge teachers' confidence in their ability to teach mathematics topics, TIMSS constructed an index of teachers' confidence in their preparation to teach mathematics (CPTM), presented in Exhibit 6.3. Teachers were asked how well prepared they felt to teach each of 12 mathematics topics (e.g., properties of geometric figures, solving linear equations and

1 Goldhaber, D.D. and Brewer, D.J. (1997), "Evaluating the Effect of Teacher Degree Level on Educational Performance" in W. Fowler (ed.), *Developments in School Finance*, 1996, NCES 97-535, Washington DC: National Center for Education Statistics; Darling-Hammond, L. (2000), *Teacher Quality and Student Achievement: A Review of State Policy Evidence*, Education Policy Analysis Archives, 8(1).

inequalities). There were three possible responses: very well prepared was assigned a value of three, somewhat prepared two, and not well prepared one. Students were assigned to the high level of the index if their teachers reported feeling very well prepared, on average, across the 12 topics (2.75 or higher). The medium level indicates that teachers reported being somewhat to well prepared (averages from 2.25 to 2.75), and the low level that they felt only somewhat prepared or less (averages less than 2.25).

The results show that average mathematics achievement is related to how well prepared teachers felt they were to teach mathematics, with higher achievement related to higher levels of teachers' confidence. On average internationally, teachers reported relatively high degrees of confidence, with 63 percent of students taught by teachers who believed they were very well prepared. Interestingly, for the United States as a whole and most Benchmarking entities, more students were taught mathematics by teachers confident about their preparation than in almost all the comparison countries. Interpreting these results should take several factors into account. For example, cultural issues may dictate that teachers in the high-scoring Asian countries are more reserved about reporting their strengths and abilities. Also, when the mathematics curriculum is more challenging, teachers may feel less confident in their academic and pedagogical preparation. Nevertheless, it appears that in relation to both high- and low-performing countries around the world, teachers in many Benchmarking entities and in the United States overall may be overconfident about their preparation to teach eighth-grade mathematics.

Exhibit R3.1 in the reference section provides the detail for the 12 topics comprising the confidence in preparation index. On average across countries, the topics having the most students (from 79 to 82 percent) taught by teachers who felt very well prepared were "fractions, decimals, and percentages;" "ratios and proportions;" "perimeter, area, and volume;" "evaluate and perform operations on algebraic expressions;" and "solving linear equations and inequalities." Teachers reported being least well prepared to teach "simple probabilities – understanding and calculations;" just more than half the students internationally (55 percent on average) were taught by teachers who felt very well prepared to teach this topic.

For the Benchmarking jurisdictions, almost all students had teachers confident in their preparation to teach the two number topics that were included in the TIMSS questionnaire: "fractions, decimals, and percentages;" and "ratios and proportions." Similarly, in algebra 90 percent or more of students in most Benchmarking entities were taught by teachers who reported being very well prepared to teach the three algebra topics: "algebraic representation;" "evaluate and perform operations on algebraic expressions;" and "solving linear equations and inequalities." Similar results were obtained for the topics "representation and interpretation of data in graphs, charts, and tables;" and "simple probabilities – understanding and calculations," even though teachers in Idaho, Massachusetts, and North Carolina were less confident about this latter topic. Teachers also appeared confident in their preparation to teach "measurement – units, instruments, and accuracy,"except in North Carolina, the Fremont/Lincoln/Westside Public Schools, Guilford County, and Rochester, where less than 80 percent of the students were taught by teachers who felt very well prepared to teach this topic. The pattern of less confidence in teaching this measurement topic was found internationally and for the United States.

Teachers in the Benchmarking entities expressed the least confidence in their preparation to teach geometry. Less than 80 percent of the students in Idaho, Oregon, the Delaware Science Coalition, and the Fremont/Lincoln/Westside Public Schools had teachers confident about their preparation in any of the three geometry topics. Across nearly all the participating states as well as in a number of the districts and consortia, teachers expressed less than full confidence in their preparation to teach "geometric figures – symmetry, motions and transformations, congruence and similarity." Interestingly, this pattern was also noted internationally and for the United States, even though these topics are included in the curriculum and taught to substantial percentages of eighth-grade students in the U.S. and abroad. Beyond those already mentioned, Benchmarking entities where less than 80 percent of students had teachers confident about their preparation to teach "coordinate geometry" were Illinois, Indiana, Missouri, the Jersey City Public Schools, and the Miami-Dade County Public Schools.

Exhibit R3.2 shows principals' opinions about the degree to which shortages of qualified mathematics teachers affect the capacity to provide instruction. On average internationally, principals reported that such shortages affect the quality of instruction some or a lot for one-third of the students. This compares with 16 percent in the United States. Benchmarking entities where principals reported that such shortages affect the capacity to provide instruction for more than one-fourth of the students were Maryland, South Carolina, Texas, Chicago, Guilford County, Jersey City, Montgomery County, and Rochester.

Teachers' beliefs about mathematics learning and instruction are to some degree related to their preparation. Exhibits R3.3 and R3.4 in the reference section show the percentages of eighth-grade students whose mathematics teachers reported certain beliefs about mathematics, the way mathematics should be taught, and the importance of various cognitive skills in achieving success in the discipline. In general, more students in the Benchmarking entities than internationally were taught by teachers agreeing that mathematics is primarily a formal way of representing the real world. Conversely, more students internationally than in the Benchmarking entities had teachers who agreed that some students have a natural talent for mathematics, and that an effective teaching approach is to give students having difficulty more practice by themselves during class. There was nearly complete agreement by teachers throughout the Benchmarking jurisdictions and around the world that more than one representation should be used in teaching a mathematics topic. Views varied substantially, for both the countries and the Benchmarking entities, regarding the importance of being able to remember formulas and procedures. Less than one-quarter of the students in the Delaware Science Coalition (similar to Chinese Taipei and Korea) were taught by teachers who believed remembering formulas and procedures was very important for students' success in mathematics. In contrast, more than half the students in Idaho, South Carolina, Guilford County, Jersey City, and Rochester (similar to the Russian Federation) had teachers who believed this to be the case.

How teachers spend their time in school is determined mainly by school and district policies and practices, but the perspectives they gain during their teacher preparation can also have an effect. Across countries, students' mathematics teachers spent only about 60 percent of their formally scheduled school time teaching mathematics (see Exhibit R3.5 in the reference section). Additionally, about 10 percent was spent teaching subjects other than mathematics, about 10 percent on curriculum planning, and about 20 percent on various administrative and other duties. The results for the United States as a whole and for most of the Benchmarking entities were very similar to the international profile.

Exhibit 6.1 Age and Gender of Teachers

		Percentage of Students by Age of Teachers					Percentage of Students by Gender of Teachers	
		29 Years or Under	30-39 Years	40-49 Years	50 Years or Older		Female	Male
Countries								
United States		11 (2.0)	25 (3.5)	37 (3.9)	27 (2.9)		60 (3.0)	40 (3.0)
Belgium (Flemish)		20 (2.7)	15 (2.4)	38 (3.0)	27 (3.1)		66 (4.8)	34 (4.8)
Canada		17 (2.4)	33 (2.7)	25 (3.1)	26 (3.0)		53 (3.0)	47 (3.0)
Chinese Taipei		10 (2.6)	34 (4.0)	30 (4.0)	26 (3.4)		51 (4.1)	49 (4.1)
Czech Republic		7 (2.5)	29 (4.8)	22 (5.0)	43 (5.6)		73 (4.0)	27 (4.0)
England	s	20 (2.9)	23 (3.5)	35 (3.6)	22 (2.7)	s	48 (3.8)	52 (3.8)
Hong Kong, SAR		32 (4.2)	38 (4.5)	19 (3.3)	11 (2.6)		44 (4.1)	56 (4.1)
Italy		0 (0.0)	8 (2.0)	58 (4.1)	34 (3.8)		76 (3.1)	24 (3.1)
Japan		21 (3.3)	39 (4.3)	33 (3.7)	7 (2.1)		27 (3.6)	73 (3.6)
Korea, Rep. of		19 (3.0)	53 (3.7)	15 (2.5)	13 (2.8)		59 (3.4)	41 (3.4)
Netherlands	r	15 (4.3)	17 (3.9)	41 (5.4)	26 (5.3)		28 (5.0)	72 (5.0)
Russian Federation		8 (2.0)	32 (3.7)	29 (2.9)	31 (4.0)		93 (2.6)	7 (2.6)
Singapore		37 (4.4)	25 (4.0)	15 (3.2)	23 (3.6)		75 (4.1)	25 (4.1)
States								
Connecticut	r	17 (5.9)	18 (4.1)	35 (7.4)	30 (7.6)	r	77 (6.7)	23 (6.7)
Idaho	r	7 (3.0)	28 (6.6)	43 (7.4)	22 (6.3)	r	56 (6.1)	44 (6.1)
Illinois		22 (5.7)	17 (3.8)	31 (5.9)	30 (7.1)		75 (4.7)	25 (4.7)
Indiana		26 (7.5)	18 (4.2)	26 (6.3)	30 (6.2)		57 (6.9)	43 (6.9)
Maryland	r	24 (5.0)	19 (4.1)	32 (5.7)	26 (6.0)	r	69 (4.8)	31 (4.8)
Massachusetts		17 (5.2)	18 (3.8)	27 (4.6)	38 (5.1)		57 (5.7)	43 (5.7)
Michigan		19 (3.7)	33 (5.7)	29 (5.2)	19 (4.6)		60 (5.7)	40 (5.7)
Missouri		11 (4.0)	40 (5.9)	29 (6.4)	20 (4.4)		66 (6.7)	34 (6.7)
North Carolina		29 (5.6)	23 (5.9)	35 (6.6)	13 (4.4)		75 (4.2)	25 (4.2)
Oregon		19 (3.2)	16 (4.3)	36 (6.7)	29 (6.6)		57 (5.0)	43 (5.0)
Pennsylvania		25 (6.9)	19 (4.4)	32 (5.6)	24 (5.7)		54 (5.4)	46 (5.4)
South Carolina		23 (5.7)	32 (4.8)	19 (3.5)	27 (5.7)		85 (5.1)	15 (5.1)
Texas		17 (5.0)	25 (4.3)	38 (6.1)	21 (4.1)		67 (5.6)	33 (5.6)
Districts and Consortia								
Academy School Dist. #20, CO		0 (0.0)	18 (0.3)	48 (0.4)	35 (0.3)		67 (0.4)	33 (0.4)
Chicago Public Schools, IL		9 (3.4)	25 (10.1)	39 (8.6)	27 (7.9)		70 (10.4)	30 (10.4)
Delaware Science Coalition, DE	r	22 (6.5)	27 (5.9)	26 (4.2)	26 (5.2)	r	57 (4.9)	43 (4.9)
First in the World Consort., IL		27 (6.8)	19 (8.4)	26 (9.3)	28 (5.6)		84 (4.7)	16 (4.7)
Fremont/Lincoln/WestSide PS, NE		28 (8.5)	39 (7.3)	7 (0.2)	25 (6.4)		78 (6.8)	22 (6.8)
Guilford County, NC		29 (6.7)	29 (4.8)	31 (3.6)	10 (4.5)		89 (3.5)	11 (3.5)
Jersey City Public Schools, NJ		0 (0.0)	23 (3.0)	37 (3.8)	40 (4.3)		57 (4.4)	43 (4.4)
Miami-Dade County PS, FL	s	14 (6.1)	21 (7.8)	32 (8.1)	34 (7.9)	s	68 (11.5)	32 (11.5)
Michigan Invitational Group, MI		25 (4.7)	12 (4.6)	32 (6.6)	32 (7.5)		49 (8.6)	51 (8.6)
Montgomery County, MD	s	25 (7.5)	11 (1.7)	29 (8.2)	35 (11.2)	s	84 (3.8)	16 (3.8)
Naperville Sch. Dist. #203, IL		22 (3.5)	18 (3.2)	30 (3.8)	30 (3.0)		25 (5.1)	75 (5.1)
Project SMART Consortium, OH		15 (5.1)	16 (5.0)	34 (5.8)	34 (6.3)		50 (5.4)	50 (5.4)
Rochester City Sch. Dist., NY		24 (5.2)	14 (4.2)	36 (3.8)	26 (4.5)		54 (5.4)	46 (5.4)
SW Math/Sci. Collaborative, PA		10 (2.9)	16 (5.2)	32 (6.4)	42 (5.5)		42 (5.2)	58 (5.2)
International Avg. (All Countries)		16 (0.5)	30 (0.6)	33 (0.6)	21 (0.5)		60 (0.6)	40 (0.6)

SOURCE: IEA Third International Mathematics and Science Study (TIMSS), 1998-1999.

Background data provided by teachers.

States in *italics* did not fully satisfy guidelines for sample participation rates (see Appendix A for details).

( ) Standard errors appear in parentheses. Because results are rounded to the nearest whole number, some totals may appear inconsistent.

An "r" indicates teacher response data available for 70-84% of students. An "s" indicates teacher response data available for 50-69% of students.

Exhibit 6.2 Teachers' Major Area of Study in Their BA, MA, or Teacher Training Certification Program

	Percentage of Students Whose Teachers Reported Having the Major Area of Study[1]				
	Mathematics	Mathematics Education	Science or Science Education	Education	Other
Countries					
United States	41 (3.4)	37 (3.4)	16 (2.4)	54 (3.4)	r 46 (3.6)
Belgium (Flemish)	89 (2.6)	38 (3.8)	73 (3.5)	42 (2.9)	37 (3.5)
Canada	22 (2.7)	19 (2.2)	24 (2.8)	49 (3.2)	68 (2.9)
Chinese Taipei	82 (3.7)	39 (4.2)	11 (2.1)	32 (3.6)	23 (3.9)
Czech Republic	85 (3.8)	34 (5.6)	53 (6.0)	34 (5.5)	53 (4.9)
England	47 (3.3)	32 (2.9)	s 20 (2.6)	s 44 (3.4)	s 41 (3.5)
Hong Kong, SAR	57 (4.2)	30 (3.9)	38 (4.4)	36 (3.8)	47 (4.5)
Italy	22 (3.3)	0 (0.0)	66 (3.4)	0 (0.0)	16 (3.1)
Japan	79 (3.6)	27 (3.6)	4 (1.7)	15 (3.2)	21 (3.5)
Korea, Rep. of	55 (4.2)	61 (4.0)	4 (1.5)	19 (3.2)	9 (2.2)
Netherlands	68 (4.9)	16 (4.2)	25 (5.0)	12 (4.3)	14 (4.4)
Russian Federation	89 (2.9)	83 (3.1)	39 (4.0)	81 (3.1)	67 (3.9)
Singapore	78 (3.6)	32 (4.0)	38 (4.2)	48 (4.8)	47 (4.3)
States					
Connecticut	31 (5.2)	29 (5.3)	r 6 (3.2)	r 69 (5.2)	s 40 (7.4)
Idaho	28 (5.2)	34 (7.2)	r 17 (5.3)	r 68 (5.9)	r 43 (7.4)
Illinois	61 (4.8)	55 (6.5)	13 (5.1)	71 (5.0)	43 (4.6)
Indiana	55 (7.3)	48 (5.0)	17 (5.1)	63 (5.0)	26 (5.5)
Maryland	40 (5.7)	35 (6.0)	r 8 (2.7)	r 63 (6.6)	r 37 (5.2)
Massachusetts	60 (5.1)	35 (4.9)	9 (2.9)	59 (4.7)	29 (5.5)
Michigan	51 (5.9)	53 (6.9)	32 (6.4)	64 (6.3)	52 (6.1)
Missouri	61 (6.4)	49 (5.2)	14 (5.2)	79 (4.3)	32 (5.9)
North Carolina	50 (5.0)	50 (6.6)	26 (4.1)	61 (5.6)	31 (5.0)
Oregon	39 (4.8)	39 (6.4)	21 (4.6)	66 (6.1)	49 (5.9)
Pennsylvania	58 (6.0)	53 (4.7)	8 (3.3)	61 (5.8)	r 33 (4.4)
South Carolina	53 (6.1)	45 (6.0)	6 (2.7)	61 (6.3)	25 (6.0)
Texas	50 (6.5)	29 (6.0)	r 18 (5.5)	r 47 (8.1)	r 51 (6.2)
Districts and Consortia					
Academy School Dist. #20, CO	55 (0.4)	39 (0.4)	20 (0.4)	66 (0.4)	12 (0.2)
Chicago Public Schools, IL	37 (9.4)	51 (9.8)	13 (2.8)	74 (9.5)	r 59 (9.1)
Delaware Science Coalition, DE	23 (5.2)	36 (6.5)	r 12 (4.6)	r 65 (6.6)	r 59 (7.4)
First in the World Consort., IL	73 (7.2)	75 (7.8)	21 (5.0)	77 (3.4)	38 (7.9)
Fremont/Lincoln/WestSide PS, NE	65 (3.1)	56 (6.0)	5 (0.2)	57 (9.1)	58 (5.5)
Guilford County, NC	59 (4.8)	64 (6.4)	13 (4.3)	47 (5.8)	37 (5.7)
Jersey City Public Schools, NJ	16 (4.9)	18 (2.6)	4 (2.7)	79 (5.0)	r 55 (6.4)
Miami-Dade County PS, FL	31 (7.9)	27 (8.8)	s 18 (8.7)	s 55 (9.3)	s 84 (6.0)
Michigan Invitational Group, MI	64 (7.6)	36 (8.9)	29 (4.0)	55 (10.0)	47 (8.0)
Montgomery County, MD	27 (6.1)	28 (7.3)	s 6 (1.2)	s 76 (7.1)	s 37 (6.2)
Naperville Sch. Dist. #203, IL	73 (5.4)	30 (2.8)	2 (0.5)	50 (5.9)	57 (4.8)
Project SMART Consortium, OH	67 (4.6)	61 (6.4)	11 (4.5)	61 (7.7)	47 (5.3)
Rochester City Sch. Dist., NY	70 (3.6)	58 (5.0)	6 (1.7)	56 (5.5)	38 (4.0)
SW Math/Sci. Collaborative, PA	63 (4.9)	61 (6.5)	12 (5.3)	64 (7.7)	r 31 (7.4)
International Avg. (All Countries)	71 (0.6)	31 (0.6)	35 (0.6)	32 (0.6)	32 (0.6)

SOURCE: IEA Third International Mathematics and Science Study (TIMSS), 1998-1999.

Background data provided by teachers.

1 Teachers who responded that they majored in more than one area are reflected in all categories that apply.

States in *italics* did not fully satisfy guidelines for sample participation rates (see Appendix A for details).

( ) Standard errors appear in parentheses. Because results are rounded to the nearest whole number, some totals may appear inconsistent.

An "r" indicates teacher response data available for 70-84% of students. An "s" indicates teacher response data available for 50-69% of students.

Exhibit 6.3 Index of Teachers' Confidence in Preparation to Teach Mathematics (CPTM)

8th Grade Mathematics

Index of Teachers' Confidence in Preparation to Teach Mathematics

Index based on teachers' responses to 12 questions about how prepared they feel to teach different mathematics topics (see reference exhibit R3.1) based on a 3-point scale: 1 = not well prepared; 2 = somewhat prepared; 3 = very well prepared. Average is computed across the 12 items for items for which the teacher did not respond do not teach. High level indicates average is greater than or equal to 2.75. Medium level indicates average is greater than or equal to 2.25 and less than 2.75. Low level indicates average is less than 2.25.

		High CPTM		Medium CPTM		Low CPTM	
		Percent of Students	Average Achievement	Percent of Students	Average Achievement	Percent of Students	Average Achievement
Jersey City Public Schools, NJ		97 (2.7)	479 (9.0)	3 (2.7)	351 (4.6)	0 (0.0)	~ ~
Naperville Sch. Dist. #203, IL		95 (1.9)	570 (3.0)	5 (1.9)	529 (8.9)	0 (0.0)	~ ~
Michigan Invitational Group, MI		94 (2.1)	530 (5.0)	6 (2.1)	519 (27.2)	0 (0.0)	~ ~
SW Math/Sci. Collaborative, PA		94 (3.4)	519 (8.1)	5 (3.4)	508 (20.0)	1 (0.0)	~ ~
Rochester City Sch. Dist., NY		93 (2.0)	444 (7.3)	7 (2.0)	406 (16.9)	0 (0.0)	~ ~
First in the World Consort., IL		93 (5.5)	564 (6.4)	7 (5.5)	491 (11.8)	0 (0.0)	~ ~
Academy School Dist. #20, CO		92 (0.2)	531 (1.9)	0 (0.0)	~ ~	8 (0.2)	495 (5.0)
Maryland	r	92 (3.0)	489 (5.6)	8 (3.0)	444 (28.1)	0 (0.0)	~ ~
Missouri		92 (3.3)	492 (5.8)	6 (2.6)	476 (13.2)	2 (1.6)	~ ~
South Carolina		92 (3.6)	506 (8.4)	8 (3.6)	472 (22.6)	0 (0.0)	~ ~
Pennsylvania		92 (5.0)	512 (7.2)	4 (1.7)	496 (27.7)	5 (4.7)	501 (6.7)
Michigan		91 (3.3)	525 (6.9)	8 (3.3)	479 (17.0)	1 (0.6)	~ ~
Project SMART Consortium, OH		90 (4.1)	526 (8.1)	10 (4.1)	476 (16.7)	0 (0.0)	~ ~
North Carolina		88 (4.1)	497 (7.0)	11 (4.0)	479 (13.7)	1 (0.9)	~ ~
United States		87 (2.4)	505 (4.2)	11 (2.3)	489 (7.0)	2 (1.0)	~ ~
Connecticut	r	87 (5.9)	519 (10.5)	11 (5.7)	526 (16.6)	1 (1.4)	~ ~
Illinois		87 (5.0)	516 (6.3)	12 (5.0)	479 (25.8)	1 (0.7)	~ ~
Massachusetts		87 (3.9)	513 (7.2)	10 (3.1)	535 (24.9)	3 (2.3)	486 (8.0)
Texas	r	87 (4.5)	525 (9.4)	12 (4.3)	485 (22.4)	1 (1.2)	~ ~
Chicago Public Schools, IL		87 (6.7)	470 (7.4)	13 (6.6)	452 (13.5)	1 (0.8)	~ ~
Indiana		86 (4.8)	513 (7.3)	11 (4.6)	545 (22.0)	2 (1.7)	~ ~
Miami-Dade County PS, FL	s	86 (5.2)	425 (11.9)	11 (5.2)	435 (53.0)	3 (2.5)	269 (37.9)
Guilford County, NC		85 (5.3)	517 (10.0)	13 (5.0)	490 (26.3)	2 (0.1)	~ ~
Montgomery County, MD	s	85 (6.5)	543 (5.2)	14 (6.6)	501 (9.9)	1 (0.1)	~ ~
Czech Republic		85 (3.6)	521 (5.1)	14 (3.8)	519 (9.5)	1 (1.3)	~ ~
Delaware Science Coalition, DE	r	85 (5.6)	480 (11.5)	12 (5.1)	499 (22.2)	3 (2.3)	417 (38.5)
Fremont/Lincoln/WestSide PS, NE		81 (4.9)	492 (10.6)	13 (4.9)	440 (24.3)	5 (0.2)	534 (5.0)
Netherlands		81 (6.2)	542 (7.1)	10 (3.0)	514 (22.4)	9 (5.8)	514 (58.7)
Oregon		78 (4.3)	516 (7.3)	18 (4.7)	506 (15.3)	4 (1.6)	480 (22.4)
Idaho	r	75 (4.9)	508 (8.2)	18 (6.1)	461 (12.3)	7 (3.8)	447 (34.9)
Chinese Taipei		71 (3.6)	586 (4.5)	15 (3.1)	587 (10.9)	14 (2.7)	572 (6.8)
Canada		71 (2.7)	537 (3.3)	21 (3.0)	530 (6.6)	8 (1.8)	515 (14.6)
Singapore		66 (4.2)	603 (7.1)	24 (3.7)	619 (12.0)	10 (2.8)	578 (20.8)
Belgium (Flemish)		65 (3.2)	559 (5.8)	32 (3.1)	561 (5.6)	3 (1.4)	558 (27.1)
Hong Kong, SAR		61 (4.3)	579 (5.5)	28 (3.9)	591 (8.2)	11 (2.7)	571 (12.0)
Italy		60 (3.9)	479 (5.5)	27 (3.5)	481 (7.2)	13 (2.3)	479 (12.4)
Korea, Rep. of		48 (3.9)	585 (3.2)	31 (3.8)	590 (4.1)	21 (3.0)	588 (3.5)
Japan		8 (2.1)	584 (6.1)	24 (3.6)	589 (4.2)	68 (4.0)	573 (2.6)
England		– –	– –	– –	– –	– –	– –
Russian Federation		– –	– –	– –	– –	– –	– –
International Avg. (All Countries)		63 (0.6)	489 (1.1)	23 (0.6)	481 (1.7)	14 (0.5)	473 (2.9)

SOURCE: IEA Third International Mathematics and Science Study (TIMSS), 1998-1999.

States in *italics* did not fully satisfy guidelines for sample participation rates (see Appendix A for details).

( ) Standard errors appear in parentheses. Because results are rounded to the nearest whole number, some totals may appear inconsistent.

A dash (–) indicates data are not available. A tilde (~) indicates insufficient data to report achievement.

An "r" indicates teacher response data available for 70-84% of students. An "s" indicates teacher response data available for 50-69% of students.

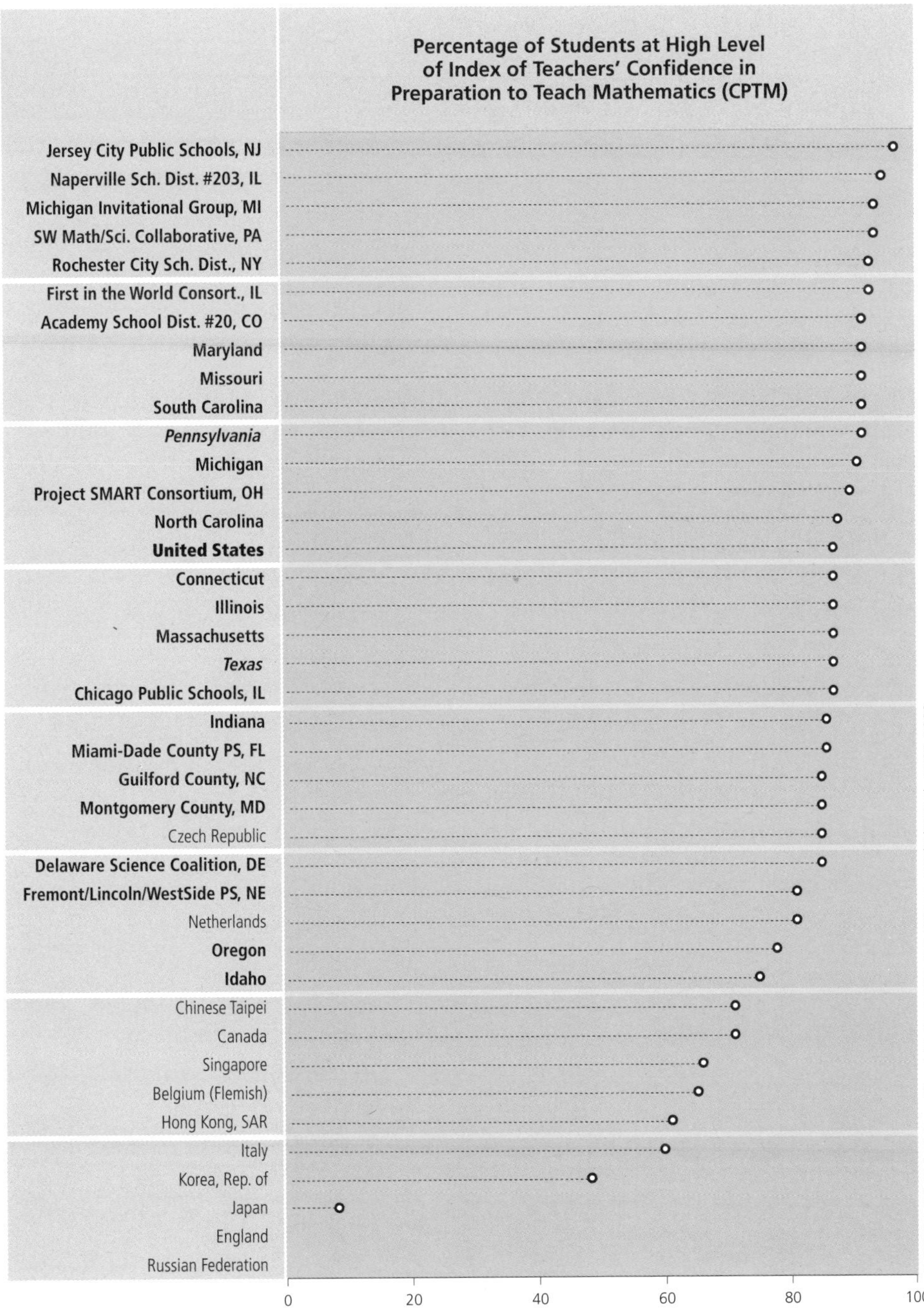

Exhibit 6.3 (Continued)

Index of Teachers' Confidence in Preparation to Teach Mathematics (CPTM)

8th Grade Mathematics

Percentage of Students at High Level of Index of Teachers' Confidence in Preparation to Teach Mathematics (CPTM)

Jersey City Public Schools, NJ
Naperville Sch. Dist. #203, IL
Michigan Invitational Group, MI
SW Math/Sci. Collaborative, PA
Rochester City Sch. Dist., NY
First in the World Consort., IL
Academy School Dist. #20, CO
Maryland
Missouri
South Carolina
Pennsylvania
Michigan
Project SMART Consortium, OH
North Carolina
United States
Connecticut
Illinois
Massachusetts
Texas
Chicago Public Schools, IL
Indiana
Miami-Dade County PS, FL
Guilford County, NC
Montgomery County, MD
Czech Republic
Delaware Science Coalition, DE
Fremont/Lincoln/WestSide PS, NE
Netherlands
Oregon
Idaho
Chinese Taipei
Canada
Singapore
Belgium (Flemish)
Hong Kong, SAR
Italy
Korea, Rep. of
Japan
England
Russian Federation

0 20 40 60 80 100

SOURCE: IEA Third International Mathematics and Science Study (TIMSS), 1998-1999.

How Much School Time Is Devoted to Mathematics Instruction?

Exhibit 6.4 presents information about the amount of mathematics instruction given to eighth-grade students in the TIMSS 1999 Benchmarking jurisdictions and the comparison countries. Since different systems have school years of different lengths (see Exhibit R3.6) and different arrangements of daily and weekly instruction, the information is given in terms of the average number of hours of mathematics instruction over the school year as reported by mathematics teachers. Canada provides 150 hours per year, on average, and the United States 144 hours, compared with the international average of 129 hours. Benchmarking entities with teachers reporting more than 150 hours of mathematics instruction per year were the Jersey City Public Schools, South Carolina, North Carolina, the Delaware Science Coalition, and the Fremont/Lincoln/Westside Public Schools. Interestingly, the teachers in the Naperville School District and the First in the World Consortium reported the least amount of mathematics instructional time (114 hours) per year. Among the reference countries, the percentage of instructional time at the eighth grade that was devoted to mathematics ranged from 17 percent in the Russian Federation to nine percent in Chinese Taipei and the Netherlands. Among the Benchmarking jurisdictions, the percentage ranged from 18 percent in North Carolina to 11 percent in Indiana, Pennsylvania, and First in the World.

As shown in Exhibit 6.5, teachers of about half the students, on average internationally, reported that mathematics classes meet for at least two hours per week but fewer than three and a half. For another one-third of students, classes meet for at least three and a half hours but fewer than five. On average, eighth graders in the United States spend more time in mathematics class per week (typically three and a half to five hours) than do their counterparts internationally. This pattern of more classroom time held for nearly all of the Benchmarking entities, with the exception of the Chicago Public Schools and Naperville (primarily two to three and a half hours), and North Carolina and the Jersey City Public Schools (primarily five hours or more).

The data, however, reveal no clear pattern between the number of in-class instructional hours and mathematics achievement either across or within participating entities. Common sense and research both support the idea that time on task is an important contributor to achievement, yet this time can be spent more or less efficiently. Time alone is not enough; it needs to be spent on high-quality mathematics instruction. Devoting extensive class time to remedial activities can deprive students of this. Also, instructional time can be spent out of school in various tutoring programs; low-performing students may be receiving additional instruction.

Videotapes of mathematics classes in the United States and Japan in TIMSS 1995 revealed that outside interruptions like those for announcements or to conduct administrative tasks can affect the flow of the lesson and detract from instructional time.[2] As shown in Exhibit 6.6, on average internationally about one-fifth of the students (21 percent) were in mathematics classes that were interrupted pretty often or almost always, and 28 percent were in classes that were never interrupted. In Japan and Korea, more than half the students were in mathematics classes that were never interrupted – compared with only 10 percent in the United States. In the United States, nearly one-third of the eighth graders were in mathematics classes that were interrupted pretty often or almost always. If anything, the teachers in most of the Benchmarking jurisdictions reported even more interruptions than did teachers in the U.S. nationally. The jurisdictions with more than 15 percent of students in classrooms that were never interrupted were Illinois, the First in the World Consortium, Montgomery County, and Naperville. Conversely, the jurisdictions with the highest percentages of students in classrooms almost always interrupted (17 to 18 percent) were the public school systems of Chicago, Jersey City, Miami-Dade, and Rochester. Students in mathematics classrooms that were frequently interrupted had substantially lower achievement than their counterparts in classrooms with fewer interruptions.

2 Stigler, J.W., Gonzales, P., Kawanaka, T., Knoll, S., and Serrano, A. (1999), *The TIMSS Videotape Classroom Study: Methods and Findings from an Exploratory Research Project on Eighth-Grade Mathematics Instruction in Germany, Japan, and the United States*, NCES 1999-074, Washington, DC: National Center for Education Statistics.

Exhibit 6.4 Mathematics Instructional Time at Grade 8

		Students' Average Yearly Mathematics Instructional Time in Hours		Mathematics Instructional Time as a Percent of Total Instructional Time[1]
Jersey City Public Schools, NJ	s	238 (9.8)		x x
South Carolina	r	189 (10.6)		x x
North Carolina	r	182 (8.9)	s	18 (1.1)
Delaware Science Coalition, DE	s	167 (17.8)		x x
Fremont/Lincoln/WestSide PS, NE	s	152 (15.3)		x x
Canada	r	150 (2.3)	r	15 (0.2)
Hong Kong, SAR	r	149 (5.4)	s	15 (0.5)
Oregon	r	148 (10.5)	s	14 (0.6)
Guilford County, NC	s	146 (6.2)	s	13 (0.4)
United States	s	144 (4.5)		x x
Texas	s	143 (10.3)		x x
Rochester City Sch. Dist., NY	r	143 (9.1)	s	13 (1.0)
Russian Federation	r	142 (3.3)	s	17 (0.6)
Missouri	r	142 (8.2)	s	14 (0.7)
Massachusetts	s	141 (4.9)		x x
Maryland	s	141 (6.9)	s	13 (0.6)
Czech Republic		139 (2.4)		15 (0.2)
Academy School Dist. #20, CO		138 (0.2)		x x
Indiana	r	135 (10.7)	s	11 (0.5)
Michigan	r	135 (5.3)		x x
Idaho	s	135 (9.0)		x x
Italy		130 (3.2)		12 (0.3)
Michigan Invitational Group, MI	r	129 (8.6)	s	13 (0.8)
Illinois	r	128 (8.7)	s	13 (1.0)
Japan		127 (1.8)		12 (0.2)
Chicago Public Schools, IL	s	127 (8.2)		x x
Chinese Taipei		126 (1.9)		9 (0.1)
Singapore		126 (3.8)		15 (0.5)
Project SMART Consortium, OH	r	124 (5.5)		x x
Pennsylvania	r	122 (7.6)	s	11 (0.7)
SW Math/Sci. Collaborative, PA		119 (3.8)	r	12 (0.4)
Korea, Rep. of		118 (3.5)		11 (0.3)
Belgium (Flemish)		116 (3.5)		12 (0.4)
England	s	115 (2.7)	s	12 (0.3)
Naperville Sch. Dist. #203, IL		114 (0.3)		12 (0.0)
First in the World Consort., IL	s	114 (9.8)	s	11 (1.2)
Netherlands	s	94 (1.6)	s	9 (0.1)
Connecticut		x x		x x
Miami-Dade County PS, FL		x x		x x
Montgomery County, MD		x x		x x
International Avg. (All Countries)		129 (0.7)		13 (0.1)

SOURCE: IEA Third International Mathematics and Science Study (TIMSS), 1998-1999.

Mathematics instructional time provided by teachers, and total instructional time provided by schools.

1 Computed as the ratio of mathematics instructional time to total instructional time averaged across students.

States in *italics* did not fully satisfy guidelines for sample participation rates (see Appendix A for details).

( ) Standard errors appear in parentheses. Because results are rounded to the nearest whole number, some totals may appear inconsistent.

An "r" indicates school and/or teacher response data available for 70-84% of students. An "s" indicates school and/or teacher response data available for 50-69% of students. An "x" indicates school and/or teacher response data available for <50% of students.

Exhibit 6.5 Number of Hours Mathematics Is Taught Weekly

	5 Hours or More		3.5 Hours to < 5		2 Hours to < 3.5		Less Than 2 Hours	
	Percent of Students	Average Achievement	Percent of Students	Average Achievement	Percent of Students	Average Achievement	Percent of Students	Average Achievement
Countries								
United States	16 (2.2)	490 (9.2)	56 (3.4)	501 (4.9)	17 (2.6)	528 (11.6)	11 (2.3)	491 (14.5)
Belgium (Flemish)	4 (1.0)	590 (11.7)	40 (2.8)	595 (4.1)	43 (3.8)	544 (7.7)	13 (3.4)	502 (18.9)
Canada r	17 (2.2)	520 (6.4)	55 (3.2)	544 (3.9)	26 (2.7)	523 (6.1)	3 (0.9)	503 (6.3)
Chinese Taipei	1 (1.1)	~ ~	48 (4.4)	592 (5.8)	51 (4.5)	577 (5.5)	0 (0.0)	~ ~
Czech Republic	4 (2.1)	600 (28.1)	52 (4.4)	517 (5.3)	44 (4.4)	517 (6.4)	0 (0.0)	~ ~
England s	2 (1.2)	~ ~	3 (1.4)	481 (10.2)	95 (2.0)	512 (5.3)	0 (0.2)	~ ~
Hong Kong, SAR	9 (2.3)	579 (15.2)	71 (4.0)	583 (5.6)	17 (3.1)	587 (11.1)	3 (1.5)	553 (16.7)
Italy	9 (2.1)	469 (11.5)	55 (3.8)	483 (5.3)	29 (4.0)	475 (7.4)	6 (1.8)	484 (10.3)
Japan	1 (1.3)	~ ~	2 (1.3)	~ ~	95 (2.0)	577 (2.1)	2 (0.9)	~ ~
Korea, Rep. of	2 (0.9)	~ ~	3 (1.1)	602 (9.6)	93 (1.8)	587 (2.1)	3 (1.1)	587 (11.7)
Netherlands	0 (0.0)	~ ~	0 (0.0)	~ ~	100 (0.5)	537 (7.2)	0 (0.0)	~ ~
Russian Federation	11 (2.5)	553 (13.4)	57 (4.1)	528 (7.7)	32 (3.8)	513 (8.5)	0 (0.0)	~ ~
Singapore	9 (2.3)	592 (24.7)	37 (3.8)	586 (11.2)	48 (4.0)	623 (7.5)	5 (2.0)	608 (20.0)
States								
Connecticut r	5 (2.5)	534 (14.7)	58 (6.1)	515 (11.1)	36 (6.7)	532 (15.2)	1 (0.1)	~ ~
Idaho r	13 (4.4)	488 (18.4)	65 (7.6)	499 (9.2)	13 (4.4)	512 (13.6)	10 (4.8)	454 (15.2)
Illinois	6 (2.2)	500 (9.6)	44 (6.6)	522 (9.3)	38 (6.5)	489 (11.0)	12 (5.1)	540 (8.5)
Indiana	7 (3.5)	565 (33.6)	55 (7.5)	509 (8.6)	26 (7.8)	517 (16.4)	12 (4.0)	517 (8.2)
Maryland r	17 (5.3)	474 (16.2)	60 (6.4)	489 (6.9)	10 (4.1)	504 (16.7)	13 (4.0)	472 (18.3)
Massachusetts r	12 (4.7)	513 (8.9)	69 (6.1)	511 (7.8)	15 (4.7)	522 (14.4)	3 (2.2)	549 (26.4)
Michigan	8 (3.1)	512 (18.6)	64 (5.4)	525 (9.5)	15 (4.0)	521 (13.3)	14 (2.9)	528 (11.4)
Missouri	7 (3.2)	479 (43.4)	65 (6.0)	491 (6.8)	22 (5.1)	493 (11.0)	6 (3.2)	502 (13.1)
North Carolina	48 (5.2)	493 (8.8)	37 (5.6)	498 (13.7)	7 (2.9)	492 (12.3)	8 (2.1)	491 (42.5)
Oregon	9 (3.8)	545 (13.7)	64 (6.6)	519 (6.5)	19 (4.7)	483 (18.4)	8 (2.4)	510 (30.5)
Pennsylvania	11 (5.1)	515 (11.1)	47 (5.0)	518 (9.9)	29 (3.8)	504 (7.3)	13 (5.5)	496 (17.7)
South Carolina	40 (6.2)	512 (7.6)	41 (5.5)	494 (15.0)	13 (4.7)	523 (24.2)	6 (2.5)	469 (36.9)
Texas r	16 (6.2)	530 (19.4)	59 (6.6)	528 (10.6)	12 (3.8)	520 (27.5)	12 (3.3)	488 (20.9)
Districts and Consortia								
Academy School Dist. #20, CO	9 (0.2)	527 (4.6)	75 (0.3)	535 (2.1)	8 (0.2)	529 (5.5)	8 (0.2)	513 (4.7)
Chicago Public Schools, IL	6 (3.6)	460 (32.4)	19 (7.8)	465 (14.0)	69 (7.6)	469 (7.9)	5 (3.0)	430 (23.3)
Delaware Science Coalition, DE r	20 (6.9)	507 (27.1)	56 (7.3)	464 (13.3)	21 (5.1)	510 (11.7)	3 (2.3)	417 (25.8)
First in the World Consort., IL	2 (2.4)	~ ~	60 (1.5)	564 (7.0)	26 (4.3)	539 (6.2)	12 (5.1)	559 (21.5)
Fremont/Lincoln/WestSide PS, NE	8 (5.2)	493 (28.5)	77 (3.8)	494 (10.1)	12 (1.2)	477 (4.5)	3 (0.1)	323 (9.2)
Guilford County, NC	15 (3.8)	500 (16.6)	64 (5.2)	513 (11.3)	6 (3.7)	524 (25.9)	15 (3.7)	502 (22.1)
Jersey City Public Schools, NJ r	69 (6.0)	467 (5.7)	31 (6.0)	495 (22.3)	0 (0.0)	~ ~	0 (0.0)	~ ~
Miami-Dade County PS, FL s	20 (7.6)	371 (26.6)	45 (10.7)	443 (18.3)	16 (8.1)	415 (22.3)	19 (7.0)	442 (33.6)
Michigan Invitational Group, MI	10 (2.6)	519 (4.7)	64 (7.8)	532 (7.6)	7 (1.1)	516 (28.3)	19 (6.9)	552 (10.7)
Montgomery County, MD s	3 (1.1)	598 (17.6)	67 (12.6)	539 (7.1)	20 (11.8)	533 (13.1)	11 (7.3)	503 (11.6)
Naperville Sch. Dist. #203, IL	2 (0.1)	~ ~	0 (0.0)	~ ~	89 (0.4)	571 (3.2)	9 (0.4)	549 (3.6)
Project SMART Consortium, OH	7 (2.1)	536 (40.9)	51 (6.0)	519 (11.2)	31 (5.5)	525 (13.1)	11 (3.2)	505 (10.1)
Rochester City Sch. Dist., NY	6 (3.3)	509 (31.2)	59 (3.9)	427 (7.7)	35 (2.9)	454 (12.9)	0 (0.0)	~ ~
SW Math/Sci. Collaborative, PA	5 (3.2)	511 (29.8)	41 (6.9)	524 (12.1)	44 (7.6)	505 (10.6)	10 (3.3)	551 (27.2)
International Avg. (All Countries)	9 (0.3)	481 (3.5)	34 (0.5)	492 (2.3)	53 (0.5)	490 (1.9)	4 (0.3)	485 (4.7)

SOURCE: IEA Third International Mathematics and Science Study (TIMSS), 1998-1999.

Background data provided by teachers.

States in *italics* did not fully satisfy guidelines for sample participation rates (see Appendix A for details).

( ) Standard errors appear in parentheses. Because results are rounded to the nearest whole number, some totals may appear inconsistent.

A tilde (~) indicates insufficient data to report achievement.

An "r" indicates teacher response data available for 70-84% of students. An "s" indicates teacher response data available for 50-69% of students.

	Never		Once in a While		Pretty Often		Almost Always	
	Percent of Students	Average Achievement	Percent of Students	Average Achievement	Percent of Students	Average Achievement	Percent of Students	Average Achievement
Countries								
United States	10 (0.4)	494 (8.2)	59 (0.9)	522 (3.9)	20 (0.5)	488 (3.9)	11 (0.6)	455 (5.1)
Belgium (Flemish)	24 (1.1)	557 (5.9)	62 (1.1)	566 (2.9)	9 (0.7)	562 (6.8)	5 (0.8)	505 (20.3)
Canada	9 (0.4)	528 (4.2)	64 (1.0)	540 (2.4)	18 (0.7)	517 (3.9)	9 (0.7)	502 (7.8)
Chinese Taipei	22 (1.1)	580 (6.1)	56 (1.0)	594 (4.4)	17 (0.9)	580 (5.4)	6 (0.6)	563 (9.0)
Czech Republic	33 (1.7)	520 (4.0)	59 (1.3)	524 (4.7)	4 (0.5)	517 (11.4)	4 (0.8)	472 (13.7)
England	10 (0.8)	508 (9.5)	66 (1.2)	509 (4.2)	19 (1.1)	474 (6.0)	6 (0.6)	437 (8.9)
Hong Kong, SAR	36 (1.0)	585 (4.4)	54 (0.8)	588 (4.0)	8 (0.6)	552 (8.9)	2 (0.2)	~ ~
Italy	16 (1.0)	480 (5.5)	54 (1.2)	488 (4.0)	18 (1.0)	477 (5.3)	11 (0.8)	450 (7.6)
Japan	53 (1.4)	580 (2.7)	42 (1.3)	581 (2.5)	4 (0.3)	559 (5.9)	1 (0.2)	~ ~
Korea, Rep. of	57 (0.9)	581 (2.0)	38 (0.8)	598 (3.0)	4 (0.2)	579 (7.5)	1 (0.1)	~ ~
Netherlands	39 (1.3)	539 (7.7)	55 (1.3)	544 (8.3)	4 (0.5)	524 (14.0)	2 (0.4)	~ ~
Russian Federation	17 (1.5)	538 (11.1)	64 (1.5)	533 (5.2)	10 (0.9)	506 (7.5)	9 (0.7)	497 (6.9)
Singapore	16 (0.8)	592 (8.9)	64 (1.0)	614 (5.9)	14 (0.6)	585 (7.4)	6 (0.4)	579 (9.5)
States								
Connecticut	10 (1.1)	529 (12.6)	59 (2.3)	529 (8.7)	18 (1.7)	488 (9.1)	12 (1.3)	471 (12.3)
Idaho	11 (0.9)	484 (14.8)	60 (1.7)	510 (6.0)	18 (1.1)	475 (8.9)	11 (1.0)	463 (9.1)
Illinois	16 (1.2)	521 (9.7)	61 (1.5)	519 (7.0)	15 (1.1)	487 (8.2)	9 (0.9)	472 (7.8)
Indiana	10 (1.2)	511 (9.8)	66 (1.6)	527 (7.3)	16 (1.1)	495 (7.5)	7 (0.8)	471 (11.3)
Maryland	12 (0.9)	494 (9.8)	60 (1.6)	513 (5.4)	17 (1.0)	475 (7.2)	11 (1.0)	465 (9.8)
Massachusetts	11 (0.7)	521 (10.0)	62 (1.3)	526 (5.8)	19 (1.2)	495 (6.8)	8 (0.8)	464 (7.7)
Michigan	11 (1.3)	509 (12.2)	61 (2.0)	534 (6.4)	18 (1.6)	501 (8.1)	11 (1.3)	476 (6.7)
Missouri	10 (0.8)	483 (9.3)	58 (1.8)	500 (5.6)	20 (0.9)	489 (6.4)	12 (1.3)	454 (9.4)
North Carolina	7 (0.5)	474 (13.6)	60 (2.0)	513 (7.2)	21 (1.1)	485 (6.6)	12 (1.3)	448 (7.5)
Oregon	11 (0.9)	491 (8.0)	59 (1.6)	532 (5.9)	19 (0.9)	499 (6.5)	11 (0.8)	486 (9.0)
Pennsylvania	13 (1.4)	506 (10.4)	59 (1.7)	522 (6.1)	18 (1.0)	494 (5.2)	10 (1.0)	462 (10.7)
South Carolina	9 (1.1)	482 (11.4)	56 (2.2)	523 (7.7)	23 (2.1)	485 (8.3)	12 (1.0)	461 (9.8)
Texas	12 (0.8)	497 (17.0)	55 (2.1)	536 (9.0)	22 (1.5)	517 (8.1)	11 (1.0)	485 (11.2)
Districts and Consortia								
Academy School Dist. #20, CO	4 (0.6)	504 (12.0)	57 (1.2)	536 (2.6)	26 (1.3)	531 (4.6)	12 (1.1)	506 (6.6)
Chicago Public Schools, IL	7 (1.0)	435 (14.8)	49 (4.3)	478 (6.4)	27 (2.6)	456 (8.3)	17 (2.8)	447 (10.8)
Delaware Science Coalition, DE	11 (0.9)	466 (9.2)	59 (2.6)	500 (9.9)	17 (1.2)	472 (8.2)	13 (1.5)	453 (10.8)
First in the World Consort., IL	17 (1.3)	559 (12.1)	66 (1.5)	568 (5.9)	14 (1.4)	530 (10.3)	4 (0.6)	521 (12.0)
Fremont/Lincoln/WestSide PS, NE	8 (1.1)	484 (16.5)	56 (2.1)	513 (9.3)	20 (2.2)	471 (9.0)	15 (1.4)	430 (11.5)
Guilford County, NC	10 (0.8)	498 (10.9)	65 (1.5)	525 (8.0)	19 (1.2)	499 (9.2)	6 (0.8)	473 (18.8)
Jersey City Public Schools, NJ	5 (0.8)	467 (13.5)	51 (2.0)	489 (7.8)	27 (1.9)	475 (11.3)	18 (1.3)	450 (12.0)
Miami-Dade County PS, FL	11 (1.0)	411 (15.6)	49 (1.7)	449 (10.3)	23 (0.8)	411 (9.8)	17 (1.4)	394 (16.0)
Michigan Invitational Group, MI	11 (1.4)	550 (6.9)	64 (2.6)	543 (6.4)	18 (1.9)	511 (8.1)	8 (1.1)	487 (12.5)
Montgomery County, MD	16 (1.2)	547 (9.3)	60 (1.7)	550 (3.8)	15 (1.6)	509 (8.3)	9 (0.9)	500 (8.7)
Naperville Sch. Dist. #203, IL	22 (1.3)	570 (5.9)	66 (1.5)	575 (3.2)	8 (0.8)	552 (8.1)	4 (0.5)	521 (9.6)
Project SMART Consortium, OH	10 (1.0)	511 (7.7)	60 (1.9)	533 (8.1)	20 (1.5)	507 (9.8)	10 (0.9)	495 (12.1)
Rochester City Sch. Dist., NY r	11 (0.9)	428 (12.6)	52 (2.5)	479 (7.4)	19 (1.8)	444 (9.8)	18 (1.7)	417 (8.5)
SW Math/Sci. Collaborative, PA	15 (2.1)	517 (10.7)	66 (1.7)	524 (6.3)	13 (1.2)	505 (12.4)	6 (1.0)	475 (15.4)
International Avg. (All Countries)	28 (0.2)	487 (1.2)	52 (0.2)	499 (0.8)	13 (0.1)	474 (1.4)	8 (0.1)	442 (1.8)

SOURCE: IEA Third International Mathematics and Science Study (TIMSS), 1998-1999.

Background data provided by students.

States in *italics* did not fully satisfy guidelines for sample participation rates (see Appendix A for details).

( ) Standard errors appear in parentheses. Because results are rounded to the nearest whole number, some totals may appear inconsistent.

A tilde (~) indicates insufficient data to report achievement.

An "r" indicates a 70-84% student response rate.

What Activities Do Students Do in Their Mathematics Lessons?

Because it can affect pedagogical strategies, class size is shown in Exhibit 6.7. Teachers' reports on the size of their eighth-grade mathematics class reveal that across countries the average was 31 students, but there was considerable variation even among the higher-performing countries – from 42 students in Korea to 19 in Belgium (Flemish). Average class size was relatively uniform across all of the Benchmarking entities, ranging from 22 to 30 students. The relationship between class size and achievement is difficult to disentangle, given the variety of policies and practices and the fact that smaller classes can be used for both advanced and remedial learning. It makes sense, however, that teachers may have an easier time managing and conducting more student-centered instructional activities with smaller classes.

Extensive research about class size in relation to achievement indicates that the existence of such a relationship is dependent on the situation.[3] Dramatic reductions in class size can be related to gains in achievement, but the chief effects of smaller classes often are in relation to teacher attitudes and instructional behaviors. Also, the research is more consistent in suggesting that reductions in class size have the potential to help students in the primary grades. The TIMSS 1999 data support the complexity of this issue. The five highest-performing countries – Singapore, Korea, Chinese Taipei, Hong Kong, and Japan – were among those with the largest mathematics classes. Within countries, several show little or no relationship between achievement and class size, often because students are mostly all in classes of similar size. Within other countries, there appears to be a curvilinear relationship, or those students with higher achievement appear to be in larger classes. In some countries, larger classes may represent the more usual situation for mathematics teaching, with smaller classes used primarily for students needing remediation or for those students in the less-advanced tracks.

Exhibit 6.8 presents a profile of the activities most commonly encountered in mathematics classes around the world, as reported by mathematics teachers. As can be seen from the international averages, the two predominant activities, accounting for nearly half of class time on average, were teacher lecture (23 percent of class time) and teacher-guided student practice (22 percent). In general for the United States overall and the Benchmarking entities, teachers' reports on the frequency of these activities matched the international profile. According to U.S. mathematics teachers, class time is spent as follows:

3 Mayer, D.P., Mullens, J.E., and Moore, M.T. (2000), *Monitoring School Quality: An Indicators Report*, NCES 2001-030, Washington, DC: National Center for Education Statistics.

15 percent on homework review; 20 percent on lecture style teacher presentation; 35 percent on teacher-guided or independent student practice; 12 percent on re-teaching and clarification; 11 percent on tests and quizzes, six percent on administrative tasks; and four percent on other activities. One noteworthy exception is 26 percent of class time in Naperville spent on homework review, compared with 15 percent for the United States.

As shown in Exhibit 6.9, most students internationally (86 percent on average) agreed with teachers' reports about the prevalence of teacher-guided activities, saying that their teachers frequently showed them how to do mathematics problems. Just as found in the 1995 videotapes, it appears that in the U.S. the teacher states the problem, demonstrates the solution, and then asks the students to practice. Ninety-four percent of U.S. eighth graders reported that their teachers showed them how to do mathematics problems almost always or pretty often during mathematics lessons. More than 90 percent of the students in each of the Benchmarking entities reported this also.

Compared with their counterparts internationally (59 percent), more U.S. students reported that working independently on worksheets or textbooks occurred almost always or pretty often (86 percent). Working on their own on worksheets or textbooks was also quite pervasive throughout the Benchmarking entities, where more than 80 percent of the students in each jurisdiction reported doing this activity that frequently.

As for working on mathematics projects, the Benchmarking states typically were below the international average (36 percent), ranging from 22 to 33 percent. There was considerable variation across the districts and consortia. Less than one-fifth of the students reported frequent project work in the Academy School District, the First in the World Consortium, and Naperville. At the other end of the continuum, 63 percent so reported in Jersey City, followed by 34 to 38 percent in Chicago, the Fremont/Lincoln/Westside Public Schools, Miami-Dade, and Rochester.

Compared with students internationally, eighth graders in each of the Benchmarking jurisdictions and in the United States overall reported an unusually large amount of classroom time devoted to working on homework. Internationally, 55 percent of the students reported frequently discussing their completed homework. The figure for the United States was 79 percent, and it ranged from 70 to 91 percent for the Benchmarking jurisdictions. An even greater difference was evident for frequently beginning homework in class – 42 percent internationally compared with 74 percent for the United States. In the Benchmarking jurisdictions, from 43 to 90 percent of the students reported beginning their homework in class almost always or pretty often.

As might be anticipated, students reported that use of the board was an extremely common presentational mode in mathematics class (see Exhibit 6.10). On average internationally, 92 percent of students reported that teachers used the board at least pretty often, and 60 percent reported that students did so. Using the board seems to be less common in the United States, especially for students (37 percent). In the United States, use of an overhead projector is a popular presentational mode, especially for teachers – 59 percent compared with 19 percent internationally. This mode was used frequently for more than 80 percent of the students in Maryland, North Carolina, Oregon, the Academy School District, the Fremont/Lincoln/Westside Public Schools, Guilford County, Montgomery County, and Naperville.

Educators, parents, employers, and most of the public support the goal of improving students' capacity for mathematics problem-solving. To examine the emphasis placed on that goal, TIMSS created an index of teachers' emphasis on mathematics reasoning and problem-solving (EMRPS). As shown in Exhibit 6.11, the index is based on teachers' responses about how often they asked students to explain the reasoning behind an idea, represent and analyze relationships using tables, charts, or graphs, work on problems for which there was no immediate solution, and write equations to represent relationships. Students were placed in the high category if, on average, they were asked to do these activities in most of their lessons. The medium level represents students asked to do these activities in some to most lessons, and students in the low category did them only in some lessons or rarely.

Nearly half the Japanese students were at the high index level, compared with the international average of 15 percent. Across countries, most students (61 percent on average) were in the medium category. An emphasis on problem-solving was related to performance, with students at the high and medium levels having higher average achievement than those at the low level, both internationally and for most entities. There was tremendous variation among the Benchmarking participants on this index. From 41 to 46 percent of the students were in the high category in Jersey City, First in the World, and the Michigan Invitational Group, compared with eight to nine percent in Chicago and Oregon.

Exhibit R3.7 in the reference section shows the percentages of students asked in most or every lesson to engage in each of the activities included in the problem-solving index. For comparison purposes, the exhibit also shows the percentages of students asked to practice computational skills in most or every lesson. According to their teachers, internationally on average nearly three-fourths of the students (73 percent) were asked to practice their computational skills in most or every mathematics lesson. Nearly as many (70 percent) were asked to explain the reasoning behind an idea this frequently. The other three problem-solving activities occurred much less often. Forty-three percent of students, on average across countries, wrote equations representing relationships in most or every lesson, but only about one-fourth (26 percent) represented and analyzed relationships using tables or graphs, and about one-fifth (21 percent) worked on problems for which there was no immediately obvious method of solution. While the Benchmarking entities did not vary greatly from the international profile, there were differences. For example, twice as many students as internationally reported spending time in most or every lesson working on problems for which there was no immediately obvious method of solution in the First in the World Consortium, the Jersey Public Schools, and the Michigan Invitational Group (44 to 51 percent). More than 90 percent of the students in Jersey City and the Michigan Invitational Group were frequently asked to explain the reasoning behind an idea, and 90 percent of the Naperville students were frequently asked to write equations to represent relationships.

Teachers were not asked about the emphasis placed on using things from everyday life in solving mathematics problems, but students were (see Exhibit R3.8). In most of the countries, students reported a moderate emphasis on doing this type of problem in mathematics class. Nearly two-thirds (65 percent), on average internationally, said these activities occur once in a while or pretty often, and an additional 15 percent said they occur almost always. The figures were somewhat higher for the United States and most Benchmarking jurisdictions. More than 60 percent of the students in Maryland, North Carolina, the Academy School District, the Fremont/Lincoln/Westside Public Schools, Jersey City, and the Michigan Invitational Group reported that they use things from everyday life in solving mathematics problems almost always or pretty often.

Exhibit 6.7 Mathematics Class Size

8th Grade Mathematics

	Overall Average Class Size	1 - 20 Students		21 - 35 Students		36 or More Students	
		Percent of Students	Average Achievement	Percent of Students	Average Achievement	Percent of Students	Average Achievement
Countries							
United States r	26 (0.7)	21 (2.6)	507 (8.4)	73 (3.0)	504 (4.9)	6 (1.4)	488 (26.2)
Belgium (Flemish)	19 (0.4)	58 (3.5)	541 (6.8)	42 (3.5)	582 (4.4)	0 (0.0)	~ ~
Canada	27 (0.3)	11 (2.1)	522 (6.7)	87 (2.3)	534 (2.9)	2 (1.0)	~ ~
Chinese Taipei	39 (0.5)	0 (0.0)	~ ~	14 (2.9)	578 (11.5)	86 (3.0)	586 (4.6)
Czech Republic r	24 (0.4)	18 (4.2)	504 (6.9)	82 (4.2)	524 (6.0)	0 (0.0)	~ ~
England	x x	x x	x x	x x	x x	x x	x x
Hong Kong, SAR	37 (0.5)	7 (1.8)	521 (20.0)	15 (3.0)	530 (10.5)	78 (3.4)	597 (4.3)
Italy	20 (0.3)	55 (3.9)	472 (5.3)	44 (3.9)	489 (6.5)	1 (0.0)	~ ~
Japan	36 (0.2)	1 (0.0)	~ ~	41 (3.4)	572 (2.9)	58 (3.3)	582 (2.3)
Korea, Rep. of	42 (0.5)	0 (0.0)	~ ~	12 (2.2)	584 (6.7)	88 (2.2)	587 (2.1)
Netherlands r	25 (0.5)	13 (4.1)	459 (18.8)	87 (4.1)	546 (8.2)	0 (0.0)	~ ~
Russian Federation	24 (0.5)	19 (3.2)	492 (10.0)	81 (3.2)	534 (5.9)	0 (0.0)	~ ~
Singapore	37 (0.3)	1 (0.4)	~ ~	32 (3.8)	602 (11.6)	68 (3.8)	607 (6.4)
States							
Connecticut s	24 (1.4)	29 (6.1)	501 (16.8)	64 (7.1)	525 (11.6)	6 (5.5)	559 (3.4)
Idaho r	22 (1.7)	43 (7.0)	481 (14.3)	52 (5.8)	503 (8.8)	6 (4.4)	488 (17.8)
Illinois	24 (0.6)	24 (5.3)	511 (10.8)	76 (5.2)	513 (7.9)	1 (0.0)	~ ~
Indiana r	22 (1.3)	40 (6.8)	517 (13.7)	59 (6.7)	512 (9.6)	1 (0.1)	~ ~
Maryland s	28 (1.2)	11 (3.4)	497 (23.2)	84 (4.7)	488 (6.3)	5 (2.6)	419 (23.6)
Massachusetts r	24 (1.1)	32 (5.1)	488 (11.6)	66 (4.8)	528 (7.4)	3 (1.5)	453 (30.5)
Michigan r	27 (1.3)	17 (3.6)	519 (8.0)	80 (3.7)	526 (9.2)	3 (2.0)	536 (29.8)
Missouri	23 (0.8)	36 (5.6)	477 (8.1)	61 (5.7)	497 (6.7)	3 (2.1)	571 (22.7)
North Carolina r	24 (0.7)	22 (5.4)	482 (17.1)	77 (5.4)	497 (7.7)	1 (0.8)	~ ~
Oregon r	24 (0.4)	26 (3.9)	500 (14.8)	74 (3.9)	521 (7.5)	0 (0.0)	~ ~
Pennsylvania	23 (0.6)	31 (4.4)	498 (11.3)	68 (4.4)	513 (6.9)	1 (0.6)	~ ~
South Carolina r	24 (1.0)	35 (5.7)	484 (13.6)	64 (5.5)	513 (12.4)	2 (1.7)	~ ~
Texas r	22 (0.9)	41 (6.1)	518 (16.9)	58 (6.0)	532 (8.4)	1 (0.9)	~ ~
Districts and Consortia							
Academy School Dist. #20, CO	27 (0.0)	9 (0.2)	474 (5.6)	88 (0.2)	541 (1.7)	3 (0.1)	508 (11.8)
Chicago Public Schools, IL	26 (1.2)	16 (7.2)	478 (27.9)	80 (6.6)	464 (6.3)	4 (0.5)	444 (5.1)
Delaware Science Coalition, DE r	29 (0.9)	9 (3.7)	417 (31.9)	78 (4.4)	480 (13.1)	13 (4.2)	559 (19.9)
First in the World Consort., IL	24 (0.6)	28 (4.3)	575 (15.6)	72 (4.3)	552 (4.9)	0 (0.0)	~ ~
Fremont/Lincoln/WestSide PS, NE	24 (0.6)	22 (4.8)	455 (19.9)	78 (4.8)	499 (11.9)	0 (0.0)	~ ~
Guilford County, NC r	24 (0.5)	15 (4.1)	494 (13.5)	85 (4.1)	512 (11.3)	0 (0.0)	~ ~
Jersey City Public Schools, NJ r	28 (3.1)	17 (4.8)	440 (21.3)	71 (4.0)	482 (11.8)	12 (4.6)	524 (31.9)
Miami-Dade County PS, FL s	30 (1.6)	16 (6.6)	369 (40.3)	56 (11.0)	427 (18.3)	28 (10.6)	437 (24.6)
Michigan Invitational Group, MI	26 (0.6)	23 (4.6)	534 (16.1)	75 (4.6)	528 (5.9)	2 (0.1)	~ ~
Montgomery County, MD s	25 (0.7)	16 (3.3)	495 (15.2)	84 (3.4)	539 (4.7)	0 (0.0)	~ ~
Naperville Sch. Dist. #203, IL	28 (0.4)	6 (2.8)	508 (23.3)	94 (2.8)	572 (3.0)	0 (0.0)	~ ~
Project SMART Consortium, OH r	24 (0.7)	23 (6.2)	533 (18.3)	77 (6.2)	523 (8.2)	0 (0.0)	~ ~
Rochester City Sch. Dist., NY	24 (0.6)	22 (4.8)	452 (13.8)	78 (4.8)	439 (7.9)	0 (0.0)	~ ~
SW Math/Sci. Collaborative, PA	24 (1.2)	35 (6.3)	507 (10.1)	62 (6.4)	521 (10.5)	3 (3.0)	455 (6.5)
International Avg. (All Countries)	31 (0.1)	17 (0.4)	468 (2.4)	53 (0.6)	488 (1.4)	30 (0.4)	471 (4.3)

SOURCE: IEA Third International Mathematics and Science Study (TIMSS), 1998-1999.

Background data provided by teachers.

States in *italics* did not fully satisfy guidelines for sample participation rates (see Appendix A for details).

( ) Standard errors appear in parentheses. Because results are rounded to the nearest whole number, some totals may appear inconsistent.

A tilde (~) indicates insufficient data to report achievement.

An "r" indicates teacher response data available for 70-84% of students. An "s" indicates teacher response data available for 50-69% of students. An "x" indicates teacher response data available for <50% of students.

	Average Percentage of Class Time Spent in a Typical Month of Lessons							
	Administrative Tasks	Homework Review	Lecture-Style Presentation by Teacher	Teacher-Guided Student Practice	Re-teaching and Clarification of Content/ Procedures	Student Independent Practice	Tests and Quizzes	Other
Countries								
United States	r 6 (0.3)	r 15 (0.4)	r 20 (0.7)	r 18 (0.4)	r 12 (0.5)	r 17 (0.9)	r 11 (0.4)	r 4 (0.5)
Belgium (Flemish)	4 (0.3)	7 (0.4)	24 (1.1)	29 (1.0)	10 (0.4)	14 (0.9)	10 (0.3)	2 (0.4)
Canada	r 5 (0.2)	r 14 (0.4)	r 20 (0.9)	r 18 (0.8)	r 10 (0.3)	r 20 (0.7)	r 10 (0.3)	r 3 (0.6)
Chinese Taipei	3 (0.6)	12 (0.5)	39 (1.3)	15 (0.5)	11 (0.6)	9 (0.5)	10 (0.5)	2 (0.4)
Czech Republic	3 (0.3)	5 (0.4)	23 (0.7)	29 (1.2)	10 (0.5)	19 (1.0)	9 (0.6)	3 (0.4)
England	s 3 (0.2)	s 6 (0.5)	s 18 (0.9)	s 27 (1.2)	s 11 (0.4)	s 24 (1.5)	s 8 (0.4)	s 3 (0.7)
Hong Kong, SAR	5 (0.7)	12 (0.7)	32 (1.6)	18 (0.8)	8 (0.4)	14 (0.8)	8 (0.4)	3 (0.4)
Italy	2 (0.2)	14 (0.5)	25 (0.7)	22 (0.7)	13 (0.4)	12 (0.5)	12 (0.5)	1 (0.2)
Japan	2 (0.5)	5 (0.4)	34 (1.6)	26 (1.3)	16 (0.9)	9 (0.7)	7 (0.5)	2 (0.3)
Korea, Rep. of	3 (0.6)	6 (0.3)	33 (1.4)	22 (0.8)	14 (0.8)	14 (0.8)	7 (0.3)	3 (0.4)
Netherlands	5 (0.4)	15 (1.5)	9 (1.2)	5 (1.0)	18 (1.1)	32 (2.0)	11 (0.6)	5 (1.0)
Russian Federation	2 (0.1)	10 (0.4)	25 (0.6)	17 (0.7)	11 (0.4)	17 (0.6)	12 (0.6)	5 (0.4)
Singapore	6 (0.6)	13 (0.7)	28 (1.5)	20 (1.2)	9 (0.3)	12 (0.8)	8 (0.4)	3 (0.3)
States								
Connecticut	r 5 (0.6)	r 15 (0.8)	r 20 (1.7)	r 22 (1.7)	r 12 (1.0)	r 14 (1.4)	r 13 (1.0)	s 3 (0.9)
Idaho	r 5 (0.6)	r 12 (0.6)	r 16 (1.2)	r 17 (1.8)	r 12 (0.7)	r 23 (2.3)	r 11 (0.7)	r 3 (0.5)
Illinois	5 (0.4)	15 (0.6)	21 (1.5)	19 (1.2)	11 (0.5)	15 (0.9)	12 (0.7)	3 (0.4)
Indiana	4 (0.4)	14 (0.9)	22 (1.6)	17 (1.3)	12 (0.7)	15 (1.2)	12 (0.6)	3 (0.7)
Maryland	r 6 (0.7)	r 13 (0.8)	r 20 (1.6)	r 18 (1.2)	r 12 (1.1)	r 15 (1.1)	r 12 (0.7)	r 4 (0.6)
Massachusetts	4 (0.4)	17 (1.0)	19 (1.1)	19 (0.9)	15 (1.0)	13 (0.7)	12 (0.6)	r 4 (1.0)
Michigan	5 (0.6)	16 (0.8)	18 (1.0)	19 (1.6)	11 (1.0)	16 (1.0)	10 (0.6)	5 (1.7)
Missouri	5 (0.5)	12 (0.6)	21 (1.2)	19 (1.2)	12 (0.8)	18 (1.2)	10 (0.6)	3 (0.7)
North Carolina	5 (0.4)	14 (1.0)	20 (1.2)	20 (1.2)	12 (0.5)	16 (1.0)	11 (0.6)	3 (0.4)
Oregon	5 (0.5)	12 (1.0)	19 (1.3)	17 (1.2)	11 (0.6)	21 (1.2)	9 (0.6)	5 (1.7)
Pennsylvania	4 (0.3)	16 (0.9)	24 (1.5)	19 (1.1)	10 (0.5)	13 (1.1)	10 (0.6)	3 (0.4)
South Carolina	5 (0.6)	13 (0.8)	23 (1.7)	19 (1.2)	12 (0.8)	15 (1.0)	11 (0.7)	3 (0.5)
Texas	r 7 (0.7)	r 12 (0.8)	r 17 (1.4)	r 21 (1.2)	r 12 (0.7)	r 17 (1.2)	r 12 (0.7)	r 4 (0.7)
Districts and Consortia								
Academy School Dist. #20, CO	5 (0.0)	18 (0.0)	20 (0.1)	14 (0.0)	12 (0.0)	16 (0.0)	13 (0.1)	r 3 (0.0)
Chicago Public Schools, IL	6 (0.7)	11 (1.1)	20 (2.2)	20 (2.0)	13 (1.0)	16 (1.7)	12 (1.1)	3 (1.0)
Delaware Science Coalition, DE	r 5 (0.5)	r 13 (0.8)	r 21 (1.1)	r 22 (1.8)	r 10 (0.6)	r 13 (1.1)	r 10 (0.5)	r 6 (1.3)
First in the World Consort., IL	3 (0.4)	17 (1.2)	24 (1.6)	16 (1.1)	11 (0.4)	12 (1.2)	11 (0.7)	7 (2.7)
Fremont/Lincoln/WestSide PS, NE	6 (0.7)	19 (1.7)	19 (2.7)	18 (1.9)	10 (0.6)	16 (1.0)	11 (1.1)	2 (0.7)
Guilford County, NC	5 (0.4)	13 (0.5)	18 (1.5)	20 (1.1)	11 (0.7)	16 (1.0)	11 (0.7)	5 (1.2)
Jersey City Public Schools, NJ	5 (0.7)	9 (0.5)	18 (1.2)	17 (0.5)	13 (0.7)	21 (1.2)	10 (0.4)	r 7 (0.7)
Miami-Dade County PS, FL	s 5 (0.8)	s 14 (1.2)	s 19 (1.7)	s 19 (1.4)	s 12 (1.1)	s 13 (1.1)	s 12 (1.1)	s 5 (0.7)
Michigan Invitational Group, MI	3 (0.3)	18 (2.3)	16 (1.8)	18 (2.4)	11 (1.2)	17 (1.0)	13 (0.7)	6 (2.1)
Montgomery County, MD	s 5 (0.4)	s 14 (1.0)	s 18 (0.6)	s 20 (1.5)	s 14 (0.8)	s 14 (0.8)	s 12 (0.8)	s 4 (0.8)
Naperville Sch. Dist. #203, IL	5 (0.5)	26 (0.7)	22 (0.8)	14 (0.7)	9 (0.3)	12 (0.9)	12 (0.5)	1 (0.3)
Project SMART Consortium, OH	5 (0.5)	15 (1.2)	21 (1.1)	19 (1.0)	11 (0.6)	16 (1.2)	11 (0.5)	2 (0.4)
Rochester City Sch. Dist., NY	5 (0.4)	14 (0.8)	22 (0.8)	17 (0.8)	13 (0.9)	15 (0.5)	10 (0.6)	3 (0.5)
SW Math/Sci. Collaborative, PA	5 (0.8)	15 (1.3)	24 (2.3)	17 (1.3)	11 (0.6)	14 (1.2)	12 (0.8)	2 (0.5)
International Avg. (All Countries)	5 (0.1)	12 (0.1)	23 (0.2)	22 (0.2)	13 (0.1)	15 (0.2)	11 (0.1)	4 (0.1)

SOURCE: IEA Third International Mathematics and Science Study (TIMSS), 1998-1999.

Background data provided by teachers.

States in *italics* did not fully satisfy guidelines for sample participation rates (see Appendix A for details).

( ) Standard errors appear in parentheses. Because results are rounded to the nearest whole number, some totals may appear inconsistent.

An "r" indicates teacher response data available for 70-84% of students. An "s" indicates teacher response data available for 50-69% of students.

Exhibit 6.9 Students Doing Various Activities in Mathematics Class

	Percentage of Students Reporting Almost Always or Pretty Often				
	We Discuss Our Completed Homework	Teacher Shows Us How to Do Mathematics Problems	We Work on Worksheets or Textbooks on Our Own	We Work on Mathematics Projects	We Begin Our Homework
Countries					
United States	79 (1.2)	94 (0.6)	86 (0.7)	29 (1.3)	74 (1.6)
Belgium (Flemish)	43 (1.4)	69 (0.9)	64 (1.0)	16 (1.1)	20 (1.4)
Canada	62 (1.4)	92 (0.5)	92 (0.5)	28 (1.1)	82 (1.2)
Chinese Taipei	55 (1.0)	91 (0.5)	59 (1.2)	55 (1.2)	34 (1.0)
Czech Republic	42 (1.8)	86 (1.1)	51 (2.4)	8 (0.6)	16 (1.6)
England	62 (1.5)	93 (0.7)	88 (1.5)	35 (1.4)	27 (1.6)
Hong Kong, SAR	35 (1.1)	91 (0.6)	69 (1.2)	67 (1.4)	40 (1.1)
Italy	64 (1.4)	80 (1.2)	34 (1.2)	22 (1.3)	39 (2.3)
Japan	19 (1.2)	88 (0.7)	38 (1.5)	6 (0.7)	20 (1.3)
Korea, Rep. of	10 (0.5)	85 (0.8)	29 (0.7)	46 (1.2)	17 (0.7)
Netherlands	68 (3.7)	70 (2.7)	92 (1.1)	3 (0.7)	89 (1.5)
Russian Federation	53 (1.4)	78 (1.2)	62 (1.3)	19 (0.9)	10 (0.8)
Singapore	61 (1.0)	97 (0.4)	75 (0.9)	15 (1.1)	60 (1.9)
States					
Connecticut	87 (1.3)	94 (1.3)	88 (1.0)	33 (3.0)	67 (2.2)
Idaho	70 (2.4)	94 (1.1)	88 (1.2)	31 (1.9)	89 (1.3)
Illinois	78 (2.2)	97 (0.5)	87 (1.0)	31 (2.2)	82 (2.5)
Indiana	80 (1.7)	95 (1.1)	88 (0.8)	30 (2.5)	84 (2.6)
Maryland	81 (1.9)	93 (1.0)	87 (1.1)	28 (2.1)	57 (3.1)
Massachusetts	82 (2.2)	94 (0.9)	85 (1.1)	22 (1.6)	63 (3.4)
Michigan	84 (1.9)	95 (0.7)	89 (0.8)	28 (2.3)	83 (2.4)
Missouri	74 (2.5)	92 (1.1)	90 (1.2)	30 (2.2)	85 (2.1)
North Carolina	89 (1.4)	98 (0.5)	90 (0.8)	31 (1.9)	79 (2.1)
Oregon	74 (2.4)	93 (1.1)	90 (1.2)	34 (2.2)	90 (1.8)
Pennsylvania	85 (1.8)	95 (0.9)	83 (1.2)	24 (2.0)	71 (3.2)
South Carolina	84 (2.0)	95 (0.9)	87 (1.6)	30 (2.2)	79 (2.2)
Texas	75 (2.9)	94 (1.3)	84 (1.4)	25 (2.1)	78 (2.4)
Districts and Consortia					
Academy School Dist. #20, CO	82 (0.9)	92 (0.9)	90 (0.9)	19 (0.9)	72 (1.1)
Chicago Public Schools, IL	74 (4.3)	96 (1.1)	81 (1.4)	34 (3.3)	53 (4.6)
Delaware Science Coalition, DE	85 (1.6)	95 (0.9)	88 (1.3)	25 (1.8)	74 (2.0)
First in the World Consort., IL	91 (1.5)	94 (1.5)	92 (1.6)	18 (2.8)	63 (3.6)
Fremont/Lincoln/WestSide PS, NE	83 (1.5)	91 (1.0)	91 (1.2)	38 (3.7)	83 (2.9)
Guilford County, NC	88 (1.4)	96 (1.0)	93 (0.8)	24 (2.2)	80 (2.5)
Jersey City Public Schools, NJ	76 (2.0)	97 (0.6)	85 (2.2)	63 (2.3)	43 (2.7)
Miami-Dade County PS, FL	71 (4.7)	92 (2.2)	83 (2.4)	34 (2.8)	58 (3.3)
Michigan Invitational Group, MI	86 (1.3)	92 (1.2)	86 (1.7)	22 (1.3)	84 (3.0)
Montgomery County, MD	83 (1.4)	93 (1.2)	92 (0.9)	24 (2.4)	69 (1.5)
Naperville Sch. Dist. #203, IL	91 (0.9)	96 (0.7)	92 (0.9)	15 (1.8)	87 (1.6)
Project SMART Consortium, OH	84 (1.9)	93 (1.5)	88 (1.2)	25 (1.8)	84 (2.5)
Rochester City Sch. Dist., NY	r 82 (1.8)	r 95 (0.8)	r 86 (1.2)	r 35 (2.9)	r 68 (3.0)
SW Math/Sci. Collaborative, PA	85 (2.1)	95 (1.0)	83 (1.9)	22 (2.2)	79 (3.3)
International Avg. (All Countries)	55 (0.2)	86 (0.2)	59 (0.2)	36 (0.2)	42 (0.2)

SOURCE: IEA Third International Mathematics and Science Study (TIMSS), 1998-1999.

Background data provided by students.

States in *italics* did not fully satisfy guidelines for sample participation rates (see Appendix A for details).

( ) Standard errors appear in parentheses. Because results are rounded to the nearest whole number, some totals may appear inconsistent.

An "r" indicates a 70-84% student response rate.

	Percentage of Students Reporting Almost Always or Pretty Often				
	Teacher Uses the Board	Teacher Uses an Overhead Projector	Teacher Uses a Computer to Demonstrate Ideas in Mathematics	Students Use the Board	Students Use an Overhead Projector
Countries					
United States	80 (1.9)	59 (3.3)	9 (0.7)	37 (1.9)	16 (1.0)
Belgium (Flemish)	96 (0.7)	11 (1.7)	2 (0.5)	42 (1.8)	2 (0.8)
Canada	91 (0.9)	42 (2.7)	5 (0.5)	25 (1.2)	7 (0.8)
Chinese Taipei	96 (0.4)	4 (0.4)	2 (0.2)	48 (1.6)	2 (0.3)
Czech Republic	97 (0.4)	9 (1.6)	2 (0.4)	91 (1.7)	4 (0.5)
England	94 (1.5)	19 (2.6)	6 (0.8)	13 (1.0)	3 (0.6)
Hong Kong, SAR	96 (0.4)	9 (0.8)	3 (0.4)	46 (1.7)	3 (0.4)
Italy	94 (0.5)	8 (0.9)	5 (0.6)	84 (1.1)	7 (0.6)
Japan	99 (0.2)	4 (0.8)	1 (0.4)	50 (2.5)	1 (0.3)
Korea, Rep. of	93 (0.5)	10 (0.8)	7 (0.9)	38 (1.7)	3 (0.3)
Netherlands	90 (1.6)	7 (1.4)	2 (0.3)	9 (1.2)	2 (0.3)
Russian Federation	96 (0.4)	7 (1.0)	1 (0.2)	92 (0.6)	4 (0.5)
Singapore	96 (1.3)	75 (2.1)	11 (1.2)	52 (2.0)	21 (1.1)
States					
Connecticut	85 (3.4)	57 (4.5)	8 (1.4)	43 (3.4)	18 (2.7)
Idaho	81 (2.9)	59 (4.1)	9 (1.5)	30 (2.7)	12 (1.2)
Illinois	75 (5.2)	64 (5.5)	8 (1.3)	37 (4.8)	16 (1.9)
Indiana	78 (3.8)	61 (5.4)	8 (1.1)	42 (3.7)	16 (1.7)
Maryland	74 (3.2)	86 (2.5)	10 (1.0)	44 (3.8)	32 (1.9)
Massachusetts	87 (2.4)	47 (5.1)	7 (1.3)	46 (3.4)	17 (2.3)
Michigan	77 (3.6)	64 (4.6)	7 (1.1)	30 (2.2)	18 (2.1)
Missouri	81 (3.2)	55 (5.0)	8 (0.9)	39 (3.0)	15 (2.0)
North Carolina	76 (2.5)	84 (2.7)	10 (1.2)	51 (3.0)	33 (2.6)
Oregon	63 (3.3)	83 (3.0)	9 (0.9)	22 (1.8)	28 (2.5)
Pennsylvania	92 (1.8)	44 (3.5)	5 (0.6)	65 (2.9)	16 (2.6)
South Carolina	63 (3.8)	80 (4.3)	10 (1.4)	32 (3.0)	16 (1.7)
Texas	71 (3.2)	72 (3.7)	9 (1.5)	32 (3.4)	22 (1.9)
Districts and Consortia					
Academy School Dist. #20, CO	70 (1.0)	85 (0.8)	6 (0.7)	30 (1.0)	23 (1.1)
Chicago Public Schools, IL	79 (6.2)	41 (9.2)	10 (1.9)	50 (4.5)	18 (4.5)
Delaware Science Coalition, DE	80 (2.8)	72 (4.2)	10 (1.1)	38 (2.8)	27 (2.7)
First in the World Consort., IL	79 (5.8)	70 (2.6)	5 (1.3)	43 (6.2)	28 (3.6)
Fremont/Lincoln/WestSide PS, NE	61 (3.7)	92 (1.0)	15 (1.4)	23 (2.7)	29 (2.8)
Guilford County, NC	67 (3.3)	89 (2.5)	6 (0.9)	35 (2.5)	25 (2.2)
Jersey City Public Schools, NJ	93 (1.9)	65 (2.8)	17 (1.9)	50 (3.2)	22 (2.5)
Miami-Dade County PS, FL	80 (4.9)	63 (6.8)	16 (1.9)	46 (6.4)	19 (2.3)
Michigan Invitational Group, MI	84 (3.4)	74 (2.8)	7 (0.9)	35 (2.2)	26 (3.3)
Montgomery County, MD	60 (3.3)	92 (1.7)	9 (0.9)	32 (2.7)	32 (2.8)
Naperville Sch. Dist. #203, IL	73 (2.3)	90 (0.6)	5 (0.7)	43 (1.9)	25 (1.6)
Project SMART Consortium, OH	80 (2.7)	66 (4.1)	11 (1.5)	45 (3.2)	25 (2.9)
Rochester City Sch. Dist., NY	r 64 (2.8)	r 74 (4.0)	r 16 (2.1)	r 35 (3.0)	r 36 (3.0)
SW Math/Sci. Collaborative, PA	95 (1.7)	40 (5.1)	5 (0.7)	57 (4.2)	10 (2.0)
International Avg. (All Countries)	92 (0.1)	19 (0.3)	5 (0.1)	60 (0.2)	9 (0.1)

SOURCE: IEA Third International Mathematics and Science Study (TIMSS), 1998-1999.

Background data provided by students.

States in *italics* did not fully satisfy guidelines for sample participation rates (see Appendix A for details).

( ) Standard errors appear in parentheses. Because results are rounded to the nearest whole number, some totals may appear inconsistent.

An "r" indicates a 70-84% student response rate.

Exhibit 6.11

Exhibit 6.11 Index of Teachers' Emphasis on Mathematics Reasoning and Problem-Solving (EMRPS)

8th Grade Mathematics

Index of Teachers' Emphasis on Mathematics Reasoning and Problem-Solving

Index based on teachers' responses to four questions about how often they ask students to: 1) explain the reasoning behind an idea; 2) represent and analyze relationships using tables, charts, or graphs; 3) work on problems for which there is no immediately obvious method of solution; 4) write equations to represent relationships (see reference exhibit R3.7). Average is computed across the four items based on a 4-point scale: 1 = never or almost never; 2 = some lessons; 3 = most lessons; 4 = every lesson. High level indicates average is greater than or equal to 3. Medium level indicates average is greater than or equal to 2.25 and less than 3. Low level indicates average is less than 2.25.

		High EMRPS		Medium EMRPS		Low EMRPS	
		Percent of Students	Average Achievement	Percent of Students	Average Achievement	Percent of Students	Average Achievement
Japan		49 (4.1)	584 (2.6)	45 (4.1)	574 (2.5)	7 (2.1)	562 (6.2)
Jersey City Public Schools, NJ	r	46 (6.4)	481 (11.1)	50 (6.0)	482 (15.3)	4 (2.5)	372 (7.2)
First in the World Consort., IL		42 (8.8)	536 (8.1)	54 (8.8)	581 (10.4)	4 (3.0)	492 (12.6)
Michigan Invitational Group, MI		41 (9.6)	521 (5.0)	52 (10.2)	549 (9.4)	7 (3.5)	484 (17.2)
Italy		30 (3.1)	484 (6.9)	58 (3.6)	479 (5.7)	12 (2.6)	472 (8.7)
Naperville Sch. Dist. #203, IL		29 (4.9)	569 (9.9)	67 (4.8)	571 (5.1)	4 (2.6)	524 (15.0)
Academy School Dist. #20, CO		26 (0.3)	552 (3.4)	53 (0.4)	533 (2.2)	21 (0.4)	504 (3.2)
Connecticut	r	26 (5.2)	554 (23.7)	57 (6.8)	509 (10.4)	17 (5.9)	508 (17.0)
Miami-Dade County PS, FL	s	25 (8.3)	443 (29.9)	55 (8.9)	410 (13.8)	21 (6.6)	425 (31.4)
Maryland	r	25 (5.6)	491 (14.9)	55 (6.3)	491 (8.4)	20 (4.2)	460 (14.7)
Czech Republic		21 (4.2)	539 (8.4)	73 (4.6)	516 (5.6)	6 (2.6)	502 (10.3)
Guilford County, NC		21 (5.4)	521 (24.3)	66 (5.9)	503 (9.8)	13 (3.5)	527 (13.4)
Michigan		21 (4.7)	558 (16.9)	60 (5.2)	516 (7.6)	19 (4.8)	510 (11.8)
Korea, Rep. of		21 (3.0)	588 (4.0)	66 (3.3)	586 (2.6)	13 (2.4)	594 (4.6)
Texas		20 (5.5)	552 (18.2)	61 (5.2)	512 (12.8)	19 (3.9)	511 (18.9)
Delaware Science Coalition, DE	r	20 (4.2)	490 (14.5)	59 (7.4)	492 (14.5)	21 (6.7)	445 (14.6)
United States		18 (2.5)	519 (12.4)	57 (2.9)	502 (4.1)	24 (2.7)	489 (6.4)
Montgomery County, MD	s	18 (6.7)	582 (11.6)	61 (6.6)	533 (7.1)	21 (5.2)	493 (7.1)
Indiana		17 (4.6)	512 (12.8)	64 (5.2)	524 (9.1)	19 (5.4)	491 (11.8)
SW Math/Sci. Collaborative, PA		17 (4.9)	517 (19.0)	62 (6.0)	527 (10.6)	21 (5.7)	492 (8.4)
Massachusetts		15 (4.2)	543 (15.7)	70 (6.5)	506 (7.1)	15 (4.9)	506 (14.8)
South Carolina		15 (3.3)	545 (26.8)	62 (5.5)	505 (8.6)	24 (4.2)	474 (17.4)
Idaho	r	14 (5.1)	511 (14.9)	52 (5.0)	500 (9.1)	34 (5.6)	479 (15.3)
Chinese Taipei		13 (2.4)	571 (7.5)	58 (4.2)	594 (6.0)	29 (3.8)	573 (6.9)
Project SMART Consortium, OH		13 (2.0)	540 (13.6)	60 (5.8)	516 (10.2)	27 (5.6)	522 (16.6)
Illinois		13 (3.6)	522 (19.6)	56 (5.8)	513 (9.2)	31 (6.8)	505 (9.8)
Canada		13 (2.0)	550 (8.1)	62 (3.4)	537 (3.5)	26 (3.0)	518 (4.9)
Fremont/Lincoln/WestSide PS, NE	r	13 (1.1)	491 (21.9)	66 (1.7)	498 (12.8)	22 (1.1)	459 (21.4)
Netherlands		12 (3.5)	561 (12.7)	60 (6.1)	528 (10.3)	28 (5.2)	547 (9.5)
Russian Federation		11 (2.5)	557 (12.8)	74 (3.9)	523 (6.6)	15 (3.6)	518 (10.5)
Pennsylvania		10 (3.3)	512 (21.2)	67 (5.4)	518 (9.0)	22 (5.8)	489 (9.2)
Missouri		10 (3.9)	503 (26.1)	55 (5.9)	495 (6.8)	35 (5.4)	483 (10.3)
Rochester City Sch. Dist., NY		10 (2.9)	443 (19.4)	73 (3.7)	444 (8.3)	17 (2.1)	429 (12.3)
North Carolina		10 (2.7)	522 (19.0)	69 (4.6)	493 (8.7)	21 (4.3)	476 (13.8)
Chicago Public Schools, IL		9 (5.7)	447 (9.3)	67 (8.5)	476 (6.5)	23 (9.1)	448 (13.0)
Oregon		8 (2.7)	561 (16.1)	64 (5.0)	518 (6.0)	28 (4.9)	494 (12.8)
Singapore		7 (2.1)	617 (25.9)	47 (4.0)	607 (8.8)	47 (4.4)	599 (8.2)
Hong Kong, SAR		6 (2.2)	597 (13.7)	56 (3.6)	591 (5.7)	38 (3.7)	570 (8.1)
England		3 (1.4)	533 (24.8)	66 (3.5)	519 (7.2)	31 (3.4)	490 (7.6)
Belgium (Flemish)		1 (0.4)	~ ~	39 (3.1)	592 (4.9)	61 (3.1)	540 (5.4)
International Avg. (All Countries)		15 (0.5)	493 (3.5)	61 (0.7)	490 (1.0)	24 (0.6)	479 (1.5)

SOURCE: IEA Third International Mathematics and Science Study (TIMSS), 1998-1999.

States in *italics* did not fully satisfy guidelines for sample participation rates (see Appendix A for details).

( ) Standard errors appear in parentheses. Because results are rounded to the nearest whole number, some totals may appear inconsistent.

A tilde (~) indicates insufficient data to report achievement.

An "r" indicates teacher response data available for 70-84% of students. An "s" indicates teacher response data available for 50-69% of students.

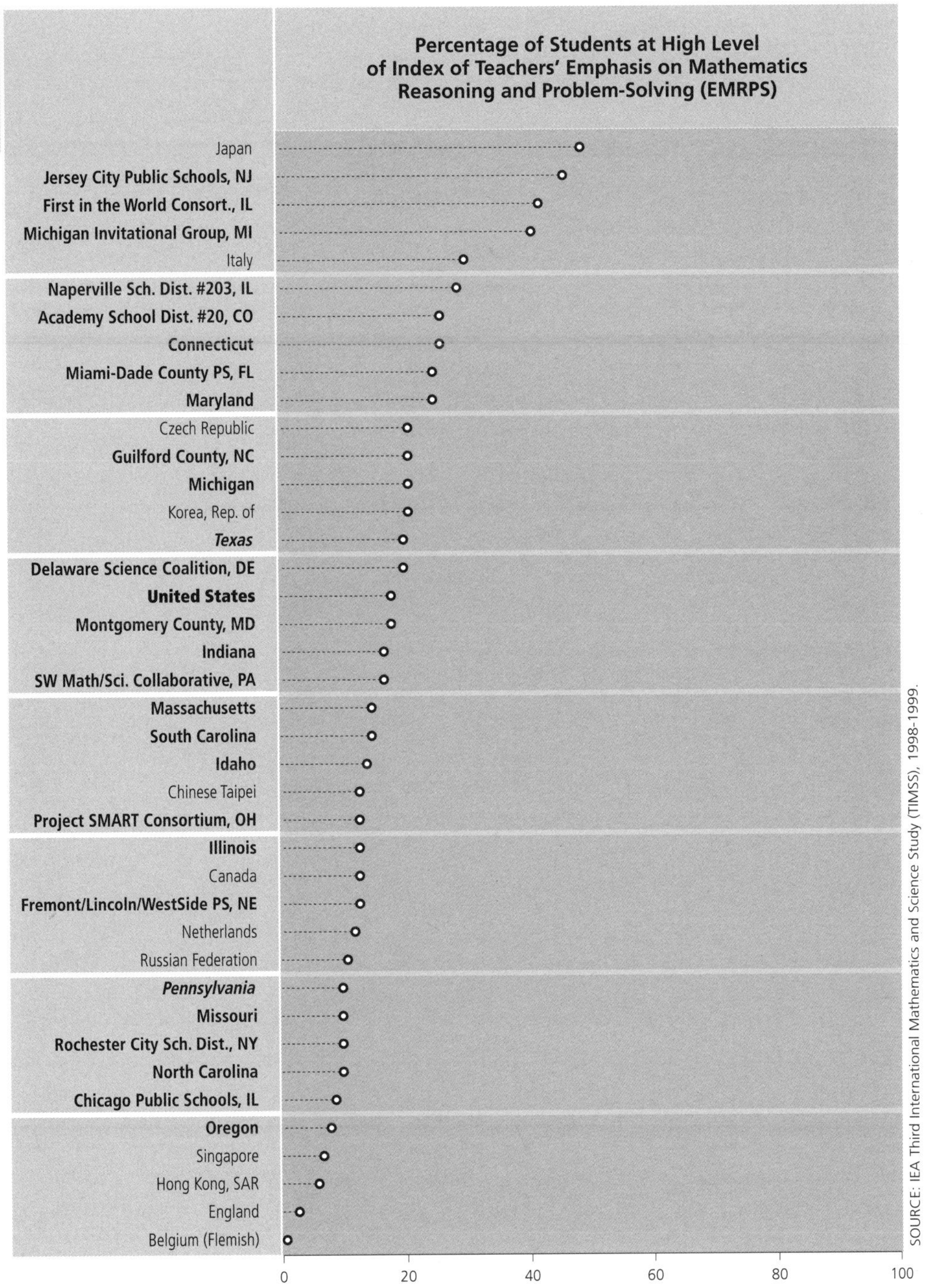

Percentage of Students at High Level of Index of Teachers' Emphasis on Mathematics Reasoning and Problem-Solving (EMRPS)

Japan
Jersey City Public Schools, NJ
First in the World Consort., IL
Michigan Invitational Group, MI
Italy
Naperville Sch. Dist. #203, IL
Academy School Dist. #20, CO
Connecticut
Miami-Dade County PS, FL
Maryland
Czech Republic
Guilford County, NC
Michigan
Korea, Rep. of
Texas
Delaware Science Coalition, DE
United States
Montgomery County, MD
Indiana
SW Math/Sci. Collaborative, PA
Massachusetts
South Carolina
Idaho
Chinese Taipei
Project SMART Consortium, OH
Illinois
Canada
Fremont/Lincoln/WestSide PS, NE
Netherlands
Russian Federation
Pennsylvania
Missouri
Rochester City Sch. Dist., NY
North Carolina
Chicago Public Schools, IL
Oregon
Singapore
Hong Kong, SAR
England
Belgium (Flemish)

0 20 40 60 80 100

SOURCE: IEA Third International Mathematics and Science Study (TIMSS), 1998-1999.

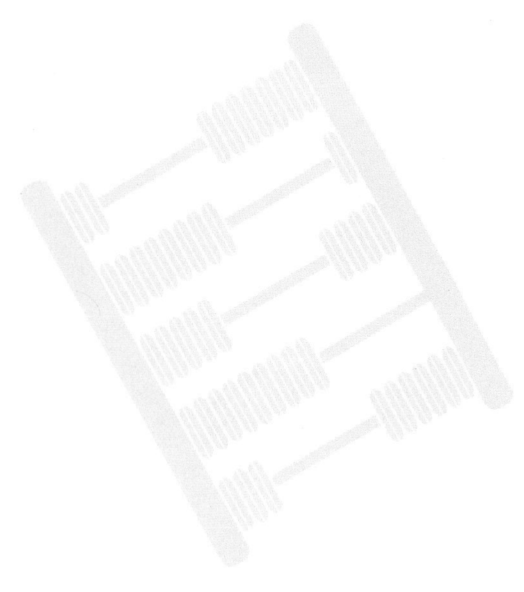

How Are Calculators and Computers Used?

Exhibit 6.12 shows data on students' access to calculators for use in mathematics class and on policies on their use for those with access. When all 38 TIMSS 1999 countries were considered, teachers in 14 countries reported that nearly all students (more than 90 percent) had access to calculators in class. In addition to the United States, the countries with this high degree of access were Australia, Belgium (Flemish), Canada, the Czech Republic, England, Finland, Hong Kong, Israel, Lithuania, the Netherlands, New Zealand, Singapore, and the Slovak Republic. For students in classes with access to calculators, most teachers reported some type of restricted use (for about two-thirds of the students on average internationally). Corresponding to the results for the United States, most students in the Benchmarking entities (83 to 100 percent) had access to calculators. The policies regarding use varied dramatically, however. Whereas use was restricted for only about one-third or less of the students in some jurisdictions – the First in the World Consortium, the Jersey City Public Schools, the Michigan Invitational Group, and Montgomery County – more than 80 percent of the students were subject to some restrictions in South Carolina, the Chicago Public Schools, and the Rochester City School District.

TIMSS combined students' and teachers' reports on the frequency of calculator use to create an index of emphasis on calculators in mathematics class (ECMC), presented in Exhibit 6.13. Students were placed in the high category if they reported using calculators in class almost always or pretty often and their teachers reported calculator use of at least once or twice a week. At the other end of the spectrum, students were placed in the low category if they reported using calculators only once in a while or never and their teachers reported asking students to use calculators never or hardly ever. There was enormous variation in the results across countries. For example, the Netherlands and Singapore had more than four-fifths of their students (95 and 85 percent, respectively) in the high category. In contrast, a number of countries had half or more of their students in the low category, including Chinese Taipei, Korea, and Japan. Since several high-performing countries have restricted calculator use and large percentages of students are in the low-use category, the relationship between calculator use and performance is difficult to interpret. Although on average internationally the relationship is unclear, in most of the countries where emphasis on calculator use was high, there was a positive association between calculator use and mathematics achievement.

Exhibit R3.9 in the reference section shows the detailed results for students' reports on the frequency of their calculator use. In the Netherlands, 67 percent of students reported almost always using calculators in their mathematics lessons. Countries with the next highest level of use included the United States (42 percent) and Canada (44 percent). The Benchmarking jurisdictions with the greatest percentages of students reporting almost always using calculators were the Academy School District (68 percent), Jersey City (68 percent), and Naperville (71 percent). Benchmarking entities with the lowest percentages of students (25 percent or less) reporting this level of calculator use were South Carolina, Texas, Chicago, Miami-Dade, and Rochester.

The percentages of students asked to use calculators for various activities at least once or twice a week are shown in Exhibit R3.10. According to teachers internationally, they asked the most students to use calculators at least weekly for checking answers, performing routine computations, and solving complex problems (43 to 44 percent each). About one-fourth of the students across countries were asked to explore number concepts and one-fifth to use calculators on their tests. Across the Benchmarking entities, students used calculators for each of the activities asked about by TIMSS, although in varying degrees.

Students' reports on the frequency of their computer use in mathematics class are presented in Exhibit 6.14. Across countries, the vast majority of students (80 percent on average internationally) reported never using computers in mathematics class. Even though more students in the Benchmarking entities than internationally used computers in mathematics class, the percentages using computers almost always or pretty often were still relatively low, ranging from 24 percent in the Jersey City Public Schools to seven percent in Idaho, Guilford County, and the Michigan Invitational Group.

Because the Internet provides a wealth of opportunities for students to collect and analyze information, TIMSS began asking about students' access to the Internet and whether they used the World Wide Web to access information for mathematics projects. The data in Exhibit 6.15 indicate great variation in Internet access across countries and across the Benchmarking participants. Still, the international averages show about one-quarter of the students with access to the Internet at school. The international average for using the Internet to access information for mathematics class on even a monthly basis was 10 percent (less than half those reporting access). For the Benchmarking jurisdictions, Internet access at school ranged from 31 to 32 percent in Rochester and Chicago to 98 percent in First in the World and Naperville. Still, the only jurisdictions reporting 20 percent or more of the students accessing information for mathematics class on even a monthly basis were Connecticut, the Delaware Science Coalition, Jersey City, and Miami-Dade.

	Percentage of Students Having Access to Calculators in Class	Policy on Use of Calculators During Mathematics Lessons for Students Having Access					
		Unrestricted Use		Restricted Use		Calculators Not Permitted	
		Percent of Students	Average Achievement	Percent of Students	Average Achievement	Percent of Students	Average Achievement
Countries							
United States	96 (1.2)	34 (3.3)	524 (6.7)	66 (3.3)	493 (4.5)	0 (0.2)	~ ~
Belgium (Flemish)	94 (2.6)	13 (2.3)	580 (8.7)	87 (2.4)	560 (5.6)	1 (0.4)	~ ~
Canada	96 (1.1)	40 (3.3)	537 (4.5)	60 (3.3)	531 (4.5)	0 (0.0)	~ ~
Chinese Taipei	51 (4.6)	13 (3.9)	576 (13.0)	85 (4.3)	577 (5.7)	3 (2.0)	599 (76.8)
Czech Republic	94 (2.4)	7 (2.7)	517 (13.4)	91 (3.1)	522 (4.7)	2 (1.5)	~ ~
England	s 100 (0.3)	s 14 (2.2)	547 (16.0)	86 (2.2)	504 (5.2)	0 (0.0)	~ ~
Hong Kong, SAR	99 (0.5)	67 (4.3)	579 (5.2)	32 (4.2)	590 (6.6)	1 (0.0)	~ ~
Italy	87 (2.0)	10 (2.6)	467 (12.0)	84 (3.1)	482 (4.6)	6 (1.6)	465 (16.9)
Japan	34 (4.3)	13 (3.9)	579 (5.4)	85 (4.4)	579 (5.1)	2 (0.2)	~ ~
Korea, Rep. of	28 (3.4)	5 (3.3)	601 (9.0)	77 (6.3)	589 (4.6)	18 (5.7)	586 (9.0)
Netherlands	100 (0.0)	85 (4.1)	540 (7.8)	15 (4.1)	522 (18.5)	0 (0.0)	~ ~
Russian Federation	– –	12 (2.5)	547 (16.2)	78 (3.4)	520 (6.2)	10 (2.3)	546 (8.7)
Singapore	100 (0.0)	31 (4.7)	622 (11.0)	69 (4.7)	597 (6.2)	0 (0.0)	~ ~
States							
Connecticut	r 96 (2.2)	r 37 (7.4)	548 (13.2)	63 (7.4)	512 (9.7)	0 (0.0)	~ ~
Idaho	r 90 (5.1)	r 23 (6.5)	510 (13.4)	75 (6.6)	490 (10.4)	2 (0.2)	~ ~
Illinois	94 (3.9)	34 (5.4)	529 (8.8)	65 (5.4)	510 (7.2)	0 (0.0)	~ ~
Indiana	94 (2.6)	22 (5.2)	519 (10.7)	75 (5.6)	519 (9.2)	3 (2.0)	492 (7.4)
Maryland	r 100 (0.1)	r 42 (6.2)	509 (7.9)	58 (6.2)	468 (7.4)	0 (0.0)	~ ~
Massachusetts	97 (2.0)	36 (6.2)	537 (9.2)	64 (6.2)	498 (6.5)	0 (0.0)	~ ~
Michigan	99 (0.7)	55 (6.3)	530 (7.3)	45 (6.3)	517 (11.2)	0 (0.0)	~ ~
Missouri	95 (3.2)	45 (6.6)	492 (8.4)	55 (6.6)	494 (6.9)	0 (0.0)	~ ~
North Carolina	99 (0.8)	29 (6.2)	485 (14.1)	70 (6.3)	496 (7.1)	1 (0.8)	~ ~
Oregon	100 (0.3)	52 (6.2)	526 (8.9)	48 (6.2)	502 (6.7)	0 (0.0)	~ ~
Pennsylvania	89 (5.9)	32 (4.6)	554 (9.9)	66 (4.8)	495 (8.0)	2 (0.2)	~ ~
South Carolina	89 (4.8)	12 (3.5)	539 (29.9)	83 (4.9)	504 (8.1)	5 (2.9)	457 (26.6)
Texas	93 (2.8)	19 (4.0)	562 (16.1)	77 (5.1)	514 (11.2)	5 (2.7)	475 (52.8)
Districts and Consortia							
Academy School Dist. #20, CO	99 (0.2)	57 (0.4)	560 (2.0)	43 (0.4)	497 (2.8)	0 (0.0)	~ ~
Chicago Public Schools, IL	94 (3.9)	6 (3.6)	473 (29.3)	91 (4.7)	468 (6.9)	3 (0.3)	473 (3.3)
Delaware Science Coalition, DE	r 95 (3.5)	r 39 (6.0)	458 (18.1)	59 (6.3)	497 (12.4)	2 (0.1)	~ ~
First in the World Consort., IL	100 (0.0)	65 (4.7)	569 (6.6)	35 (4.7)	538 (8.9)	0 (0.0)	~ ~
Fremont/Lincoln/WestSide PS, NE	100 (0.0)	26 (9.5)	470 (12.6)	74 (9.5)	493 (11.9)	0 (0.0)	~ ~
Guilford County, NC	97 (0.6)	22 (3.9)	547 (12.4)	78 (3.9)	497 (10.7)	0 (0.0)	~ ~
Jersey City Public Schools, NJ	100 (0.0)	93 (5.0)	469 (7.0)	7 (5.0)	601 (5.3)	0 (0.0)	~ ~
Miami-Dade County PS, FL	s 88 (7.8)	s 25 (7.4)	446 (33.3)	75 (7.4)	404 (16.3)	0 (0.0)	~ ~
Michigan Invitational Group, MI	98 (1.7)	68 (6.5)	535 (6.7)	32 (6.5)	533 (7.5)	0 (0.0)	~ ~
Montgomery County, MD	s 100 (0.0)	s 69 (5.8)	547 (8.2)	31 (5.8)	505 (10.9)	0 (0.0)	~ ~
Naperville Sch. Dist. #203, IL	100 (0.0)	60 (3.1)	572 (5.2)	40 (3.1)	563 (6.7)	0 (0.0)	~ ~
Project SMART Consortium, OH	88 (4.8)	25 (5.6)	567 (21.0)	70 (6.3)	517 (8.6)	5 (3.3)	478 (10.1)
Rochester City Sch. Dist., NY	83 (3.1)	12 (5.1)	521 (24.6)	83 (6.1)	431 (5.6)	5 (3.8)	533 (8.2)
SW Math/Sci. Collaborative, PA	100 (0.0)	45 (7.1)	541 (9.8)	55 (7.1)	498 (10.7)	0 (0.0)	~ ~
International Avg. (All Countries)	73 (0.5)	21 (0.5)	490 (2.2)	67 (0.7)	488 (1.2)	12 (0.6)	464 (3.5)

SOURCE: IEA Third International Mathematics and Science Study (TIMSS), 1998-1999.

Background data provided by teachers.

* The use of calculators on TIMSS was not allowed in 1995 or in 1999.

States in *italics* did not fully satisfy guidelines for sample participation rates (see Appendix A for details).

( ) Standard errors appear in parentheses. Because results are rounded to the nearest whole number, some totals may appear inconsistent.

A dash (–) indicates data are not available. A tilde (~) indicates insufficient data to report achievement.

An "r" indicates teacher response data available for 70-84% of students. An "s" indicates teacher response data available for 50-69% of students.

Exhibit 6.13 Index of Emphasis on Calculators in Mathematics Class (ECMC)*

Index of Emphasis on Calculators in Mathematics Class

Index based on students' reports of the frequency of using calculators in mathematics lessons and teachers' reports of students' use of calculators in mathematics class for five activities: checking answers; tests and exams; routine computation; solving complex problems; and exploring number concepts (see reference exhibits R3.9-R3.10). High level indicates the student reported using calculators in mathematics lessons always or pretty often, and the teacher reported students use calculators at least once or twice a week for any of the tasks. Low level indicates the student reported using calculators once in a while or never, and the teacher reported students use calculators never or hardly ever for all of the tasks. Medium level includes all other possible combinations of responses.

		High ECMC		Medium ECMC		Low ECMC	
		Percent of Students	Average Achievement	Percent of Students	Average Achievement	Percent of Students	Average Achievement
Netherlands		95 (1.1)	538 (7.2)	5 (1.1)	512 (23.5)	0 (0.0)	~ ~
Jersey City Public Schools, NJ	r	93 (0.8)	485 (9.8)	7 (0.8)	432 (12.3)	0 (0.0)	~ ~
Naperville Sch. Dist. #203, IL		92 (0.8)	570 (2.8)	8 (0.8)	549 (14.2)	0 (0.0)	~ ~
Montgomery County, MD	s	90 (3.6)	540 (7.5)	10 (3.6)	484 (17.8)	0 (0.0)	~ ~
Academy School Dist. #20, CO		90 (0.8)	540 (1.8)	8 (0.8)	461 (5.7)	1 (0.3)	~ ~
Michigan Invitational Group, MI		90 (3.2)	536 (5.0)	9 (2.8)	506 (8.8)	2 (0.1)	~ ~
Oregon		87 (2.3)	521 (5.2)	13 (2.2)	485 (9.1)	0 (0.0)	~ ~
First in the World Consort., IL		86 (2.4)	560 (5.8)	14 (2.4)	547 (17.7)	0 (0.0)	~ ~
Singapore		85 (1.6)	611 (6.3)	15 (1.6)	567 (7.1)	0 (0.0)	~ ~
Fremont/Lincoln/WestSide PS, NE		83 (4.2)	492 (12.0)	17 (4.2)	463 (9.8)	0 (0.0)	~ ~
England	s	80 (2.3)	524 (5.7)	19 (2.2)	462 (6.5)	1 (0.7)	~ ~
North Carolina		79 (3.6)	500 (5.7)	20 (3.6)	480 (11.8)	1 (0.6)	~ ~
Canada	r	79 (1.9)	537 (3.0)	18 (1.7)	523 (4.7)	3 (0.9)	548 (6.8)
Michigan		78 (3.3)	530 (6.8)	21 (3.1)	507 (7.6)	1 (0.9)	~ ~
Missouri		78 (4.1)	497 (5.4)	17 (4.5)	476 (14.9)	5 (3.1)	461 (77.6)
Connecticut	r	76 (5.1)	528 (9.1)	19 (3.7)	505 (14.6)	5 (2.0)	497 (43.9)
Hong Kong, SAR		75 (1.9)	586 (4.4)	25 (1.8)	577 (6.3)	0 (0.2)	~ ~
Guilford County, NC		73 (5.5)	506 (9.6)	25 (5.4)	512 (15.5)	2 (0.4)	~ ~
Illinois		72 (4.7)	526 (6.2)	22 (3.4)	487 (7.8)	7 (3.8)	436 (7.8)
SW Math/Sci. Collaborative, PA		70 (5.4)	528 (7.6)	29 (5.1)	499 (11.1)	1 (0.7)	~ ~
Maryland	r	66 (5.3)	503 (4.7)	33 (5.3)	459 (9.3)	1 (0.5)	~ ~
United States	r	65 (3.2)	515 (4.5)	31 (2.9)	489 (6.4)	5 (1.2)	476 (10.8)
Massachusetts		64 (5.3)	518 (7.5)	33 (4.9)	505 (8.2)	3 (1.8)	497 (84.9)
Pennsylvania		63 (6.1)	521 (8.3)	25 (3.6)	497 (8.5)	12 (5.7)	492 (8.5)
Idaho	r	61 (6.2)	499 (9.6)	30 (3.4)	488 (13.8)	9 (4.6)	495 (12.5)
Delaware Science Coalition, DE	r	58 (4.1)	486 (11.9)	39 (3.8)	484 (14.3)	4 (2.6)	527 (29.9)
Indiana		56 (4.8)	523 (8.4)	39 (4.2)	513 (9.1)	5 (2.4)	492 (20.5)
Italy		52 (2.4)	486 (4.6)	37 (2.3)	474 (5.7)	11 (1.8)	483 (12.0)
Project SMART Consortium, OH		50 (2.9)	545 (11.6)	39 (4.3)	502 (8.3)	10 (3.5)	483 (8.9)
Miami-Dade County PS, FL	s	46 (7.6)	419 (16.1)	43 (5.3)	420 (12.5)	11 (7.3)	475 (56.9)
South Carolina		45 (5.2)	525 (10.4)	43 (4.6)	491 (12.4)	12 (3.4)	477 (21.9)
Belgium (Flemish)		39 (2.7)	571 (6.3)	54 (2.7)	562 (6.9)	7 (2.6)	532 (27.9)
Texas	r	37 (4.4)	550 (10.7)	52 (4.7)	504 (13.0)	12 (4.5)	519 (17.2)
Czech Republic		35 (3.2)	528 (7.1)	60 (3.5)	517 (4.7)	5 (2.0)	507 (26.2)
Chicago Public Schools, IL		32 (4.6)	471 (8.4)	53 (6.3)	471 (8.6)	15 (8.3)	446 (10.8)
Russian Federation		29 (2.3)	522 (9.3)	60 (2.1)	528 (6.3)	12 (2.4)	539 (13.3)
Rochester City Sch. Dist., NY	r	24 (4.9)	458 (19.4)	60 (4.4)	449 (6.3)	16 (3.6)	448 (16.9)
Chinese Taipei		2 (0.4)	~ ~	48 (4.0)	576 (4.8)	50 (4.2)	598 (5.4)
Korea, Rep. of		0 (0.3)	~ ~	29 (3.3)	587 (4.0)	71 (3.3)	587 (2.4)
Japan		0 (0.1)	~ ~	21 (3.2)	573 (6.4)	79 (3.2)	579 (2.2)
International Avg. (All Countries)		32 (0.3)	481 (1.8)	42 (0.5)	484 (1.2)	26 (0.5)	481 (3.3)

SOURCE: IEA Third International Mathematics and Science Study (TIMSS), 1998-1999.

* The use of calculators on TIMSS was not allowed in 1995 or in 1999.

States in *italics* did not fully satisfy guidelines for sample participation rates (see Appendix A for details).

( ) Standard errors appear in parentheses. Because results are rounded to the nearest whole number, some totals may appear inconsistent.

A tilde (~) indicates insufficient data to report achievement.

An "r" indicates teacher and/or student response data available for 70-84% of students. An "s" indicates teacher and/or student response data available for 50-69% of students.

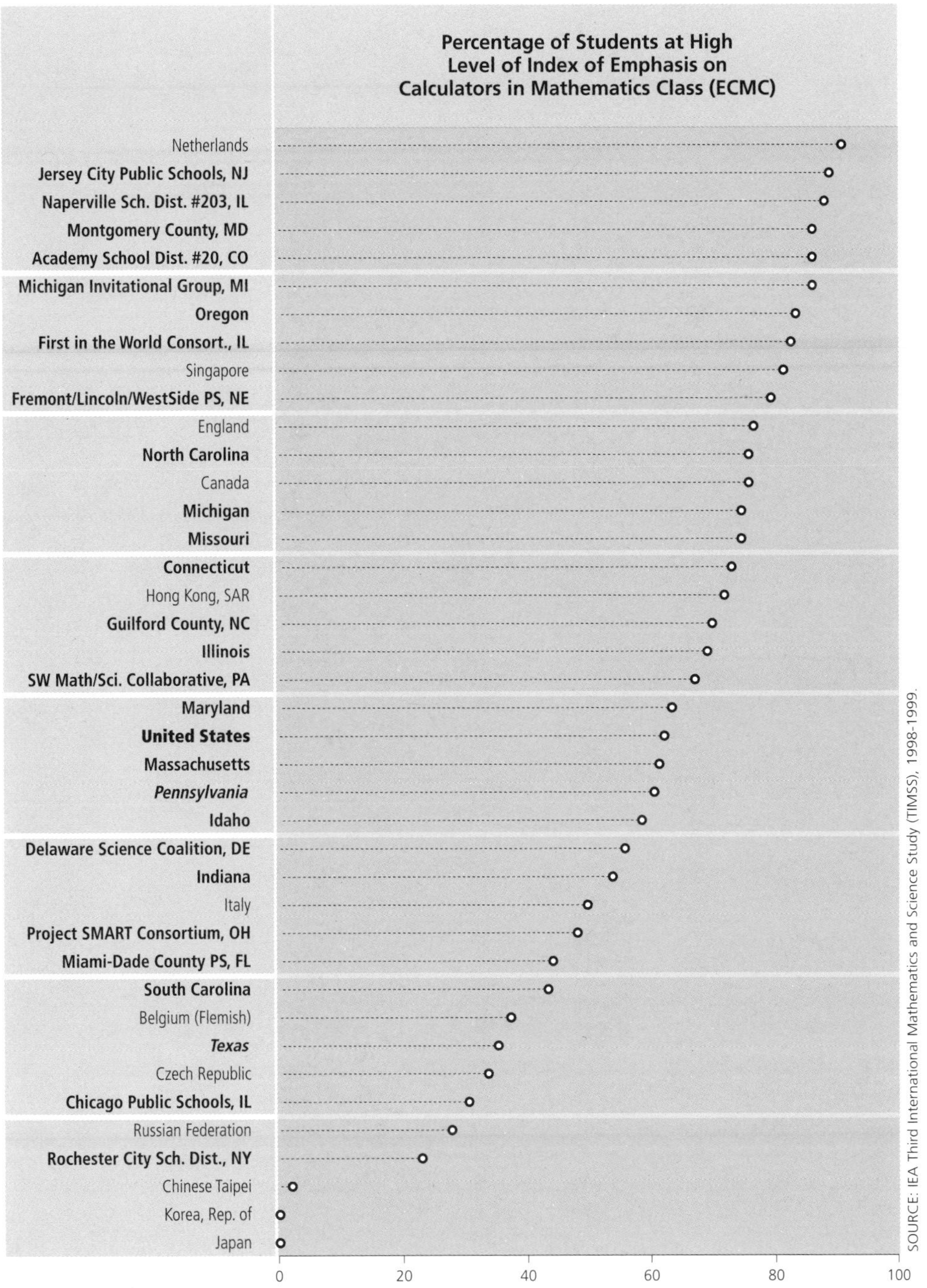

Percentage of Students at High Level of Index of Emphasis on Calculators in Mathematics Class (ECMC)

Netherlands
Jersey City Public Schools, NJ
Naperville Sch. Dist. #203, IL
Montgomery County, MD
Academy School Dist. #20, CO
Michigan Invitational Group, MI
Oregon
First in the World Consort., IL
Singapore
Fremont/Lincoln/WestSide PS, NE
England
North Carolina
Canada
Michigan
Missouri
Connecticut
Hong Kong, SAR
Guilford County, NC
Illinois
SW Math/Sci. Collaborative, PA
Maryland
United States
Massachusetts
Pennsylvania
Idaho
Delaware Science Coalition, DE
Indiana
Italy
Project SMART Consortium, OH
Miami-Dade County PS, FL
South Carolina
Belgium (Flemish)
Texas
Czech Republic
Chicago Public Schools, IL
Russian Federation
Rochester City Sch. Dist., NY
Chinese Taipei
Korea, Rep. of
Japan

0 20 40 60 80 100

SOURCE: IEA Third International Mathematics and Science Study (TIMSS), 1998-1999.

	Almost Always or Pretty Often		Once in a While		Never	
	Percent of Students	Average Achievement	Percent of Students	Average Achievement	Percent of Students	Average Achievement
Countries						
United States	12 (1.1)	463 (7.3)	27 (2.0)	520 (5.2)	61 (2.7)	506 (4.0)
Belgium (Flemish)	1 (0.4)	~ ~	5 (1.2)	536 (17.4)	93 (1.3)	562 (3.1)
Canada	8 (0.7)	507 (7.1)	25 (1.5)	534 (3.8)	67 (1.6)	534 (2.5)
Chinese Taipei	13 (0.6)	548 (7.5)	21 (0.6)	564 (5.2)	66 (0.9)	601 (3.8)
Czech Republic	2 (0.7)	~ ~	14 (2.4)	526 (8.4)	84 (2.6)	520 (3.8)
England	11 (1.7)	466 (10.4)	43 (2.2)	512 (5.1)	46 (2.7)	492 (5.2)
Hong Kong, SAR	8 (0.5)	561 (9.5)	18 (0.8)	577 (6.2)	75 (1.1)	587 (4.1)
Italy	11 (1.3)	464 (7.4)	17 (1.6)	489 (5.5)	72 (2.3)	482 (4.0)
Japan	2 (0.5)	~ ~	21 (2.3)	576 (3.7)	76 (2.7)	581 (2.0)
Korea, Rep. of	3 (0.3)	567 (7.9)	13 (0.7)	596 (3.9)	83 (0.8)	587 (2.2)
Netherlands	1 (0.2)	~ ~	19 (3.2)	543 (9.6)	80 (3.2)	541 (8.2)
Russian Federation	1 (0.2)	~ ~	3 (0.4)	513 (11.1)	97 (0.4)	530 (5.7)
Singapore	11 (0.8)	590 (11.0)	43 (2.5)	625 (6.8)	46 (2.7)	589 (6.1)
States						
Connecticut	12 (1.9)	483 (9.6)	31 (2.9)	529 (9.7)	57 (3.8)	513 (9.9)
Idaho	7 (0.9)	434 (15.0)	17 (1.5)	507 (8.5)	76 (2.1)	498 (7.1)
Illinois	12 (1.8)	474 (7.7)	36 (2.8)	521 (8.6)	52 (4.0)	510 (7.9)
Indiana	10 (1.8)	479 (16.5)	25 (3.6)	517 (9.9)	65 (5.1)	522 (7.0)
Maryland	13 (1.7)	447 (11.1)	36 (2.0)	504 (7.5)	51 (2.5)	507 (6.8)
Massachusetts	13 (2.7)	488 (9.5)	24 (2.7)	530 (7.5)	64 (4.3)	513 (5.7)
Michigan	9 (1.3)	467 (9.6)	28 (3.4)	540 (10.6)	63 (3.6)	518 (6.8)
Missouri	9 (1.7)	453 (7.7)	20 (2.6)	489 (7.5)	71 (3.4)	496 (6.1)
North Carolina	13 (2.2)	456 (10.0)	34 (2.4)	500 (8.0)	53 (3.6)	503 (7.6)
Oregon	12 (1.3)	482 (11.1)	26 (1.9)	534 (6.4)	62 (2.5)	515 (5.8)
Pennsylvania	8 (1.0)	465 (11.3)	22 (2.4)	524 (7.7)	70 (3.0)	509 (6.5)
South Carolina	11 (1.5)	444 (8.4)	25 (2.4)	514 (10.5)	64 (3.5)	509 (7.6)
Texas	14 (3.0)	489 (16.3)	33 (3.1)	533 (10.3)	52 (4.8)	522 (10.2)
Districts and Consortia						
Academy School Dist. #20, CO	9 (0.9)	506 (10.1)	32 (1.2)	547 (3.4)	59 (1.4)	523 (2.9)
Chicago Public Schools, IL	15 (3.4)	437 (12.8)	28 (4.1)	469 (8.4)	58 (7.1)	467 (6.8)
Delaware Science Coalition, DE	9 (1.0)	415 (9.0)	16 (1.7)	495 (16.3)	75 (1.9)	492 (9.0)
First in the World Consort., IL	8 (1.4)	518 (23.4)	44 (3.8)	571 (6.0)	48 (4.3)	556 (8.5)
Fremont/Lincoln/WestSide PS, NE	13 (1.2)	463 (13.0)	37 (3.5)	513 (13.9)	51 (4.1)	478 (6.8)
Guilford County, NC	7 (0.9)	478 (11.7)	43 (1.7)	526 (8.1)	50 (2.1)	510 (10.0)
Jersey City Public Schools, NJ	24 (2.5)	462 (15.5)	41 (1.7)	483 (7.7)	35 (2.8)	480 (11.9)
Miami-Dade County PS, FL	14 (2.1)	361 (16.3)	16 (2.0)	428 (17.3)	70 (3.3)	439 (7.6)
Michigan Invitational Group, MI	7 (0.9)	502 (20.5)	24 (1.9)	543 (6.2)	69 (2.2)	533 (5.9)
Montgomery County, MD	10 (0.9)	488 (9.9)	37 (2.2)	546 (6.2)	53 (2.4)	542 (5.3)
Naperville Sch. Dist. #203, IL	8 (0.7)	549 (9.9)	44 (2.5)	579 (5.2)	48 (2.9)	565 (4.6)
Project SMART Consortium, OH	17 (2.6)	494 (9.7)	36 (3.2)	536 (10.2)	47 (3.9)	521 (8.6)
Rochester City Sch. Dist., NY r	14 (1.6)	444 (6.2)	14 (1.9)	450 (14.2)	72 (2.8)	457 (7.3)
SW Math/Sci. Collaborative, PA	8 (1.6)	486 (17.9)	28 (4.3)	530 (11.1)	64 (4.9)	516 (7.6)
International Avg. (All Countries)	5 (0.1)	455 (2.8)	14 (0.2)	488 (1.5)	80 (0.3)	498 (0.7)

SOURCE: IEA Third International Mathematics and Science Study (TIMSS), 1998-1999.

Background data provided by students.

States in *italics* did not fully satisfy guidelines for sample participation rates (see Appendix A for details).

( ) Standard errors appear in parentheses. Because results are rounded to the nearest whole number, some totals may appear inconsistent.

A tilde (~) indicates insufficient data to report achievement.

An "r" indicates a 70-84% student response rate.

Exhibit 6.15 Access to the Internet and Use of the Internet for Mathematics Projects

	Percentage of Students				
	Have Access to the Internet			Use the Internet for Mathematics Projects at Least Once a Month	
	At Home	At School	Elsewhere	Use E-mail to Work with Students in Other Schools	Use the World Wide Web to Access Information
Countries					
United States	59 (1.7)	76 (3.2)	81 (0.9)	13 (0.5)	17 (0.8)
Belgium (Flemish)	27 (0.9)	44 (2.7)	64 (1.1)	5 (0.5)	9 (0.9)
Canada	57 (1.3)	87 (1.5)	84 (0.8)	8 (0.4)	12 (0.5)
Chinese Taipei	32 (1.1)	61 (3.2)	41 (0.8)	10 (0.4)	12 (0.5)
Czech Republic	7 (0.7)	16 (2.6)	39 (1.6)	3 (0.4)	5 (0.4)
England	36 (1.1)	65 (3.1)	53 (1.3)	8 (0.7)	18 (0.9)
Hong Kong, SAR	34 (1.1)	26 (2.2)	34 (0.8)	10 (0.6)	11 (0.6)
Italy	13 (0.7)	20 (2.2)	27 (1.1)	7 (0.6)	8 (0.7)
Japan	r 13 (0.9)	6 (1.6)	s 2 (0.3)	8 (0.8)	7 (0.8)
Korea, Rep. of	23 (0.7)	6 (1.2)	36 (1.0)	4 (0.3)	6 (0.3)
Netherlands	41 (1.8)	53 (5.4)	74 (1.8)	6 (0.7)	6 (0.9)
Russian Federation	3 (0.3)	1 (0.4)	17 (0.9)	3 (0.3)	4 (0.4)
Singapore	47 (1.9)	48 (3.2)	39 (0.9)	9 (0.7)	15 (0.8)
States					
Connecticut	71 (2.5)	85 (2.3)	85 (0.8)	14 (1.2)	20 (1.5)
Idaho	53 (2.7)	84 (4.1)	78 (1.4)	11 (0.9)	12 (1.0)
Illinois	56 (2.3)	79 (3.6)	79 (1.5)	12 (0.8)	16 (1.2)
Indiana	59 (2.0)	70 (5.8)	85 (1.5)	10 (1.0)	13 (1.1)
Maryland	66 (1.8)	77 (3.2)	83 (0.8)	13 (0.8)	18 (1.0)
Massachusetts	68 (2.1)	78 (3.6)	83 (1.3)	14 (1.0)	18 (1.1)
Michigan	61 (2.4)	80 (3.7)	83 (1.2)	10 (0.9)	12 (1.1)
Missouri	49 (1.5)	77 (5.3)	82 (1.0)	11 (0.8)	15 (0.7)
North Carolina	51 (2.0)	80 (2.7)	82 (0.9)	13 (0.9)	19 (1.3)
Oregon	61 (2.1)	85 (4.4)	82 (1.7)	11 (0.6)	14 (1.1)
Pennsylvania	64 (2.7)	69 (4.0)	82 (0.9)	11 (0.8)	16 (1.5)
South Carolina	52 (2.2)	92 (1.5)	81 (1.3)	12 (0.9)	19 (1.3)
Texas	54 (3.5)	82 (3.5)	79 (2.2)	14 (1.1)	19 (1.2)
Districts and Consortia					
Academy School Dist. #20, CO	84 (1.1)	93 (0.7)	78 (1.2)	12 (0.9)	17 (1.1)
Chicago Public Schools, IL	35 (2.4)	32 (6.8)	72 (1.9)	10 (1.2)	16 (1.6)
Delaware Science Coalition, DE	66 (2.3)	88 (1.5)	84 (1.0)	17 (1.3)	20 (1.7)
First in the World Consort., IL	82 (1.0)	98 (0.6)	86 (1.7)	13 (1.1)	19 (1.3)
Fremont/Lincoln/WestSide PS, NE	61 (1.9)	91 (1.4)	85 (1.6)	11 (1.3)	16 (1.8)
Guilford County, NC	64 (1.9)	89 (1.0)	89 (1.1)	12 (1.2)	19 (1.5)
Jersey City Public Schools, NJ	38 (2.2)	92 (1.2)	71 (2.1)	19 (1.4)	33 (2.3)
Miami-Dade County PS, FL	47 (3.1)	59 (6.7)	73 (2.4)	20 (2.5)	22 (1.8)
Michigan Invitational Group, MI	62 (2.1)	90 (1.3)	83 (1.4)	7 (0.8)	14 (1.4)
Montgomery County, MD	77 (1.8)	92 (1.0)	74 (2.2)	13 (1.2)	18 (1.2)
Naperville Sch. Dist. #203, IL	86 (1.0)	98 (0.4)	87 (0.8)	10 (0.8)	14 (1.3)
Project SMART Consortium, OH	63 (1.8)	83 (1.1)	91 (0.7)	12 (1.2)	15 (0.9)
Rochester City Sch. Dist., NY	31 (2.3)	31 (1.6)	74 (2.0)	13 (1.7)	15 (1.0)
SW Math/Sci. Collaborative, PA	58 (2.7)	80 (4.7)	83 (1.6)	10 (0.8)	14 (1.3)
International Avg. (All Countries)	19 (0.2)	27 (0.4)	43 (0.2)	8 (0.1)	10 (0.1)

SOURCE: IEA Third International Mathematics and Science Study (TIMSS), 1998-1999.

Background data provided by students.

States in *italics* did not fully satisfy guidelines for sample participation rates (see Appendix A for details).

( ) Standard errors appear in parentheses. Because results are rounded to the nearest whole number, some totals may appear inconsistent.

An "r" indicates a 70-84% student response rate. An "s" indicates a 50-69% student response rate.

What Are the Roles of Homework and Assessment?

The amount of time students spend on homework assignments is an important consideration in examining their opportunity to learn mathematics. Exhibit 6.16 presents the index of teachers' emphasis on mathematics homework (EMH). Students in the high category had teachers who reported giving relatively long homework assignments (more than 30 minutes) on a relatively frequent basis (at least once or twice a week). Those in the low category had teachers who gave short assignments (less than 30 minutes) relatively infrequently (less than once a week or never). The medium level includes all other combinations of responses. Details from teachers' reports about the length and frequency of their homework assignments are found in the reference section in Exhibit R3.11.

The results show substantial variation across countries and Benchmarking entities in the emphasis placed on homework. Together with Italy, Singapore, and the Russian Federation among the comparison countries, the Academy School District had more than half its students in the high category. For the remaining Benchmarking participants, the majority of students were in the medium category. Very few students were in the low category. One notable exception is Japan (34 percent in the low category), where students were more likely to spend extra time in tutoring and special schools than doing homework.[4] There was little relationship between the amount of homework assigned and students' performance. Again, lower-performing students may need more homework assignments for remedial reasons.

Since problem-solving activities will potentially be more beneficial if they can be extended to out-of-class-situations and stretched over a longer time, TIMSS asked teachers how often they assigned homework based on projects and investigations. The data in Exhibit R3.12 in the reference section show that most students (82 percent on average internationally) had teachers that never or rarely gave such homework. Even though teachers in some of the Benchmarking entities reported giving project-based homework more frequently than did teachers internationally, such assignments did not appear to be made very often. The Benchmarking entities where approximately one-third or more of the students were given projects to do as homework at least sometimes were Connecticut, Massachusetts, Oregon, South Carolina, the Jersey City Public Schools, the Miami-Dade County Public Schools, Montgomery County, and the Project SMART Consortium.

4 Robitaille, D.F., (1997), *National Contexts for Mathematics and Science Education: An Encyclopedia of the Education Systems Participating in TIMSS*, Vancouver, BC: Pacific Educational Press.

One theme in recommendations for educational reform is to make assessment a continuous process that relies on a variety of methods and sources of data, rather than on a few high-stakes tests. Exhibit 6.17 shows teachers' reports about the weight given to various types of assessment. Teachers in the United States as a whole and in most of the Benchmarking jurisdictions reported placing less weight on informal assessment approaches than did teachers internationally. On average internationally, the most emphasis was placed on students' responses in class, which were given quite a lot or a great deal of weight for 77 percent of the students. The next heaviest weight internationally was given to teacher-made tests requiring explanations (67 percent of students on average) and to observations of students (64 percent). While the use of teacher-made tests requiring explanations was similar to the international average in many Benchmarking jurisdictions, students' responses in class and observations of students were given less weight in the United States as a whole and in most Benchmarking entities (generally for about half the students or less). Exceptions included Jersey City and Miami-Dade, as well as Chicago to some extent.

Internationally, the least weight reportedly was given to external standardized tests, teacher-made objective tests, and projects or practical exercises. On average across countries, about two-fifths of the students (from 37 to 42 percent) had mathematics teachers who reported giving quite a lot or a great deal of weight to such assessments. Across the Benchmarking entities, generally even less weight than internationally was given to external standardized tests. The jurisdictions more similar to the international average were Michigan, North Carolina, Texas, the Academy School District, and Jersey City.

As shown in Exhibit R3.13 in the reference section, eighth-grade students reported substantial variation in the frequency of testing in mathematics class. On average internationally, students were split about in half, with 57 percent reporting having a quiz or test in class almost always or pretty often and 43 percent reporting such testing only once in a while or never. At least three-fourths of the students reported frequent testing in Belgium (Flemish), Canada, the Russian Federation, and the United States. Across the Benchmarking jurisdictions about 80 to 90 percent of the students reported frequent testing. In contrast, about half or more reported infrequent testing in the Czech Republic, Hong Kong, Italy, Japan, and Korea. Within participating entities, there was a tendency for the most frequent testing to be associated with lower-achieving students. One could argue that these students can least afford time diverted from their instructional program. However, teachers may provide shorter lessons and follow-up quizzes for lower-achieving students to monitor their grasp of the subject matter more closely.

Index of Teachers' Emphasis on Mathematics Homework

Index based on teachers' responses to two questions about how often they usually assign mathematics homework and how many minutes of mathematics homework they usually assign students (see reference exhibit R3.11). High level indicates the assignment of more than 30 minutes of homework at least once or twice a week. Low level indicates the assignment of less than 30 minutes of homework less than once a week or never assigning homework. Medium level includes all other possible combinations of responses.

	High EMH		Medium EMH		Low EMH	
	Percent of Students	Average Achievement	Percent of Students	Average Achievement	Percent of Students	Average Achievement
Italy	80 (3.0)	479 (4.9)	20 (2.9)	479 (7.9)	0 (0.0)	~ ~
Academy School Dist. #20, CO	73 (0.4)	546 (1.6)	25 (0.4)	483 (4.0)	2 (0.1)	~ ~
Singapore	66 (4.6)	613 (6.9)	34 (4.6)	587 (10.6)	0 (0.0)	~ ~
Russian Federation	57 (4.6)	527 (6.7)	43 (4.6)	525 (7.8)	0 (0.0)	~ ~
Chinese Taipei	48 (3.6)	593 (6.4)	50 (3.7)	580 (5.5)	2 (1.1)	~ ~
Hong Kong, SAR	41 (4.3)	580 (5.9)	57 (4.4)	585 (5.8)	2 (1.2)	~ ~
Jersey City Public Schools, NJ	40 (5.7)	492 (16.0)	60 (5.7)	464 (8.3)	0 (0.0)	~ ~
First in the World Consort., IL	37 (5.1)	595 (12.0)	63 (5.1)	533 (7.2)	0 (0.0)	~ ~
Chicago Public Schools, IL	37 (9.1)	472 (12.9)	63 (9.1)	457 (7.5)	0 (0.0)	~ ~
Texas	35 (6.2)	546 (16.3)	63 (6.7)	500 (9.0)	2 (1.5)	~ ~
Massachusetts	35 (6.5)	525 (9.9)	65 (6.5)	506 (6.9)	0 (0.0)	~ ~
SW Math/Sci. Collaborative, PA	34 (5.3)	552 (13.5)	65 (5.3)	501 (8.8)	1 (0.9)	~ ~
Michigan	32 (4.3)	549 (15.0)	68 (4.3)	502 (7.0)	0 (0.0)	~ ~
Naperville Sch. Dist. #203, IL	29 (2.3)	588 (3.5)	68 (2.3)	559 (4.1)	2 (0.1)	~ ~
South Carolina	29 (6.2)	527 (14.1)	71 (6.2)	491 (8.8)	0 (0.0)	~ ~
Michigan Invitational Group, MI	28 (6.9)	570 (14.9)	72 (6.9)	517 (5.3)	0 (0.0)	~ ~
England	28 (2.9)	529 (8.2)	71 (3.0)	485 (4.7)	1 (0.5)	~ ~
Guilford County, NC	27 (6.0)	539 (13.1)	71 (6.5)	504 (11.0)	2 (0.1)	~ ~
Illinois	26 (5.4)	530 (11.6)	74 (5.4)	502 (7.6)	0 (0.0)	~ ~
Project SMART Consortium, OH	25 (5.7)	567 (16.1)	75 (5.7)	505 (6.8)	0 (0.0)	~ ~
Montgomery County, MD	25 (4.1)	569 (10.5)	74 (4.1)	526 (3.4)	0 (0.1)	~ ~
Missouri	25 (5.7)	498 (15.8)	74 (5.6)	487 (5.7)	1 (1.1)	~ ~
United States	25 (2.1)	528 (9.6)	75 (2.0)	495 (3.8)	1 (0.6)	~ ~
Korea, Rep. of	25 (3.4)	587 (4.2)	62 (3.6)	586 (2.9)	14 (2.6)	593 (4.4)
Pennsylvania	24 (5.2)	535 (12.6)	76 (5.2)	499 (6.3)	0 (0.0)	~ ~
Connecticut	22 (5.1)	545 (20.3)	78 (5.1)	503 (9.3)	0 (0.0)	~ ~
North Carolina	21 (5.1)	534 (13.1)	75 (5.0)	486 (6.8)	4 (2.2)	463 (27.7)
Oregon	21 (4.5)	558 (12.0)	76 (4.8)	506 (6.0)	3 (2.0)	453 (68.7)
Fremont/Lincoln/WestSide PS, NE	20 (2.9)	541 (29.6)	80 (2.9)	475 (7.1)	0 (0.0)	~ ~
Rochester City Sch. Dist., NY	20 (5.1)	502 (11.5)	80 (5.1)	430 (6.4)	0 (0.0)	~ ~
Indiana	18 (4.8)	560 (11.2)	82 (4.8)	504 (7.4)	0 (0.0)	~ ~
Miami-Dade County PS, FL	18 (4.6)	411 (15.3)	82 (4.6)	424 (10.5)	0 (0.0)	~ ~
Canada	16 (2.3)	527 (6.2)	83 (2.4)	532 (2.8)	1 (0.6)	~ ~
Idaho	14 (3.2)	516 (20.7)	83 (3.4)	492 (7.1)	3 (1.0)	476 (38.3)
Delaware Science Coalition, DE	14 (4.4)	528 (18.5)	86 (4.4)	472 (9.4)	0 (0.0)	~ ~
Maryland	14 (2.5)	524 (16.6)	85 (2.8)	491 (6.5)	2 (1.5)	~ ~
Japan	11 (2.5)	578 (3.9)	55 (4.3)	580 (2.8)	34 (4.3)	574 (5.3)
Netherlands	11 (2.6)	555 (14.6)	88 (2.6)	538 (8.0)	1 (0.5)	~ ~
Belgium (Flemish)	10 (2.0)	582 (8.6)	73 (3.6)	557 (5.5)	17 (3.2)	548 (15.0)
Czech Republic	2 (1.2)	~ ~	85 (3.8)	520 (4.8)	13 (3.6)	513 (9.9)
International Avg. (All Countries)	35 (0.6)	491 (1.8)	62 (0.6)	485 (1.0)	4 (0.2)	484 (4.0)

SOURCE: IEA Third International Mathematics and Science Study (TIMSS), 1998-1999.

States in *italics* did not fully satisfy guidelines for sample participation rates (see Appendix A for details).

A tilde (~) indicates insufficient data to report achievement.

( ) Standard errors appear in parentheses. Because results are rounded to the nearest whole number, some totals may appear inconsistent.

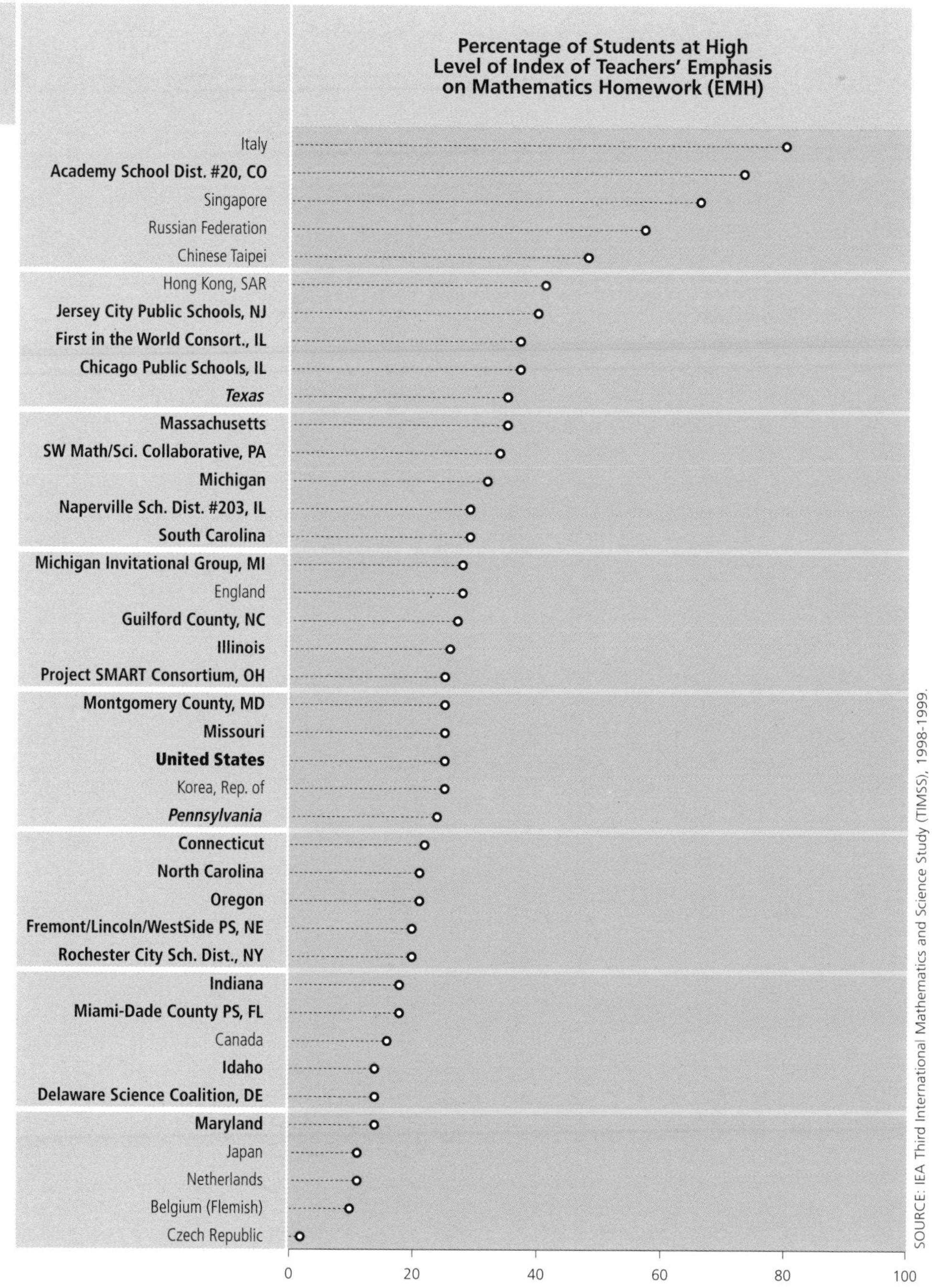

Percentage of Students at High Level of Index of Teachers' Emphasis on Mathematics Homework (EMH)

Italy
Academy School Dist. #20, CO
Singapore
Russian Federation
Chinese Taipei
Hong Kong, SAR
Jersey City Public Schools, NJ
First in the World Consort., IL
Chicago Public Schools, IL
Texas
Massachusetts
SW Math/Sci. Collaborative, PA
Michigan
Naperville Sch. Dist. #203, IL
South Carolina
Michigan Invitational Group, MI
England
Guilford County, NC
Illinois
Project SMART Consortium, OH
Montgomery County, MD
Missouri
United States
Korea, Rep. of
Pennsylvania
Connecticut
North Carolina
Oregon
Fremont/Lincoln/WestSide PS, NE
Rochester City Sch. Dist., NY
Indiana
Miami-Dade County PS, FL
Canada
Idaho
Delaware Science Coalition, DE
Maryland
Japan
Netherlands
Belgium (Flemish)
Czech Republic

0 20 40 60 80 100

SOURCE: IEA Third International Mathematics and Science Study (TIMSS), 1998-1999.

	Percentage of Students by Type of Assessment						
	External Standardized Tests	Teacher-Made Tests Requiring Explanations	Teacher-Made Objective Tests	Homework Assignments	Projects or Practical Exercises	Observations of Students	Students' Responses in Class
Countries							
United States	28 (3.0)	55 (3.3)	28 (3.5)	56 (4.3)	33 (3.5)	40 (3.2)	41 (3.6)
Belgium (Flemish)	12 (3.0)	94 (1.4)	11 (2.4)	23 (3.0)	12 (2.1)	17 (3.4)	52 (4.4)
Canada	21 (3.1)	61 (3.0)	r 26 (2.8)	r 51 (3.8)	r 38 (2.7)	r 34 (3.2)	42 (3.4)
Chinese Taipei	36 (4.0)	43 (4.0)	76 (3.4)	81 (3.2)	17 (3.4)	68 (3.1)	72 (3.6)
Czech Republic	53 (5.4)	97 (1.8)	9 (2.6)	26 (5.0)	23 (5.2)	80 (4.2)	98 (1.5)
England	s 51 (4.1)	s 35 (3.6)	s 7 (1.4)	s 81 (2.2)	s 41 (3.4)	s 78 (2.9)	s 78 (2.7)
Hong Kong, SAR	17 (3.2)	52 (4.2)	47 (3.6)	44 (4.0)	10 (2.6)	38 (4.3)	44 (4.3)
Italy	22 (3.2)	92 (2.2)	63 (3.8)	67 (3.6)	75 (3.1)	96 (1.4)	99 (0.6)
Japan	15 (2.9)	55 (4.4)	25 (3.9)	47 (4.0)	41 (4.0)	67 (4.1)	65 (4.3)
Korea, Rep. of	37 (3.8)	48 (3.7)	45 (3.7)	32 (3.6)	43 (3.3)	50 (4.1)	61 (4.1)
Netherlands	29 (5.5)	96 (1.8)	20 (5.8)	18 (4.7)	8 (2.6)	28 (4.7)	27 (5.4)
Russian Federation	– –	98 (1.0)	54 (4.4)	68 (3.7)	59 (3.8)	91 (2.2)	86 (2.5)
Singapore	36 (4.2)	22 (3.9)	5 (2.0)	61 (4.5)	37 (4.2)	46 (4.6)	52 (4.2)
States							
Connecticut	s 11 (3.7)	s 56 (7.3)	s 21 (6.8)	s 45 (5.6)	s 61 (8.5)	s 49 (8.6)	s 53 (7.3)
Idaho	r 25 (5.1)	r 37 (6.1)	r 21 (5.7)	r 79 (5.7)	r 27 (6.3)	r 29 (6.9)	r 33 (7.5)
Illinois	24 (4.4)	47 (5.9)	32 (5.7)	60 (5.9)	28 (5.5)	23 (4.6)	27 (5.4)
Indiana	28 (6.6)	61 (4.9)	27 (5.8)	60 (5.6)	23 (4.3)	33 (5.8)	29 (6.2)
Maryland	r 26 (6.0)	r 61 (5.5)	r 19 (4.9)	r 47 (6.0)	r 28 (3.5)	r 41 (6.4)	r 42 (6.4)
Massachusetts	19 (4.6)	64 (4.7)	20 (4.1)	56 (6.2)	41 (5.2)	53 (6.1)	57 (5.8)
Michigan	36 (7.3)	48 (5.8)	27 (6.2)	54 (6.0)	33 (5.4)	25 (5.1)	32 (5.2)
Missouri	21 (4.5)	60 (5.7)	24 (4.7)	73 (5.2)	45 (5.7)	42 (5.9)	36 (5.2)
North Carolina	39 (6.1)	44 (5.0)	48 (5.3)	58 (6.4)	34 (5.0)	46 (5.8)	48 (4.4)
Oregon	14 (3.9)	60 (6.4)	27 (6.4)	76 (6.0)	33 (6.2)	44 (6.3)	40 (6.1)
Pennsylvania	18 (4.3)	58 (5.3)	20 (5.4)	47 (6.5)	24 (5.1)	39 (6.7)	42 (6.6)
South Carolina	13 (2.6)	66 (7.3)	44 (5.6)	36 (5.4)	35 (6.8)	48 (5.9)	42 (5.4)
Texas	42 (6.0)	r 49 (6.1)	55 (6.9)	53 (6.9)	r 33 (5.9)	52 (6.6)	52 (6.1)
Districts and Consortia							
Academy School Dist. #20, CO	43 (0.4)	33 (0.3)	6 (0.2)	72 (0.3)	38 (0.4)	39 (0.4)	43 (0.4)
Chicago Public Schools, IL	26 (8.6)	51 (10.2)	60 (10.6)	59 (10.0)	41 (12.8)	56 (12.6)	71 (10.7)
Delaware Science Coalition, DE	r 23 (5.7)	r 64 (6.7)	r 13 (4.9)	r 41 (6.9)	r 37 (5.0)	r 41 (7.1)	r 43 (6.1)
First in the World Consort., IL	r 10 (3.5)	r 77 (4.9)	r 35 (7.4)	r 17 (4.4)	r 38 (5.3)	r 26 (8.2)	r 31 (4.8)
Fremont/Lincoln/WestSide PS, NE	8 (5.5)	42 (9.7)	37 (8.6)	49 (9.2)	20 (5.3)	29 (1.7)	r 19 (3.3)
Guilford County, NC	22 (4.1)	57 (5.2)	47 (5.9)	57 (6.7)	39 (7.1)	46 (5.9)	39 (6.5)
Jersey City Public Schools, NJ	63 (6.5)	96 (3.8)	58 (6.0)	40 (5.0)	82 (4.5)	82 (3.7)	82 (3.7)
Miami-Dade County PS, FL	s 21 (6.1)	s 66 (8.2)	s 35 (8.9)	s 67 (9.5)	s 51 (7.6)	s 67 (9.7)	s 77 (8.3)
Michigan Invitational Group, MI	11 (2.6)	74 (4.7)	9 (6.3)	59 (7.9)	41 (6.6)	41 (8.9)	35 (7.6)
Montgomery County, MD	s 24 (7.0)	s 77 (3.1)	s 16 (5.6)	s 40 (6.0)	s 28 (6.6)	s 26 (7.5)	s 21 (5.8)
Naperville Sch. Dist. #203, IL	16 (2.8)	54 (4.5)	16 (4.5)	48 (3.7)	33 (3.9)	39 (6.0)	29 (5.7)
Project SMART Consortium, OH	21 (5.4)	62 (6.5)	28 (6.5)	47 (6.2)	41 (6.6)	45 (7.6)	45 (7.2)
Rochester City Sch. Dist., NY	1 (0.0)	60 (4.2)	36 (6.6)	50 (5.8)	29 (5.7)	30 (5.2)	34 (6.2)
SW Math/Sci. Collaborative, PA	22 (5.7)	59 (6.8)	17 (5.0)	44 (7.6)	23 (5.8)	42 (4.6)	49 (5.7)
International Avg. (All Countries)	37 (0.6)	67 (0.6)	39 (0.6)	60 (0.6)	42 (0.6)	64 (0.6)	77 (0.5)

SOURCE: IEA Third International Mathematics and Science Study (TIMSS), 1998-1999.

Background data provided by teachers.

States in *italics* did not fully satisfy guidelines for sample participation rates (see Appendix A for details).

( ) Standard errors appear in parentheses. Because results are rounded to the nearest whole number, some totals may appear inconsistent.

A dash (–) indicates data are not available.

An "r" indicates teacher response data available for 70-84% of students. An "s" indicates teacher response data available for 50-69% of students.

In What Types of Professional Development Activities Do U.S. Mathematics Teachers Participate?

As a TIMSS 1999 national option, the United States asked mathematics teachers to describe their professional development during the 1998-99 school year, defined as June 1998 to May 1999. Since no other countries asked these questions, cross-country comparisons are not possible. Comparisons, however, can be made to the United States as a whole and among the Benchmarking jurisdictions. Teachers were asked both how often they observed and were observed by other teachers (see Exhibit 6.18). In the U.S. overall, these observations of and by teachers were reported by the mathematics teachers of 25 and 35 percent of the students, respectively. Among the Benchmarking states, the results for classroom observation as a professional development approach resembled the national results. Among districts and consortia, observations were used most extensively in the First in the World Consortium and Montgomery County with more than half the students having teachers who reported both observing and being observed by other teachers.

The professional development activities teachers were asked about include the following school- and district-based activities: immersion or internship activities; receiving mentoring, coaching, lead teaching, or observation; teacher resource centers; committees or task forces; and teacher study groups. As shown in Exhibit 6.19, participation on committees or task forces was the most frequently used of these activities. It was reported nationally by the mathematics teachers of more than half the eighth graders (55 percent), and was similarly popular among the Benchmarking participants.

Mathematics teachers were asked about their participation in several types of workshops, conferences, and networks, including within-district workshops and institutes; out-of-district workshops and institutes; teacher collaborative or networks; out-of-district conferences; and other forms of organized professional development (see Exhibit 6.20). They were also asked about individual activities, including taking courses for college credit; individual research projects; individual learning; and other individual professional development activities (see Exhibit 6.21). Of all of the professional development activities, within-district workshops or institutes (79 percent of the students) and individual learning (84 percent) were generally the most frequent activities in which mathematics teachers of U.S. eighth-grade students participated during the 1998-99 school year. Even though there was considerable variation, these activities were also widely reported by teachers in the Benchmarking jurisdictions.

Teachers' reports about the topics heavily emphasized in their professional development are presented in Exhibit 6.22. Nationally, mathematics teachers of 63 percent of eighth graders reported that curriculum was emphasized quite a lot or a great deal. The next greatest emphasis was on general pedagogy, mathematics pedagogy, and instructional technology (45 to 47 percent of the students). Teachers reported the least emphasis on content knowledge (28 percent) and leadership development (15 percent). Again, although there was variation across the Benchmarking participants, the national pattern held in many jurisdictions.

The most interesting result about professional development may be the limited emphasis on content knowledge in relation to the other topics. Further detail about the types of content emphasized is provided in Exhibit 6.23. Nationally, teachers reported that the five content areas (fractions and number sense; measurement; data representation, analysis, and probability; geometry; and algebra) were emphasized relatively equally (from 45 to 56 percent). In general, the pattern of relatively equal emphasis was also found in the Benchmarking states. There was more variation within some districts and consortia. For example, the Academy School District focused relatively less emphasis on professional development in geometry (17 percent) than in the other four areas (28 to 42 percent). Montgomery County placed relatively less emphasis on measurement (18 percent) and more emphasis on data representation, analysis, and probability (72 percent). The First in the World Consortium placed relatively more emphasis on geometry (77 percent) and relatively less on data representation, analysis, and probability (37 percent).

Teachers in the United States reported a relatively heavy focus on curriculum in their professional development activities. Their reports about familiarity with various curriculum documents are presented in Exhibit 6.24. Nationally, teachers of most students (91 percent) reported that they were fairly or very familiar with the curriculum guides for their school and their school district, and this held across most of the Benchmarking jurisdictions. U.S. mathematics teachers of 82 percent of the eighth-grade students reported being very familiar with the NCTM *Professional Standards for Teaching Mathematics.* For the Benchmarking states, this ranged from 71 percent in Idaho to 98 percent in South Carolina. For districts and consortia, it ranged from 62 percent in the Chicago Public Schools to 97 percent in the Fremont/Lincoln/Westside Public Schools.

Fewer teachers than might be anticipated reported being at least fairly familiar with their state curriculum guides. Nationally, 74 percent of the eighth graders had mathematics teachers who so reported. Among states the figure ranged from 57 percent in Pennsylvania to 98 percent in South Carolina, and among districts and consortia from 54 percent in the Southwest Pennsylvania Math and Science Collaborative to 100 percent in the Academy School District.

Exhibit 6.18 Students Taught by Teachers Who Participated in Professional Development – Classroom Observation

8th Grade Mathematics

	Observation of Other Teachers[1]		Observation by Other Teachers[2]	
	Percent of Students	Number of Class Periods Observed Averaged Across Students[3]	Percent of Students	Number of Class Periods Observed Averaged Across Students[3]
States				
Connecticut	r 29 (6.9)	5 (1.1)	r 51 (8.0)	5 (1.7)
Idaho	r 12 (4.8)	2 (0.2)	r 34 (8.5)	7 (2.6)
Illinois	9 (3.5)	3 (0.4)	23 (5.5)	10 (3.1)
Indiana	10 (3.4)	11 (4.8)	33 (6.2)	7 (1.9)
Maryland	r 29 (5.1)	6 (1.9)	r 45 (6.1)	4 (0.5)
Massachusetts	24 (5.1)	4 (0.8)	34 (5.4)	8 (2.8)
Michigan	14 (4.0)	6 (1.2)	26 (5.4)	10 (3.3)
Missouri	19 (5.2)	4 (1.9)	25 (6.0)	4 (1.5)
North Carolina	31 (6.4)	5 (1.0)	47 (7.7)	4 (0.7)
Oregon	23 (4.0)	5 (1.7)	23 (5.1)	5 (2.6)
Pennsylvania	25 (4.6)	4 (0.5)	42 (5.7)	5 (1.3)
South Carolina	28 (5.6)	3 (0.4)	47 (5.7)	4 (0.6)
Texas	r 39 (5.3)	6 (0.9)	r 51 (6.1)	4 (0.9)
Districts and Consortia				
Academy School Dist. #20, CO	18 (0.3)	2 (0.0)	40 (0.4)	10 (0.1)
Chicago Public Schools, IL	2 (2.2)	~ ~	31 (12.0)	10 (3.0)
Delaware Science Coalition, DE	r 16 (5.5)	5 (1.3)	r 23 (4.7)	9 (3.5)
First in the World Consort., IL	66 (4.5)	11 (0.9)	59 (3.4)	12 (2.4)
Fremont/Lincoln/WestSide PS, NE	27 (8.4)	17 (4.9)	51 (10.5)	20 (3.4)
Guilford County, NC	52 (6.3)	4 (0.5)	41 (5.9)	8 (2.3)
Jersey City Public Schools, NJ	5 (1.5)	3 (0.4)	22 (2.3)	5 (0.4)
Miami-Dade County PS, FL	s 33 (6.3)	3 (0.8)	s 35 (7.5)	2 (0.4)
Michigan Invitational Group, MI	18 (7.3)	3 (0.5)	17 (5.9)	3 (0.7)
Montgomery County, MD	s 51 (5.7)	8 (1.0)	s 85 (5.0)	4 (0.6)
Naperville Sch. Dist. #203, IL	21 (3.5)	5 (1.0)	34 (4.4)	4 (0.7)
Project SMART Consortium, OH	37 (6.0)	9 (2.5)	47 (6.7)	9 (2.1)
Rochester City Sch. Dist., NY	s 14 (1.8)	2 (0.5)	s 47 (6.9)	11 (1.7)
SW Math/Sci. Collaborative, PA	25 (4.8)	4 (1.0)	37 (7.2)	7 (2.1)
United States	25 (3.0)	4 (0.8)	35 (3.3)	5 (1.0)

SOURCE: IEA Third International Mathematics and Science Study (TIMSS), 1998-1999.

Background data provided by teachers.

1 Based on complete class periods teachers observed other teachers in their school teach mathematics from the beginning of the 1998-99 school year until the time of testing.

2 Based on complete class periods teachers were observed while teaching mathematics by other teachers in their school from the beginning of the 1998-99 school year until the time of testing.

3 Teachers who did not participate in the professional development activity were not included in the average.

States in *italics* did not fully satisfy guidelines for sample participation rates (see Appendix A for details).

( ) Standard errors appear in parentheses. Because results are rounded to the nearest whole number, some totals may appear inconsistent.

A tilde (~) indicates insufficient data to report average number of class periods.

An "r" indicates teacher response data available for 70-84% of students. An "s" indicates teacher response data available for 50-69% of students.

Exhibit 6.19 Students Taught by Teachers Who Participated in Professional Development – School- and District-Based Activities*

8th Grade Mathematics

	Immersion or Internship Activities		Receipt of Mentoring or Observation		Teacher Resource Center		Committees or Task Forces		Teacher Study Groups	
	Percent of Students	Teacher Hours Averaged Across Students[1]	Percent of Students	Teacher Hours Averaged Across Students[1]	Percent of Students	Teacher Hours Averaged Across Students[1]	Percent of Students	Teacher Hours Averaged Across Students[1]	Percent of Students	Teacher Hours Averaged Across Students[1]
States										
Connecticut	r 3 (0.3)	2 (0.0)	r 32 (7.5)	11 (3.3)	r 9 (5.0)	3 (0.8)	r 55 (6.6)	9 (1.5)	r 26 (4.8)	8 (1.3)
Idaho	r 3 (2.6)	17 (3.3)	r 24 (5.3)	8 (3.3)	r 8 (5.0)	10 (9.6)	r 51 (6.9)	15 (2.1)	r 26 (6.4)	4 (1.0)
Illinois	4 (1.8)	5 (1.4)	20 (4.4)	11 (3.5)	14 (4.1)	12 (3.9)	55 (6.5)	16 (2.9)	23 (6.0)	9 (1.7)
Indiana	5 (3.3)	45 (20.6)	14 (5.4)	10 (7.1)	3 (1.7)	3 (1.1)	61 (5.9)	9 (1.2)	21 (5.8)	7 (2.4)
Maryland	r 6 (3.3)	18 (17.6)	r 33 (6.4)	4 (0.7)	r 21 (4.7)	7 (2.8)	r 35 (6.9)	14 (3.0)	r 22 (6.1)	12 (3.9)
Massachusetts	7 (2.2)	14 (8.4)	32 (5.9)	5 (0.7)	16 (4.7)	4 (0.8)	61 (5.9)	12 (1.4)	46 (7.8)	10 (1.9)
Michigan	0 (0.0)	~ ~	21 (4.7)	4 (0.7)	11 (3.7)	6 (3.7)	54 (7.0)	12 (2.0)	18 (4.9)	12 (1.7)
Missouri	6 (2.6)	23 (9.8)	27 (5.7)	4 (0.7)	6 (3.3)	4 (0.6)	60 (6.3)	10 (1.9)	20 (4.5)	5 (1.3)
North Carolina	2 (1.8)	~ ~	41 (5.4)	11 (2.7)	14 (3.9)	7 (1.2)	56 (5.3)	7 (0.9)	29 (5.8)	12 (3.6)
Oregon	5 (2.3)	7 (3.5)	35 (5.2)	7 (2.5)	11 (3.7)	10 (3.8)	68 (3.3)	15 (3.0)	29 (5.4)	11 (2.3)
Pennsylvania	14 (3.1)	10 (2.3)	30 (5.6)	8 (2.4)	15 (3.6)	9 (2.7)	58 (6.2)	10 (1.3)	20 (4.3)	6 (0.9)
South Carolina	4 (2.4)	14 (8.7)	23 (5.5)	12 (3.9)	25 (5.1)	9 (2.6)	46 (6.6)	14 (2.6)	21 (5.4)	10 (1.6)
Texas	r 18 (6.6)	12 (4.5)	r 39 (6.7)	13 (4.9)	r 24 (4.3)	5 (0.8)	r 61 (6.8)	13 (2.0)	r 42 (6.7)	16 (4.5)
Districts and Consortia										
Academy School Dist. #20, CO	18 (0.3)	9 (0.1)	49 (0.4)	7 (0.1)	15 (0.3)	3 (0.0)	48 (0.4)	16 (0.1)	40 (0.4)	7 (0.1)
Chicago Public Schools, IL	9 (5.3)	3 (0.9)	25 (8.8)	21 (9.9)	29 (9.8)	12 (2.0)	34 (9.5)	11 (2.3)	22 (7.8)	15 (6.0)
Delaware Science Coalition, DE	r 0 (0.0)	~ ~	r 28 (6.7)	11 (2.8)	r 36 (5.5)	4 (0.7)	r 71 (5.6)	10 (1.2)	r 24 (5.4)	9 (2.6)
First in the World Consort., IL	5 (0.3)	5 (0.0)	r 51 (5.5)	24 (2.9)	23 (6.3)	5 (1.0)	82 (7.9)	10 (1.6)	30 (9.7)	15 (3.4)
Fremont/Lincoln/WestSide PS, NE	0 (0.0)	~ ~	33 (8.5)	22 (7.6)	12 (4.0)	4 (0.5)	49 (7.0)	6 (1.6)	22 (4.1)	5 (0.6)
Guilford County, NC	6 (0.9)	10 (0.0)	47 (5.6)	18 (2.3)	43 (5.4)	9 (1.2)	58 (6.8)	15 (2.6)	31 (5.0)	15 (3.3)
Jersey City Public Schools, NJ	3 (0.2)	15 (0.0)	r 35 (3.5)	10 (0.7)	14 (2.8)	4 (0.3)	r 45 (4.9)	11 (0.8)	30 (4.5)	25 (5.3)
Miami-Dade County PS, FL	s 9 (5.3)	17 (6.3)	s 24 (4.9)	8 (5.2)	s 42 (10.2)	8 (2.7)	s 56 (10.3)	15 (3.5)	s 54 (10.3)	19 (5.0)
Michigan Invitational Group, MI	0 (0.0)	~ ~	25 (8.6)	6 (1.4)	5 (0.2)	2 (0.0)	59 (7.0)	10 (0.9)	32 (4.7)	11 (0.8)
Montgomery County, MD	s 8 (3.5)	11 (2.6)	s 50 (6.3)	3 (0.8)	s 22 (7.2)	5 (0.9)	s 57 (6.7)	19 (2.2)	s 12 (2.4)	25 (17.7)
Naperville Sch. Dist. #203, IL	0 (0.0)	~ ~	26 (5.6)	6 (1.1)	17 (2.8)	4 (0.1)	64 (4.5)	46 (2.7)	25 (2.8)	30 (9.4)
Project SMART Consortium, OH	6 (2.8)	33 (14.7)	25 (6.8)	5 (0.7)	23 (6.5)	4 (0.7)	64 (6.5)	13 (1.4)	19 (4.7)	6 (0.8)
Rochester City Sch. Dist., NY	s 0 (0.0)	~ ~	s 34 (6.8)	8 (1.6)	s 47 (8.2)	6 (0.8)	s 39 (7.8)	10 (2.3)	s 31 (7.1)	8 (0.8)
SW Math/Sci. Collaborative, PA	5 (3.3)	4 (0.7)	18 (5.2)	12 (7.5)	11 (4.8)	6 (1.2)	42 (6.9)	11 (1.7)	16 (4.5)	7 (0.8)
United States	6 (2.1)	14 (3.6)	27 (3.2)	5 (0.6)	12 (2.4)	5 (1.5)	55 (3.2)	12 (1.5)	30 (3.4)	11 (2.5)

SOURCE: IEA Third International Mathematics and Science Study (TIMSS), 1998-1999.

Background data provided by teachers.

* Based on participation in professional development activities from June 1998 until the time of testing.

1 Teachers who did not participate in the professional development activity were not included in the average.

States in *italics* did not fully satisfy guidelines for sample participation rates (see Appendix A for details).

( ) Standard errors appear in parentheses. Because results are rounded to the nearest whole number, some totals may appear inconsistent.

A tilde (~) indicates insufficient data to report average hours.

An "r" indicates teacher response data available for 70-84% of students. An "s" indicates teacher response data available for 50-69% of students.

Students Taught by Teachers Who Participated in Professional Development – Workshops, Conferences, and Networks*

8th Grade Mathematics

	Within-District Workshops/ Institutes		Out-of-District Workshops/ Institutes		Teacher Collaborative or Networks		Out-of-District Conferences		Other Organized Professional Development	
	Percent of Students	Teacher Hours Averaged Across Students[1]	Percent of Students	Teacher Hours Averaged Across Students[1]	Percent of Students	Teacher Hours Averaged Across Students[1]	Percent of Students	Teacher Hours Averaged Across Students[1]	Percent of Students	Teacher Hours Averaged Across Students[1]
States										
Connecticut	r 82 (5.8)	14 (1.7)	r 33 (7.2)	15 (2.0)	r 30 (6.9)	12 (3.3)	r 41 (7.5)	12 (2.2)	r 11 (5.1)	6 (1.7)
Idaho	r 64 (5.7)	12 (1.4)	r 34 (4.9)	25 (5.0)	r 14 (4.2)	7 (1.0)	r 37 (7.5)	15 (2.4)	r 12 (3.7)	6 (1.8)
Illinois	81 (5.0)	10 (1.3)	53 (6.1)	9 (1.6)	12 (3.1)	7 (1.4)	38 (6.5)	11 (2.1)	22 (6.3)	10 (3.1)
Indiana	76 (7.5)	11 (1.3)	33 (6.8)	9 (1.4)	18 (4.2)	6 (0.9)	30 (7.1)	8 (0.9)	15 (3.9)	18 (9.1)
Maryland	r 79 (4.8)	18 (1.7)	r 30 (5.6)	13 (2.8)	r 30 (5.6)	12 (3.0)	r 23 (5.5)	12 (3.2)	r 23 (4.8)	9 (1.2)
Massachusetts	82 (4.7)	14 (2.0)	45 (5.4)	11 (2.1)	23 (5.8)	7 (1.3)	35 (6.1)	8 (1.5)	r 39 (6.1)	11 (3.3)
Michigan	70 (6.3)	15 (1.6)	32 (6.1)	13 (2.3)	13 (3.4)	6 (1.3)	30 (5.3)	10 (1.9)	13 (4.4)	7 (1.7)
Missouri	76 (6.1)	12 (2.0)	41 (6.6)	13 (3.5)	19 (4.7)	5 (1.1)	49 (6.7)	16 (2.7)	17 (3.6)	9 (2.4)
North Carolina	87 (3.5)	14 (1.5)	27 (4.2)	17 (6.3)	27 (5.7)	12 (3.8)	37 (5.2)	10 (1.6)	r 19 (4.6)	15 (5.7)
Oregon	83 (4.2)	13 (1.5)	42 (5.9)	10 (1.2)	23 (5.6)	7 (0.9)	39 (5.5)	16 (1.7)	19 (4.3)	15 (3.6)
Pennsylvania	75 (4.8)	13 (1.9)	47 (6.2)	8 (1.2)	20 (4.5)	10 (1.6)	29 (5.5)	11 (2.8)	19 (4.7)	11 (3.3)
South Carolina	75 (4.0)	19 (2.4)	27 (6.4)	15 (2.9)	16 (4.6)	5 (1.2)	35 (4.7)	19 (4.7)	26 (5.0)	13 (3.5)
Texas	r 94 (3.0)	26 (4.1)	r 62 (5.8)	20 (2.9)	r 27 (7.3)	14 (5.3)	r 39 (6.6)	21 (3.9)	r 32 (5.3)	22 (4.7)
Districts and Consortia										
Academy School Dist. #20, CO	67 (0.4)	10 (0.1)	37 (0.4)	13 (0.1)	r 0 (0.0)	~ ~	24 (0.3)	7 (0.0)	6 (0.2)	8 (0.0)
Chicago Public Schools, IL	67 (11.4)	11 (2.5)	22 (7.9)	8 (2.8)	30 (11.8)	8 (2.2)	23 (8.7)	11 (2.4)	16 (8.2)	7 (2.0)
Delaware Science Coalition, DE	r 79 (4.6)	15 (1.4)	r 39 (6.5)	11 (3.1)	r 29 (6.0)	8 (1.5)	r 33 (5.7)	11 (4.0)	r 16 (4.9)	11 (4.3)
First in the World Consort., IL	68 (4.7)	12 (2.1)	64 (6.0)	12 (1.7)	69 (6.0)	13 (3.9)	54 (8.7)	14 (1.9)	r 24 (6.2)	10 (1.5)
Fremont/Lincoln/WestSide PS, NE	97 (0.2)	13 (2.1)	29 (5.3)	15 (3.1)	15 (1.8)	2 (0.0)	35 (8.6)	15 (2.2)	r 34 (6.1)	12 (1.3)
Guilford County, NC	78 (4.8)	23 (3.1)	16 (3.4)	23 (10.3)	26 (6.3)	6 (1.4)	29 (5.1)	10 (1.5)	r 15 (4.7)	9 (1.3)
Jersey City Public Schools, NJ	85 (2.7)	11 (0.3)	41 (4.4)	16 (0.8)	16 (2.2)	22 (2.7)	26 (4.1)	11 (1.0)	45 (3.3)	7 (0.1)
Miami-Dade County PS, FL	s 88 (6.2)	24 (3.2)	s 16 (8.5)	5 (0.8)	s 35 (12.3)	8 (1.8)	s 11 (7.6)	3 (0.8)	s 33 (8.5)	12 (4.8)
Michigan Invitational Group, MI	74 (4.9)	12 (2.4)	39 (8.0)	18 (4.5)	33 (5.0)	8 (2.3)	27 (8.7)	10 (2.3)	10 (6.0)	6 (0.6)
Montgomery County, MD	s 86 (5.1)	27 (1.7)	s 34 (6.7)	13 (3.9)	s 29 (5.8)	20 (9.9)	s 28 (6.9)	8 (0.7)	s 25 (6.2)	7 (2.2)
Naperville Sch. Dist. #203, IL	72 (5.7)	24 (1.1)	45 (3.9)	6 (0.2)	18 (3.6)	11 (1.0)	38 (4.5)	7 (0.2)	20 (2.6)	7 (0.1)
Project SMART Consortium, OH	83 (6.0)	15 (1.3)	53 (5.8)	7 (0.8)	29 (5.5)	8 (1.7)	30 (6.3)	11 (2.6)	16 (6.5)	8 (3.4)
Rochester City Sch. Dist., NY	s 97 (3.5)	11 (1.9)	s 44 (8.2)	19 (3.3)	s 43 (5.8)	12 (1.8)	s 2 (0.2)	~ ~	s 27 (6.5)	10 (1.2)
SW Math/Sci. Collaborative, PA	74 (7.4)	16 (2.0)	42 (7.6)	10 (1.4)	24 (6.4)	12 (2.8)	20 (4.8)	10 (3.6)	6 (3.5)	5 (0.9)
United States	79 (3.1)	15 (1.3)	37 (3.2)	16 (1.9)	21 (2.7)	10 (1.6)	34 (2.7)	13 (1.6)	r 18 (2.5)	11 (1.7)

SOURCE: IEA Third International Mathematics and Science Study (TIMSS), 1998-1999.

Background data provided by teachers.

* Based on participation in professional development activities from June 1998 until the time of testing.

1 Teachers who did not participate in the professional development activity were not included in the average.

States in *italics* did not fully satisfy guidelines for sample participation rates (see Appendix A for details).

( ) Standard errors appear in parentheses. Because results are rounded to the nearest whole number, some totals may appear inconsistent.

A tilde (~) indicates insufficient data to report average hours.

An "r" indicates teacher response data available for 70-84% of students. An "s" indicates teacher response data available for 50-69% of students.

Exhibit 6.21 Students Taught by Teachers Who Participated in Professional Development – Individual Activities*

8th Grade Mathematics

	Courses for College Credit[1]		Individual Research Projects		Individual Learning		Other Individual Professional Development	
	Percent of Students	Teacher Hours Averaged Across Students[2]	Percent of Students	Teacher Hours Averaged Across Students[2]	Percent of Students	Teacher Hours Averaged Across Students[2]	Percent of Students	Teacher Hours Averaged Across Students[2]
States								
Connecticut	r 15 (4.4)	27 (7.2)	r 35 (6.3)	23 (4.5)	r 84 (5.6)	25 (2.7)	s 31 (6.4)	18 (5.3)
Idaho	r 54 (8.2)	27 (2.9)	r 22 (4.1)	23 (5.4)	r 68 (5.6)	27 (3.9)	r 29 (7.1)	31 (8.8)
Illinois	36 (7.0)	24 (5.7)	33 (7.3)	23 (6.4)	88 (4.0)	23 (3.9)	19 (5.3)	21 (9.2)
Indiana	21 (4.5)	40 (9.1)	21 (4.7)	13 (2.8)	84 (5.5)	19 (1.7)	20 (4.8)	19 (5.5)
Maryland	r 31 (4.5)	40 (6.9)	r 25 (5.2)	26 (5.9)	r 79 (6.1)	23 (2.2)	r 26 (6.0)	24 (4.6)
Massachusetts	27 (5.5)	43 (4.2)	36 (6.3)	19 (3.6)	84 (4.0)	26 (3.4)	r 37 (7.4)	21 (5.0)
Michigan	17 (4.7)	22 (5.9)	37 (6.1)	15 (4.2)	85 (4.2)	18 (2.9)	r 39 (6.3)	16 (4.7)
Missouri	23 (4.3)	19 (6.5)	20 (4.6)	43 (11.6)	83 (4.7)	20 (2.4)	15 (4.7)	17 (4.4)
North Carolina	17 (4.8)	30 (7.6)	39 (5.6)	18 (3.7)	80 (3.5)	16 (2.1)	r 20 (4.8)	19 (4.5)
Oregon	28 (4.2)	28 (5.6)	36 (4.6)	18 (4.2)	86 (3.6)	24 (2.6)	34 (5.5)	28 (8.1)
Pennsylvania	31 (5.5)	34 (6.2)	36 (6.6)	12 (2.5)	93 (3.0)	23 (3.2)	r 23 (4.6)	13 (1.9)
South Carolina	47 (6.3)	33 (5.8)	36 (6.3)	17 (5.4)	86 (3.8)	25 (3.6)	24 (4.4)	17 (5.7)
Texas	r 16 (4.1)	36 (9.7)	r 34 (6.4)	22 (2.8)	r 81 (2.9)	28 (3.5)	r 41 (7.0)	19 (3.6)
Districts and Consortia								
Academy School Dist. #20, CO	40 (0.4)	18 (0.7)	r 44 (0.4)	17 (0.1)	92 (0.2)	25 (0.3)	r 11 (0.3)	2 (0.0)
Chicago Public Schools, IL	28 (10.7)	16 (7.1)	25 (8.8)	27 (7.7)	75 (8.9)	22 (5.3)	r 17 (9.2)	10 (2.9)
Delaware Science Coalition, DE	r 28 (6.1)	46 (9.4)	r 41 (6.0)	19 (3.5)	r 81 (4.0)	31 (5.1)	r 36 (6.2)	23 (3.6)
First in the World Consort., IL	11 (3.5)	12 (3.8)	42 (6.0)	28 (9.5)	100 (0.0)	26 (5.0)	s 18 (4.6)	8 (1.4)
Fremont/Lincoln/WestSide PS, NE	31 (7.2)	52 (6.3)	40 (9.1)	14 (4.4)	91 (1.2)	25 (3.0)	r 35 (3.4)	21 (2.6)
Guilford County, NC	14 (4.6)	29 (5.3)	30 (4.6)	22 (7.6)	74 (3.5)	23 (2.3)	23 (3.0)	12 (1.0)
Jersey City Public Schools, NJ	r 13 (3.7)	33 (5.3)	39 (3.7)	20 (2.0)	85 (2.4)	35 (1.8)	r 31 (5.6)	13 (2.1)
Miami-Dade County PS, FL	s 39 (8.4)	18 (7.6)	s 56 (8.5)	15 (5.1)	s 78 (5.3)	20 (4.3)	x x	x x
Michigan Invitational Group, MI	23 (1.3)	20 (1.2)	19 (5.1)	5 (0.6)	76 (2.9)	22 (2.6)	7 (2.6)	33 (8.8)
Montgomery County, MD	s 39 (5.8)	39 (6.7)	s 46 (6.7)	29 (3.6)	s 90 (3.4)	25 (2.5)	s 34 (6.5)	19 (6.1)
Naperville Sch. Dist. #203, IL	22 (2.5)	56 (10.3)	39 (2.6)	24 (1.5)	85 (4.2)	23 (1.3)	21 (2.6)	9 (0.1)
Project SMART Consortium, OH	38 (5.5)	24 (7.0)	34 (6.6)	25 (4.8)	81 (5.4)	26 (1.4)	25 (5.5)	14 (3.6)
Rochester City Sch. Dist., NY	s 19 (3.6)	90 (0.0)	s 45 (8.2)	10 (1.5)	s 92 (0.9)	23 (4.8)	s 44 (6.7)	10 (1.5)
SW Math/Sci. Collaborative, PA	10 (4.6)	50 (7.2)	27 (6.4)	23 (8.4)	83 (5.4)	24 (2.7)	23 (6.6)	21 (5.4)
United States	27 (2.9)	35 (4.8)	r 33 (3.7)	21 (2.2)	84 (2.3)	26 (2.3)	r 25 (3.7)	18 (2.1)

SOURCE: IEA Third International Mathematics and Science Study (TIMSS), 1998-1999.

Background data provided by teachers.

* Based on participation in professional development activities from June 1998 until the time of testing.

1 The response range had a maximum of 90 hours spent in courses for college credit.

2 Teachers who did not participate in the professional development activity were not included in the average.

States in *italics* did not fully satisfy guidelines for sample participation rates (see Appendix A for details).

( ) Standard errors appear in parentheses. Because results are rounded to the nearest whole number, some totals may appear inconsistent.

An "r" indicates teacher response data available for 70-84% of students. An "s" indicates teacher response data available for 50-69% of students. An "x" indicates teacher response data available for <50% of students.

8th Grade Mathematics

	Percentage of Students Whose Teachers Reported That the Topic is Emphasized Quite a Lot or A Great Deal in Their Professional Development[1]						
	Content Knowledge	Curriculum	General Instruction/ Pedagogy	Subject-Specific Instruction/ Pedagogy	Assessment	Instructional Technology	Leadership Development
States							
Connecticut	r 22 (6.4)	r 57 (7.2)	r 43 (6.1)	r 37 (7.7)	r 35 (6.4)	r 48 (7.0)	r 11 (4.2)
Idaho	r 28 (6.1)	r 37 (5.3)	r 41 (6.3)	r 32 (6.0)	r 26 (4.9)	r 42 (6.7)	r 11 (2.8)
Illinois	20 (5.3)	62 (5.8)	50 (5.8)	33 (5.6)	45 (7.2)	60 (6.8)	14 (5.5)
Indiana	9 (4.1)	56 (6.9)	35 (5.8)	29 (5.6)	23 (5.7)	27 (6.1)	13 (5.2)
Maryland	r 28 (4.2)	r 55 (6.1)	r 55 (4.9)	r 45 (5.9)	r 42 (5.4)	r 63 (4.9)	r 12 (3.3)
Massachusetts	32 (5.0)	66 (5.8)	52 (5.2)	50 (7.3)	35 (5.4)	43 (5.4)	20 (5.0)
Michigan	24 (5.5)	57 (5.5)	60 (5.0)	41 (6.2)	33 (5.8)	35 (6.5)	15 (4.3)
Missouri	14 (3.1)	58 (6.9)	50 (5.5)	44 (6.4)	48 (5.8)	34 (6.4)	8 (2.8)
North Carolina	19 (3.8)	64 (7.3)	57 (4.6)	45 (4.8)	34 (5.0)	62 (5.4)	19 (5.2)
Oregon	23 (5.2)	64 (4.7)	42 (6.1)	30 (6.0)	57 (5.3)	16 (5.7)	17 (4.4)
Pennsylvania	26 (5.8)	63 (6.3)	44 (6.0)	39 (5.4)	34 (4.7)	42 (5.2)	24 (4.1)
South Carolina	24 (5.0)	78 (4.9)	43 (6.7)	55 (6.9)	31 (5.7)	44 (7.0)	21 (5.8)
Texas	r 26 (6.1)	r 77 (6.1)	r 66 (5.8)	r 57 (6.9)	r 41 (6.9)	r 64 (6.0)	r 25 (5.9)
Districts and Consortia							
Academy School Dist. #20, CO	26 (0.3)	52 (0.4)	30 (0.3)	46 (0.4)	30 (0.3)	54 (0.4)	9 (0.2)
Chicago Public Schools, IL	r 30 (8.7)	r 63 (8.9)	r 73 (12.0)	r 44 (10.4)	r 49 (9.3)	r 44 (9.1)	r 24 (9.4)
Delaware Science Coalition, DE	r 23 (5.3)	r 79 (6.2)	r 32 (7.4)	r 54 (8.3)	r 28 (6.1)	r 46 (7.0)	r 14 (5.5)
First in the World Consort., IL	42 (8.8)	87 (5.3)	70 (4.6)	51 (4.9)	34 (7.5)	53 (7.7)	7 (1.0)
Fremont/Lincoln/WestSide PS, NE	39 (5.4)	72 (7.3)	38 (3.3)	45 (8.5)	45 (7.4)	28 (4.2)	22 (3.5)
Guilford County, NC	31 (6.5)	76 (5.3)	79 (4.7)	49 (5.6)	46 (6.6)	46 (5.8)	24 (4.3)
Jersey City Public Schools, NJ	49 (3.7)	57 (5.0)	70 (5.7)	59 (5.1)	53 (2.4)	51 (4.3)	15 (1.7)
Miami-Dade County PS, FL	s 56 (9.2)	s 65 (9.9)	s 58 (9.7)	s 64 (7.3)	s 49 (8.2)	s 68 (7.5)	s 32 (7.6)
Michigan Invitational Group, MI	19 (7.9)	57 (3.4)	30 (5.8)	41 (7.4)	26 (9.0)	28 (6.5)	19 (7.8)
Montgomery County, MD	s 24 (4.9)	s 77 (4.7)	s 52 (6.1)	s 47 (6.7)	s 56 (8.5)	s 85 (5.2)	s 23 (4.9)
Naperville Sch. Dist. #203, IL	0 (0.0)	49 (4.4)	34 (3.6)	23 (2.6)	35 (4.3)	57 (3.8)	19 (2.5)
Project SMART Consortium, OH	19 (5.0)	52 (4.7)	53 (5.9)	49 (7.3)	43 (5.9)	46 (6.5)	20 (4.9)
Rochester City Sch. Dist., NY	s 35 (8.0)	s 69 (5.4)	s 53 (7.9)	s 62 (6.6)	s 62 (8.2)	s 18 (6.9)	s 29 (6.0)
SW Math/Sci. Collaborative, PA	21 (5.2)	59 (6.8)	39 (7.4)	31 (6.1)	30 (7.0)	39 (7.2)	11 (4.4)
United States	r 28 (3.3)	63 (3.3)	45 (3.1)	47 (3.9)	r 33 (3.1)	45 (3.7)	r 15 (2.5)

SOURCE: IEA Third International Mathematics and Science Study (TIMSS), 1998-1999.

Background data provided by teachers.

1 Based on participation in professional development activities from June 1998 until the time of testing. Does not include students whose teachers reported that they do not teach the topic.

States in *italics* did not fully satisfy guidelines for sample participation rates (see Appendix A for details).

( ) Standard errors appear in parentheses. Because results are rounded to the nearest whole number, some totals may appear inconsistent.

An "r" indicates teacher response data available for 70-84% of students. An "s" indicates teacher response data available for 50-69% of students.

	Percentage of Students Whose Teachers Reported That the Content Area is Focused On in Their Professional Development[1]									
		Fractions and Number Sense		Measurement		Data Representation, Analysis, and Probability		Geometry		Algebra
States										
Connecticut	r	32 (7.6)	r	29 (7.2)	r	42 (6.5)	r	32 (7.4)	r	44 (7.1)
Idaho	r	40 (6.9)	r	34 (6.5)	r	33 (5.4)	r	24 (5.8)	r	37 (5.5)
Illinois		46 (5.7)		39 (6.5)		49 (6.8)		39 (5.6)		46 (5.4)
Indiana		40 (6.2)		32 (6.2)		37 (6.8)		26 (6.0)		41 (6.0)
Maryland	r	46 (6.5)	r	41 (7.3)	r	65 (5.7)	r	40 (6.0)	r	58 (6.8)
Massachusetts		52 (5.6)		52 (6.4)		52 (5.0)		43 (5.7)		53 (5.5)
Michigan		39 (5.6)		29 (5.3)		44 (7.0)		38 (6.8)		48 (6.6)
Missouri		47 (6.4)		51 (6.3)		54 (6.1)		47 (5.2)		52 (4.7)
North Carolina		53 (6.6)		53 (6.7)		53 (5.9)		53 (7.1)		56 (5.9)
Oregon		42 (7.0)		41 (5.8)		46 (5.1)		38 (5.6)		45 (5.5)
Pennsylvania	r	37 (5.6)	r	35 (5.6)	r	41 (6.6)	r	24 (4.4)	r	37 (6.2)
South Carolina		52 (6.8)		45 (5.5)		56 (7.2)		42 (6.0)		58 (6.4)
Texas	r	59 (7.0)	r	47 (7.0)	r	56 (6.8)	r	45 (7.1)	r	64 (7.0)
Districts and Consortia										
Academy School Dist. #20, CO		42 (0.4)		28 (0.4)		30 (0.4)		17 (0.4)		37 (0.4)
Chicago Public Schools, IL	r	41 (11.0)	r	37 (9.5)	r	41 (12.2)	r	34 (9.0)	r	40 (11.0)
Delaware Science Coalition, DE	r	61 (6.5)	r	63 (7.1)	r	59 (6.1)	r	52 (6.3)	r	64 (6.7)
First in the World Consort., IL		46 (6.4)		52 (9.0)		37 (6.5)		77 (6.7)		66 (9.7)
Fremont/Lincoln/WestSide PS, NE		52 (8.6)		33 (5.4)		55 (8.0)		39 (1.4)		52 (8.6)
Guilford County, NC		45 (6.3)		36 (6.4)		34 (6.3)		40 (6.2)		51 (5.7)
Jersey City Public Schools, NJ		53 (5.2)		58 (5.2)		46 (3.9)		50 (4.3)		54 (5.7)
Miami-Dade County PS, FL	s	57 (8.5)	s	66 (7.6)	s	68 (7.7)	s	60 (8.4)	s	59 (7.1)
Michigan Invitational Group, MI		39 (4.8)		34 (4.7)		45 (4.6)		35 (8.0)		48 (4.1)
Montgomery County, MD	s	34 (6.3)	s	18 (3.6)	s	72 (9.1)	s	48 (7.3)	s	64 (9.1)
Naperville Sch. Dist. #203, IL		26 (2.8)		17 (2.8)		47 (5.2)		22 (0.7)		40 (4.5)
Project SMART Consortium, OH		36 (5.9)		41 (4.6)		47 (5.7)		34 (4.4)		46 (6.6)
Rochester City Sch. Dist., NY	s	76 (5.5)	s	86 (6.9)	s	84 (6.3)	s	76 (6.2)	s	81 (6.8)
SW Math/Sci. Collaborative, PA		30 (4.8)		34 (5.7)		38 (7.0)		36 (5.6)		36 (7.9)
United States		54 (3.3)		45 (3.3)	r	50 (3.0)	r	45 (2.4)	r	56 (3.1)

SOURCE: IEA Third International Mathematics and Science Study (TIMSS), 1998-1999.

Background data provided by teachers.

1 Content areas are focused on in professional development if 80% or more of the TIMSS topics in the content area are reported by teachers to have been focused on in their professional development from June 1998 until the time of testing.

States in *italics* did not fully satisfy guidelines for sample participation rates (see Appendix A for details).

( ) Standard errors appear in parentheses. Because results are rounded to the nearest whole number, some totals may appear inconsistent.

An "r" indicates teacher response data available for 70-84% of students. An "s" indicates teacher response data available for 50-69% of students.

	Percentage of Students Whose Teachers Reported Being Fairly Familiar or Very Familiar with the Curriculum Document											
		National Council of Teachers of Mathematics (NCTM) *Professional Standards for Teaching Mathematics*		State Education Department Curriculum Guide		School District Curriculum Guide		School Curriculum Guide		National Assessment of Educational Progress (NAEP) Assessment Frameworks/ Specifications		State Education Department Assessment Specifications
States												
Connecticut	r	96 (2.3)	r	73 (5.5)	r	95 (2.6)	r	98 (1.2)	r	38 (6.7)	r	64 (7.1)
Idaho	r	71 (4.0)	r	60 (5.7)	r	84 (5.4)	r	87 (4.4)	r	8 (3.9)	r	39 (7.6)
Illinois		84 (3.8)		58 (7.5)		95 (2.7)		82 (3.2)		14 (3.0)	r	56 (8.7)
Indiana		92 (3.9)		92 (3.3)		98 (1.7)		97 (2.2)		12 (3.8)		59 (6.4)
Maryland	r	94 (3.0)	r	63 (7.0)	r	96 (3.0)	s	89 (2.5)	r	35 (4.6)	s	62 (5.6)
Massachusetts		85 (4.4)		86 (4.2)		94 (2.2)		94 (2.9)		40 (5.5)		74 (5.9)
Michigan		90 (3.6)		72 (5.3)		94 (2.9)		90 (4.1)		12 (3.9)		57 (6.8)
Missouri		90 (3.1)		73 (5.1)		97 (2.5)		96 (3.2)		46 (6.0)		76 (5.9)
North Carolina		87 (3.5)		98 (1.3)		97 (1.8)		91 (2.6)		28 (4.2)		46 (5.7)
Oregon		78 (3.8)		93 (2.2)		92 (3.9)		92 (3.1)		16 (4.3)		82 (5.0)
Pennsylvania		88 (5.5)		57 (4.0)		87 (5.9)		78 (3.8)		29 (4.2)	r	56 (4.3)
South Carolina		98 (1.3)		98 (2.3)		100 (0.0)		97 (0.4)		62 (5.7)		76 (4.6)
Texas	r	79 (5.8)	r	62 (7.2)	r	97 (2.0)	r	94 (3.3)	r	29 (6.9)	r	69 (6.4)
Districts and Consortia												
Academy School Dist. #20, CO		88 (0.4)		100 (0.0)		100 (0.0)		100 (0.0)		17 (0.3)		64 (0.4)
Chicago Public Schools, IL		62 (10.1)		70 (9.3)		90 (5.9)	r	100 (0.0)		22 (8.2)		33 (5.9)
Delaware Science Coalition, DE	r	92 (4.6)	r	88 (4.3)	r	91 (3.0)	r	91 (3.7)	r	40 (6.9)	r	65 (6.5)
First in the World Consort., IL		95 (5.1)		80 (6.7)		96 (2.7)		98 (1.8)		36 (10.6)		59 (10.3)
Fremont/Lincoln/WestSide PS, NE		97 (0.1)		76 (4.5)		97 (3.0)		100 (0.0)		30 (5.9)		41 (7.4)
Guilford County, NC		84 (3.3)		99 (1.4)		96 (3.1)		97 (3.3)		32 (3.6)		66 (4.9)
Jersey City Public Schools, NJ		97 (0.4)		97 (3.0)		100 (0.0)		100 (0.0)		63 (4.4)		82 (5.1)
Miami-Dade County PS, FL	s	86 (4.9)	s	90 (5.0)	s	85 (7.6)	s	95 (4.0)	s	39 (10.5)	s	59 (10.5)
Michigan Invitational Group, MI		91 (2.5)		61 (5.5)		95 (0.2)		92 (0.5)		25 (3.0)		62 (7.6)
Montgomery County, MD	s	91 (3.5)	s	76 (4.6)	s	98 (2.1)		x x	s	39 (7.2)	s	67 (6.8)
Naperville Sch. Dist. #203, IL		90 (3.7)		62 (3.7)		92 (0.9)		95 (1.1)		32 (4.1)		62 (4.0)
Project SMART Consortium, OH		94 (2.0)		68 (5.4)		95 (0.3)		97 (2.8)		10 (4.3)		40 (4.6)
Rochester City Sch. Dist., NY		82 (1.6)		68 (4.5)		100 (0.0)		89 (4.9)		19 (4.7)		61 (5.0)
SW Math/Sci. Collaborative, PA		90 (4.9)		54 (7.7)		85 (5.5)		86 (5.6)		16 (4.6)		66 (7.7)
United States		82 (2.6)		74 (3.8)		91 (2.2)		91 (2.1)		27 (3.0)		51 (3.8)

SOURCE: IEA Third International Mathematics and Science Study (TIMSS), 1998-1999.

Background data provided by teachers.

States in *italics* did not fully satisfy guidelines for sample participation rates (see Appendix A for details).

( ) Standard errors appear in parentheses. Because results are rounded to the nearest whole number, some totals may appear inconsistent.

An "r" indicates teacher response data available for 70-84% of students. An "s" indicates teacher response data available for 50-69% of students. An "x" indicates teacher response data available for <50% of students.

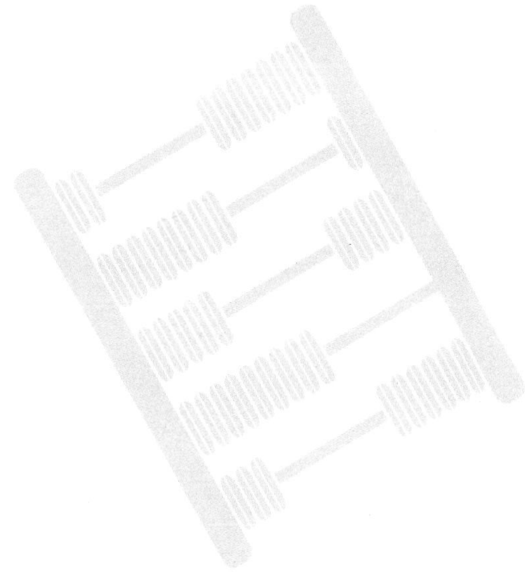

CHAPTER 7

School Contexts for Learning and Instruction

Chapter 7 presents findings about the school contexts for learning and instruction in mathematics, including school characteristics, policies, and practices. Information is presented about the percentage of students eligible for free or reduced-price lunch for each Benchmarking participant, and about the extent of school resources, including computers and Internet access, for the Benchmarking participants and for selected reference countries. Data are also provided on the role of the school principal and on issues related to school climate and environment, including attendance problems and school safety.

7

What Is the Economic Composition of the Student Body?

There is considerable evidence that student achievement is greater in schools with higher proportions of students from advantaged socio-economic backgrounds.[1] To provide information on the composition of the student body, schools' reports on the percentage of their students that are eligible to receive free or reduced-price lunch are summarized in Exhibit 7.1 for each of the Benchmarking participants.[2] The Benchmarking participants span almost the complete range on this factor, from the Naperville School District and the Academy School District, with just a few percent of low-income students, to the Jersey City Public Schools, where almost all students (89 percent) were eligible to receive free or reduced-price lunch. Although mathematics achievement was not perfectly correlated with the percentage of students eligible for free or reduced-price lunch, it is noticeable that several high-performing jurisdictions had low percentages of eligible students, and that three of the four lowest-performing[3] – the Chicago Public Schools, the Rochester City School District, and the Jersey City Public Schools – had the highest percentages of such students.

1 Data on this issue from TIMSS 1995 are presented in Martin, M.O., Mullis, I.V.S., Gregory, K.D., Hoyle, C.D., and Shen, C. (2000), *Effective Schools in Science and Mathematics: IEA's Third International Mathematics and Science Study*, Chestnut Hill, MA: Boston College.

2 These data were collected only in the United States and in the Benchmarking jurisdictions.

3 The response rate from schools in the Miami-Dade County Public Schools was insufficient for reliable reporting.

Exhibit 7.1 Students Eligible to Receive Free/Reduced Price Lunch

8th Grade Mathematics

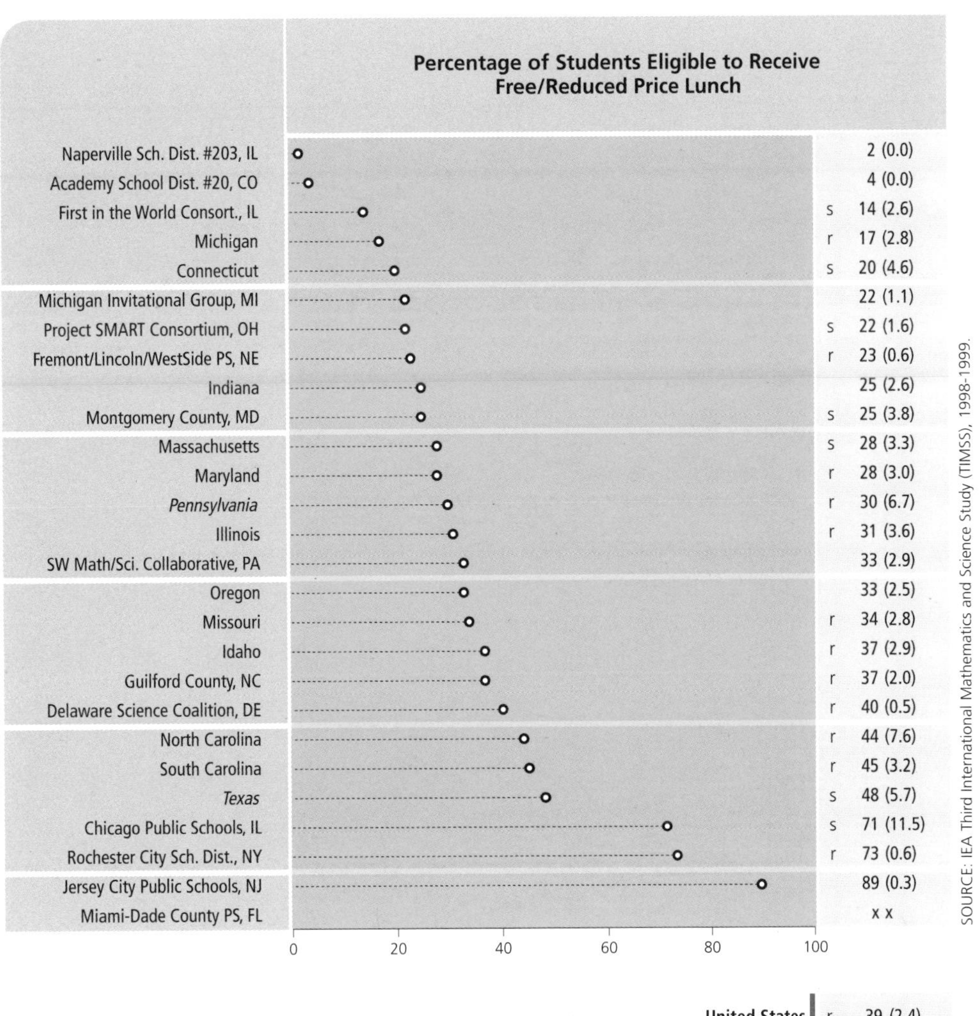

United States r 39 (2.4)

Background data provided by schools.

States in *italics* did not fully satisfy guidelines for sample participation rates (see Appendix A for details).

( ) Standard errors appear in parentheses. Because results are rounded to the nearest whole number, some totals may appear inconsistent.

An "r" indicates school response data available for 70-84% of students. An "s" indicates school response data available for 50-69% of students. An "x" indicates school response data available for <50% of students.

What School Resources Are Available to Support Mathematics Learning?

Some school resources are specific to mathematics, but many are general resources that improve learning opportunities across the curriculum. All the available resources can work together to support mathematics learning and instruction. TIMSS collected data on a range of school resources, including those of a general nature such as buildings and infrastructure, as well as equipment and materials specifically related to mathematics learning.

To measure the extent of school resources in each participating entity, TIMSS created an index of availability of school resources for mathematics instruction (ASRMI). As described in Exhibit 7.2, the index is based on schools' average response to five questions about shortages that affect their general capacity to provide instruction and five questions about shortages that affect mathematics instruction in particular. Students were placed in the high category if principals reported that shortages, both general and for mathematics in particular, had no or little effect on instructional capacity. The medium level indicates that one type of shortage affects instruction some or a lot, and the low level that both shortages affect it some or a lot.

Schools in the United States appear to be fairly well-resourced in comparison with the TIMSS 1999 countries. Across the United States as a whole, 37 percent of students were in schools reporting that resource shortages had little effect on instruction, compared with 19 percent on average internationally. Of the reference countries, only Belgium (Flemish), Singapore, the Czech Republic, and the Netherlands reported higher percentages in this category. Across the Benchmarking participants, reports varied widely. In the Academy School District, the First in the World Consortium, and Naperville, more than 75 percent of students were in well-resourced schools, whereas in North Carolina and Oregon 17 percent or less were in such schools.

In many of the Benchmarking jurisdictions and TIMSS 1999 countries, students in schools in the high category had higher average mathematics achievement than those in the low category. For example, in the United States 37 percent of the students were in the high category with an average mathematics achievement of 516, compared with four percent in the low category with an average of 480. However, the relationship between a country's average mathematics achievement and availability of instructional resources is complex. For example, in some countries that performed significantly above the international average, including Korea, Chinese Taipei, and the Russian Federation, few students (six percent or less) were in schools with high availability of resources for mathematics instruction. In contrast, in other high-performing countries such as Belgium (Flemish) and the Netherlands, no students were in schools with low availability of resources.

Exhibit R4.1 in the reference section shows the results for each of the types of facilities and materials summarized in the general capacity part of the index. There was substantial variation across countries, but internationally on average, nearly half the students were in schools where mathematics instruction was negatively affected by shortages or inadequacies in instructional materials, the budget for supplies, school buildings, and instructional space. Generally, the Benchmarking participants reported fewer students in schools where mathematics instruction was negatively affected by resource shortages, but again the situation varied widely across jurisdictions. Shortage of instructional space was a problem in Oregon, the Fremont/Lincoln/Westside Public Schools, Jersey City, Miami-Dade, and Montgomery County, where more than half of the eighth-grade students were affected. Inadequate school buildings or grounds were also a problem in Miami-Dade, and Oregon had more than half its students in schools that reported shortages of instructional materials and budget for supplies.

Exhibit R4.2, also in the reference section, shows the results for each of the types of equipment and materials summarized in the mathematics instructional capacity part of the index. More than half the students, on average across all the TIMSS 1999 countries, were in schools where shortages or inadequacies in computers and computer software affected the capacity to provide mathematics instruction. Although the Benchmarking entities generally reported fewer students affected by such shortages, Idaho, Missouri, North Carolina, and the Delaware Science Coalition had a majority of their students affected by shortages of both computers and computer software, and many other jurisdictions came close. No participants reported a majority of students affected by shortages in calculators or library materials, and only Chicago had a majority affected by shortages in audio-visual resources.

Exhibits R4.3 and R4.4 in the reference section present more data on access to computers and the Internet for instructional purposes. Benchmarking participants appear to be relatively well equipped with computers, compared with countries internationally, as almost all students were in schools with fewer than 15 students per computer. Internet access was also widespread across Benchmarking entities. In all states except Indiana, Missouri, and Pennsylvania, more than 90 percent of students were in schools with Internet access. School districts with relatively low levels of Internet access were those in Rochester (69 percent) and Chicago (just 44 percent).

Index of Availability of School Resources for Mathematics Instruction

Index based on schools' average response to five questions about shortages that affect general capacity to provide instruction (instructional materials; budget for supplies; school buildings and grounds; heating/cooling and lighting systems; instructional space), and the average response to five questions about shortages that affect mathematics instruction (computers; computer software; calculators; library materials; audio-visual resources) (see reference exhibits R4.1-R4.2). High level indicates that both shortages, on average, affect instructional capacity none or a little. Medium level indicates that one shortage affects instructional capacity none or a little and the other shortage affects instructional capacity some or a lot. Low level indicates that both shortages affect instructional capacity some or a lot.

		High ASRMI		Medium ASRMI		Low ASRMI	
		Percent of Students	Average Achievement	Percent of Students	Average Achievement	Percent of Students	Average Achievement
Academy School Dist. #20, CO		83 (0.4)	529 (1.8)	17 (0.4)	524 (4.9)	0 (0.0)	~ ~
First in the World Consort., IL	r	79 (1.0)	564 (7.8)	21 (1.0)	531 (15.9)	0 (0.0)	~ ~
Naperville Sch. Dist. #203, IL		76 (1.5)	569 (3.5)	24 (1.5)	569 (5.0)	0 (0.0)	~ ~
Belgium (Flemish)		54 (4.6)	556 (7.2)	46 (4.6)	558 (10.0)	0 (0.0)	~ ~
Singapore		50 (4.0)	603 (8.4)	46 (4.1)	608 (8.8)	4 (1.4)	589 (16.2)
Czech Republic		50 (3.6)	525 (6.7)	49 (3.9)	516 (5.8)	2 (1.5)	~ ~
Connecticut	s	47 (9.4)	528 (17.6)	50 (9.5)	523 (8.2)	3 (0.3)	479 (10.1)
Texas	r	44 (5.0)	523 (17.8)	52 (5.9)	517 (12.6)	4 (3.9)	500 (4.7)
Montgomery County, MD	s	43 (13.6)	540 (7.7)	57 (13.6)	535 (6.9)	0 (0.0)	~ ~
SW Math/Sci. Collaborative, PA		43 (9.1)	518 (11.8)	52 (9.6)	519 (11.0)	5 (3.4)	498 (4.3)
Michigan		43 (7.6)	540 (11.1)	52 (8.0)	517 (7.4)	5 (3.2)	505 (11.4)
Pennsylvania		43 (6.2)	522 (10.6)	54 (6.5)	504 (7.6)	3 (1.9)	520 (22.2)
Fremont/Lincoln/WestSide PS, NE	r	43 (1.7)	491 (15.3)	46 (1.5)	472 (9.8)	11 (1.3)	568 (58.7)
Illinois		42 (5.4)	526 (8.3)	57 (5.4)	508 (8.4)	1 (0.9)	~ ~
Rochester City Sch. Dist., NY	r	40 (1.6)	467 (12.2)	44 (1.6)	423 (9.7)	16 (0.5)	436 (18.0)
Netherlands	r	40 (6.2)	539 (10.5)	60 (6.2)	552 (10.5)	0 (0.0)	~ ~
United States	r	37 (3.8)	516 (6.9)	59 (3.6)	493 (5.2)	4 (1.5)	480 (14.2)
Japan		36 (4.3)	582 (3.9)	61 (4.2)	578 (2.6)	3 (1.5)	562 (5.5)
Indiana		36 (7.8)	515 (12.3)	62 (7.7)	514 (8.2)	2 (1.8)	~ ~
Guilford County, NC	s	36 (1.3)	496 (13.0)	64 (1.3)	523 (14.9)	0 (0.0)	~ ~
Massachusetts	s	36 (7.4)	522 (13.3)	64 (7.4)	516 (7.8)	0 (0.0)	~ ~
Project SMART Consortium, OH		35 (1.6)	536 (15.2)	61 (1.5)	507 (8.0)	4 (0.5)	516 (43.0)
Idaho	r	32 (7.9)	481 (12.9)	63 (8.7)	505 (9.2)	4 (3.5)	472 (17.6)
Delaware Science Coalition, DE	r	32 (1.5)	447 (15.7)	59 (1.9)	484 (14.0)	9 (1.8)	496 (48.0)
Miami-Dade County PS, FL	s	31 (12.2)	458 (10.1)	57 (13.5)	426 (16.2)	11 (7.8)	399 (4.4)
Canada		31 (2.5)	547 (4.9)	64 (2.7)	523 (3.1)	5 (1.1)	528 (12.8)
Maryland	r	30 (6.8)	470 (11.1)	52 (7.6)	506 (8.9)	18 (5.8)	473 (11.3)
Missouri		30 (6.1)	501 (10.0)	68 (6.3)	483 (7.6)	3 (1.8)	482 (56.0)
Michigan Invitational Group, MI		29 (1.4)	530 (16.3)	66 (1.5)	537 (5.2)	5 (1.2)	497 (12.4)
Italy		28 (3.4)	484 (8.4)	66 (4.0)	478 (4.6)	6 (2.0)	473 (8.6)
England	r	26 (4.2)	535 (10.1)	72 (4.4)	486 (5.4)	2 (1.5)	~ ~
Chicago Public Schools, IL	s	25 (12.0)	472 (13.4)	65 (11.6)	456 (6.0)	10 (6.7)	467 (33.9)
Jersey City Public Schools, NJ		25 (0.8)	461 (16.2)	66 (1.1)	485 (12.8)	9 (0.7)	473 (7.5)
Hong Kong, SAR		22 (4.1)	585 (12.8)	67 (4.4)	586 (5.8)	10 (2.7)	567 (11.1)
South Carolina		21 (7.0)	501 (15.5)	74 (6.4)	498 (9.4)	6 (4.3)	532 (25.6)
North Carolina	r	17 (6.1)	465 (10.2)	76 (6.0)	501 (5.4)	6 (4.4)	523 (12.0)
Oregon		11 (5.0)	525 (21.6)	77 (6.4)	517 (7.9)	12 (5.5)	500 (14.1)
Chinese Taipei		6 (1.9)	580 (14.2)	78 (3.2)	587 (4.8)	16 (2.7)	577 (10.7)
Korea, Rep. of		4 (1.6)	594 (12.1)	81 (3.5)	588 (2.1)	16 (3.1)	583 (4.1)
Russian Federation		1 (0.9)	~ ~	47 (4.0)	536 (8.4)	52 (3.9)	518 (6.6)
International Avg. (All Countries)		19 (0.5)	497 (2.5)	63 (0.7)	486 (1.0)	18 (0.5)	476 (2.0)

SOURCE: IEA Third International Mathematics and Science Study (TIMSS), 1998-1999.

States in *italics* did not fully satisfy guidelines for sample participation rates (see Appendix A for details).

( ) Standard errors appear in parentheses. Because results are rounded to the nearest whole number, some totals may appear inconsistent.

A tilde (~) indicates insufficient data to report achievement.

An "r" indicates school response data available for 70-84% of students. An "s" indicates school response data available for 50-69% of students.

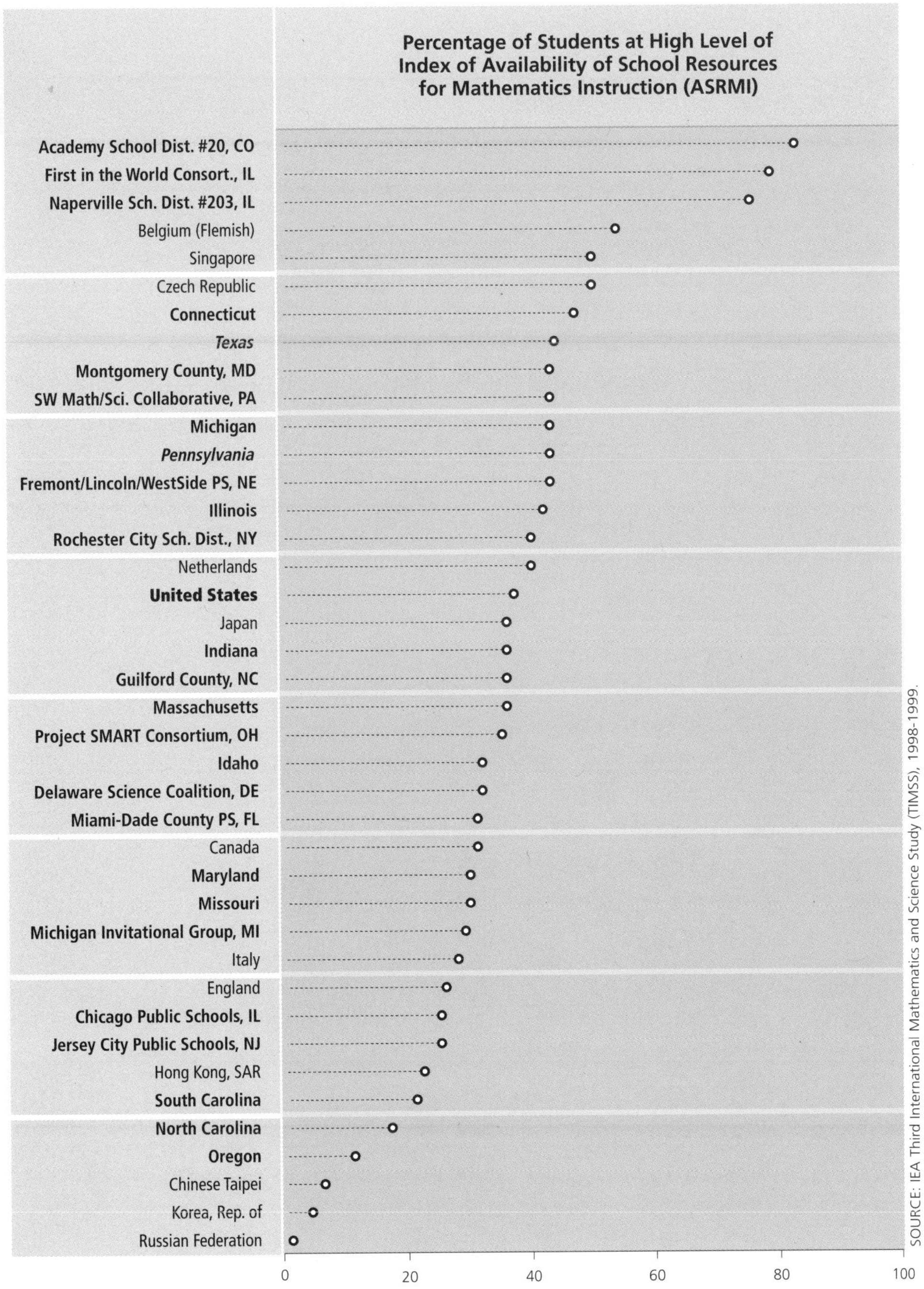

Percentage of Students at High Level of Index of Availability of School Resources for Mathematics Instruction (ASRMI)

Academy School Dist. #20, CO
First in the World Consort., IL
Naperville Sch. Dist. #203, IL
Belgium (Flemish)
Singapore
Czech Republic
Connecticut
Texas
Montgomery County, MD
SW Math/Sci. Collaborative, PA
Michigan
Pennsylvania
Fremont/Lincoln/WestSide PS, NE
Illinois
Rochester City Sch. Dist., NY
Netherlands
United States
Japan
Indiana
Guilford County, NC
Massachusetts
Project SMART Consortium, OH
Idaho
Delaware Science Coalition, DE
Miami-Dade County PS, FL
Canada
Maryland
Missouri
Michigan Invitational Group, MI
Italy
England
Chicago Public Schools, IL
Jersey City Public Schools, NJ
Hong Kong, SAR
South Carolina
North Carolina
Oregon
Chinese Taipei
Korea, Rep. of
Russian Federation

0 20 40 60 80 100

SOURCE: IEA Third International Mathematics and Science Study (TIMSS), 1998-1999.

What Is the Role of the School Principal?

To better understand the roles and responsibilities of schools across countries, TIMSS asked school principals how much time per month they spend on various school-related activities. Specifically, they were asked how much time they spend on instructional leadership activities, including discussing educational objectives with teachers, initiating curriculum revisions and planning, training teachers, and engaging in professional development activities. They were also asked how much time they spend talking with parents, counseling and disciplining students, and responding to requests from local, regional, or national education officials. Further, they responded to questions about how much time they spend on administrative duties, including hiring teachers, representing the school in the community and at official meetings, and doing internal tasks (e.g., regulations, school budget, timetable). Finally, they were asked how much time they spend teaching.

The results presented in Exhibit 7.3 show that principals reported spending per month, on average across all the TIMSS 1999 countries, 51 hours on administrative duties, 35 hours communicating with various constituents, 33 hours on instructional leadership activities, and 16 hours teaching.[4] Compared with the international profile, principals in the United States reported spending more time communicating with students, parents, and education officials (over 50 hours per month, on average), and very little time teaching. Reports from principals in the Benchmarking jurisdictions generally resembled those of the United States overall. It is interesting to note that principals in Jersey City and Rochester reported spending 72 hours per month communicating with students, parents, and education officials, while principals in Indiana and the Michigan Invitational Group reported spending 74 hours per month on administrative duties.

A number of the comparison countries, such as Canada, Chinese Taipei, Hong Kong, and Singapore, have patterns of principals' use of time similar to that of the United States. For example, unlike in most European countries (e.g., the Czech Republic and Russian Federation among comparison countries), principals in these countries spend relatively little time teaching, and most of it on administrative duties, communicating with constituents, and engaging in instructional leadership activities.

4 Activities reported by principals are not necessarily exclusive; principals may have reported engaging in more than one activity at the same time.

	Average Total Hours Per Month Spent on Activities[1]			
	Instructional Leadership Activities[2]	Communicating with Students, Parents, and Education Officials[3]	Administrative Duties[4]	Teaching (including preparation)
Countries				
United States	r 34 (1.9)	r 52 (2.4)	r 56 (3.2)	r 3 (0.6)
Belgium (Flemish)	29 (2.3)	27 (2.1)	56 (2.5)	0 (0.1)
Canada	25 (1.1)	54 (1.4)	54 (2.1)	5 (0.9)
Chinese Taipei	24 (1.4)	34 (1.7)	86 (4.1)	4 (0.6)
Czech Republic	32 (1.9)	33 (1.8)	44 (2.4)	36 (1.8)
England	– –	– –	– –	– –
Hong Kong, SAR	r 43 (3.2)	r 29 (1.8)	r 75 (4.2)	r 3 (0.6)
Italy	36 (1.4)	44 (2.1)	45 (1.7)	– –
Japan	33 (2.0)	19 (1.3)	69 (3.6)	1 (0.8)
Korea, Rep. of	30 (2.1)	22 (1.6)	46 (3.6)	3 (0.5)
Netherlands	r 42 (4.0)	r 20 (2.0)	r 49 (5.6)	r 7 (1.7)
Russian Federation	r 44 (1.9)	r 33 (1.7)	r 65 (3.1)	r 46 (2.1)
Singapore	45 (2.2)	46 (1.9)	56 (3.1)	3 (0.6)
States				
Connecticut	s 38 (5.6)	s 55 (4.9)	s 51 (6.0)	s 1 (0.4)
Idaho	r 33 (2.2)	r 41 (3.3)	r 53 (6.1)	r 2 (0.9)
Illinois	r 36 (2.1)	r 49 (3.5)	r 61 (4.9)	r 2 (1.0)
Indiana	37 (3.9)	53 (5.8)	74 (6.0)	3 (1.0)
Maryland	r 38 (2.8)	r 60 (4.0)	r 56 (3.9)	r 1 (0.3)
Massachusetts	s 32 (3.1)	s 48 (4.1)	s 56 (6.6)	s 1 (0.4)
Michigan	35 (2.8)	53 (4.8)	61 (5.2)	3 (1.4)
Missouri	34 (3.3)	55 (4.9)	57 (4.9)	1 (0.5)
North Carolina	r 43 (3.7)	r 66 (6.5)	r 54 (5.0)	r 2 (0.8)
Oregon	38 (4.3)	51 (5.1)	58 (5.2)	2 (0.7)
Pennsylvania	r 27 (2.1)	r 57 (4.1)	r 59 (6.0)	r 2 (0.6)
South Carolina	r 35 (3.6)	r 62 (4.8)	r 53 (5.3)	r 2 (1.1)
Texas	s 35 (4.5)	s 57 (5.3)	s 64 (6.0)	s 2 (0.6)
Districts and Consortia				
Academy School Dist. #20, CO	25 (0.1)	45 (0.1)	46 (0.1)	1 (0.0)
Chicago Public Schools, IL	s 46 (9.0)	s 51 (5.5)	s 58 (8.9)	s 2 (0.8)
Delaware Science Coalition, DE	s 37 (1.2)	s 60 (1.3)	s 53 (2.4)	s 0 (0.0)
First in the World Consort., IL	r 32 (0.5)	r 48 (0.3)	r 47 (0.9)	r 1 (0.1)
Fremont/Lincoln/WestSide PS, NE	s 27 (0.3)	s 56 (0.5)	s 42 (0.5)	s 1 (0.1)
Guilford County, NC	r 41 (0.4)	r 65 (0.5)	r 56 (0.7)	r 1 (0.0)
Jersey City Public Schools, NJ	r 34 (0.7)	r 72 (0.6)	r 36 (0.7)	r 3 (0.1)
Miami-Dade County PS, FL	x x	x x	x x	x x
Michigan Invitational Group, MI	31 (0.5)	63 (1.0)	74 (1.4)	1 (0.0)
Montgomery County, MD	s 35 (6.2)	s 46 (4.3)	s 48 (6.4)	s 1 (0.4)
Naperville Sch. Dist. #203, IL	36 (0.7)	37 (0.7)	67 (0.8)	0 (0.0)
Project SMART Consortium, OH	r 31 (0.6)	r 58 (1.0)	r 54 (1.2)	r 1 (0.1)
Rochester City Sch. Dist., NY	r 35 (0.4)	r 72 (0.8)	r 51 (0.7)	r 8 (0.4)
SW Math/Sci. Collaborative, PA	33 (3.6)	62 (5.8)	40 (4.6)	4 (1.6)
International Avg. (All Countries)	33 (0.3)	35 (0.3)	51 (0.5)	16 (0.2)

SOURCE: IEA Third International Mathematics and Science Study (TIMSS), 1998-1999.

Background data provided by schools.

1 Total hours reported for activities in each category averaged across schools. Activities are not necessarily exclusive; principals may have reported engaging in more than one activity at the same time.

2 Includes discussing educational objectives with teachers; initiating curriculum revision and/or planning; training teachers; and professional development activities.

3 Includes talking with parents, counseling and disciplining of students and responding to requests from local, regional, or national education officials.

4 Includes hiring teachers; representing the school in the community; representing the school at official meetings; internal administrative tasks (e.g., regulations, school budget, timetable).

States in *italics* did not fully satisfy guidelines for sample participation rates (see Appendix A for details).

( ) Standard errors appear in parentheses. Because results are rounded to the nearest whole number, some totals may appear inconsistent.

A dash (–) indicates data are not available.

An "r" indicates school response data available for 70-84% of students. An "s" indicates school response data available for 50-69% of students. An "x" indicates school response data available for <50% of students.

What Are the Schools' Expectations of Parents?

Schools' expectations for parental involvement are shown in Exhibit 7.4. Clearly schools expect help from parents. On average across all the TIMSS 1999 countries, 85 percent of the students attended schools expecting parents to ensure that their children complete their homework, and 79 percent attended schools expecting parents to volunteer for school projects or field trips. About half the students were in schools expecting parents to help raise funds and to serve on committees. Only 28 percent were in schools expecting parents to help as aides in the classroom.

In the United States, almost all students were in schools that expected parents to ensure that their children completed their homework and to volunteer for school projects, programs, or field trips. Parents generally were not often expected to serve as teacher aides (with the notable exception of the Chicago Public Schools, where 34 percent of students were in such schools), but were more often expected to serve on committees and to raise funds for the school. Schools in the Benchmarking jurisdictions generally resembled those in the United States overall, with few major differences.

	Percentage of Students Whose Schools Reported That They Expect Parents to Be Involved in the School-Related Activity									
		Be Sure Child Completes Homework		Serve as Teacher Aides in Classroom		Volunteer for School Projects, Programs, or Field Trips		Raise Funds for the School		Serve on Committees[1]
Countries										
United States	r	99 (0.7)	r	15 (3.0)	r	94 (1.7)	r	55 (4.7)	r	68 (4.1)
Belgium (Flemish)		94 (2.1)		19 (3.7)		39 (4.3)		9 (2.7)		10 (2.7)
Canada		99 (0.6)		15 (1.7)		82 (2.2)		52 (3.4)		55 (2.7)
Chinese Taipei		97 (1.3)		58 (4.2)		90 (2.5)		41 (4.2)		56 (4.4)
Czech Republic		91 (3.1)		7 (2.7)		80 (3.8)		32 (4.7)		35 (4.9)
England		– –		– –		– –		– –		– –
Hong Kong, SAR		96 (1.8)		30 (4.2)		77 (3.8)		60 (4.6)		21 (3.7)
Italy		91 (2.3)		9 (2.2)		70 (3.4)		25 (3.1)		42 (3.7)
Japan		43 (4.4)		5 (2.0)		81 (2.8)		6 (2.0)		8 (2.2)
Korea, Rep. of		64 (3.9)		33 (4.1)		71 (3.8)		31 (3.8)		44 (4.2)
Netherlands	r	81 (5.6)	r	46 (6.2)	r	61 (6.2)	r	16 (5.2)	r	46 (6.5)
Russian Federation		78 (3.1)		36 (3.3)		91 (1.7)		59 (2.8)		59 (4.1)
Singapore		95 (1.8)		6 (2.2)		44 (4.5)		51 (4.3)		41 (4.3)
States										
Connecticut	s	100 (0.0)	s	7 (4.4)	s	83 (6.6)	s	54 (8.6)	s	42 (8.9)
Idaho	r	97 (0.3)	r	7 (4.2)	r	86 (5.3)	r	20 (6.9)	r	43 (8.8)
Illinois		97 (2.5)		13 (4.4)		85 (6.5)		41 (6.8)		47 (6.9)
Indiana		100 (0.0)		8 (4.1)		87 (4.3)		50 (7.6)		42 (6.9)
Maryland	r	95 (3.5)	r	16 (5.4)	r	93 (4.0)	r	68 (7.8)	r	60 (7.8)
Massachusetts	s	100 (0.0)	s	8 (4.5)	s	91 (5.3)	s	65 (7.9)	s	86 (6.2)
Michigan		98 (1.8)		13 (5.0)		98 (1.6)		47 (7.6)		63 (6.6)
Missouri		96 (3.1)		5 (3.5)		73 (7.7)		33 (8.2)		50 (8.5)
North Carolina	r	100 (0.0)	r	22 (7.5)	r	95 (3.2)	r	76 (7.4)	r	61 (7.8)
Oregon		98 (2.3)		22 (8.0)		91 (3.4)		58 (7.6)		72 (6.1)
Pennsylvania		100 (0.0)		14 (6.3)		84 (5.3)		52 (6.5)		34 (6.2)
South Carolina		100 (0.0)		27 (7.5)		100 (0.0)		77 (7.2)		91 (4.4)
Texas	r	97 (2.7)	r	9 (5.1)	r	94 (3.9)	r	36 (8.7)	r	65 (6.9)
Districts and Consortia										
Academy School Dist. #20, CO		100 (0.0)		0 (0.0)		100 (0.0)		46 (0.4)		75 (0.3)
Chicago Public Schools, IL	r	100 (0.0)	r	34 (8.8)	r	94 (6.0)	r	68 (11.8)	r	80 (8.9)
Delaware Science Coalition, DE	r	98 (0.1)	r	9 (0.5)	r	90 (0.5)	r	53 (1.9)	r	60 (2.0)
First in the World Consort., IL	r	100 (0.0)	r	20 (1.5)	r	98 (0.1)	r	56 (1.2)	r	37 (1.3)
Fremont/Lincoln/WestSide PS, NE	r	100 (0.0)	r	0 (0.0)	r	72 (1.9)	r	33 (1.2)	r	48 (1.6)
Guilford County, NC	r	100 (0.0)	s	0 (0.0)	r	100 (0.0)	r	88 (1.0)	r	77 (0.7)
Jersey City Public Schools, NJ		100 (0.0)		6 (0.2)		90 (0.6)		54 (1.4)		77 (0.8)
Miami-Dade County PS, FL		x x		x x		x x		x x		x x
Michigan Invitational Group, MI		85 (1.5)		4 (0.3)		73 (1.2)		34 (1.3)		76 (1.4)
Montgomery County, MD	s	100 (0.0)	s	20 (11.3)	s	100 (0.0)	s	88 (2.3)	s	59 (12.3)
Naperville Sch. Dist. #203, IL		100 (0.0)		0 (0.0)		81 (0.6)		36 (1.8)		36 (1.8)
Project SMART Consortium, OH		93 (1.0)		14 (0.5)		80 (1.4)		45 (1.4)		52 (1.4)
Rochester City Sch. Dist., NY	r	100 (0.0)	r	19 (1.3)	r	90 (0.9)	r	57 (1.6)	r	100 (0.0)
SW Math/Sci. Collaborative, PA		100 (0.0)		7 (4.0)		88 (6.2)		48 (8.0)		41 (8.2)
International Avg. (All Countries)		85 (0.5)		28 (0.6)		79 (0.5)		51 (0.6)		47 (0.6)

SOURCE: IEA Third International Mathematics and Science Study (TIMSS), 1998-1999.

Background data provided by schools.

1 Serve on committees which select school personnel or review school finances.

States in *italics* did not fully satisfy guidelines for sample participation rates (see Appendix A for details).

( ) Standard errors appear in parentheses. Because results are rounded to the nearest whole number, some totals may appear inconsistent.

A dash (–) indicates data are not available.

An "r" indicates school response data available for 70-84% of students. An "s" indicates school response data available for 50-69% of students. An "x" indicates school response data available for <50% of students.

How Serious Are School Attendance Problems?

In some countries, schools are confronted with high rates of absenteeism, which can influence instructional continuity and reduce the time for learning. In general, research has shown that greater truancy is related to less serious attitudes towards school and lower academic achievement. To examine this issue, TIMSS developed an index of good school and class attendance (SCA) based on schools' responses to three questions about the seriousness of students' absenteeism, arriving late at school, and skipping class. The high index level indicates that schools reported that all three types of behavior are not a problem. The low level indicates that two or more are a serious problem, or that two are minor problems and one a serious problem. The medium category includes all other combinations of responses.

The results of the index are presented in Exhibit 7.5. Sixty percent of students on average across all the TIMSS 1999 countries were in the medium category, where principals had judged their schools to have a moderate attendance problem. Exactly one-fifth of the students were in schools at the high level of the index, and another 19 percent were in schools at the low level. Although countries varied considerably, there was a modest positive relationship between good attendance and mathematics achievement on average across countries.

The results for the United States resemble the international averages, and also show a positive relationship between attendance and mathematics achievement. Across the Benchmarking entities, the situation varied considerably. Participants with the highest percentages of students in schools with good attendance included Naperville and the Academy School District, with more than 40 percent of the students in this category. Jurisdictions with less than 10 percent of students in this category included Pennsylvania, Jersey City, Oregon, the Delaware Science Coalition, and Rochester.

The information used to compute this index appears in Exhibit 7.6, together with data showing the percentages of students in schools where the behavior occurs at least weekly. Arriving late and absenteeism were more common in the United States than in the TIMSS 1999 countries generally, but were not usually considered to be serious problems. Among Benchmarking participants, Naperville had the fewest students in schools that reported attendance problems. In contrast, Rochester reported the most problems, with almost all students in schools where tardiness, absenteeism, and skipping class are frequent occurrences and sometimes constitute serious problems.

Exhibit 7.5 Index of Good School and Class Attendance (SCA)

8th Grade Mathematics

Index of Good School and Class Attendance

Index based on schools' responses to three questions about the seriousness of attendance problems in school: arriving late at school; absenteeism; skipping class (see exhibit 7.6). High level indicates that all three behaviors are reported to be not a problem. Low level indicates that two or more behaviors are reported to be a serious problem, or two behaviors are reported to be minor problems and the third a serious problem. Medium level includes all other possible combinations of responses.

		High SCA		Medium SCA		Low SCA	
		Percent of Students	Average Achievement	Percent of Students	Average Achievement	Percent of Students	Average Achievement
Naperville Sch. Dist. #203, IL		55 (1.5)	564 (4.0)	45 (1.5)	576 (3.6)	0 (0.0)	~ ~
Belgium (Flemish)		52 (4.4)	579 (7.1)	45 (4.5)	536 (7.4)	3 (1.0)	535 (9.3)
Academy School Dist. #20, CO		42 (0.4)	524 (3.3)	58 (0.4)	531 (1.8)	0 (0.0)	~ ~
Czech Republic		36 (5.8)	526 (9.9)	56 (6.0)	516 (4.4)	8 (2.3)	539 (20.2)
Michigan Invitational Group, MI		34 (1.4)	533 (11.0)	66 (1.4)	532 (7.0)	0 (0.0)	~ ~
Italy		33 (3.3)	497 (5.8)	58 (3.6)	481 (5.1)	9 (2.4)	424 (12.4)
Singapore		32 (4.1)	630 (11.9)	64 (4.0)	592 (7.0)	3 (1.6)	597 (19.3)
Korea, Rep. of		31 (3.7)	585 (3.7)	61 (4.0)	588 (2.4)	9 (2.4)	595 (5.4)
Netherlands	r	30 (7.3)	524 (14.5)	46 (7.3)	555 (6.6)	24 (7.5)	519 (27.9)
First in the World Consort., IL	r	28 (1.4)	568 (18.2)	72 (1.4)	549 (8.6)	0 (0.0)	~ ~
Chinese Taipei		28 (3.7)	616 (7.6)	61 (3.6)	570 (4.0)	11 (2.7)	591 (10.1)
Michigan	r	28 (6.7)	529 (6.3)	69 (6.2)	526 (10.1)	3 (2.5)	496 (57.7)
Chicago Public Schools, IL	s	27 (13.5)	486 (15.6)	65 (13.2)	456 (9.5)	8 (1.2)	442 (20.9)
Indiana		27 (7.8)	544 (9.2)	66 (8.4)	506 (9.3)	7 (3.7)	503 (6.0)
Hong Kong, SAR		25 (3.9)	603 (7.4)	68 (4.3)	582 (6.8)	7 (2.5)	540 (13.3)
Project SMART Consortium, OH	s	25 (1.2)	537 (23.8)	71 (1.2)	507 (9.8)	4 (0.2)	477 (16.0)
Illinois		22 (6.5)	519 (12.6)	73 (6.7)	510 (6.5)	5 (0.4)	540 (10.4)
Connecticut	s	22 (6.6)	551 (28.7)	78 (6.6)	512 (10.9)	0 (0.0)	~ ~
United States	r	19 (3.0)	534 (11.5)	68 (3.4)	498 (5.2)	13 (2.5)	470 (9.3)
Fremont/Lincoln/WestSide PS, NE	s	18 (0.6)	507 (19.1)	69 (1.5)	470 (12.2)	13 (1.5)	568 (58.7)
Canada		18 (2.2)	530 (7.1)	73 (3.0)	530 (3.0)	9 (2.0)	535 (7.9)
Texas	s	15 (7.0)	544 (17.6)	81 (7.3)	516 (12.0)	4 (2.8)	454 (13.8)
Montgomery County, MD	s	15 (11.0)	566 (9.6)	85 (11.0)	531 (4.2)	0 (0.0)	~ ~
Massachusetts	s	14 (5.1)	537 (14.2)	74 (6.2)	515 (7.1)	11 (5.4)	513 (8.1)
Idaho	r	14 (6.7)	498 (14.7)	78 (7.6)	499 (8.6)	8 (3.6)	469 (24.9)
SW Math/Sci. Collaborative, PA		13 (3.6)	545 (10.2)	78 (6.2)	522 (8.9)	9 (4.6)	448 (18.3)
Guilford County, NC	r	13 (0.6)	545 (18.7)	79 (1.0)	515 (12.0)	8 (0.9)	448 (19.6)
South Carolina	r	11 (4.0)	484 (29.5)	75 (5.4)	507 (7.8)	13 (4.0)	485 (27.2)
Maryland	r	11 (4.5)	514 (9.2)	80 (6.1)	490 (6.4)	10 (5.1)	452 (23.1)
Russian Federation		10 (1.7)	535 (12.0)	70 (3.8)	532 (6.4)	20 (3.4)	500 (8.2)
Missouri		10 (5.0)	511 (13.2)	80 (7.0)	491 (6.5)	10 (5.1)	424 (24.7)
North Carolina	r	10 (4.2)	483 (16.5)	84 (5.7)	502 (6.1)	6 (4.0)	452 (8.6)
Pennsylvania		9 (5.1)	525 (12.0)	83 (6.6)	514 (7.0)	8 (4.1)	471 (18.0)
Japan		7 (2.4)	590 (12.2)	47 (4.1)	579 (2.6)	46 (3.9)	576 (2.4)
Jersey City Public Schools, NJ	r	7 (0.3)	517 (9.1)	90 (0.4)	472 (10.8)	3 (0.1)	442 (16.4)
Oregon		4 (3.0)	487 (2.3)	84 (5.9)	515 (7.5)	12 (4.8)	504 (13.9)
Delaware Science Coalition, DE	r	0 (0.0)	~ ~	88 (2.0)	462 (10.9)	12 (2.0)	534 (38.5)
Rochester City Sch. Dist., NY	s	0 (0.0)	~ ~	50 (1.5)	448 (11.8)	50 (1.5)	433 (10.8)
Miami-Dade County PS, FL		x x	x x	x x	x x	x x	x x
England		– –	– –	– –	– –	– –	– –
International Avg. (All Countries)		20 (0.6)	497 (2.8)	60 (0.7)	488 (1.0)	19 (0.5)	474 (2.0)

SOURCE: IEA Third International Mathematics and Science Study (TIMSS), 1998-1999.

States in *italics* did not fully satisfy guidelines for sample participation rates (see Appendix A for details).

( ) Standard errors appear in parentheses. Because results are rounded to the nearest whole number, some totals may appear inconsistent.

A dash (–) indicates data are not available. A tilde (~) indicates insufficient data to report achievement.

An "r" indicates school response data available for 70-84% of students. An "s" indicates school response data available for 50-69% of students. An "x" indicates school response data available for <50% of students.

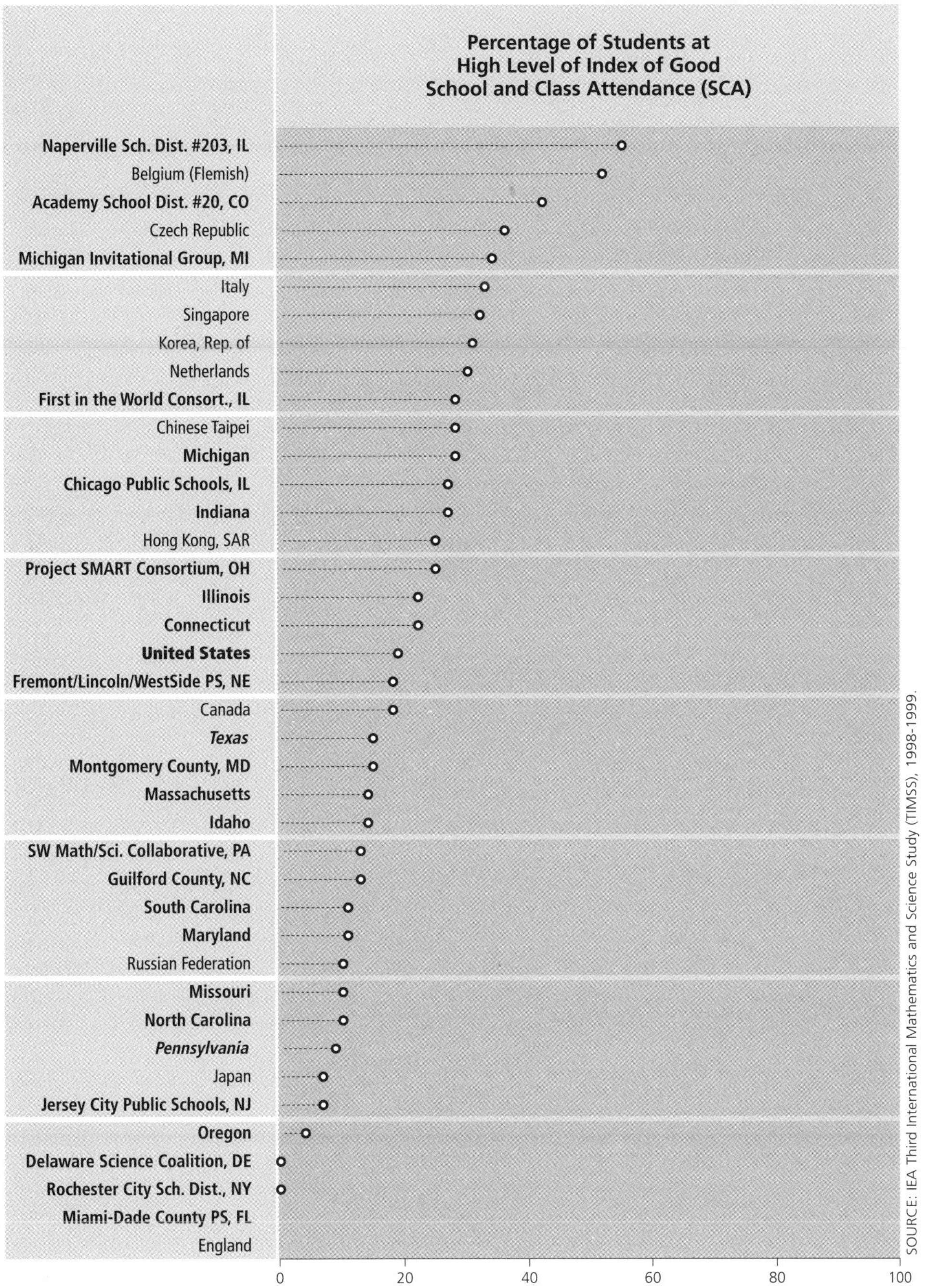

Percentage of Students at High Level of Index of Good School and Class Attendance (SCA)

Naperville Sch. Dist. #203, IL
Belgium (Flemish)
Academy School Dist. #20, CO
Czech Republic
Michigan Invitational Group, MI
Italy
Singapore
Korea, Rep. of
Netherlands
First in the World Consort., IL
Chinese Taipei
Michigan
Chicago Public Schools, IL
Indiana
Hong Kong, SAR
Project SMART Consortium, OH
Illinois
Connecticut
United States
Fremont/Lincoln/WestSide PS, NE
Canada
Texas
Montgomery County, MD
Massachusetts
Idaho
SW Math/Sci. Collaborative, PA
Guilford County, NC
South Carolina
Maryland
Russian Federation
Missouri
North Carolina
Pennsylvania
Japan
Jersey City Public Schools, NJ
Oregon
Delaware Science Coalition, DE
Rochester City Sch. Dist., NY
Miami-Dade County PS, FL
England

0 20 40 60 80 100

SOURCE: IEA Third International Mathematics and Science Study (TIMSS), 1998-1999.

Exhibit 7.6 Frequency and Seriousness of Student Attendance Problems

8th Grade Mathematics

	Percentage of Students Whose Schools Reported the Behavior					
	Arriving Late		Absenteeism		Skipping Class	
	Occurs at Least Weekly	Is a Serious Problem	Occurs at Least Weekly	Is a Serious Problem	Occurs at Least Weekly	Is a Serious Problem
Countries						
United States	r 71 (3.7)	r 12 (2.3)	r 60 (4.2)	r 12 (2.7)	r 29 (3.6)	r 4 (1.8)
Belgium (Flemish)	44 (4.7)	3 (1.4)	11 (2.4)	4 (1.8)	4 (1.3)	2 (1.0)
Canada	58 (2.7)	7 (1.7)	45 (3.1)	7 (1.6)	22 (2.3)	3 (1.0)
Chinese Taipei	43 (4.1)	2 (1.1)	32 (4.0)	10 (2.7)	30 (3.8)	11 (2.8)
Czech Republic	21 (3.8)	0 (0.3)	9 (2.8)	8 (2.5)	5 (2.2)	8 (2.4)
England	– –	– –	– –	– –	– –	– –
Hong Kong, SAR	r 61 (4.8)	9 (2.8)	r 34 (4.5)	3 (1.6)	r 10 (2.8)	r 1 (0.9)
Italy	32 (3.6)	4 (1.6)	11 (2.2)	9 (2.3)	8 (2.2)	7 (2.0)
Japan	55 (4.1)	20 (3.4)	63 (4.1)	76 (3.9)	14 (3.2)	27 (3.8)
Korea, Rep. of	32 (4.0)	1 (1.0)	31 (4.1)	12 (2.9)	21 (3.6)	5 (1.8)
Netherlands	r 76 (4.9)	r 18 (6.8)	r 35 (5.9)	r 12 (6.4)	r 44 (6.5)	r 15 (7.1)
Russian Federation	41 (3.8)	14 (3.5)	22 (2.9)	12 (2.2)	32 (4.2)	10 (2.2)
Singapore	51 (4.8)	3 (1.6)	40 (4.4)	3 (1.5)	23 (4.0)	0 (0.0)
States						
Connecticut	s 67 (9.4)	s 0 (0.0)	s 48 (9.5)	s 4 (0.5)	s 20 (6.7)	s 0 (0.0)
Idaho	r 72 (8.9)	r 5 (2.7)	r 67 (8.5)	r 8 (3.6)	r 31 (7.3)	r 1 (0.1)
Illinois	57 (8.4)	5 (3.0)	42 (7.4)	7 (1.2)	r 9 (4.0)	0 (0.0)
Indiana	64 (7.9)	7 (3.5)	55 (7.9)	9 (4.2)	20 (4.5)	0 (0.0)
Maryland	r 63 (7.1)	r 10 (5.1)	r 51 (6.9)	r 10 (5.1)	r 21 (6.0)	r 0 (0.0)
Massachusetts	s 59 (8.9)	s 16 (7.5)	s 62 (7.6)	s 14 (6.1)	s 17 (6.6)	s 0 (0.0)
Michigan	48 (7.1)	r 1 (1.0)	37 (7.3)	r 5 (3.4)	11 (4.5)	r 0 (0.0)
Missouri	76 (6.0)	2 (1.7)	69 (6.7)	13 (5.6)	33 (6.5)	r 9 (5.0)
North Carolina	r 54 (8.3)	r 3 (0.2)	r 52 (9.0)	r 11 (5.0)	r 16 (6.2)	r 0 (0.0)
Oregon	81 (6.5)	r 8 (3.0)	75 (7.6)	19 (5.3)	43 (8.1)	5 (1.8)
Pennsylvania	73 (7.2)	8 (4.1)	50 (6.7)	8 (4.1)	17 (5.0)	1 (0.0)
South Carolina	r 73 (6.5)	r 10 (4.9)	r 67 (7.8)	r 20 (5.1)	16 (4.4)	r 0 (0.0)
Texas	r 81 (7.3)	s 4 (2.8)	r 68 (7.6)	s 1 (1.4)	r 39 (6.1)	s 0 (0.0)
Districts and Consortia						
Academy School Dist. #20, CO	54 (0.4)	0 (0.0)	29 (0.4)	0 (0.0)	46 (0.4)	0 (0.0)
Chicago Public Schools, IL	s 66 (8.3)	s 8 (1.2)	s 49 (11.4)	s 10 (7.8)	s 14 (6.1)	r 0 (0.0)
Delaware Science Coalition, DE	r 84 (2.0)	r 0 (0.0)	r 90 (0.6)	r 12 (2.0)	s 54 (1.7)	r 0 (0.0)
First in the World Consort., IL	r 62 (1.4)	r 0 (0.0)	r 15 (0.4)	r 0 (0.0)	r 0 (0.0)	r 0 (0.0)
Fremont/Lincoln/WestSide PS, NE	r 68 (1.1)	s 0 (0.0)	r 58 (1.4)	s 13 (1.5)	r 48 (1.7)	s 0 (0.0)
Guilford County, NC	r 77 (0.9)	r 0 (0.0)	r 88 (0.6)	r 8 (0.9)	r 36 (1.1)	r 0 (0.0)
Jersey City Public Schools, NJ	66 (1.0)	r 12 (0.8)	50 (1.4)	r 0 (0.0)	0 (0.0)	r 0 (0.0)
Miami-Dade County PS, FL	x x	x x	x x	x x	x x	x x
Michigan Invitational Group, MI	48 (1.5)	9 (0.8)	40 (1.6)	0 (0.0)	31 (1.5)	0 (0.0)
Montgomery County, MD	s 83 (9.6)	s 0 (0.0)	s 61 (12.2)	s 0 (0.0)	s 12 (7.2)	s 0 (0.0)
Naperville Sch. Dist. #203, IL	39 (1.9)	0 (0.0)	15 (2.1)	0 (0.0)	0 (0.0)	0 (0.0)
Project SMART Consortium, OH	r 73 (1.1)	s 4 (0.2)	r 47 (1.6)	s 4 (0.2)	r 33 (1.6)	s 0 (0.0)
Rochester City Sch. Dist., NY	r 100 (0.0)	s 19 (0.6)	r 100 (0.0)	s 19 (0.6)	r 84 (0.5)	s 30 (1.5)
SW Math/Sci. Collaborative, PA	68 (7.7)	9 (4.6)	62 (6.2)	7 (4.3)	26 (8.7)	3 (2.9)
International Avg. (All Countries)	49 (0.6)	11 (0.4)	38 (0.6)	17 (0.5)	27 (0.6)	13 (0.5)

SOURCE: IEA Third International Mathematics and Science Study (TIMSS), 1998-1999.

Background data provided by schools.

States in *italics* did not fully satisfy guidelines for sample participation rates (see Appendix A for details).

( ) Standard errors appear in parentheses. Because results are rounded to the nearest whole number, some totals may appear inconsistent.

A dash (–) indicates data are not available.

An "r" indicates school response data available for 70-84% of students. An "s" indicates school response data available for 50-69% of students. An "x" indicates school response data available for <50% of students.

How Safe and Orderly Are Schools?

Discipline that maintains an orderly atmosphere conducive to learning is very important to school quality, and research indicates that urban schools have conditions less conducive to learning than non-urban schools.[5] For example, urban schools report more crime against students and teachers at school and that physical conflict among students is a serious or moderate problem. Among the Benchmarking participants there was considerable variation in principals' reports about the seriousness of a variety of potential discipline problems.

The frequency and seriousness of student behavior threatening an orderly school environment are presented in Exhibit 7.7. The three types of behavior are violating the dress code, creating a classroom disturbance, and cheating. Violation of dress code is likely to reflect, at least partially, whether there is a uniform requirement. For many countries, violating the dress code was not reported to be a serious problem; on average internationally only six percent of the students were in schools where it was a serious problem. Dress code violations were more frequently reported in the United States, where 42 percent of students were in schools where this occurs at least weekly, compared with 24 percent internationally. This was also a frequent problem in Texas and in Rochester, with 79 and 59 percent of students, respectively, in such schools.

Classroom disturbance was a more frequent problem in schools in the United States, as well as a more serious one. More than two-thirds of U.S. eighth-grade students were in schools where disturbances occur at least weekly, and 11 percent where these are a serious problem. Benchmarking jurisdictions where classroom disturbances were both more frequent and more serious than in the United States generally included Maryland, Missouri, North Carolina, Pennsylvania, the Delaware Science Coalition, Guilford County, the Michigan Invitational Group, Montgomery County, and Rochester.

The frequency and seriousness of student behavior threatening a safe school environment are shown in Exhibit 7.8. The five types of behavior are vandalism, theft, physical injury to other students, intimidation or verbal abuse of other students, and intimidation or verbal abuse of teachers or staff. As in other reports of student behavior, cross-national comparisons are difficult because of differing perceptions of what constitutes a serious problem. However, with only a few exceptions, the overwhelming majority of students attend schools judged to have few serious problems. The incidence of such student behavior was

5 Mayer, D.P., Mullens, J.E., and Moore, M.T. (2000), *Monitoring School Quality: An Indicators Report*, NCES 2001-030, Washington, DC: National Center for Education Statistics; Kaufman, P., Chen, X., Choy, S.P., Ruddy, S.A., Miller, A.K., Fleury, J.K., Chandler, K.A., Rand, M.R., Klaus, P., and Planty, M.G. (2000), *Indicators of School Crime and Safety, 2000*, NCES 2001-017/NCJ-184176, Washington, DC: U.S. Departments of Education and Justice.

generally low in most countries. The exception was intimidation or verbal abuse of other students. Some countries had relatively high percentages of students in schools where this occurs at least weekly; in Canada, the Netherlands, and the United States, more than 40 percent of the students were in such schools. Among Benchmarking participants, intimidation or verbal abuse of other students was a frequent and serious problem in Idaho, Maryland, Oregon, Pennsylvania, the Delaware Science Coalition, the Fremont/Lincoln/Westside Public Schools, the Project SMART Consortium, and Rochester. Vandalism was a frequent and serious problem in Rochester.

Frequency and Seriousness of Student Behavior Threatening an Orderly School Environment

	Percentage of Students Whose Schools Reported the Behavior					
	Violating Dress Code		Classroom Disturbance		Cheating	
	Occurs at Least Weekly	Is a Serious Problem	Occurs at Least Weekly	Is a Serious Problem	Occurs at Least Weekly	Is a Serious Problem
Countries						
United States	r 42 (4.0)	r 3 (1.2)	r 69 (4.3)	r 11 (2.6)	r 12 (2.8)	r 1 (0.0)
Belgium (Flemish)	6 (2.1)	0 (0.0)	40 (5.4)	7 (2.5)	14 (2.7)	1 (0.0)
Canada	22 (1.8)	2 (0.8)	60 (2.6)	21 (2.3)	4 (1.4)	2 (0.9)
Chinese Taipei	41 (4.1)	3 (1.5)	30 (3.8)	4 (1.6)	9 (2.1)	8 (2.3)
Czech Republic	3 (1.7)	0 (0.0)	63 (4.7)	21 (4.4)	9 (4.3)	11 (3.5)
England	– –	– –	– –	– –	– –	– –
Hong Kong, SAR	r 42 (4.6)	r 7 (2.5)	36 (4.7)	r 9 (2.9)	4 (1.7)	r 4 (1.9)
Italy	– –	– –	47 (4.0)	32 (3.6)	13 (2.7)	5 (1.4)
Japan	30 (4.0)	18 (3.5)	5 (1.5)	23 (3.7)	2 (1.1)	13 (2.8)
Korea, Rep. of	37 (4.3)	3 (1.4)	43 (4.2)	7 (1.8)	3 (1.3)	8 (2.5)
Netherlands	r 10 (4.2)	r 0 (0.0)	r 76 (5.5)	r 14 (5.4)	r 60 (6.5)	r 1 (0.8)
Russian Federation	7 (2.2)	0 (0.0)	13 (2.8)	4 (1.6)	1 (0.5)	2 (1.2)
Singapore	36 (4.8)	2 (1.3)	32 (3.9)	3 (1.7)	3 (1.4)	0 (0.0)
States						
Connecticut	s 22 (7.5)	s 0 (0.0)	s 71 (10.3)	s 11 (5.8)	s 8 (4.9)	s 7 (4.6)
Idaho	r 21 (8.2)	r 0 (0.0)	r 76 (6.8)	r 8 (3.9)	r 15 (5.4)	r 0 (0.0)
Illinois	16 (5.9)	2 (1.1)	65 (8.0)	6 (3.4)	10 (3.9)	0 (0.0)
Indiana	19 (6.2)	3 (0.2)	70 (5.5)	11 (4.8)	12 (5.0)	1 (1.2)
Maryland	r 36 (7.4)	r 4 (3.0)	r 84 (5.8)	r 26 (7.9)	r 9 (4.3)	r 0 (0.0)
Massachusetts	s 15 (5.5)	s 0 (0.0)	s 73 (8.4)	s 11 (4.4)	s 8 (4.8)	s 3 (2.6)
Michigan	16 (6.2)	r 2 (0.2)	68 (6.7)	r 7 (3.6)	5 (2.8)	r 0 (0.0)
Missouri	33 (7.6)	r 0 (0.0)	83 (5.1)	r 13 (4.7)	12 (4.1)	r 0 (0.0)
North Carolina	r 31 (8.6)	r 0 (0.0)	r 86 (5.7)	r 15 (6.3)	r 8 (4.4)	r 0 (0.0)
Oregon	21 (6.3)	0 (0.0)	77 (6.3)	6 (3.7)	4 (2.9)	0 (0.0)
Pennsylvania	34 (5.2)	6 (5.9)	82 (4.7)	15 (7.5)	5 (2.2)	1 (0.1)
South Carolina	r 47 (8.8)	r 5 (3.3)	86 (6.5)	r 10 (4.6)	13 (5.8)	r 1 (1.4)
Texas	r 79 (3.7)	s 11 (6.6)	r 79 (6.0)	s 8 (5.2)	r 12 (6.1)	s 0 (0.0)
Districts and Consortia						
Academy School Dist. #20, CO	0 (0.0)	0 (0.0)	100 (0.0)	0 (0.0)	0 (0.0)	0 (0.0)
Chicago Public Schools, IL	r 40 (9.7)	r 10 (7.5)	s 62 (9.0)	s 0 (0.0)	s 19 (10.2)	s 0 (0.0)
Delaware Science Coalition, DE	r 39 (2.0)	r 6 (0.5)	r 96 (0.4)	r 23 (1.8)	r 18 (0.8)	r 0 (0.0)
First in the World Consort., IL	r 0 (0.0)	r 0 (0.0)	r 44 (1.1)	r 0 (0.1)	r 0 (0.1)	r 0 (0.0)
Fremont/Lincoln/WestSide PS, NE	r 43 (1.8)	s 0 (0.0)	r 65 (1.3)	s 9 (0.5)	r 13 (0.9)	s 0 (0.0)
Guilford County, NC	r 42 (1.2)	r 0 (0.0)	r 88 (1.0)	r 17 (0.9)	r 19 (1.2)	s 0 (0.0)
Jersey City Public Schools, NJ	r 19 (1.1)	r 6 (0.9)	44 (1.6)	r 9 (0.8)	11 (1.0)	r 0 (0.0)
Miami-Dade County PS, FL	x x	x x	x x	x x	x x	x x
Michigan Invitational Group, MI	31 (1.5)	0 (0.0)	84 (1.4)	15 (1.5)	25 (1.2)	2 (0.1)
Montgomery County, MD	s 38 (12.6)	s 0 (0.0)	s 86 (9.8)	s 13 (8.1)	s 7 (1.1)	s 0 (0.0)
Naperville Sch. Dist. #203, IL	0 (0.0)	0 (0.0)	15 (2.1)	0 (0.0)	21 (1.0)	0 (0.0)
Project SMART Consortium, OH	r 27 (1.3)	s 0 (0.0)	r 65 (1.4)	s 14 (0.8)	r 0 (0.0)	s 0 (0.0)
Rochester City Sch. Dist., NY	r 59 (1.5)	s 0 (0.0)	r 100 (0.0)	s 50 (1.7)	s 0 (0.0)	s 0 (0.0)
SW Math/Sci. Collaborative, PA	47 (9.1)	2 (2.1)	67 (7.2)	11 (5.4)	7 (2.9)	0 (0.0)
International Avg. (All Countries)	24 (0.6)	6 (0.3)	39 (0.6)	13 (0.5)	11 (0.4)	7 (0.3)

SOURCE: IEA Third International Mathematics and Science Study (TIMSS), 1998-1999.

Background data provided by schools.

States in *italics* did not fully satisfy guidelines for sample participation rates (see Appendix A for details).

( ) Standard errors appear in parentheses. Because results are rounded to the nearest whole number, some totals may appear inconsistent.

A dash (–) indicates data are not available.

An "r" indicates school response data available for 70-84% of students. An "s" indicates school response data available for 50-69% of students. An "x" indicates school response data available for <50% of students.

Frequency and Seriousness of Student Behavior Threatening a Safe School Environment

8th Grade Mathematics

	Percentage of Students Whose Schools Reported the Behavior					
	Vandalism		Theft		Physical Injury to Other Students	
	Occurs at Least Weekly	Is a Serious Problem	Occurs at Least Weekly	Is a Serious Problem	Occurs at Least Weekly	Is a Serious Problem
Countries						
United States	r 11 (2.3)	r 1 (0.8)	r 10 (2.5)	r 2 (1.1)	r 10 (2.4)	r 3 (1.8)
Belgium (Flemish)	8 (2.4)	9 (2.6)	7 (2.2)	9 (2.5)	8 (1.9)	6 (2.1)
Canada	15 (1.5)	6 (2.0)	7 (1.4)	6 (1.9)	6 (1.8)	4 (1.5)
Chinese Taipei	14 (3.1)	11 (2.5)	7 (2.2)	16 (2.9)	8 (2.3)	21 (3.2)
Czech Republic	13 (2.7)	21 (3.6)	3 (1.9)	17 (3.8)	2 (1.7)	17 (3.7)
England	– –	– –	– –	– –	– –	– –
Hong Kong, SAR	18 (3.7)	r 6 (2.3)	8 (2.6)	r 5 (2.2)	5 (2.1)	r 3 (1.6)
Italy	7 (1.9)	18 (2.8)	4 (1.4)	16 (2.8)	9 (2.1)	19 (3.0)
Japan	3 (1.3)	23 (3.5)	1 (0.9)	25 (3.7)	1 (0.9)	22 (3.6)
Korea, Rep. of	12 (2.8)	10 (2.5)	9 (2.5)	13 (3.0)	10 (2.6)	9 (2.6)
Netherlands	r 45 (7.6)	r 28 (7.4)	r 22 (5.9)	r 19 (6.4)	r 2 (1.3)	r 4 (2.0)
Russian Federation	0 (0.4)	3 (1.5)	1 (0.5)	6 (2.0)	2 (1.1)	4 (1.3)
Singapore	5 (1.8)	2 (1.3)	5 (2.0)	2 (1.4)	1 (0.7)	0 (0.0)
States						
Connecticut	s 12 (6.0)	s 0 (0.0)	s 12 (6.0)	s 0 (0.0)	s 25 (8.2)	s 13 (6.1)
Idaho	r 15 (5.6)	r 0 (0.0)	r 17 (5.9)	r 4 (3.2)	r 25 (8.2)	r 0 (0.0)
Illinois	3 (0.9)	2 (0.1)	5 (2.4)	0 (0.0)	9 (3.8)	4 (3.0)
Indiana	2 (0.1)	0 (0.0)	6 (3.7)	2 (2.2)	8 (4.0)	2 (2.2)
Maryland	r 7 (3.7)	r 3 (0.2)	r 6 (3.4)	r 0 (0.0)	r 33 (8.3)	r 9 (5.1)
Massachusetts	s 6 (3.5)	s 0 (0.0)	s 6 (3.8)	s 3 (2.4)	s 9 (4.5)	s 0 (0.0)
Michigan	6 (3.2)	r 2 (0.2)	3 (2.1)	r 2 (0.1)	6 (2.7)	r 4 (2.7)
Missouri	9 (5.0)	r 2 (2.2)	7 (3.9)	r 7 (3.9)	8 (4.9)	r 5 (3.6)
North Carolina	r 20 (7.3)	r 0 (0.0)	r 20 (7.1)	r 3 (2.5)	r 8 (4.4)	r 0 (0.0)
Oregon	7 (3.9)	2 (1.7)	12 (4.9)	0 (0.0)	7 (4.4)	2 (2.3)
Pennsylvania	7 (2.9)	r 1 (0.9)	6 (2.9)	r 2 (1.8)	9 (3.6)	5 (3.1)
South Carolina	5 (3.6)	r 0 (0.0)	18 (5.9)	r 0 (0.0)	8 (4.6)	r 3 (2.5)
Texas	r 12 (6.2)	s 0 (0.0)	r 16 (7.3)	s 0 (0.0)	r 9 (5.1)	s 0 (0.0)
Districts and Consortia						
Academy School Dist. #20, CO	0 (0.0)	0 (0.0)	0 (0.0)	0 (0.0)	0 (0.0)	r 0 (0.0)
Chicago Public Schools, IL	s 6 (1.0)	s 0 (0.0)	s 6 (1.0)	s 0 (0.0)	s 6 (1.0)	s 0 (0.0)
Delaware Science Coalition, DE	r 6 (0.5)	r 6 (0.5)	r 5 (2.1)	r 0 (0.0)	s 28 (2.6)	r 6 (0.5)
First in the World Consort., IL	r 13 (0.4)	r 0 (0.0)	r 13 (0.4)	r 0 (0.0)	r 0 (0.0)	r 0 (0.0)
Fremont/Lincoln/WestSide PS, NE	r 0 (0.0)	s 0 (0.0)	r 25 (1.4)	s 0 (0.0)	r 25 (1.4)	s 13 (1.5)
Guilford County, NC	r 0 (0.0)	r 0 (0.0)	r 0 (0.0)	s 0 (0.0)	r 7 (0.4)	s 0 (0.0)
Jersey City Public Schools, NJ	11 (0.9)	r 0 (0.0)	0 (0.0)	r 6 (0.4)	10 (0.3)	r 9 (0.8)
Miami-Dade County PS, FL	x x	x x	x x	x x	x x	x x
Michigan Invitational Group, MI	19 (1.3)	0 (0.0)	0 (0.0)	0 (0.0)	11 (0.8)	0 (0.0)
Montgomery County, MD	s 12 (7.2)	s 0 (0.0)	s 7 (1.1)	s 0 (0.0)	s 0 (0.0)	s 0 (0.0)
Naperville Sch. Dist. #203, IL	0 (0.0)	0 (0.0)	0 (0.0)	0 (0.0)	0 (0.0)	0 (0.0)
Project SMART Consortium, OH	r 16 (1.2)	s 0 (0.0)	r 23 (1.5)	s 0 (0.0)	r 16 (0.8)	s 10 (0.8)
Rochester City Sch. Dist., NY	r 60 (1.6)	s 36 (1.7)	r 19 (1.8)	s 0 (0.0)	r 30 (1.3)	s 0 (0.0)
SW Math/Sci. Collaborative, PA	14 (5.8)	4 (0.4)	14 (4.7)	4 (0.4)	17 (6.7)	2 (2.1)
International Avg. (All Countries)	11 (0.4)	13 (0.5)	6 (0.3)	12 (0.5)	6 (0.3)	10 (0.4)

SOURCE: IEA Third International Mathematics and Science Study (TIMSS), 1998-1999.

Background data provided by schools.

States in *italics* did not fully satisfy guidelines for sample participation rates (see Appendix A for details).

( ) Standard errors appear in parentheses. Because results are rounded to the nearest whole number, some totals may appear inconsistent.

A dash (–) indicates data are not available.

An "r" indicates school response data available for 70-84% of students. An "s" indicates school response data available for 50-69% of students. An "x" indicates school response data available for <50% of students.

	Percentage of Students Whose Schools Reported the Behavior			
	Intimidation or Verbal Abuse of Other Students		Intimidation or Verbal Abuse of Teachers or Staff	
	Occurs at Least Weekly	Is a Serious Problem	Occurs at Least Weekly	Is a Serious Problem
Countries				
United States	r 46 (4.3)	r 16 (3.6)	r 7 (2.0)	r 3 (1.5)
Belgium (Flemish)	23 (3.4)	15 (3.7)	5 (1.5)	3 (1.2)
Canada	42 (3.0)	22 (2.5)	4 (1.2)	3 (1.1)
Chinese Taipei	11 (2.7)	18 (3.1)	1 (1.0)	17 (3.0)
Czech Republic	5 (1.5)	17 (3.6)	0 (0.0)	9 (2.6)
England	– –	– –	– –	– –
Hong Kong, SAR	r 8 (2.7)	r 4 (1.8)	r 3 (1.5)	r 2 (1.3)
Italy	14 (2.3)	23 (3.0)	4 (1.7)	13 (2.7)
Japan	3 (1.5)	25 (3.8)	2 (1.2)	23 (3.7)
Korea, Rep. of	12 (2.9)	12 (2.8)	8 (2.3)	9 (2.5)
Netherlands	r 49 (7.3)	r 23 (6.9)	r 17 (6.6)	r 16 (6.4)
Russian Federation	3 (1.3)	7 (2.1)	1 (0.5)	1 (0.6)
Singapore	7 (2.3)	2 (1.2)	1 (0.7)	1 (0.9)
States				
Connecticut	s 53 (11.3)	s 14 (6.2)	s 5 (3.9)	s 6 (4.5)
Idaho	r 62 (9.7)	r 29 (7.3)	r 13 (3.5)	r 2 (0.1)
Illinois	42 (7.2)	11 (4.6)	6 (3.3)	3 (2.6)
Indiana	35 (7.1)	7 (2.0)	2 (0.1)	0 (0.0)
Maryland	r 66 (7.1)	r 25 (7.3)	r 36 (6.5)	r 16 (6.1)
Massachusetts	s 52 (9.2)	s 15 (7.2)	s 9 (4.4)	s 4 (2.7)
Michigan	46 (5.1)	r 16 (5.4)	0 (0.0)	r 2 (0.1)
Missouri	49 (7.7)	r 13 (3.9)	21 (5.9)	r 5 (3.4)
North Carolina	r 49 (6.8)	r 18 (5.8)	r 12 (5.1)	r 0 (0.1)
Oregon	67 (7.8)	23 (7.9)	4 (2.7)	2 (2.3)
Pennsylvania	53 (8.2)	21 (7.3)	13 (4.0)	9 (4.9)
South Carolina	47 (8.9)	r 9 (4.3)	8 (4.6)	r 3 (2.5)
Texas	r 43 (5.1)	s 12 (6.3)	r 2 (2.5)	s 0 (0.0)
Districts and Consortia				
Academy School Dist. #20, CO	25 (0.3)	0 (0.0)	0 (0.0)	0 (0.0)
Chicago Public Schools, IL	s 30 (12.5)	s 0 (0.0)	s 0 (0.0)	s 0 (0.0)
Delaware Science Coalition, DE	r 83 (0.9)	r 13 (0.7)	r 16 (1.9)	r 10 (0.6)
First in the World Consort., IL	r 37 (1.0)	r 0 (0.1)	r 0 (0.1)	r 0 (0.1)
Fremont/Lincoln/WestSide PS, NE	r 51 (1.6)	s 24 (1.1)	r 43 (1.8)	s 0 (0.0)
Guilford County, NC	r 46 (1.2)	s 6 (0.5)	r 9 (0.4)	s 10 (0.5)
Jersey City Public Schools, NJ	36 (1.3)	r 19 (1.0)	35 (1.3)	r 9 (0.8)
Miami-Dade County PS, FL	x x	x x	x x	x x
Michigan Invitational Group, MI	50 (1.5)	14 (0.7)	12 (0.8)	0 (0.0)
Montgomery County, MD	s 48 (8.8)	s 23 (11.1)	s 28 (14.9)	x x
Naperville Sch. Dist. #203, IL	21 (1.0)	0 (0.0)	0 (0.0)	0 (0.0)
Project SMART Consortium, OH	r 61 (1.6)	s 26 (1.0)	r 16 (0.8)	s 18 (0.9)
Rochester City Sch. Dist., NY	r 100 (0.0)	s 36 (1.7)	r 50 (1.7)	s 0 (0.0)
SW Math/Sci. Collaborative, PA	52 (9.4)	14 (6.3)	22 (7.7)	4 (3.3)
International Avg. (All Countries)	16 (0.5)	14 (0.5)	4 (0.3)	9 (0.4)

SOURCE: IEA Third International Mathematics and Science Study (TIMSS), 1998-1999.

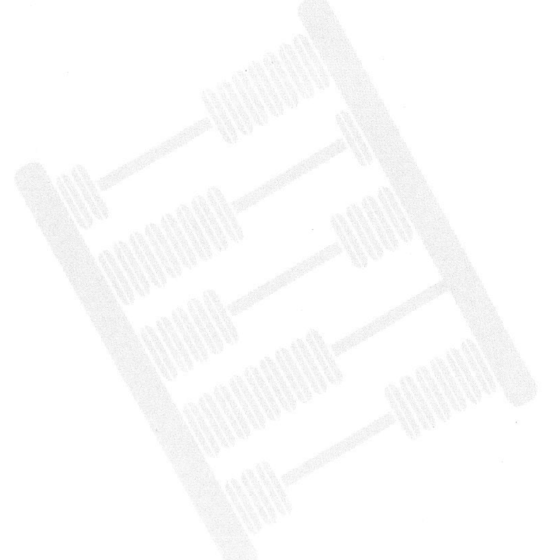

REFERENCE

1 Students' Backgrounds and Attitudes Towards Mathematics

Exhibit R1.1 Educational Aids in the Home: Dictionary, Study Desk/Table, and Computer

8th Grade Mathematics

	Have All Three Educational Aids		Do Not Have All Three Educational Aids		Percentage of Students		
	Percent of Students	Average Achievement	Percent of Students	Average Achievement	Have Dictionary	Have Study Desk/Table for Own Use	Have Computer
Countries							
United States	74 (1.3)	518 (3.7)	26 (1.3)	463 (4.3)	97 (0.3)	90 (0.5)	80 (1.2)
Belgium (Flemish)	82 (1.2)	567 (3.3)	18 (1.2)	520 (8.6)	98 (0.7)	96 (0.6)	86 (1.0)
Canada	78 (0.8)	537 (2.6)	22 (0.8)	510 (3.1)	98 (0.2)	91 (0.6)	85 (0.8)
Chinese Taipei	61 (1.1)	608 (3.8)	39 (1.1)	551 (4.4)	98 (0.2)	94 (0.4)	63 (1.0)
Czech Republic	43 (1.2)	541 (4.5)	57 (1.2)	504 (4.6)	94 (0.8)	91 (0.7)	47 (1.2)
England	79 (0.9)	507 (4.0)	21 (0.9)	461 (6.0)	98 (0.3)	92 (0.6)	85 (0.8)
Hong Kong, SAR	57 (1.3)	592 (4.1)	43 (1.3)	571 (4.9)	99 (0.1)	75 (0.9)	72 (1.3)
Italy	59 (1.1)	492 (4.0)	41 (1.1)	461 (4.2)	98 (0.3)	93 (0.6)	63 (1.0)
Japan	52 (1.0)	592 (2.3)	48 (1.0)	566 (2.3)	99 (0.1)	97 (0.2)	52 (0.9)
Korea, Rep. of	65 (0.9)	602 (1.7)	35 (0.9)	561 (3.0)	99 (0.2)	96 (0.2)	67 (0.9)
Netherlands	94 (1.0)	543 (7.2)	6 (1.0)	509 (8.7)	100 (0.2)	99 (0.2)	96 (1.0)
Russian Federation	19 (1.2)	537 (6.6)	81 (1.2)	524 (6.3)	88 (1.3)	92 (0.8)	22 (1.2)
Singapore	75 (1.4)	615 (6.1)	25 (1.4)	573 (7.1)	99 (0.2)	92 (0.5)	80 (1.3)
States							
Connecticut	82 (2.0)	523 (8.5)	18 (2.0)	466 (10.2)	97 (0.3)	92 (0.9)	88 (1.7)
Idaho	75 (2.3)	507 (6.6)	25 (2.3)	464 (9.4)	94 (0.9)	90 (0.9)	82 (2.1)
Illinois	75 (2.1)	523 (6.9)	25 (2.1)	470 (6.1)	98 (0.5)	91 (0.8)	80 (2.1)
Indiana	74 (2.0)	525 (7.3)	26 (2.0)	487 (7.8)	97 (0.4)	90 (1.2)	81 (1.5)
Maryland	80 (1.6)	506 (5.8)	20 (1.6)	452 (7.6)	98 (0.3)	91 (0.9)	86 (1.4)
Massachusetts	82 (1.8)	522 (5.7)	18 (1.8)	475 (8.0)	98 (0.3)	93 (0.7)	87 (1.6)
Michigan	79 (1.9)	527 (6.5)	21 (1.9)	481 (9.8)	98 (0.3)	90 (0.9)	85 (1.7)
Missouri	69 (2.0)	504 (5.2)	31 (2.0)	461 (5.9)	96 (0.6)	90 (0.6)	76 (1.8)
North Carolina	68 (2.0)	510 (7.1)	32 (2.0)	463 (6.6)	97 (0.4)	89 (0.9)	74 (1.8)
Oregon	79 (2.0)	526 (5.1)	21 (2.0)	473 (7.7)	97 (0.6)	91 (1.0)	86 (1.7)
Pennsylvania	78 (2.4)	518 (6.0)	22 (2.4)	472 (6.4)	98 (0.7)	91 (1.1)	83 (2.0)
South Carolina	67 (2.2)	519 (7.1)	33 (2.2)	468 (7.6)	97 (0.4)	89 (1.0)	75 (2.2)
Texas	65 (3.6)	546 (6.7)	35 (3.6)	469 (9.6)	95 (0.7)	86 (1.7)	73 (3.3)
Districts and Consortia							
Academy School Dist. #20, CO	92 (0.8)	532 (2.0)	8 (0.8)	495 (8.4)	99 (0.3)	96 (0.6)	96 (0.5)
Chicago Public Schools, IL	54 (1.9)	474 (7.2)	46 (1.9)	450 (5.9)	98 (0.5)	85 (1.5)	61 (1.7)
Delaware Science Coalition, DE	76 (2.1)	492 (9.3)	24 (2.1)	443 (7.9)	97 (0.6)	90 (1.1)	82 (1.6)
First in the World Consort., IL	91 (1.2)	564 (5.8)	9 (1.2)	520 (12.7)	98 (0.3)	95 (1.2)	96 (0.6)
Fremont/Lincoln/WestSide PS, NE	77 (1.8)	503 (8.6)	23 (1.8)	439 (9.9)	96 (0.9)	92 (1.0)	81 (1.6)
Guilford County, NC	76 (1.8)	527 (7.1)	24 (1.8)	470 (9.8)	98 (0.5)	92 (1.1)	81 (1.6)
Jersey City Public Schools, NJ	49 (2.8)	494 (11.0)	51 (2.8)	460 (6.1)	96 (0.7)	81 (1.4)	58 (2.3)
Miami-Dade County PS, FL	58 (3.0)	445 (10.3)	42 (3.0)	391 (8.7)	95 (0.8)	84 (1.4)	66 (2.8)
Michigan Invitational Group, MI	82 (1.2)	540 (5.5)	18 (1.2)	500 (9.6)	97 (0.5)	91 (1.0)	89 (1.6)
Montgomery County, MD	86 (1.9)	548 (4.1)	14 (1.9)	476 (7.8)	99 (0.4)	93 (0.9)	91 (1.4)
Naperville Sch. Dist. #203, IL	96 (0.6)	570 (2.7)	4 (0.6)	551 (13.8)	99 (0.3)	97 (0.5)	98 (0.4)
Project SMART Consortium, OH	76 (1.5)	530 (8.1)	24 (1.5)	492 (5.9)	98 (0.6)	91 (1.1)	83 (1.2)
Rochester City Sch. Dist., NY	52 (2.5)	458 (9.1)	48 (2.5)	436 (6.2)	94 (0.7)	83 (1.4)	61 (2.3)
SW Math/Sci. Collaborative, PA	75 (2.1)	530 (6.6)	25 (2.1)	476 (10.0)	98 (0.4)	90 (0.9)	82 (1.9)
International Avg. (All Countries)	41 (0.2)	516 (1.2)	59 (0.2)	471 (0.8)	90 (0.1)	86 (0.1)	45 (0.2)

SOURCE: IEA Third International Mathematics and Science Study (TIMSS), 1998-1999.

Background data provided by students.

States in *italics* did not fully satisfy guidelines for sample participation rates (see Appendix A for details).

( ) Standard errors appear in parentheses. Because results are rounded to the nearest whole number, some totals may appear inconsistent.

Exhibit R1.2 Number of Books in the Home

8th Grade Mathematics

	Three or More Bookcases (More Than 200 Books)		About Two Bookcases (101-200 Books)		About One Bookcase (26-100 Books)		About One Shelf (11-25 Books)		None or Very Few (0-10 Books)	
	Percent of Students	Average Achievement	Percent of Students	Average Achievement	Percent of Students	Average Achievement	Percent of Students	Average Achievement	Percent of Students	Average Achievement
Countries										
United States	28 (1.2)	537 (5.5)	22 (0.6)	523 (3.5)	29 (0.8)	495 (3.1)	14 (0.7)	461 (5.0)	8 (0.6)	439 (5.2)
Belgium (Flemish)	14 (0.8)	580 (5.1)	14 (0.6)	578 (7.3)	31 (1.3)	569 (6.1)	21 (0.7)	549 (5.1)	19 (1.3)	523 (4.9)
Canada	31 (0.9)	543 (3.5)	24 (0.8)	536 (3.5)	28 (0.7)	527 (3.3)	11 (0.5)	507 (5.1)	5 (0.3)	510 (5.9)
Chinese Taipei	16 (0.8)	637 (5.8)	12 (0.5)	629 (5.7)	31 (0.7)	599 (4.4)	23 (0.7)	563 (4.0)	17 (0.9)	513 (4.4)
Czech Republic	28 (1.4)	539 (5.0)	30 (1.4)	532 (5.6)	34 (1.1)	506 (5.1)	7 (0.8)	472 (7.2)	1 (0.2)	~ ~
England	26 (1.2)	537 (6.1)	23 (0.8)	505 (4.9)	32 (1.1)	488 (3.8)	13 (0.8)	456 (5.8)	7 (0.7)	438 (9.7)
Hong Kong, SAR	8 (0.5)	588 (8.8)	10 (0.5)	590 (7.4)	27 (0.7)	592 (4.3)	27 (0.7)	584 (5.0)	28 (0.9)	568 (4.3)
Italy	20 (0.9)	505 (5.6)	15 (0.7)	495 (4.9)	28 (0.9)	487 (4.5)	25 (0.9)	460 (5.4)	12 (0.8)	444 (7.0)
Japan	18 (0.7)	605 (4.6)	18 (0.6)	598 (4.0)	31 (0.7)	577 (2.7)	19 (0.6)	565 (4.3)	14 (0.6)	549 (4.8)
Korea, Rep. of	20 (0.8)	625 (2.9)	23 (0.6)	605 (3.1)	36 (0.7)	581 (2.2)	10 (0.5)	550 (3.8)	10 (0.4)	527 (5.1)
Netherlands	24 (1.8)	564 (8.5)	23 (1.2)	551 (8.1)	31 (1.1)	540 (8.2)	15 (1.4)	512 (9.6)	8 (1.4)	499 (11.1)
Russian Federation	23 (1.5)	556 (6.3)	29 (1.1)	539 (5.5)	31 (1.3)	517 (5.3)	13 (1.0)	485 (9.0)	4 (0.5)	460 (16.2)
Singapore	12 (0.6)	618 (8.3)	14 (0.7)	627 (9.0)	40 (1.1)	613 (6.1)	22 (1.0)	586 (6.8)	12 (0.8)	572 (7.2)
States										
Connecticut	35 (2.7)	544 (8.4)	23 (0.9)	519 (8.5)	25 (1.3)	507 (7.8)	10 (1.4)	461 (9.6)	8 (1.4)	436 (14.6)
Idaho	32 (1.6)	518 (8.0)	23 (1.1)	511 (6.1)	27 (1.4)	491 (7.2)	11 (1.2)	453 (9.7)	7 (1.0)	412 (11.2)
Illinois	29 (2.5)	542 (7.6)	23 (0.9)	522 (7.1)	30 (1.6)	499 (6.2)	12 (1.1)	464 (7.5)	6 (0.8)	452 (6.0)
Indiana	30 (2.2)	542 (7.8)	23 (1.0)	527 (7.8)	28 (1.2)	508 (6.6)	11 (1.3)	479 (6.7)	8 (1.0)	447 (9.5)
Maryland	31 (1.8)	531 (6.1)	23 (0.8)	508 (6.1)	27 (1.0)	481 (5.9)	13 (0.8)	452 (8.9)	7 (0.8)	433 (10.7)
Massachusetts	32 (1.9)	543 (6.2)	23 (1.1)	518 (5.7)	27 (1.1)	508 (5.8)	11 (1.1)	476 (7.6)	7 (1.1)	445 (8.6)
Michigan	36 (1.9)	544 (7.4)	24 (1.0)	523 (7.0)	26 (0.9)	508 (7.1)	10 (1.1)	467 (8.5)	5 (0.7)	461 (9.7)
Missouri	26 (1.6)	511 (6.9)	21 (1.3)	506 (5.3)	31 (1.2)	490 (6.1)	13 (0.8)	461 (9.3)	10 (0.8)	443 (8.4)
North Carolina	23 (1.8)	525 (8.3)	24 (0.9)	517 (8.0)	32 (1.3)	486 (6.4)	15 (1.1)	463 (6.1)	7 (0.7)	439 (8.6)
Oregon	33 (2.1)	546 (6.9)	23 (1.0)	528 (5.2)	27 (1.1)	499 (5.2)	10 (1.4)	472 (9.2)	6 (0.8)	436 (11.0)
Pennsylvania	28 (2.2)	537 (8.6)	25 (0.8)	521 (5.9)	30 (1.7)	496 (5.7)	11 (1.0)	468 (8.1)	6 (0.7)	448 (7.3)
South Carolina	23 (1.3)	540 (10.4)	21 (1.1)	524 (8.1)	30 (1.1)	499 (7.3)	16 (0.9)	460 (8.8)	9 (0.9)	439 (4.4)
Texas	20 (2.1)	575 (7.0)	19 (1.5)	543 (6.7)	30 (1.6)	524 (7.9)	16 (1.4)	479 (9.1)	15 (2.1)	448 (9.0)
Districts and Consortia										
Academy School Dist. #20, CO	46 (1.2)	544 (2.2)	25 (1.2)	526 (4.1)	21 (1.1)	516 (4.9)	5 (0.5)	497 (10.2)	3 (0.5)	461 (13.5)
Chicago Public Schools, IL	17 (2.6)	474 (9.2)	18 (1.6)	481 (6.9)	35 (1.8)	463 (7.0)	21 (1.8)	451 (5.7)	10 (1.2)	435 (10.1)
Delaware Science Coalition, DE	28 (2.1)	518 (9.2)	21 (1.5)	502 (11.0)	27 (1.5)	481 (7.1)	14 (1.3)	435 (8.3)	10 (1.3)	399 (10.6)
First in the World Consort., IL	41 (2.2)	570 (8.5)	28 (2.0)	570 (7.6)	23 (1.7)	555 (8.2)	5 (0.9)	493 (11.8)	3 (0.9)	478 (20.1)
Fremont/Lincoln/WestSide PS, NE	32 (1.7)	504 (9.6)	23 (1.0)	513 (10.3)	27 (2.2)	485 (8.0)	8 (0.8)	436 (13.5)	10 (1.2)	436 (12.0)
Guilford County, NC	29 (2.3)	555 (8.0)	25 (1.1)	523 (8.8)	29 (1.7)	493 (8.4)	12 (1.8)	478 (10.4)	5 (0.9)	439 (13.5)
Jersey City Public Schools, NJ	12 (1.4)	488 (17.4)	16 (1.3)	497 (11.1)	33 (1.9)	489 (8.5)	23 (1.8)	466 (8.7)	16 (1.8)	434 (7.2)
Miami-Dade County PS, FL	14 (2.6)	470 (19.3)	14 (1.3)	456 (9.8)	31 (1.2)	428 (8.5)	25 (2.1)	403 (10.7)	17 (1.8)	378 (11.7)
Michigan Invitational Group, MI	37 (2.7)	545 (7.1)	26 (2.0)	535 (6.4)	27 (1.8)	525 (7.4)	6 (0.8)	513 (11.8)	4 (0.7)	481 (13.2)
Montgomery County, MD	41 (2.3)	563 (5.0)	21 (1.8)	554 (8.3)	24 (1.2)	520 (4.7)	8 (1.2)	486 (8.6)	6 (0.9)	454 (11.3)
Naperville Sch. Dist. #203, IL	49 (1.4)	579 (4.5)	28 (1.2)	566 (3.6)	18 (1.1)	560 (5.1)	4 (0.5)	542 (7.2)	1 (0.3)	~ ~
Project SMART Consortium, OH	26 (2.3)	544 (11.4)	24 (1.3)	532 (9.5)	32 (1.3)	519 (7.2)	11 (1.4)	493 (6.9)	8 (0.9)	454 (7.6)
Rochester City Sch. Dist., NY	17 (2.1)	478 (13.9)	15 (1.0)	471 (10.1)	28 (1.6)	452 (7.3)	21 (1.9)	423 (7.1)	19 (1.5)	419 (8.0)
SW Math/Sci. Collaborative, PA	28 (2.5)	548 (7.6)	23 (1.2)	532 (7.3)	31 (1.9)	510 (6.9)	11 (1.3)	474 (10.4)	6 (1.3)	433 (12.1)
International Avg. (All Countries)	18 (0.2)	515 (1.3)	16 (0.1)	509 (1.1)	29 (0.2)	492 (0.8)	22 (0.1)	464 (0.9)	14 (0.2)	443 (1.6)

SOURCE: IEA Third International Mathematics and Science Study (TIMSS), 1998-1999.

Background data provided by students.

States in *italics* did not fully satisfy guidelines for sample participation rates (see Appendix A for details).

( ) Standard errors appear in parentheses. Because results are rounded to the nearest whole number, some totals may appear inconsistent.

A tilde (~) indicates insufficient data to report achievement.

Exhibit R1.3 Highest Level of Education of Either Parent*

8th Grade Mathematics

	Finished University[1]		Finished Upper Secondary School But Not University[2]		Finished Primary School But Not Upper Secondary School[3]		Did Not Finish Primary School[4]		Do Not Know	
	Percent of Students	Average Achievement	Percent of Students	Average Achievement	Percent of Students	Average Achievement	Percent of Students	Average Achievement	Percent of Students	Average Achievement
Countries										
United States	35 (1.7)	535 (4.9)	46 (1.3)	496 (3.2)	5 (0.4)	456 (5.4)	1 (0.2)	~ ~	13 (0.7)	468 (5.6)
Belgium (Flemish)	16 (1.0)	595 (6.9)	45 (0.9)	568 (3.8)	10 (0.7)	540 (6.3)	0 (0.1)	~ ~	29 (1.0)	534 (4.2)
Canada	45 (1.3)	543 (3.5)	34 (1.0)	530 (2.8)	6 (0.5)	509 (8.1)	0 (0.1)	~ ~	15 (0.7)	506 (4.0)
Chinese Taipei	15 (1.0)	635 (6.5)	64 (0.8)	590 (3.8)	14 (0.7)	550 (4.8)	1 (0.1)	~ ~	7 (0.4)	527 (7.5)
Czech Republic	22 (1.2)	555 (6.2)	46 (1.3)	527 (4.5)	21 (1.2)	503 (5.5)	0 (0.0)	~ ~	11 (0.9)	480 (7.1)
England	– –	– –	– –	– –	– –	– –	– –	– –	– –	– –
Hong Kong, SAR	7 (0.7)	607 (8.5)	38 (1.0)	591 (4.3)	32 (0.9)	583 (4.2)	9 (0.7)	558 (7.6)	13 (0.6)	568 (5.7)
Italy	10 (0.8)	513 (6.4)	45 (1.3)	499 (3.6)	40 (1.5)	455 (4.5)	2 (0.3)	~ ~	3 (0.4)	453 (12.0)
Japan	– –	– –	– –	– –	– –	– –	– –	– –	– –	– –
Korea, Rep. of	25 (1.0)	620 (2.5)	48 (0.8)	587 (2.7)	14 (0.5)	566 (3.2)	5 (0.4)	559 (6.1)	8 (0.4)	545 (4.3)
Netherlands	12 (1.1)	572 (9.8)	53 (2.4)	553 (6.8)	7 (1.0)	518 (12.6)	1 (0.5)	~ ~	27 (2.1)	516 (9.1)
Russian Federation	33 (1.4)	551 (6.4)	47 (1.2)	522 (6.4)	5 (0.5)	488 (10.2)	1 (0.2)	~ ~	14 (0.9)	504 (6.1)
Singapore	11 (1.0)	652 (8.4)	51 (1.0)	608 (5.8)	23 (1.0)	589 (6.9)	4 (0.3)	579 (9.2)	12 (0.6)	588 (7.0)
States										
Connecticut	41 (2.8)	540 (10.8)	42 (2.1)	498 (7.3)	4 (0.7)	458 (12.7)	0 (0.2)	~ ~	13 (0.9)	492 (10.5)
Idaho	31 (2.1)	517 (7.3)	46 (1.9)	500 (6.0)	6 (1.0)	446 (11.2)	1 (0.2)	~ ~	16 (0.7)	469 (11.6)
Illinois	34 (2.8)	542 (8.1)	47 (2.1)	502 (6.2)	5 (0.8)	461 (8.3)	0 (0.2)	~ ~	14 (1.1)	476 (7.3)
Indiana	36 (2.8)	543 (8.6)	48 (2.9)	505 (6.2)	5 (0.8)	474 (9.9)	0 (0.1)	~ ~	11 (1.1)	487 (11.1)
Maryland	39 (2.0)	523 (7.2)	43 (1.8)	484 (5.9)	4 (0.5)	444 (6.9)	0 (0.1)	~ ~	14 (0.8)	465 (8.3)
Massachusetts	38 (2.2)	541 (6.4)	43 (1.3)	503 (6.6)	4 (0.7)	472 (10.6)	1 (0.2)	~ ~	14 (1.0)	486 (7.7)
Michigan	40 (3.2)	543 (8.1)	47 (2.7)	510 (6.1)	2 (0.4)	~ ~	0 (0.0)	~ ~	11 (0.9)	479 (8.7)
Missouri	29 (1.7)	513 (8.1)	50 (1.9)	486 (5.8)	6 (1.1)	468 (9.5)	0 (0.1)	~ ~	14 (0.9)	468 (7.3)
North Carolina	25 (3.1)	525 (15.6)	59 (4.1)	492 (2.8)	5 (0.3)	445 (23.4)	0 (0.1)	~ ~	10 (0.9)	466 (5.2)
Oregon	39 (2.5)	544 (5.2)	46 (2.3)	506 (6.4)	5 (0.6)	447 (12.9)	1 (0.2)	~ ~	9 (0.7)	478 (9.3)
Pennsylvania	34 (2.4)	531 (11.2)	49 (2.0)	501 (5.1)	3 (0.5)	460 (12.8)	0 (0.2)	~ ~	14 (1.1)	487 (7.4)
South Carolina	30 (2.1)	535 (8.8)	52 (1.9)	493 (7.0)	6 (0.7)	467 (8.6)	0 (0.0)	~ ~	12 (1.1)	481 (8.1)
Texas	37 (2.3)	558 (4.5)	38 (0.9)	510 (18.6)	9 (1.4)	476 (15.2)	1 (0.4)	~ ~	15 (1.4)	481 (16.6)
Districts and Consortia										
Academy School Dist. #20, CO	59 (1.7)	543 (2.7)	28 (1.3)	511 (4.7)	1 (0.2)	~ ~	0 (0.1)	~ ~	12 (1.0)	508 (6.5)
Chicago Public Schools, IL	24 (3.3)	470 (12.1)	47 (2.3)	470 (5.4)	11 (1.6)	454 (8.4)	2 (0.6)	~ ~	17 (1.4)	438 (7.2)
Delaware Science Coalition, DE	35 (2.6)	513 (13.4)	48 (2.0)	472 (8.7)	4 (0.7)	435 (13.1)	1 (0.4)	~ ~	12 (1.1)	438 (13.5)
First in the World Consort., IL	58 (4.0)	577 (8.6)	28 (2.4)	548 (11.3)	3 (0.7)	518 (15.4)	1 (0.4)	~ ~	11 (1.4)	513 (11.2)
Fremont/Lincoln/WestSide PS, NE	39 (2.1)	509 (10.9)	40 (2.5)	488 (9.5)	4 (0.8)	429 (11.4)	0 (0.1)	~ ~	17 (2.2)	456 (13.9)
Guilford County, NC	39 (3.4)	543 (14.3)	49 (2.9)	498 (9.4)	4 (0.7)	467 (13.5)	0 (0.2)	~ ~	9 (1.0)	497 (14.7)
Jersey City Public Schools, NJ	23 (2.0)	487 (13.1)	48 (2.0)	481 (8.1)	9 (0.9)	461 (10.6)	1 (0.4)	~ ~	19 (1.3)	468 (9.5)
Miami-Dade County PS, FL	28 (2.5)	446 (13.5)	42 (1.7)	430 (8.4)	8 (0.7)	389 (11.5)	1 (0.2)	~ ~	21 (1.4)	393 (8.8)
Michigan Invitational Group, MI	41 (2.7)	549 (9.3)	47 (2.0)	528 (7.5)	1 (0.3)	~ ~	0 (0.2)	~ ~	11 (1.3)	500 (10.3)
Montgomery County, MD	54 (2.6)	566 (5.5)	27 (1.9)	510 (6.2)	4 (0.9)	463 (11.1)	1 (0.2)	~ ~	14 (1.2)	513 (8.1)
Naperville Sch. Dist. #203, IL	71 (1.6)	580 (5.6)	19 (1.3)	550 (4.0)	1 (0.2)	~ ~	0 (0.2)	~ ~	9 (0.9)	533 (8.6)
Project SMART Consortium, OH	36 (2.5)	541 (10.7)	46 (2.1)	520 (7.7)	3 (0.7)	471 (14.4)	0 (0.2)	~ ~	14 (1.4)	485 (9.0)
Rochester City Sch. Dist., NY	22 (1.7)	458 (13.2)	48 (2.1)	449 (9.6)	8 (0.9)	435 (10.2)	1 (0.2)	~ ~	21 (2.0)	436 (9.0)
SW Math/Sci. Collaborative, PA	37 (2.8)	545 (10.4)	48 (2.3)	507 (7.2)	3 (0.5)	464 (10.9)	0 (0.0)	~ ~	13 (0.9)	485 (9.8)
International Avg. (All Countries)	20 (0.2)	525 (1.4)	41 (0.2)	492 (0.8)	21 (0.2)	460 (1.1)	6 (0.1)	418 (3.0)	12 (0.1)	463 (1.3)

SOURCE: IEA Third International Mathematics and Science Study (TIMSS), 1998-1999.

Background data provided by students

* Response categories were defined by each country to conform to their own educational system and may not be strictly comparable across countries. See Reference Exhibit R1.4 for country modifications to the definitions of educational levels.

1 In most countries, defined as completion of at least a 4-year degree program at a university or an equivalent institute of higher education.

2 Finished upper secondary school with or without some tertiary education not equivalent to a university degree. In most countries, finished secondary corresponds to completion of an upper-secondary track terminating after 11 to 13 years of schooling (ISCED level 3 vocational, apprenticeship or academic tracks).

3 Finished primary school or attended some secondary school not equivalent to completion of upper secondary.

4 Some primary school or did not go to school.

States in *italics* did not fully satisfy guidelines for sample participation rates (see Appendix A for details).

( ) Standard errors appear in parentheses. Because results are rounded to the nearest whole number, some totals may appear inconsistent.

A dash (–) indicates data are not available. A tilde (~) indicates insufficient data to report achievement.

Exhibit R1.4 Country Modifications to the Definitions of Educational Levels for Parents' Education or Students' Expectations for Finishing School*

8th Grade Mathematics

	Finished University	Finished Upper Secondary School But Not University	
		Post-Secondary Level	Upper-Secondary Level[1]
Internationally Defined Level	Finished University	Some Vocational-Technical Education After Secondary School or Some University	Finished Secondary School
United States (P) ‡	Completed Bachelor's Degree at College or University	Some Vocational-Technical Education After Secondary School or Some Community College, College or University Courses	Finish High School
United States (S) §	Finish community college, college or university	Some Vocational-Technical Education After Secondary School or Some Community College, College or University Courses	Finish High School
Australia §			
Belgium (Flemish) §		Post-Secondary Tertiary Higher Education Outside University or Some Years of University	Finish Higher Secondary School
Canada	Finish University or College	Some Vocational-Technical Education After Secondary School or Some University or College	
Chile			
Cyprus §	University Degree		Finish Upper Secondary
Czech Republic (P) §‡	Finish University (4-5 years university study)	Some Vocational-Technical Education After Secondary School or Some University	Vocational Training or Secondary With Maturita
Czech Republic (S)	Finish University (4-5 years university study)	Medium-cycle higher education or bachelor studies (3 years university study or special higher education)	Vocational Training or Secondary With Maturita
Finland			Finish secondary school (about 12 years)
Hungary §	University or College Degree	Not Included	Apprenticeship (3-year trade school) or Final Exam in Secondary School (4-year academic/vocational)
Indonesia	Completed University Degree (Sarjana 1/2/3)	Academy (3 years or less of higher education outside university - Diploma D1/D2/D3) or Some University (Did Not Complete Degree)	Finish Secondary (SMP, SMA, SMEA, STM, etc.)
Italy §	Finish University (Laurea o Dottorato di Ricerca 4-6 Year)	Vocational/Professional Course After Secondary Diploma or Some University (2-3 Year Short-Course Diploma)	Finish Secondary School With Maturita (Classical/Technical) or Vocational Training Diploma
Japan (S) 3	University or Graduate School	Vocational/Technical Education After Secondary or 2-year college	Upper secondary
Korea, Rep. of §			
Latvia (LSS) §	Higher Education (5 years)	Vocational School (Post-Secondary) or Technikum (3 years) or Some Higher Education	Finish Secondary or Vocational School (11 years)
Lithuania §	University or Other Higher Education	Vocational or Agricultural School or College (Technical, Art, Music)	
Netherlands	University With Diploma	Vocational/Technical Education After Secondary (bv.heao, hts, pedagogical academy) or Some Years At University (Without Diploma)	Finish Secondary School With Diploma
New Zealand (P) ‡	University or Teachers' College (College of Education)	Vocational/Polytechnic Education After Secondary School or Some University	Complete Form 6 or Form 7
New Zealand (S) §	University, College of Education (teacher training) or degree or national diploma course at polytech	Certificate course at polytech (e.g., trade certificate) or some university	Finish secondary school (complete Form 6 or Form 7)
Philippines §	Finish College/University	Some Vocational/Technical Education After High School or Some College/University	Finish High School
Romania §	Finish University (facultate)	Post-Secondary Technical School or Did Not Complete University	Finish Senior Secondary (liceu)
Singapore §		Finish JC/Pre-U or Polytechnic or Some Other Vocational/Technical Education After Secondary (e.g., ITE, VITB)' [includes GCE 'A' level, which is 2 years additional schooling beyond completion of secondary.]	Finish Secondary School
Slovenia (S) §‡			Finish gymnasium or secondary school
South Africa §		Finish Technikon or Some University	Finish Secondary
Thailand §	Graduate level (Finish Tertiary Education, 4 years)	Diploma/Undergraduate Level (higher certificate, 2 years)	Finish Academic or Vocational/Technical Upper-Secondary Track
Tunisia	Bachelor's Degree (BA)		

SOURCE: IEA Third International Mathematics and Science Study (TIMSS), 1998-1999.

National educational level is the same as the internationally-defined level

* Educational levels were translated and defined in most countries to be comparable to the internationally-defined levels. Countries that used modified response options to conform to their national education systems are indicated to aid in the interpretation of the reporting categories in Exhibits 4.5 and R1.3. National modifications pertain to both the parents' education and student's expectations questions unless otherwise indicated.

1 Upper-secondary corresponds to ISCED level 3 tracks terminating after 11 to 13 years in most countries. (Education at a Glance, OECD, 1995.)

2 Primary school or lower educational levels were included only in the parents' education question.

3 Japan administered the question pertaining to students' expectations but not the question pertaining to parents' education.

§ Some educational levels modified from 1995.

‡ Educational levels differ for the parent's education (P) question and the students' expectations (S) question.

Exhibit R1.4 (Continued) | Country Modifications to the Definitions of Educational Levels for Parents' Education or Students' Expectations for Finishing School*

Finished Primary School But Not Upper Secondary School		Did Not Finish Primary School[2]	
Lower-Secondary Level	**Primary Level[2]**		
Finished Some Secondary School	Finished Primary School	Some Primary School or Did Not Go to School	**Internationally Defined Level**
Some High School	Finish Elementary School	Finish elementary school or did not go to school	United States (P)
Some High School			United States (S)
		Less Than Year 6 in Primary School	Australia
Finish Lower Secondary School	Finish Basic School	Some Years of Basic School or Did Not Go to School	Belgium (Flemish)
			Canada
	Finish Primary School (grade 8)		Chile
Finish Lower Secondary (Gymnasium - grade 9)			Cyprus
Vocational Training or Secondary School Without Maturita		Not Included	Czech Republic (P)
Vocational Training or Secondary School Without Maturita			Czech Republic (S)
Some Secondary School (10 - 11 years)	Finish Primary School (about 9 years)	Did Not Go to School, Primary School or Part of Lower Secondary (< 9 years)	Finland
Finish General School (grade 8)	Some General School	Not Included	Hungary
	Finish Primary School (SD)		Indonesia
Finish Middle School			Italy
Lower Secondary			Japan (S)
Some High School	Finish Middle School	Some middle school or did not go to school	Korea, Rep. of
			Latvia (LSS)
	Finish Basic School (grade 10)	Some Basic School or Did Not Go to School	Lithuania
Some Years of Secondary School (mavo, havo, vwo) without Diploma	Finish Primary School (grade 8)		Netherlands
			New Zealand (P)
			New Zealand (S)
Some High School	Finish Elementary School	Some Elementary School or Did Not Go to School	Philippines
Did Not Complete Senior Secondary	Finish Junior Secondary (Gymnasium - grade 8)	Did Not Finish Grade 8 or Did Not Go to School	Romania
			Singapore
			Slovenia (S)
			South Africa
Finish Lower Secondary School	Finish Upper Primary School	Finish Lower Primary School or Did Not Go to School	Thailand
			Tunisia

SOURCE: IEA Third International Mathematics and Science Study (TIMSS), 1998-1999.

National educational level is the same as the internationally-defined level

Exhibit R1.5 Students' Perception of the Importance of Various Activities

8th Grade Mathematics

	Percentage of Students Agreeing That It Is Important to Do Each Activity				
	Do Well in Mathematics	Do Well in Science	Do Well in Language	Have Time to Have Fun	Be Good at Sports
Countries					
United States	97 (0.3)	96 (0.3)	96 (0.3)	99 (0.2)	84 (0.6)
Belgium (Flemish)	98 (0.3)	91 (0.8)	96 (0.4)	98 (0.4)	77 (0.9)
Canada	98 (0.2)	95 (0.4)	97 (0.5)	99 (0.2)	82 (0.6)
Chinese Taipei	89 (0.5)	89 (0.5)	89 (0.5)	99 (0.1)	94 (0.3)
Czech Republic	98 (0.3)	93 (0.6)	97 (0.4)	97 (0.4)	82 (1.0)
England	99 (0.2)	97 (0.3)	99 (0.2)	98 (0.3)	79 (0.9)
Hong Kong, SAR	95 (0.4)	86 (0.7)	96 (0.4)	97 (0.3)	84 (0.6)
Italy	97 (0.4)	94 (0.5)	97 (0.3)	98 (0.3)	89 (0.6)
Japan	88 (0.5)	83 (0.7)	89 (0.6)	99 (0.2)	82 (0.6)
Korea, Rep. of	90 (0.4)	87 (0.5)	89 (0.4)	92 (0.3)	88 (0.5)
Netherlands	98 (0.3)	94 (0.9)	99 (0.3)	98 (0.3)	76 (1.5)
Russian Federation	97 (0.4)	96 (0.3)	97 (0.4)	98 (0.3)	90 (0.6)
Singapore	99 (0.2)	98 (0.2)	100 (0.1)	93 (0.6)	90 (0.5)
States					
Connecticut	97 (0.5)	96 (0.6)	97 (0.4)	99 (0.3)	82 (1.0)
Idaho	96 (0.4)	94 (0.5)	95 (0.6)	99 (0.2)	86 (0.8)
Illinois	98 (0.3)	96 (0.5)	97 (0.4)	99 (0.2)	83 (1.0)
Indiana	97 (0.4)	96 (0.5)	96 (0.6)	99 (0.2)	82 (0.8)
Maryland	97 (0.4)	95 (0.5)	96 (0.4)	98 (0.3)	84 (0.8)
Massachusetts	97 (0.4)	96 (0.5)	96 (0.5)	99 (0.2)	82 (0.9)
Michigan	97 (0.4)	96 (0.4)	96 (0.5)	99 (0.2)	84 (1.2)
Missouri	97 (0.5)	95 (0.6)	95 (0.5)	98 (0.4)	85 (1.0)
North Carolina	99 (0.2)	97 (0.4)	99 (0.3)	99 (0.2)	87 (0.6)
Oregon	97 (0.5)	95 (0.7)	95 (0.6)	98 (0.3)	84 (1.2)
Pennsylvania	96 (1.0)	94 (0.7)	95 (0.9)	99 (0.3)	83 (0.9)
South Carolina	98 (0.4)	97 (0.4)	97 (0.3)	98 (0.3)	84 (0.8)
Texas	97 (0.4)	95 (0.6)	95 (0.5)	98 (0.7)	85 (1.1)
Districts and Consortia					
Academy School Dist. #20, CO	97 (0.4)	95 (0.6)	95 (0.6)	99 (0.3)	85 (1.0)
Chicago Public Schools, IL	99 (0.4)	95 (0.7)	97 (0.9)	95 (1.1)	83 (1.3)
Delaware Science Coalition, DE	97 (0.4)	94 (0.8)	96 (0.4)	98 (0.4)	85 (1.1)
First in the World Consort., IL	97 (0.8)	96 (0.8)	97 (0.7)	100 (0.2)	81 (1.2)
Fremont/Lincoln/WestSide PS, NE	95 (0.4)	93 (0.4)	94 (0.5)	99 (0.3)	82 (1.2)
Guilford County, NC	99 (0.4)	98 (0.3)	99 (0.4)	99 (0.4)	84 (1.5)
Jersey City Public Schools, NJ	99 (0.3)	98 (0.3)	99 (0.4)	96 (0.8)	84 (1.2)
Miami-Dade County PS, FL	97 (0.7)	97 (0.8)	98 (0.6)	97 (0.6)	85 (1.2)
Michigan Invitational Group, MI	97 (0.6)	95 (0.7)	97 (0.5)	100 (0.2)	82 (1.5)
Montgomery County, MD	97 (0.8)	94 (0.8)	96 (0.8)	99 (0.3)	83 (1.1)
Naperville Sch. Dist. #203, IL	97 (0.3)	96 (0.4)	96 (0.4)	99 (0.3)	84 (0.9)
Project SMART Consortium, OH	98 (0.5)	96 (0.6)	97 (0.4)	99 (0.3)	85 (0.8)
Rochester City Sch. Dist., NY	99 (0.5)	98 (0.7)	98 (0.5)	98 (0.4)	85 (1.7)
SW Math/Sci. Collaborative, PA	98 (0.5)	96 (0.7)	95 (0.6)	99 (0.3)	83 (1.3)
International Avg. (All Countries)	96 (0.1)	92 (0.1)	96 (0.1)	92 (0.1)	87 (0.1)

SOURCE: IEA Third International Mathematics and Science Study (TIMSS), 1998-1999.

Background data provided by students.

States in *italics* did not fully satisfy guidelines for sample participation rates (see Appendix A for details).

( ) Standard errors appear in parentheses. Because results are rounded to the nearest whole number, some totals may appear inconsistent.

Students' Perception of Their Mothers' View of the Importance of Various Activities

	Percentage of Students Agreeing That Their Mothers Think It Is Important to Do Each Activity				
	Do Well in Mathematics	Do Well in Science	Do Well in Language	Have Time to Have Fun	Be Good at Sports
Countries					
United States	98 (0.2)	98 (0.2)	98 (0.2)	93 (0.4)	76 (0.6)
Belgium (Flemish)	97 (0.4)	92 (0.6)	97 (0.5)	96 (0.5)	66 (1.6)
Canada	99 (0.1)	98 (0.3)	99 (0.2)	96 (0.4)	76 (0.8)
Chinese Taipei	95 (0.5)	95 (0.4)	93 (0.4)	95 (0.3)	91 (0.4)
Czech Republic	99 (0.2)	96 (0.5)	99 (0.3)	90 (0.7)	72 (1.1)
England	99 (0.2)	98 (0.3)	99 (0.2)	94 (0.5)	74 (1.0)
Hong Kong, SAR	96 (0.3)	87 (0.7)	97 (0.3)	82 (0.7)	73 (0.9)
Italy	99 (0.3)	97 (0.3)	99 (0.2)	95 (0.4)	84 (0.8)
Japan	92 (0.5)	87 (0.6)	92 (0.5)	94 (0.4)	82 (0.6)
Korea, Rep. of	95 (0.3)	90 (0.4)	92 (0.4)	66 (0.7)	78 (0.6)
Netherlands	98 (0.3)	94 (0.8)	98 (0.3)	97 (0.5)	59 (1.9)
Russian Federation	96 (0.4)	96 (0.4)	97 (0.4)	92 (0.4)	86 (0.7)
Singapore	99 (0.2)	98 (0.2)	98 (0.2)	76 (0.9)	80 (0.7)
States					
Connecticut	98 (0.3)	98 (0.4)	98 (0.3)	93 (0.7)	75 (1.2)
Idaho	98 (0.4)	97 (0.5)	97 (0.4)	94 (0.5)	82 (1.2)
Illinois	99 (0.2)	97 (0.4)	98 (0.3)	92 (0.9)	74 (1.2)
Indiana	99 (0.4)	98 (0.5)	98 (0.4)	95 (0.5)	74 (0.8)
Maryland	98 (0.3)	97 (0.3)	98 (0.3)	93 (0.4)	76 (1.1)
Massachusetts	98 (0.3)	98 (0.3)	98 (0.3)	93 (0.6)	73 (0.9)
Michigan	98 (0.3)	98 (0.4)	98 (0.3)	94 (0.4)	76 (1.5)
Missouri	98 (0.4)	98 (0.4)	98 (0.4)	93 (0.6)	78 (1.1)
North Carolina	99 (0.3)	98 (0.2)	99 (0.3)	94 (0.6)	80 (0.9)
Oregon	98 (0.4)	97 (0.5)	97 (0.5)	93 (0.6)	78 (1.4)
Pennsylvania	98 (0.6)	98 (0.9)	98 (0.7)	94 (0.5)	77 (1.3)
South Carolina	98 (0.4)	98 (0.4)	98 (0.3)	93 (0.8)	76 (1.3)
Texas	97 (0.4)	97 (0.5)	97 (0.5)	91 (1.1)	80 (1.3)
Districts and Consortia					
Academy School Dist. #20, CO	98 (0.3)	98 (0.4)	97 (0.4)	94 (0.7)	77 (1.1)
Chicago Public Schools, IL	98 (0.5)	96 (0.9)	97 (0.8)	85 (1.2)	72 (1.8)
Delaware Science Coalition, DE	97 (0.6)	96 (0.9)	97 (0.5)	90 (0.7)	77 (1.1)
First in the World Consort., IL	99 (0.4)	98 (0.4)	98 (0.5)	94 (0.6)	66 (2.3)
Fremont/Lincoln/WestSide PS, NE	97 (0.5)	97 (1.0)	97 (1.0)	95 (1.2)	71 (1.8)
Guilford County, NC	99 (0.3)	99 (0.3)	99 (0.3)	94 (0.6)	77 (1.4)
Jersey City Public Schools, NJ	99 (0.3)	98 (0.4)	98 (0.3)	88 (1.3)	78 (1.2)
Miami-Dade County PS, FL	97 (0.6)	98 (0.4)	98 (0.5)	88 (1.3)	79 (1.9)
Michigan Invitational Group, MI	99 (0.4)	98 (0.4)	98 (0.4)	94 (0.8)	75 (1.4)
Montgomery County, MD	98 (0.6)	97 (0.8)	98 (0.6)	92 (0.8)	74 (1.1)
Naperville Sch. Dist. #203, IL	99 (0.2)	99 (0.3)	99 (0.3)	95 (0.6)	75 (1.5)
Project SMART Consortium, OH	97 (0.5)	98 (0.5)	98 (0.4)	94 (0.8)	77 (1.8)
Rochester City Sch. Dist., NY	97 (0.7)	96 (0.7)	97 (0.8)	91 (1.0)	79 (1.9)
SW Math/Sci. Collaborative, PA	98 (0.3)	98 (0.4)	98 (0.3)	93 (0.7)	77 (1.5)
International Avg. (All Countries)	96 (0.1)	93 (0.1)	96 (0.1)	85 (0.1)	81 (0.1)

SOURCE: IEA Third International Mathematics and Science Study (TIMSS), 1998-1999.

Background data provided by students.

States in *italics* did not fully satisfy guidelines for sample participation rates (see Appendix A for details).

( ) Standard errors appear in parentheses. Because results are rounded to the nearest whole number, some totals may appear inconsistent.

Students' Perception of Their Friends' View of the Importance of Various Activities

8th Grade Mathematics

	Percentage of Students Agreeing That Their Friends Think It Is Important To Do Each Activity				
	Do Well in Mathematics	Do Well in Science	Do Well in Language	Have Time to Have Fun	Be Good at Sports
Countries					
United States	79 (0.8)	72 (0.8)	76 (1.0)	98 (0.2)	86 (0.5)
Belgium (Flemish)	81 (1.1)	66 (1.2)	77 (1.4)	98 (0.5)	76 (1.1)
Canada	84 (0.6)	72 (0.9)	82 (0.7)	99 (0.1)	84 (0.9)
Chinese Taipei	84 (0.7)	82 (0.7)	84 (0.6)	98 (0.2)	94 (0.4)
Czech Republic	84 (0.9)	68 (1.0)	83 (0.8)	97 (0.4)	83 (0.9)
England	90 (0.8)	84 (1.0)	90 (0.7)	99 (0.2)	80 (1.0)
Hong Kong, SAR	84 (0.7)	66 (1.0)	87 (0.8)	96 (0.3)	83 (0.8)
Italy	80 (0.9)	66 (1.3)	84 (0.7)	98 (0.3)	94 (0.5)
Japan	85 (0.6)	78 (0.8)	85 (0.8)	99 (0.2)	80 (0.7)
Korea, Rep. of	77 (0.7)	72 (0.8)	73 (0.8)	93 (0.3)	80 (0.8)
Netherlands	88 (1.0)	79 (1.2)	90 (0.9)	98 (0.4)	70 (1.9)
Russian Federation	89 (0.6)	83 (0.7)	89 (0.6)	97 (0.4)	87 (0.8)
Singapore	96 (0.3)	94 (0.6)	97 (0.3)	93 (0.6)	88 (0.6)
States					
Connecticut	78 (1.5)	71 (2.1)	76 (1.7)	98 (0.4)	84 (1.1)
Idaho	77 (1.8)	71 (2.2)	74 (1.5)	98 (0.4)	87 (1.1)
Illinois	80 (1.7)	70 (2.1)	75 (2.0)	98 (0.3)	86 (1.1)
Indiana	79 (1.3)	73 (1.5)	76 (1.3)	99 (0.3)	86 (0.9)
Maryland	76 (1.1)	69 (1.3)	75 (1.2)	98 (0.3)	85 (0.9)
Massachusetts	74 (1.5)	69 (1.8)	72 (1.4)	99 (0.2)	85 (0.9)
Michigan	79 (1.0)	75 (1.3)	75 (1.4)	98 (0.3)	87 (1.0)
Missouri	76 (1.3)	71 (1.4)	73 (1.3)	98 (0.4)	85 (1.2)
North Carolina	85 (1.3)	78 (1.5)	84 (1.3)	99 (0.2)	89 (1.0)
Oregon	76 (1.6)	70 (1.9)	74 (1.7)	98 (0.3)	87 (1.1)
Pennsylvania	77 (1.2)	70 (1.2)	74 (1.2)	99 (0.3)	87 (0.8)
South Carolina	83 (1.0)	74 (1.3)	82 (0.8)	98 (0.4)	87 (0.8)
Texas	77 (1.3)	70 (1.7)	74 (1.5)	98 (0.6)	87 (1.0)
Districts and Consortia					
Academy School Dist. #20, CO	77 (1.1)	74 (1.2)	75 (1.2)	99 (0.3)	86 (0.9)
Chicago Public Schools, IL	88 (1.3)	65 (2.4)	78 (2.2)	96 (0.9)	85 (1.2)
Delaware Science Coalition, DE	73 (1.6)	67 (1.8)	74 (1.3)	98 (0.6)	87 (1.1)
First in the World Consort., IL	77 (1.8)	71 (1.4)	74 (1.7)	99 (0.5)	82 (1.3)
Fremont/Lincoln/WestSide PS, NE	75 (1.4)	69 (1.1)	70 (1.1)	97 (1.1)	83 (1.6)
Guilford County, NC	88 (1.3)	82 (1.5)	87 (1.3)	99 (0.3)	87 (1.2)
Jersey City Public Schools, NJ	89 (1.3)	76 (1.6)	88 (1.2)	97 (0.7)	88 (1.0)
Miami-Dade County PS, FL	80 (1.4)	73 (1.4)	80 (1.0)	97 (0.5)	84 (1.1)
Michigan Invitational Group, MI	76 (1.6)	72 (1.8)	73 (1.4)	98 (0.8)	83 (1.8)
Montgomery County, MD	78 (1.6)	69 (1.8)	75 (1.6)	99 (0.4)	85 (1.1)
Naperville Sch. Dist. #203, IL	84 (1.1)	79 (1.2)	82 (1.1)	99 (0.3)	83 (1.0)
Project SMART Consortium, OH	76 (1.2)	73 (1.3)	74 (1.5)	99 (0.3)	85 (1.1)
Rochester City Sch. Dist., NY	82 (1.5)	79 (1.5)	79 (1.6)	97 (0.8)	85 (1.6)
SW Math/Sci. Collaborative, PA	79 (1.2)	72 (1.4)	75 (1.0)	99 (0.2)	86 (1.6)
International Avg. (All Countries)	86 (0.1)	77 (0.2)	86 (0.1)	92 (0.1)	85 (0.1)

SOURCE: IEA Third International Mathematics and Science Study (TIMSS), 1998-1999.

Background data provided by students.

States in *italics* did not fully satisfy guidelines for sample participation rates (see Appendix A for details).

( ) Standard errors appear in parentheses. Because results are rounded to the nearest whole number, some totals may appear inconsistent.

	Percentage of Students Reporting								
	To Get Desired Job			To Please Parents			To Get Into Desired Secondary School or University		
	Strongly Agree	Agree	Disagree/ Strongly Disagree	Strongly Agree	Agree	Disagree/ Strongly Disagree	Strongly Agree	Agree	Disagree/ Strongly Disagree
Countries									
United States	41 (0.8)	40 (0.7)	18 (0.6)	34 (0.8)	47 (0.7)	19 (0.6)	58 (1.2)	36 (1.0)	6 (0.3)
Belgium (Flemish)	18 (1.2)	40 (1.4)	42 (1.1)	15 (0.6)	52 (1.2)	33 (1.1)	24 (0.8)	46 (1.3)	30 (1.2)
Canada	43 (1.1)	40 (0.9)	17 (0.6)	25 (0.5)	46 (0.6)	30 (0.6)	57 (0.8)	36 (0.6)	7 (0.5)
Chinese Taipei	27 (0.7)	50 (0.8)	23 (0.9)	29 (0.8)	50 (0.6)	20 (0.7)	42 (0.9)	46 (0.7)	11 (0.5)
Czech Republic	32 (1.2)	48 (1.2)	20 (1.0)	22 (1.1)	56 (1.0)	22 (1.0)	46 (1.3)	39 (1.0)	15 (0.9)
England	36 (1.2)	41 (1.0)	23 (1.0)	21 (1.0)	41 (0.9)	38 (1.0)	43 (1.3)	42 (1.1)	15 (0.9)
Hong Kong, SAR	28 (0.8)	53 (0.8)	19 (0.6)	26 (0.7)	55 (0.7)	19 (0.7)	29 (0.8)	49 (0.8)	22 (0.8)
Italy	30 (0.7)	45 (1.0)	24 (0.8)	27 (1.0)	51 (1.0)	22 (0.9)	33 (0.7)	46 (1.0)	20 (0.9)
Japan	12 (0.5)	39 (0.7)	49 (1.0)	6 (0.4)	25 (0.7)	69 (0.8)	34 (0.8)	54 (0.7)	11 (0.7)
Korea, Rep. of	10 (0.5)	34 (0.6)	56 (0.7)	12 (0.5)	50 (0.7)	38 (0.7)	31 (0.7)	54 (0.7)	15 (0.5)
Netherlands	18 (1.2)	37 (0.9)	45 (1.3)	7 (0.8)	36 (1.2)	57 (1.3)	20 (1.2)	45 (1.1)	35 (1.6)
Russian Federation	42 (1.1)	42 (1.0)	16 (0.8)	20 (0.7)	40 (1.0)	39 (1.3)	40 (1.0)	48 (1.0)	12 (0.6)
Singapore	40 (1.0)	46 (0.8)	13 (0.6)	26 (0.8)	46 (0.6)	28 (0.8)	54 (1.1)	41 (1.0)	5 (0.4)
States									
Connecticut	36 (1.4)	43 (1.1)	21 (1.3)	32 (1.5)	47 (1.4)	20 (0.9)	54 (1.8)	41 (1.9)	6 (0.6)
Idaho	42 (1.4)	42 (1.2)	16 (0.9)	36 (1.0)	48 (1.1)	17 (0.8)	57 (1.5)	37 (1.3)	6 (0.8)
Illinois	40 (1.1)	42 (0.9)	18 (0.6)	29 (1.2)	49 (1.4)	22 (0.9)	58 (1.3)	37 (1.2)	5 (0.8)
Indiana	45 (2.2)	41 (2.3)	15 (0.8)	34 (1.6)	49 (1.4)	17 (0.9)	59 (2.3)	36 (2.0)	5 (0.5)
Maryland	43 (0.9)	40 (0.8)	16 (0.7)	36 (1.1)	46 (1.0)	18 (1.0)	59 (1.2)	35 (1.1)	6 (0.6)
Massachusetts	33 (1.1)	42 (1.2)	25 (1.4)	31 (0.8)	47 (1.1)	22 (1.1)	50 (1.2)	43 (1.1)	7 (0.5)
Michigan	38 (1.1)	44 (1.0)	18 (1.1)	30 (1.1)	49 (1.1)	20 (1.2)	55 (1.4)	40 (1.3)	5 (0.5)
Missouri	39 (1.5)	44 (1.3)	17 (1.2)	37 (1.1)	45 (1.3)	19 (1.1)	53 (1.2)	38 (1.2)	9 (0.6)
North Carolina	49 (1.2)	37 (1.2)	14 (0.8)	42 (1.5)	43 (1.1)	15 (1.1)	67 (1.5)	29 (1.4)	4 (0.3)
Oregon	39 (1.7)	43 (1.5)	18 (0.9)	32 (1.4)	52 (2.0)	17 (1.1)	52 (1.6)	41 (1.7)	7 (0.8)
Pennsylvania	36 (2.1)	44 (1.1)	20 (1.3)	30 (1.3)	49 (1.2)	21 (1.2)	52 (1.9)	40 (1.1)	8 (1.1)
South Carolina	46 (1.4)	40 (1.2)	14 (0.7)	39 (0.8)	44 (1.1)	18 (0.9)	66 (1.2)	30 (1.2)	3 (0.5)
Texas	43 (1.2)	41 (1.2)	16 (1.1)	36 (1.5)	46 (0.9)	18 (1.6)	60 (1.2)	34 (1.1)	6 (0.7)
Districts and Consortia									
Academy School Dist. #20, CO	42 (1.3)	41 (1.4)	17 (1.1)	37 (1.1)	47 (1.5)	16 (1.1)	60 (1.6)	36 (1.5)	4 (0.5)
Chicago Public Schools, IL	43 (3.1)	43 (2.2)	14 (1.5)	26 (2.0)	44 (1.9)	30 (2.0)	60 (3.4)	35 (2.9)	4 (0.9)
Delaware Science Coalition, DE	45 (1.0)	40 (1.5)	16 (1.1)	33 (1.3)	47 (1.8)	20 (1.3)	55 (1.2)	40 (1.2)	6 (0.7)
First in the World Consort., IL	35 (1.0)	40 (2.2)	26 (2.0)	28 (2.3)	50 (2.0)	22 (1.2)	55 (1.6)	40 (1.3)	5 (0.9)
Fremont/Lincoln/WestSide PS, NE	43 (2.1)	41 (2.0)	16 (1.4)	35 (1.3)	49 (1.9)	17 (1.6)	54 (1.8)	40 (1.9)	6 (1.1)
Guilford County, NC	44 (2.0)	38 (1.6)	18 (1.6)	39 (1.7)	45 (2.2)	16 (1.4)	66 (2.2)	30 (1.8)	4 (0.7)
Jersey City Public Schools, NJ	48 (1.6)	36 (1.5)	16 (1.1)	35 (1.2)	43 (1.5)	22 (1.0)	64 (1.7)	31 (1.6)	5 (0.6)
Miami-Dade County PS, FL	49 (2.0)	36 (1.4)	15 (1.4)	37 (1.5)	44 (1.4)	19 (1.7)	61 (1.7)	33 (1.0)	6 (1.0)
Michigan Invitational Group, MI	36 (1.5)	45 (1.9)	19 (1.3)	29 (1.6)	50 (1.8)	22 (1.8)	53 (2.2)	41 (1.6)	6 (1.0)
Montgomery County, MD	38 (1.4)	44 (1.5)	19 (1.7)	40 (2.0)	46 (2.0)	14 (1.2)	59 (1.5)	36 (1.5)	5 (0.9)
Naperville Sch. Dist. #203, IL	38 (1.6)	42 (2.0)	20 (1.4)	35 (1.3)	48 (1.5)	17 (1.3)	59 (1.5)	37 (1.6)	4 (0.6)
Project SMART Consortium, OH	38 (1.8)	44 (1.3)	18 (1.2)	30 (1.2)	51 (1.4)	19 (1.1)	52 (1.9)	42 (1.9)	6 (1.0)
Rochester City Sch. Dist., NY	r 51 (2.6)	36 (2.2)	13 (1.2)	r 34 (2.2)	42 (2.1)	24 (1.7)	r 61 (2.2)	33 (2.0)	6 (0.7)
SW Math/Sci. Collaborative, PA	37 (1.5)	42 (1.6)	20 (0.9)	31 (1.1)	51 (1.4)	17 (1.0)	54 (1.6)	39 (1.4)	7 (0.8)
International Avg. (All Countries)	39 (0.2)	42 (0.2)	19 (0.1)	29 (0.1)	42 (0.1)	30 (0.2)	46 (0.2)	41 (0.2)	13 (0.1)

SOURCE: IEA Third International Mathematics and Science Study (TIMSS), 1998-1999.

Background data provided by students.

States in *italics* did not fully satisfy guidelines for sample participation rates (see Appendix A for details).

( ) Standard errors appear in parentheses. Because results are rounded to the nearest whole number, some totals may appear inconsistent.

An "r" indicates a 70-84% student response rate.

Exhibit R1.9 Students' Daily Out-of-School Study Time

8th Grade Mathematics

	Average Hours Spent Each Day Studying or Doing Homework[1]				Percentage of Students Reporting Spending Some Time Studying All Three Subjects: Mathematics, Science, and Other
	Mathematics	Science	Other School Subjects	Total	
Countries					
United States	0.8 (0.02)	0.6 (0.01)	0.9 (0.02)	2.1 (0.04)	72 (1.6)
Belgium (Flemish)	1.1 (0.03)	0.8 (0.03)	1.4 (0.04)	2.9 (0.05)	86 (1.2)
Canada	0.8 (0.02)	0.6 (0.01)	1.0 (0.02)	2.2 (0.04)	78 (1.0)
Chinese Taipei	0.7 (0.02)	0.6 (0.02)	1.0 (0.02)	2.0 (0.05)	55 (1.3)
Czech Republic	0.7 (0.02)	0.6 (0.02)	0.7 (0.02)	1.9 (0.04)	74 (1.4)
England	– –	– –	– –	– –	– –
Hong Kong, SAR	0.7 (0.02)	0.5 (0.01)	0.7 (0.02)	1.6 (0.04)	53 (1.3)
Italy	1.3 (0.03)	1.0 (0.02)	1.9 (0.03)	3.6 (0.04)	91 (0.8)
Japan	0.6 (0.01)	0.4 (0.01)	0.8 (0.02)	1.7 (0.04)	59 (1.4)
Korea, Rep. of	0.6 (0.02)	0.4 (0.01)	0.7 (0.02)	1.6 (0.03)	50 (0.9)
Netherlands	0.6 (0.02)	0.6 (0.02)	1.0 (0.02)	2.2 (0.04)	89 (1.1)
Russian Federation	1.1 (0.03)	1.5 (0.03)	1.2 (0.04)	3.1 (0.05)	89 (0.7)
Singapore	1.3 (0.02)	1.2 (0.02)	1.7 (0.03)	3.5 (0.04)	90 (0.8)
States					
Connecticut	0.8 (0.02)	0.7 (0.02)	1.0 (0.02)	2.2 (0.05)	83 (1.8)
Idaho	0.7 (0.02)	0.6 (0.02)	0.8 (0.02)	1.9 (0.04)	65 (2.7)
Illinois	0.8 (0.02)	0.6 (0.02)	1.0 (0.03)	2.2 (0.05)	77 (1.6)
Indiana	0.7 (0.03)	0.5 (0.02)	0.8 (0.03)	1.9 (0.06)	70 (2.2)
Maryland	0.8 (0.02)	0.6 (0.02)	0.9 (0.02)	2.0 (0.04)	76 (1.4)
Massachusetts	0.8 (0.02)	0.7 (0.02)	1.0 (0.03)	2.3 (0.06)	84 (1.4)
Michigan	0.8 (0.03)	0.6 (0.02)	0.9 (0.03)	2.0 (0.05)	75 (1.6)
Missouri	0.7 (0.03)	0.5 (0.02)	0.8 (0.03)	1.9 (0.06)	65 (1.9)
North Carolina	0.8 (0.02)	0.6 (0.02)	0.9 (0.03)	2.1 (0.05)	74 (2.1)
Oregon	0.8 (0.02)	0.5 (0.03)	0.9 (0.03)	2.0 (0.04)	68 (2.2)
Pennsylvania	0.7 (0.03)	0.6 (0.02)	0.8 (0.03)	1.9 (0.07)	72 (1.9)
South Carolina	0.8 (0.02)	0.6 (0.02)	0.9 (0.03)	2.0 (0.05)	73 (1.6)
Texas	0.8 (0.04)	0.5 (0.03)	0.8 (0.03)	1.8 (0.07)	60 (2.3)
Districts and Consortia					
Academy School Dist. #20, CO	1.0 (0.03)	0.8 (0.03)	1.1 (0.03)	2.5 (0.05)	86 (0.8)
Chicago Public Schools, IL	1.2 (0.06)	0.8 (0.03)	1.3 (0.03)	2.7 (0.07)	79 (2.0)
Delaware Science Coalition, DE	0.7 (0.03)	0.6 (0.03)	0.8 (0.03)	1.9 (0.04)	70 (2.2)
First in the World Consort., IL	0.8 (0.02)	0.6 (0.03)	1.1 (0.05)	2.3 (0.07)	84 (1.7)
Fremont/Lincoln/WestSide PS, NE	0.7 (0.05)	0.5 (0.03)	0.9 (0.04)	1.8 (0.09)	65 (1.5)
Guilford County, NC	0.9 (0.03)	0.6 (0.02)	0.9 (0.03)	2.3 (0.05)	82 (1.6)
Jersey City Public Schools, NJ	1.1 (0.05)	0.8 (0.03)	1.3 (0.05)	2.7 (0.09)	76 (2.5)
Miami-Dade County PS, FL	0.9 (0.03)	0.7 (0.04)	0.9 (0.04)	2.2 (0.08)	69 (2.3)
Michigan Invitational Group, MI	0.7 (0.03)	0.6 (0.01)	0.8 (0.03)	2.0 (0.06)	76 (1.5)
Montgomery County, MD	0.9 (0.04)	0.7 (0.03)	1.0 (0.03)	2.4 (0.04)	81 (1.4)
Naperville Sch. Dist. #203, IL	0.8 (0.02)	0.6 (0.02)	1.0 (0.03)	2.3 (0.04)	85 (1.4)
Project SMART Consortium, OH	0.6 (0.02)	0.5 (0.02)	0.8 (0.03)	1.8 (0.04)	71 (1.8)
Rochester City Sch. Dist., NY	0.8 (0.05)	0.7 (0.04)	0.9 (0.05)	2.1 (0.07)	74 (2.4)
SW Math/Sci. Collaborative, PA	0.7 (0.03)	0.5 (0.02)	0.8 (0.02)	1.9 (0.05)	72 (2.1)
International Avg. (All Countries)	1.1 (0.00)	1.0 (0.00)	1.3 (0.01)	2.8 (0.01)	80 (0.2)

SOURCE: IEA Third International Mathematics and Science Study (TIMSS), 1998-1999.

Background data provided by students.

1 Average hours based on: No time=0; less than 1 hour=.5; 1-2 hours=1.5; 3-5 hours=4; more than 5 hours=7.

States in *italics* did not fully satisfy guidelines for sample participation rates (see Appendix A for details).

( ) Standard errors appear in parentheses. Because results are rounded to the nearest whole number, some totals may appear inconsistent.

A dash (–) indicates data are not available.

	Average Hours Spent Each Day[1]					
	Watching Television or Videos	Playing Computer Games	Playing or Talking With Friends	Doing Jobs at Home	Playing Sports	Reading a Book for Enjoyment
Countries						
United States	2.5 (0.06)	0.9 (0.02)	2.4 (0.05)	1.1 (0.03)	1.9 (0.03)	0.6 (0.02)
Belgium (Flemish)	2.1 (0.04)	0.9 (0.04)	1.8 (0.05)	1.0 (0.04)	1.8 (0.07)	0.6 (0.02)
Canada	2.2 (0.03)	0.8 (0.02)	2.1 (0.04)	1.1 (0.03)	1.9 (0.03)	0.7 (0.04)
Chinese Taipei	2.0 (0.04)	0.9 (0.03)	1.3 (0.03)	1.0 (0.02)	1.2 (0.02)	0.9 (0.02)
Czech Republic	2.3 (0.05)	0.9 (0.06)	3.0 (0.07)	1.2 (0.03)	2.0 (0.05)	1.0 (0.04)
England	2.6 (0.05)	1.2 (0.04)	2.5 (0.08)	0.8 (0.02)	1.6 (0.04)	0.6 (0.02)
Hong Kong, SAR	2.4 (0.04)	1.0 (0.03)	1.3 (0.04)	0.6 (0.01)	1.0 (0.03)	0.8 (0.02)
Italy	1.8 (0.03)	1.0 (0.03)	2.7 (0.05)	1.1 (0.03)	1.7 (0.03)	0.7 (0.02)
Japan	3.1 (0.05)	0.9 (0.03)	1.8 (0.04)	0.5 (0.02)	1.1 (0.03)	0.8 (0.02)
Korea, Rep. of	2.9 (0.04)	0.8 (0.03)	1.3 (0.03)	0.6 (0.01)	0.6 (0.02)	0.6 (0.01)
Netherlands	2.4 (0.10)	0.9 (0.04)	2.6 (0.09)	0.8 (0.04)	1.8 (0.06)	0.7 (0.04)
Russian Federation	2.6 (0.05)	0.7 (0.03)	3.0 (0.05)	1.5 (0.03)	1.3 (0.03)	1.2 (0.03)
Singapore	2.4 (0.04)	1.1 (0.03)	1.5 (0.04)	0.9 (0.02)	1.5 (0.04)	1.0 (0.02)
States						
Connecticut	2.4 (0.09)	0.9 (0.04)	2.6 (0.08)	1.0 (0.06)	2.0 (0.05)	0.6 (0.03)
Idaho	2.1 (0.08)	0.8 (0.02)	2.2 (0.07)	1.2 (0.05)	2.0 (0.08)	0.7 (0.03)
Illinois	2.6 (0.09)	0.9 (0.05)	2.5 (0.09)	1.1 (0.05)	1.9 (0.04)	0.7 (0.03)
Indiana	2.4 (0.07)	0.9 (0.04)	2.4 (0.09)	1.1 (0.04)	1.9 (0.07)	0.6 (0.04)
Maryland	3.0 (0.10)	1.1 (0.04)	2.8 (0.07)	1.1 (0.04)	2.0 (0.05)	0.6 (0.02)
Massachusetts	2.3 (0.07)	1.0 (0.03)	2.6 (0.08)	0.9 (0.03)	1.9 (0.04)	0.5 (0.03)
Michigan	2.2 (0.09)	0.8 (0.04)	2.3 (0.08)	1.0 (0.06)	2.0 (0.06)	0.6 (0.03)
Missouri	2.6 (0.08)	0.9 (0.04)	2.7 (0.09)	1.3 (0.05)	1.9 (0.04)	0.5 (0.02)
North Carolina	2.9 (0.09)	0.9 (0.04)	2.5 (0.06)	1.3 (0.03)	1.9 (0.05)	0.6 (0.02)
Oregon	2.0 (0.06)	0.8 (0.04)	2.3 (0.06)	1.1 (0.04)	2.0 (0.05)	0.7 (0.03)
Pennsylvania	2.4 (0.09)	0.9 (0.04)	2.7 (0.09)	1.0 (0.04)	2.0 (0.04)	0.5 (0.03)
South Carolina	2.9 (0.09)	1.0 (0.05)	2.5 (0.06)	1.2 (0.05)	2.0 (0.06)	0.7 (0.03)
Texas	2.6 (0.09)	0.9 (0.05)	2.3 (0.09)	1.2 (0.06)	1.8 (0.06)	0.6 (0.03)
Districts and Consortia						
Academy School Dist. #20, CO	2.1 (0.06)	0.9 (0.05)	2.1 (0.05)	0.9 (0.02)	2.0 (0.05)	0.7 (0.03)
Chicago Public Schools, IL	3.3 (0.13)	1.0 (0.09)	2.7 (0.13)	1.7 (0.10)	2.0 (0.08)	1.2 (0.12)
Delaware Science Coalition, DE	2.8 (0.10)	1.0 (0.06)	2.8 (0.11)	1.1 (0.05)	2.0 (0.06)	0.6 (0.03)
First in the World Consort., IL	1.9 (0.06)	0.7 (0.05)	2.1 (0.09)	0.7 (0.02)	1.7 (0.07)	0.7 (0.04)
Fremont/Lincoln/WestSide PS, NE	2.5 (0.08)	0.9 (0.08)	2.8 (0.09)	1.0 (0.04)	2.0 (0.08)	0.7 (0.05)
Guilford County, NC	2.8 (0.08)	0.9 (0.05)	2.5 (0.08)	1.1 (0.04)	1.9 (0.07)	0.7 (0.04)
Jersey City Public Schools, NJ	3.2 (0.09)	1.0 (0.06)	2.8 (0.10)	1.4 (0.05)	1.9 (0.07)	0.9 (0.05)
Miami-Dade County PS, FL	3.1 (0.12)	1.1 (0.07)	2.5 (0.11)	1.4 (0.06)	2.1 (0.12)	0.9 (0.08)
Michigan Invitational Group, MI	2.0 (0.08)	0.8 (0.05)	2.3 (0.10)	1.0 (0.04)	1.9 (0.08)	0.6 (0.04)
Montgomery County, MD	2.5 (0.08)	0.9 (0.05)	2.3 (0.08)	0.9 (0.04)	1.8 (0.05)	0.7 (0.02)
Naperville Sch. Dist. #203, IL	1.8 (0.05)	0.7 (0.03)	2.0 (0.05)	0.7 (0.03)	2.0 (0.05)	0.8 (0.03)
Project SMART Consortium, OH	2.5 (0.08)	0.9 (0.06)	2.9 (0.10)	1.0 (0.05)	2.2 (0.09)	0.5 (0.03)
Rochester City Sch. Dist., NY	3.6 (0.11)	1.2 (0.08)	2.9 (0.10)	1.5 (0.07)	1.9 (0.07)	0.7 (0.05)
SW Math/Sci. Collaborative, PA	2.4 (0.07)	0.9 (0.04)	2.5 (0.10)	0.9 (0.04)	2.0 (0.06)	0.5 (0.03)
International Avg. (All Countries)	2.3 (0.01)	0.8 (0.01)	1.9 (0.01)	1.4 (0.01)	1.5 (0.01)	1.0 (0.00)

SOURCE: IEA Third International Mathematics and Science Study (TIMSS), 1998-1999.

Background data provided by students.

* Activities are not necessarily exclusive; students may have reported engaging in more than one activity at the same time.

1 Average hours based on: No time=0; less than 1 hour=.5; 1-2 hours=1.5; 3-5 hours=4; more than 5 hours=7.

States in *italics* did not fully satisfy guidelines for sample participation rates (see Appendix A for details).

( ) Standard errors appear in parentheses. Because results are rounded to the nearest whole number, some totals may appear inconsistent.

Students' Reports That Mathematics Is Not One of Their Strengths

8th Grade Mathematics

	Strongly Disagree		Disagree		Agree		Strongly Agree		Average[1]
	Percent of Students	Average Achievement	Percent of Students	Average Achievement	Percent of Students	Average Achievement	Percent of Students	Average Achievement	
Countries									
United States	26 (0.8)	547 (4.6)	33 (0.6)	517 (4.5)	24 (0.6)	478 (4.4)	18 (0.7)	455 (4.3)	2.3 (0.02)
Belgium (Flemish)	17 (0.8)	598 (5.4)	30 (0.9)	580 (4.8)	31 (1.0)	540 (4.5)	21 (0.9)	526 (3.8)	2.6 (0.02)
Canada	26 (1.2)	577 (3.2)	31 (0.7)	542 (3.0)	26 (1.0)	502 (4.7)	17 (0.8)	485 (3.4)	2.3 (0.03)
Chinese Taipei	16 (0.6)	646 (6.1)	28 (0.7)	623 (4.3)	34 (0.7)	564 (4.5)	23 (0.7)	533 (3.9)	2.6 (0.02)
Czech Republic	12 (0.9)	567 (6.9)	35 (1.2)	541 (5.0)	36 (1.1)	500 (4.7)	16 (0.9)	486 (5.8)	2.6 (0.03)
England	19 (0.9)	539 (6.9)	40 (1.2)	512 (4.5)	28 (1.2)	471 (4.8)	13 (0.8)	458 (6.3)	2.4 (0.02)
Hong Kong, SAR	11 (0.5)	619 (5.5)	31 (0.8)	606 (3.9)	39 (0.7)	573 (4.4)	20 (0.8)	549 (5.0)	2.7 (0.02)
Italy	22 (0.9)	529 (5.6)	32 (0.9)	495 (3.6)	27 (0.8)	456 (4.9)	18 (0.9)	435 (5.0)	2.4 (0.02)
Japan	12 (0.5)	602 (4.9)	29 (0.7)	598 (2.7)	38 (0.6)	576 (2.7)	21 (0.7)	545 (3.3)	2.7 (0.02)
Korea, Rep. of	10 (0.5)	648 (3.9)	33 (0.7)	621 (3.0)	41 (0.8)	564 (2.7)	15 (0.5)	536 (3.6)	2.6 (0.02)
Netherlands	16 (0.8)	570 (7.4)	28 (1.4)	557 (7.9)	36 (1.5)	529 (8.2)	20 (1.0)	515 (9.2)	2.6 (0.03)
Russian Federation	21 (0.9)	580 (5.8)	40 (0.9)	538 (5.6)	30 (1.2)	501 (7.3)	9 (0.6)	471 (10.0)	2.3 (0.03)
Singapore	16 (0.7)	631 (7.3)	37 (0.7)	614 (6.1)	33 (0.8)	593 (6.4)	13 (0.6)	575 (6.9)	2.4 (0.02)
States									
Connecticut	28 (1.4)	547 (10.4)	35 (1.2)	525 (9.8)	23 (1.1)	484 (8.7)	15 (1.2)	468 (8.4)	2.3 (0.04)
Idaho	23 (1.4)	535 (7.8)	33 (1.8)	512 (7.7)	25 (1.2)	468 (7.1)	20 (1.6)	460 (6.8)	2.4 (0.05)
Illinois	27 (1.5)	546 (9.2)	36 (0.9)	521 (7.8)	23 (1.1)	477 (6.8)	13 (0.9)	462 (6.2)	2.2 (0.04)
Indiana	23 (1.9)	558 (8.3)	35 (1.3)	532 (6.3)	25 (1.3)	487 (6.5)	17 (1.7)	468 (6.4)	2.4 (0.06)
Maryland	26 (1.2)	533 (5.5)	32 (0.9)	507 (6.7)	25 (0.9)	475 (7.8)	17 (1.1)	450 (7.8)	2.3 (0.04)
Massachusetts	24 (1.6)	560 (7.2)	34 (1.0)	524 (5.4)	25 (1.1)	487 (7.1)	17 (1.2)	468 (7.1)	2.4 (0.04)
Michigan	28 (1.4)	557 (8.9)	33 (1.2)	529 (7.1)	24 (1.1)	488 (7.0)	15 (0.8)	473 (6.5)	2.3 (0.03)
Missouri	22 (1.1)	527 (7.2)	31 (1.1)	507 (4.9)	27 (1.2)	469 (5.0)	20 (1.5)	455 (7.8)	2.5 (0.04)
North Carolina	27 (1.5)	535 (7.9)	37 (1.1)	504 (7.7)	22 (1.0)	465 (7.1)	14 (0.9)	448 (5.6)	2.2 (0.03)
Oregon	25 (1.5)	555 (6.9)	35 (1.3)	530 (7.3)	25 (1.1)	485 (6.7)	15 (1.6)	471 (6.1)	2.3 (0.05)
Pennsylvania	26 (1.1)	544 (10.2)	34 (1.2)	522 (5.3)	24 (0.9)	480 (6.0)	16 (1.1)	462 (7.2)	2.3 (0.03)
South Carolina	25 (1.1)	542 (8.2)	34 (1.2)	516 (7.6)	23 (0.9)	469 (8.1)	18 (1.2)	462 (7.9)	2.3 (0.04)
Texas	27 (1.6)	564 (8.6)	33 (1.6)	531 (10.3)	25 (1.1)	486 (9.6)	15 (1.0)	480 (9.4)	2.3 (0.03)
Districts and Consortia									
Academy School Dist. #20, CO	28 (1.1)	564 (3.9)	35 (1.2)	537 (3.3)	24 (1.4)	500 (4.1)	13 (1.0)	487 (6.5)	2.2 (0.03)
Chicago Public Schools, IL	27 (1.8)	499 (7.4)	35 (1.6)	470 (7.7)	26 (1.3)	436 (7.7)	12 (1.3)	419 (5.0)	2.2 (0.04)
Delaware Science Coalition, DE	25 (1.5)	529 (10.8)	30 (1.5)	492 (10.0)	27 (1.6)	457 (9.3)	18 (1.8)	442 (9.1)	2.4 (0.04)
First in the World Consort., IL	34 (2.3)	596 (7.1)	36 (2.1)	562 (7.9)	18 (1.1)	532 (7.8)	11 (1.3)	490 (8.7)	2.1 (0.04)
Fremont/Lincoln/WestSide PS, NE	26 (2.2)	535 (13.0)	31 (2.1)	515 (6.5)	26 (1.2)	463 (8.4)	17 (2.4)	412 (7.8)	2.3 (0.06)
Guilford County, NC	28 (1.8)	537 (9.4)	36 (1.6)	521 (9.0)	23 (1.6)	497 (8.0)	13 (1.0)	478 (10.5)	2.2 (0.04)
Jersey City Public Schools, NJ	26 (2.0)	526 (10.0)	33 (1.8)	485 (7.3)	24 (1.8)	450 (10.3)	18 (1.3)	427 (7.6)	2.3 (0.05)
Miami-Dade County PS, FL	20 (1.5)	473 (14.1)	32 (1.8)	436 (10.5)	29 (1.9)	401 (9.5)	20 (2.3)	389 (9.2)	2.5 (0.06)
Michigan Invitational Group, MI	24 (2.0)	579 (6.2)	35 (1.6)	543 (7.8)	25 (1.7)	507 (6.3)	16 (1.3)	480 (13.2)	2.3 (0.04)
Montgomery County, MD	31 (2.2)	571 (6.7)	34 (1.0)	545 (5.7)	22 (1.4)	513 (5.0)	13 (1.1)	488 (8.7)	2.2 (0.05)
Naperville Sch. Dist. #203, IL	35 (1.1)	606 (3.3)	32 (1.3)	568 (3.7)	20 (1.2)	540 (4.8)	13 (0.7)	520 (6.6)	2.1 (0.02)
Project SMART Consortium, OH	28 (2.2)	561 (10.6)	34 (1.9)	535 (6.9)	23 (1.8)	487 (6.4)	15 (1.6)	467 (6.7)	2.2 (0.06)
Rochester City Sch. Dist., NY	29 (1.3)	474 (7.9)	27 (1.8)	467 (7.8)	27 (1.3)	432 (6.8)	17 (1.2)	403 (9.4)	2.3 (0.03)
SW Math/Sci. Collaborative, PA	28 (1.3)	555 (9.5)	35 (1.0)	529 (7.4)	24 (0.9)	485 (8.3)	13 (0.9)	464 (8.0)	2.2 (0.03)
International Avg. (All Countries)	17 (0.1)	532 (1.0)	33 (0.2)	506 (0.8)	33 (0.2)	469 (0.8)	17 (0.1)	450 (0.9)	2.5 (0.00)

SOURCE: IEA Third International Mathematics and Science Study (TIMSS), 1998-1999.

Background data provided by students.

1 Average scale value based on: Strongly disagree=4; disagree=3; agree=2; strongly agree=1.

States in *italics* did not fully satisfy guidelines for sample participation rates (see Appendix A for details).

( ) Standard errors appear in parentheses. Because results are rounded to the nearest whole number, some totals may appear inconsistent.

Exhibit R1.12 Students' Liking Mathematics

8th Grade Mathematics

	Like a Lot		Like		Dislike		Dislike a Lot		Average[1]
	Percent of Students	Average Achievement	Percent of Students	Average Achievement	Percent of Students	Average Achievement	Percent of Students	Average Achievement	
Countries									
United States	23 (0.9)	527 (4.5)	46 (0.6)	505 (4.0)	19 (0.7)	496 (4.5)	12 (0.7)	465 (5.6)	2.8 (0.02)
Belgium (Flemish)	20 (0.9)	598 (5.1)	46 (1.3)	562 (4.2)	24 (1.3)	537 (6.8)	10 (0.7)	518 (5.1)	2.8 (0.02)
Canada	24 (1.0)	561 (3.8)	49 (1.6)	531 (2.6)	18 (0.8)	513 (4.8)	9 (0.5)	486 (4.6)	2.9 (0.02)
Chinese Taipei	15 (0.7)	654 (5.3)	41 (0.8)	617 (3.5)	33 (0.8)	546 (5.2)	12 (0.6)	502 (5.3)	2.6 (0.02)
Czech Republic	11 (0.9)	580 (7.9)	44 (1.5)	530 (4.5)	34 (1.7)	498 (5.6)	11 (0.8)	489 (7.6)	2.5 (0.03)
England	23 (1.1)	514 (6.8)	54 (1.1)	497 (4.3)	16 (0.9)	487 (5.5)	6 (0.5)	470 (8.7)	3.0 (0.02)
Hong Kong, SAR	22 (0.7)	610 (4.7)	53 (0.7)	587 (4.0)	20 (0.8)	558 (4.7)	5 (0.4)	521 (7.3)	2.9 (0.02)
Italy	30 (1.0)	517 (4.6)	38 (1.1)	482 (4.3)	22 (0.8)	446 (5.1)	10 (0.7)	433 (5.8)	2.9 (0.02)
Japan	9 (0.5)	631 (5.7)	39 (0.9)	600 (2.2)	38 (1.0)	563 (2.5)	14 (0.6)	530 (4.1)	2.4 (0.02)
Korea, Rep. of	12 (0.5)	647 (4.2)	42 (0.8)	608 (2.4)	38 (0.7)	557 (2.7)	8 (0.4)	536 (3.8)	2.6 (0.02)
Netherlands	– –	– –	– –	– –	– –	– –	– –	– –	– –
Russian Federation	22 (1.0)	562 (5.5)	56 (0.8)	529 (5.8)	19 (0.9)	498 (8.1)	3 (0.3)	460 (13.3)	3.0 (0.02)
Singapore	30 (1.0)	626 (6.6)	49 (0.8)	602 (6.4)	14 (0.6)	583 (7.3)	6 (0.4)	564 (7.5)	3.0 (0.02)
States									
Connecticut	23 (1.8)	534 (11.6)	48 (1.6)	511 (9.7)	18 (1.2)	507 (8.7)	11 (1.3)	486 (9.2)	2.8 (0.04)
Idaho	19 (1.5)	519 (8.8)	46 (1.1)	502 (6.8)	22 (1.2)	482 (8.9)	12 (1.3)	454 (8.0)	2.7 (0.05)
Illinois	27 (1.4)	534 (8.6)	48 (1.0)	507 (7.1)	17 (1.0)	502 (7.0)	8 (0.8)	463 (8.6)	2.9 (0.03)
Indiana	23 (1.7)	535 (8.7)	47 (1.0)	520 (6.8)	18 (1.2)	503 (10.0)	12 (1.1)	484 (9.6)	2.8 (0.04)
Maryland	25 (1.6)	518 (6.0)	45 (1.5)	497 (6.6)	19 (1.1)	484 (7.4)	11 (1.1)	467 (10.3)	2.8 (0.04)
Massachusetts	20 (1.5)	536 (6.7)	47 (1.3)	517 (6.5)	21 (1.2)	502 (8.4)	12 (0.9)	482 (6.7)	2.8 (0.04)
Michigan	22 (1.4)	540 (10.9)	47 (1.3)	522 (7.7)	20 (0.9)	509 (6.8)	11 (1.0)	474 (7.3)	2.8 (0.03)
Missouri	22 (1.5)	511 (7.8)	45 (1.2)	496 (4.8)	20 (1.2)	473 (6.8)	13 (1.2)	460 (8.8)	2.8 (0.04)
North Carolina	32 (1.6)	512 (8.5)	47 (1.2)	494 (7.9)	14 (0.9)	482 (8.5)	7 (0.6)	458 (8.2)	3.1 (0.03)
Oregon	20 (1.6)	536 (8.0)	49 (1.5)	519 (6.6)	21 (1.1)	503 (7.0)	10 (1.2)	477 (8.1)	2.8 (0.04)
Pennsylvania	24 (1.3)	530 (10.3)	48 (1.5)	510 (6.0)	17 (1.2)	494 (7.2)	10 (1.1)	469 (10.0)	2.9 (0.04)
South Carolina	26 (1.4)	517 (8.7)	47 (1.5)	502 (8.8)	17 (1.1)	500 (8.0)	11 (0.9)	472 (9.2)	2.9 (0.03)
Texas	23 (1.5)	536 (12.9)	50 (1.4)	522 (9.6)	18 (1.2)	512 (10.0)	9 (0.9)	479 (10.4)	2.9 (0.04)
Districts and Consortia									
Academy School Dist. #20, CO	17 (1.1)	547 (6.6)	46 (1.5)	537 (3.4)	24 (1.1)	518 (4.0)	13 (1.1)	499 (6.4)	2.7 (0.03)
Chicago Public Schools, IL	34 (2.7)	484 (7.9)	46 (2.1)	458 (6.5)	14 (1.9)	449 (7.1)	5 (0.9)	415 (8.3)	3.1 (0.05)
Delaware Science Coalition, DE	23 (1.4)	503 (10.9)	47 (1.7)	485 (9.3)	18 (1.5)	457 (10.0)	12 (1.3)	467 (10.9)	2.8 (0.03)
First in the World Consort., IL	20 (1.5)	586 (5.8)	48 (2.3)	563 (6.2)	20 (1.9)	545 (8.9)	11 (1.8)	521 (13.8)	2.8 (0.04)
Fremont/Lincoln/WestSide PS, NE	23 (1.3)	511 (14.9)	47 (1.3)	496 (8.8)	20 (1.3)	480 (11.7)	10 (0.8)	413 (10.4)	2.8 (0.03)
Guilford County, NC	30 (1.3)	515 (10.2)	46 (1.1)	514 (8.8)	17 (1.4)	518 (7.6)	7 (0.9)	500 (13.2)	3.0 (0.04)
Jersey City Public Schools, NJ	34 (2.5)	511 (10.6)	47 (2.1)	472 (6.7)	13 (0.9)	439 (11.6)	7 (1.0)	397 (13.7)	3.1 (0.04)
Miami-Dade County PS, FL	24 (1.7)	442 (13.5)	46 (1.5)	425 (10.3)	18 (1.8)	409 (10.2)	12 (1.5)	398 (8.2)	2.8 (0.06)
Michigan Invitational Group, MI	19 (1.1)	567 (7.0)	50 (1.8)	540 (6.3)	19 (1.5)	513 (7.5)	12 (1.1)	479 (13.8)	2.8 (0.04)
Montgomery County, MD	21 (1.4)	560 (7.2)	47 (1.5)	534 (3.9)	20 (1.5)	536 (4.9)	12 (1.3)	516 (8.7)	2.8 (0.04)
Naperville Sch. Dist. #203, IL	25 (1.0)	602 (3.9)	47 (1.5)	569 (3.7)	19 (1.4)	553 (6.2)	9 (0.9)	518 (6.7)	2.9 (0.02)
Project SMART Consortium, OH	20 (1.6)	559 (10.1)	48 (2.0)	524 (7.6)	20 (1.5)	508 (8.1)	12 (1.5)	464 (11.3)	2.8 (0.05)
Rochester City Sch. Dist., NY	30 (2.0)	466 (8.7)	45 (1.8)	453 (8.8)	14 (1.0)	429 (9.6)	11 (1.6)	407 (12.7)	3.0 (0.05)
SW Math/Sci. Collaborative, PA	25 (1.6)	543 (8.9)	48 (1.4)	518 (8.2)	17 (1.4)	497 (8.0)	9 (1.1)	484 (11.3)	2.9 (0.04)
International Avg. (All Countries)	24 (0.2)	518 (0.9)	48 (0.2)	489 (0.8)	21 (0.2)	466 (1.0)	7 (0.1)	456 (1.4)	2.9 (0.00)

SOURCE: IEA Third International Mathematics and Science Study (TIMSS), 1998-1999.

Background date provided by students.

1 Average scale value based on: Like a lot=4; like=3; dislike=2; dislike a lot=1.

States in *italics* did not fully satisfy guidelines for sample participation rates (see Appendix A for details).

( ) Standard errors appear in parentheses. Because results are rounded to the nearest whole number, some totals may appear inconsistent.

A dash (–) indicates data are not available.

REFERENCE

2 The Mathematics Curriculum

	Percentage of Students Whose Schools Reported Various Organizational Approaches in Mathematics Instruction to Accommodate Students with Different Abilities or Interests in Mathematics				
	All Classes Study Similar Content but at Different Levels of Difficulty	Students Are Grouped by Ability within Classes	Enrichment Mathematics Is Offered	Remedial Mathematics Is Offered	Different Classes Study Different Content
Countries					
United States	r 49 (4.7)	r 49 (4.2)	r 79 (2.8)	r 64 (3.9)	r 37 (4.2)
Belgium (Flemish)	66 (5.1)	11 (3.2)	36 (5.0)	81 (4.7)	100 (0.0)
Canada	s 77 (3.4)	s 43 (4.3)	s 66 (3.8)	s 87 (2.5)	s 17 (3.0)
Chinese Taipei	50 (4.2)	25 (3.7)	88 (2.7)	81 (3.5)	18 (3.1)
Czech Republic	68 (4.3)	44 (5.0)	29 (3.9)	62 (4.3)	7 (3.0)
England	r 78 (3.6)	r 57 (4.7)	r 48 (5.0)	r 61 (4.8)	r 0 (0.0)
Hong Kong, SAR	r 62 (4.9)	17 (3.5)	63 (4.4)	59 (4.8)	r 3 (1.7)
Italy	0 (0.0)	0 (0.0)	51 (3.8)	81 (3.0)	0 (0.0)
Japan	31 (3.9)	13 (3.1)	32 (3.5)	67 (4.3)	13 (2.9)
Korea, Rep. of	66 (3.9)	41 (4.3)	27 (3.5)	26 (3.5)	38 (4.5)
Netherlands	r 55 (6.8)	r 39 (6.9)	r 90 (3.8)	r 64 (7.5)	r 60 (6.8)
Russian Federation	32 (3.8)	47 (4.0)	90 (3.0)	53 (3.8)	25 (3.5)
Singapore	0 (0.0)	0 (0.0)	80 (3.5)	99 (0.8)	82 (3.6)
States					
Connecticut	s 56 (9.5)	s 70 (8.4)	s 98 (2.1)	s 62 (9.5)	s 65 (9.7)
Idaho	r 46 (7.0)	r 57 (9.8)	r 73 (7.7)	r 80 (6.8)	r 66 (9.7)
Illinois	50 (6.2)	r 67 (5.6)	84 (3.7)	43 (7.2)	55 (5.9)
Indiana	51 (7.8)	52 (8.9)	85 (5.3)	43 (8.4)	43 (7.4)
Maryland	r 61 (8.0)	r 86 (4.2)	r 86 (5.1)	r 69 (7.7)	r 66 (7.0)
Massachusetts	s 54 (9.8)	s 37 (8.8)	s 84 (7.0)	s 63 (9.7)	s 41 (10.0)
Michigan	36 (7.5)	62 (6.2)	79 (6.2)	57 (8.1)	58 (6.9)
Missouri	36 (7.2)	48 (5.8)	64 (5.8)	38 (7.2)	41 (6.1)
North Carolina	r 81 (5.8)	r 73 (7.2)	r 94 (3.6)	r 71 (7.1)	r 40 (7.3)
Oregon	65 (8.3)	62 (8.4)	93 (4.2)	83 (6.0)	75 (7.5)
Pennsylvania	48 (8.5)	52 (8.2)	84 (6.1)	62 (6.5)	59 (5.5)
South Carolina	74 (6.5)	46 (8.1)	98 (2.5)	r 60 (7.4)	r 51 (6.7)
Texas	r 79 (7.5)	r 39 (6.8)	r 100 (0.0)	r 56 (9.4)	r 41 (8.6)
Districts and Consortia					
Academy School Dist. #20, CO	r 35 (0.4)	75 (0.3)	100 (0.0)	83 (0.4)	r 100 (0.0)
Chicago Public Schools, IL	r 78 (7.1)	s 54 (11.5)	r 28 (12.0)	r 70 (9.3)	r 15 (7.8)
Delaware Science Coalition, DE	r 54 (2.0)	r 58 (2.1)	r 96 (0.2)	r 53 (1.9)	r 64 (1.9)
First in the World Consort., IL	r 40 (1.3)	r 58 (1.1)	r 100 (0.0)	r 35 (1.6)	r 88 (0.4)
Fremont/Lincoln/WestSide PS, NE	r 80 (2.1)	s 68 (1.3)	s 100 (0.0)	r 76 (0.9)	s 84 (0.6)
Guilford County, NC	s 56 (1.2)	s 91 (0.2)	r 82 (0.8)	r 56 (1.2)	s 94 (0.6)
Jersey City Public Schools, NJ	58 (1.3)	16 (0.7)	11 (2.1)	52 (1.4)	0 (0.0)
Miami-Dade County PS, FL	s 83 (9.9)	s 74 (13.5)	s 100 (0.0)	s 40 (15.4)	x x
Michigan Invitational Group, MI	41 (1.4)	23 (1.2)	59 (1.5)	45 (1.5)	31 (1.1)
Montgomery County, MD	s 57 (10.7)	s 82 (8.8)	s 100 (0.0)	s 78 (11.2)	s 46 (15.5)
Naperville Sch. Dist. #203, IL	45 (1.5)	15 (2.1)	100 (0.0)	76 (1.5)	57 (1.5)
Project SMART Consortium, OH	r 37 (1.3)	46 (1.5)	96 (0.5)	41 (1.4)	r 63 (1.2)
Rochester City Sch. Dist., NY	r 100 (0.0)	r 0 (0.0)	r 100 (0.0)	r 46 (1.6)	r 27 (1.6)
SW Math/Sci. Collaborative, PA	50 (7.6)	46 (8.6)	90 (5.7)	53 (8.8)	57 (8.0)
International Avg. (All Countries)	58 (0.6)	35 (0.6)	58 (0.6)	72 (0.6)	17 (0.5)

SOURCE: IEA Third International Mathematics and Science Study (TIMSS), 1998-1999.

Background data provided by schools.

States in *italics* did not fully satisfy guidelines for sample participation rates (see Appendix A for details).

( ) Standard errors appear in parentheses. Because results are rounded to the nearest whole number, some totals may appear inconsistent.

An "r" indicates school response data available for 70-84% of students. An "s" indicates school response data available for 50-69% of students. An "x" indicates school response data available for <50% of students.

Exhibit R2.2 Detailed Information About Topics in the Intended Curriculum, Up to and Including Eighth Grade – Fractions and Number Sense

8th Grade Mathematics

	Whole numbers – including place values, factorization and operations (+, –, x, ÷)	Understanding and representing common fractions	Computations with common fractions	Understanding and representing decimal fractions	Computations with decimal fractions	Relationships between common and decimal fractions, ordering of fractions	Rounding whole numbers and decimal fractions	Estimating the results of computations	Number lines
Countries									
United States	●	●	●	●	●	●	●	●	●
Belgium (Flemish)	●	●	●	●	●	●	●	●	●
Canada	●	●	●	●	●	●	●	●	●
Chinese Taipei	●	●	●	●	●	●	●	●	●
Czech Republic	●	●	●	●	●	●	●	●	●
England	●	●	◉	●	◉	◉	◉	●	●
Hong Kong, SAR	●	●	●	●	●	●	●	●	●
Italy	●	●	●	●	●	●	●	●	●
Japan	●	●	●	●	●	●	●	●	●
Korea, Rep. of	●	●	●	●	●	●	●	○	●
Netherlands	●	●	◉	●	●	◉	●	●	●
Russian Federation	●	●	●	●	●	●	●	●	●
Singapore	●	●	●	●	●	●	●	●	●
States									
Connecticut	●	●	●	●	●	●	●	●	●
Idaho	●	●	●	●	●	●	●	●	●
Illinois	●	●	●	●	●	●	●	●	●
Indiana	●	●	●	●	●	●	●	●	●
Maryland	●	●	●	●	●	●	●	●	●
Massachusetts	●	●	●	●	●	●	●	●	●
Michigan	●	●	●	●	●	●	●	●	●
Missouri	●	●	●	●	●	●	●	●	●
North Carolina	●	●	●	●	●	●	●	●	●
Oregon	●	●	●	●	●	●	●	●	●
Pennsylvania	●	●	●	●	●	●	●	●	●
South Carolina	●	●	●	●	●	●	●	●	●
Texas	●	●	●	●	●	●	●	●	●
Districts and Consortia									
Academy School Dist. #20, CO	–	–	–	–	–	–	–	–	–
Chicago Public Schools, IL	●	●	●	●	●	●	●	●	●
Delaware Science Coalition, DE	●	●	●	●	●	●	●	●	●
First in the World Consort., IL	●	●	●	●	●	●	●	●	●
Fremont/Lincoln/WestSide PS, NE	●	●	●	●	●	●	●	●	●
Guilford County, NC	●	●	●	●	●	●	●	●	●
Jersey City Public Schools, NJ	●	●	●	●	●	●	●	●	●
Miami-Dade County PS, FL	●	●	●	●	●	●	●	●	●
Michigan Invitational Group, MI	●	●	●	●	●	●	●	●	●
Montgomery County, MD	●	●	●	●	●	●	●	●	●
Naperville Sch. Dist. #203, IL	●	●	●	●	●	●	●	●	●
Project SMART Consortium, OH	●	●	●	●	●	●	●	●	●
Rochester City Sch. Dist., NY	●	●	●	●	●	●	●	●	●
SW Math/Sci. Collaborative, PA	–	–	–	–	–	–	–	–	–

SOURCE: IEA Third International Mathematics and Science Study (TIMSS), 1998-1999.

Background data provided by coordinators from participating jurisdictions.

Whole number powers of integers	Computations with percentages and problems involving percentages	Simple computations with negative numbers	Square roots (of perfect squares less than 144), small integer exponents	Prime factors, highest common factor, lowest common multiple, rules for divisibility	Sets, subsets, union, intersection, venn diagrams	Rate problems	Concepts of ratio and proportion; ratio and proportion problems	
								Countries
●	●	●	●	●	●	●	●	United States
●	●	●	●	●	●	●	●	Belgium (Flemish)
●	●	●	●	●	○	●	●	Canada
●	●	●	●	●	○	◉	◉	Chinese Taipei
●	●	●	●	●	◦	●	●	Czech Republic
⊙	◉	◉	◉	⊙	○	⊙	⊙	England
●	●	●	●	●	○	●	●	Hong Kong, SAR
●	●	●	●	●	●	●	●	Italy
●	●	●	○	○	○	●	●	Japan
○	●	●	○	●	●	●	●	Korea, Rep. of
●	◉	◉	●	○	○	◉	◉	Netherlands
●	●	●	●	●	◦	○	●	Russian Federation
●	●	●	●	●	○	●	●	Singapore
								States
◉	●	●	◉	◉	●	◉	●	Connecticut
●	●	●	●	⊙	⊙	●	●	Idaho
●	●	●	●	●	◉	●	●	Illinois
⊙	●	●	●	●	○	●	●	Indiana
●	●	●	●	●	●	●	●	Maryland
●	●	●	●	●	●	○	●	Massachusetts
●	●	●	●	●	◉	●	●	Michigan
●	●	●	●	●	●	●	●	Missouri
●	●	●	●	●	●	●	●	North Carolina
●	●	●	●	○	○	●	●	Oregon
●	●	●	●	●	●	●	●	Pennsylvania
●	●	◉	●	●	⊙	●	●	South Carolina
●	●	●	●	●	◉	●	●	Texas
								Districts and Consortia
–	–	–	–	–	–	–	–	Academy School Dist. #20, CO
●	●	●	◉	●	◉	●	●	Chicago Public Schools, IL
●	●	●	◉	●	◉	●	●	Delaware Science Coalition, DE
●	●	●	●	●	●	●	●	First in the World Consort., IL
●	●	●	●	●	●	●	●	Fremont/Lincoln/WestSide PS, NE
●	●	●	●	●	●	●	●	Guilford County, NC
●	●	●	●	●	○	●	●	Jersey City Public Schools, NJ
●	●	●	●	●	◉	●	●	Miami-Dade County PS, FL
●	●	●	●	●	◉	●	●	Michigan Invitational Group, MI
●	●	●	◉	●	◉	●	●	Montgomery County, MD
●	●	●	●	●	●	●	●	Naperville Sch. Dist. #203, IL
●	●	●	●	●	●	⊙	●	Project SMART Consortium, OH
●	●	●	●	●	●	●	●	Rochester City Sch. Dist., NY
–	–	–	–	–	–	–	–	SW Math/Sci. Collaborative, PA

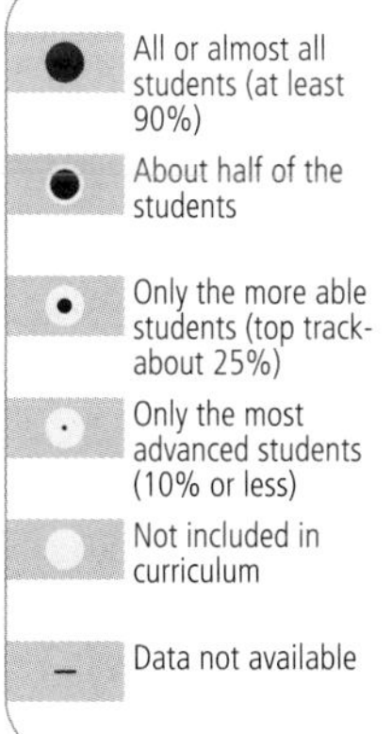

SOURCE: IEA Third International Mathematics and Science Study (TIMSS), 1998-1999.

Exhibit R2.3 Detailed Information About Topics in the Intended Curriculum, Up to and Including Eighth Grade – Measurement

TIMSS 1999 Benchmarking
Boston College

8th Grade Mathematics

	Units of measurement; standard metric units	Reading measurement instruments	Estimates of measurement; accuracy of measurement	Conversions of units between measurement systems	Perimeter and area of simple shapes – triangles, rectangles, and circles	Perimeter and area of combined shapes	Volume of rectangular solids i.e., Volume = length x width x height	Volume of other solids (e.g., pyramids, cylinders, cones, spheres)	Computing with measurements (+, –, x, ÷)	Scales applied to maps and models
Countries										
United States	●	●	●	●	●	●	●	●	●	●
Belgium (Flemish)	●	●	●	●	●	●	●	◉	●	●
Canada	●	●	●	●	●	●	●	◉	●	●
Chinese Taipei	●	◉	◉	◉	●	●	●	○	●	◉
Czech Republic	●	●	●	○	●	●	●	●	●	●
England	●	●	◉	◉	◉	◉	◉	⊙	●	◉
Hong Kong, SAR	●	●	●	○	●	●	●	○	●	●
Italy	●	●	○	●	●	●	●	●	●	○
Japan	●	●	●	●	●	●	●	●	●	●
Korea, Rep. of	●	●	●	●	●	●	●	●	●	●
Netherlands	●	●	⊙	○	◉	◉	●	◉	◉	●
Russian Federation	●	●	●	○	◉	○	●	○	●	●
Singapore	●	●	●	●	●	●	●	●	●	●
States										
Connecticut	●	●	●	⊙	●	●	●	◉	●	●
Idaho	●	●	●	⊙	●	●	⊙	⊙	●	●
Illinois	●	●	●	◉	●	●	●	◉	●	●
Indiana	●	●	●	○	●	●	●	●	●	●
Maryland	●	●	●	○	●	●	●	●	●	●
Massachusetts	●	●	●	●	●	●	●	●	●	●
Michigan	●	●	◉	○	●	●	●	◉	◉	●
Missouri	●	●	●	●	●	●	●	◉	●	●
North Carolina	●	●	●	○	●	●	●	●	●	●
Oregon	●	●	●	●	●	●	●	○	●	●
Pennsylvania	●	●	●	●	●	●	●	●	●	●
South Carolina	●	●	●	○	●	●	●	●	●	●
Texas	●	●	●	●	●	●	●	●	●	●
Districts and Consortia										
Academy School Dist. #20, CO	–	–	–	–	–	–	–	–	–	–
Chicago Public Schools, IL	●	●	●	●	●	●	●	◉	●	●
Delaware Science Coalition, DE	●	●	●	○	●	●	●	●	●	●
First in the World Consort., IL	●	●	●	●	●	●	●	●	●	●
Fremont/Lincoln/WestSide PS, NE	●	●	●	●	●	●	●	●	●	●
Guilford County, NC	●	●	●	○	●	●	●	●	●	●
Jersey City Public Schools, NJ	●	●	●	●	●	●	●	●	●	●
Miami-Dade County PS, FL	●	●	●	●	●	●	●	●	●	●
Michigan Invitational Group, MI	◉	◉	●	◉	●	●	●	●	●	●
Montgomery County, MD	●	●	◉	○	●	●	●	●	●	●
Naperville Sch. Dist. #203, IL	●	●	●	●	●	●	●	·	●	●
Project SMART Consortium, OH	●	●	●	◉	●	◉	●	●	●	●
Rochester City Sch. Dist., NY	●	●	●	●	●	●	●	●	●	●
SW Math/Sci. Collaborative, PA	–	–	–	–	–	–	–	–	–	–

● All or almost all students (at least 90%)

◉ About half of the students

⊙ Only the more able students (top track- about 25%)

· Only the most advanced students (10% or less)

○ Not included in curriculum

– Data not available

SOURCE: IEA Third International Mathematics and Science Study (TIMSS), 1998-1999.

Background data provided by coordinators from participating jurisdictions.

Detailed Information About Topics in the Intended Curriculum, Up to and Including Eighth Grade – Data Representation, Analysis, and Probability

8th Grade Mathematics

	Collecting and graphing data from a survey	Representation and interpretation of data in graphs, charts, and tables	Arithmetic mean	Median and mode	Simple probabilities – understanding and calculations
Countries					
United States	●	●	●	●	●
Belgium (Flemish)	●	●	●	◉	●
Canada	●	●	●	●	●
Chinese Taipei	●	◉	○	○	●
Czech Republic	●	●	●	●	◦
England	●	●	◉	◉	◉
Hong Kong, SAR	●	●	○	○	○
Italy	●	●	●	○	●
Japan	●	●	●	●	○
Korea, Rep. of	●	●	●	○	●
Netherlands	●	●	●	◉	◉
Russian Federation	●	●	●	●	●
Singapore	●	●	●	●	○
States					
Connecticut	●	●	●	●	●
Idaho	●	●	●	●	●
Illinois	●	●	●	●	●
Indiana	●	●	●	●	●
Maryland	●	●	●	●	●
Massachusetts	●	●	●	●	●
Michigan	●	●	●	●	●
Missouri	●	●	●	●	●
North Carolina	●	●	●	●	●
Oregon	●	●	●	●	●
Pennsylvania	●	●	●	●	●
South Carolina	●	●	●	●	●
Texas	●	●	●	●	●
Districts and Consortia					
Academy School Dist. #20, CO	–	–	–	–	–
Chicago Public Schools, IL	◉	◉	●	●	●
Delaware Science Coalition, DE	●	●	●	●	●
First in the World Consort., IL	●	●	●	●	●
Fremont/Lincoln/WestSide PS, NE	●	●	●	●	●
Guilford County, NC	●	●	●	●	●
Jersey City Public Schools, NJ	●	●	●	●	●
Miami-Dade County PS, FL	●	●	●	●	●
Michigan Invitational Group, MI	●	●	●	●	●
Montgomery County, MD	●	●	●	●	●
Naperville Sch. Dist. #203, IL	●	●	●	●	●
Project SMART Consortium, OH	●	●	●	●	●
Rochester City Sch. Dist., NY	●	●	●	●	●
SW Math/Sci. Collaborative, PA	–	–	–	–	–

● All or almost all students (at least 90%)

◉ About half of the students

⊙ Only the more able students (top track-about 25%)

◦ Only the most advanced students (10% or less)

○ Not included in curriculum

– Data not available

SOURCE: IEA Third International Mathematics and Science Study (TIMSS), 1998-1999.

Background data provided by coordinators from participating jurisdictions.

Detailed Information About Topics in the Intended Curriculum, Up to and Including Eighth Grade – Geometry

8th Grade Mathematics

	Cartesian coordinates of points in a plane	Coordinates of points on a given straight line	Simple two dimensional geometry – angles on a straight line, parallel lines, triangles and quadrilaterals	Congruence and similarity	Angles – (acute, right, supplementary, etc.)	Pythagorean theorem (without proof)	Symmetry and transformations (reflection and rotation)	Visualization of three-dimensional shapes	Geometric constructions with straight-edge and compass	Regular polygons and their properties – names (e.g., hexagon and octagon), sum of angles, etc.	Proofs (formal deductive demonstrations of geometric relationships)	Sine, cosine, and tangent in right-angle triangles	Nets of solids
Countries													
United States	●	●	●	●	●	●	●	●	●	●	◦	○	●
Belgium (Flemish)	●	◉	●	●	●	○	●	●	●	○	⊙	○	●
Canada	●	◉	●	●	●	●	●	●	●	●	○	○	●
Chinese Taipei	●	●	●	◉	●	●	⊙	◉	●	◉	○	○	○
Czech Republic	●	○	●	●	●	●	●	●	●	●	⊙	○	◉
England	●	◉	◉	⊙	◉	⊙	●	●	◉	◉	○	◦	◉
Hong Kong, SAR	●	●	●	●	●	●	○	○	●	●	●	●	○
Italy	●	●	●	●	●	●	●	●	●	●	●	○	●
Japan	●	●	●	●	●	○	●	●	●	●	●	○	●
Korea, Rep. of	○	○	●	●	●	○	○	●	●	●	●	○	○
Netherlands	●	●	●	○	●	●	◉	●	○	◉	○	○	●
Russian Federation	●	●	●	◉	●	●	◉	●	●	◉	●	○	○
Singapore	●	●	●	●	●	◉	●	●	●	●	○	◉	●
States													
Connecticut	●	●	●	●	●	◉	●	●	◉	●	⊙	◉	◉
Idaho	●	●	●	●	●	◦	●	⊙	◦	●	○	◦	◦
Illinois	●	●	●	●	●	●	●	●	◉	●	◦	⊙	●
Indiana	●	●	●	●	●	●	●	●	○	●	○	○	○
Maryland	●	●	●	●	●	●	●	●	●	●	○	○	○
Massachusetts	●	●	●	●	●	●	●	●	○	○	○	○	●
Michigan	●	●	●	●	●	●	●	●	◉	●	⊙	⊙	●
Missouri	●	●	●	●	●	◉	●	●	●	◉	◦	⊙	◦
North Carolina	●	●	●	●	●	●	●	●	○	●	⊙	○	●
Oregon	●	●	●	●	●	●	●	●	○	●	○	○	○
Pennsylvania	●	●	●	●	●	●	●	●	●	●	●	●	●
South Carolina	●	●	●	●	●	●	●	●	●	●	◦	●	●
Texas	●	●	●	●	●	●	●	●	●	●	◉	◉	●
Districts and Consortia													
Academy School Dist. #20, CO	–	–	–	–	–	–	–	–	–	–	–	–	–
Chicago Public Schools, IL	●	◉	●	◉	●	◉	●	●	◉	●	◦	○	◉
Delaware Science Coalition, DE	●	●	●	●	●	⊙	●	●	◉	●	○	○	●
First in the World Consort., IL	●	●	●	●	●	●	●	●	●	●	⊙	⊙	●
Fremont/Lincoln/WestSide PS, NE	●	●	●	●	●	●	●	●	◉	●	◦	●	◉
Guilford County, NC	●	●	●	●	●	●	●	●	○	●	◦	○	●
Jersey City Public Schools, NJ	●	●	●	●	●	●	●	●	●	●	◦	◦	●
Miami-Dade County PS, FL	●	●	●	●	●	●	●	●	●	●	○	○	●
Michigan Invitational Group, MI	●	●	●	●	●	●	●	●	●	●	○	○	◦
Montgomery County, MD	●	●	●	●	●	●	●	●	●	●	◦	○	●
Naperville Sch. Dist. #203, IL	●	●	●	●	●	●	●	●	●	●	◦	◦	●
Project SMART Consortium, OH	●	⊙	●	◉	●	●	●	●	○	●	○	●	●
Rochester City Sch. Dist., NY	●	●	●	●	●	●	●	●	●	●	◉	◉	◉
SW Math/Sci. Collaborative, PA	–	–	–	–	–	–	–	–	–	–	–	–	–

● All or almost all students (at least 90%)

◉ About half of the students

⊙ Only the more able students (top track- about 25%)

◦ Only the most advanced students (10% or less)

○ Not included in curriculum

– Data not available

SOURCE: IEA Third International Mathematics and Science Study (TIMSS), 1998-1999.

Background data provided by coordinators from participating jurisdictions.

R

Exhibit R2.6 Detailed Information About Topics in the Intended Curriculum, Up to and Including Eighth Grade – Algebra

8th Grade Mathematics

	Number patterns and simple relations	Writing expressions for general terms in number pattern sequence	Translating from verbal descriptions to symbolic expressions	Simple algebraic expressions	Evaluating simple algebraic expressions by substitution of given value of variables	Representing situations algebraically; formulas	Solving simple equations	Solving simple inequalities	Solving simultaneous equations in two variables	Interpreting linear relations	Using the graph of a relationship to interpolate/extrapolate
Countries											
United States	●	●	●	●	●	●	●	●	◦	●	◦
Belgium (Flemish)	●	●	●	●	●	●	●	○	○	○	○
Canada	●	◉	●	●	●	●	●	◦	○	⊙	◉
Chinese Taipei	●	◉	●	●	●	◉	●	⊙	●	◉	○
Czech Republic	◉	◉	●	●	●	●	●	◦	○	○	○
England	●	◉	◉	◉	◉	◉	◉	⊙	◦	○	○
Hong Kong, SAR	●	○	●	●	●	●	●	○	●	●	○
Italy	●	●	●	●	●	●	●	●	●	●	○
Japan	●	●	●	●	●	●	●	●	●	●	●
Korea, Rep. of	●	○	●	●	●	●	●	●	●	●	●
Netherlands	●	◉	◉	◉	⊙	◉	◉	○	○	●	●
Russian Federation	○	○	●	●	●	●	●	●	●	●	○
Singapore	●	●	●	●	●	●	●	◉	●	●	●
States											
Connecticut	●	●	●	●	●	●	●	◉	⊙	◉	◉
Idaho	●	●	●	●	●	●	●	⊙	⊙	⊙	◦
Illinois	●	●	●	●	●	●	●	●	●	●	●
Indiana	●	⊙	●	●	●	●	●	●	○	●	●
Maryland	●	○	●	●	●	○	●	●	○	●	○
Massachusetts	●	○	●	●	●	○	●	○	○	○	●
Michigan	●	●	●	●	●	◉	●	●	◉	◉	●
Missouri	●	●	●	●	●	◉	●	◉	⊙	⊙	◦
North Carolina	●	●	●	●	●	●	●	●	◉	◉	◉
Oregon	●	●	●	●	●	●	●	●	○	○	○
Pennsylvania	●	●	●	●	●	●	●	●	●	●	●
South Carolina	●	●	●	●	●	●	●	●	◦	●	◉
Texas	●	●	●	●	●	●	●	●	⊙	⊙	⊙
Districts and Consortia											
Academy School Dist. #20, CO	–	–	–	–	–	–	–	–	–	–	–
Chicago Public Schools, IL	●	◉	◉	◉	◉	◉	◉	⊙	◦	◦	◦
Delaware Science Coalition, DE	●	●	●	●	●	●	●	●	○	●	●
First in the World Consort., IL	●	●	●	●	●	●	●	●	●	●	⊙
Fremont/Lincoln/WestSide PS, NE	●	●	●	●	●	●	●	●	◉	◉	●
Guilford County, NC	●	●	●	●	●	●	●	●	◉	●	◉
Jersey City Public Schools, NJ	●	●	●	●	●	●	●	●	◦	⊙	◦
Miami-Dade County PS, FL	●	●	●	●	●	●	●	●	⊙	●	●
Michigan Invitational Group, MI	●	●	●	●	●	●	●	●	◦	●	◉
Montgomery County, MD	●	◉	●	●	●	●	●	●	◉	◉	●
Naperville Sch. Dist. #203, IL	●	●	●	●	●	●	●	●	⊙	●	⊙
Project SMART Consortium, OH	●	●	●	●	●	●	●	○	○	●	●
Rochester City Sch. Dist., NY	●	●	●	●	●	●	●	●	◉	●	◉
SW Math/Sci. Collaborative, PA	–	–	–	–	–	–	–	–	–	–	–

●	All or almost all students (at least 90%)
◉	About half of the students
⊙	Only the more able students (top track-about 25%)
◦	Only the most advanced students (10% or less)
○	Not included in curriculum
–	Data not available

SOURCE: IEA Third International Mathematics and Science Study (TIMSS), 1998-1999.

Background data provided by coordinators from participating jurisdictions.

		Percentage of Students					
		Taught Topics Before This Year Only		Taught Topics During This Year[1]			Not Yet Taught 50% or More of Topics
		More Than 80% of Topics	More Than 50% Up to and Including 80% of Topics	More Than 50% of Topics Each Taught More Than 5 Periods	More Than 50% of Topics Each Taught at Least 1-5 Periods	50% or Less of Topics Taught	
Countries							
United States		8 (1.4)	9 (1.4)	34 (2.8)	48 (3.2)	1 (0.7)	0 (0.1)
Belgium (Flemish)		21 (3.0)	19 (2.3)	2 (1.0)	42 (3.7)	10 (3.6)	6 (2.9)
Canada	r	1 (0.6)	9 (2.0)	27 (2.7)	63 (3.3)	1 (0.4)	0 (0.3)
Chinese Taipei		90 (2.4)	8 (2.1)	0 (0.0)	2 (1.1)	0 (0.0)	0 (0.0)
Czech Republic		53 (5.7)	25 (4.3)	5 (2.2)	16 (3.3)	1 (0.8)	0 (0.0)
England	s	8 (2.4)	19 (3.3)	3 (0.9)	63 (4.8)	6 (2.1)	1 (0.6)
Hong Kong, SAR		18 (3.0)	56 (4.5)	2 (1.2)	18 (3.6)	5 (2.0)	1 (0.8)
Italy		39 (3.9)	42 (4.1)	4 (1.3)	14 (2.9)	1 (0.5)	0 (0.0)
Japan		51 (4.9)	30 (4.3)	1 (0.0)	16 (3.3)	2 (1.2)	0 (0.0)
Korea, Rep. of		10 (2.4)	14 (2.8)	11 (2.5)	57 (4.0)	6 (2.0)	2 (1.3)
Netherlands		8 (2.3)	28 (5.8)	17 (6.3)	41 (5.8)	5 (2.7)	0 (0.0)
Russian Federation		– –	– –	– –	– –	– –	– –
Singapore		37 (4.2)	35 (4.3)	6 (2.0)	22 (3.7)	0 (0.0)	0 (0.0)
States							
Connecticut	r	16 (5.4)	17 (5.4)	33 (6.0)	32 (5.4)	2 (1.5)	0 (0.0)
Idaho	r	6 (4.0)	5 (2.4)	32 (5.2)	55 (6.0)	1 (0.1)	0 (0.3)
Illinois		6 (2.3)	16 (4.8)	31 (5.3)	44 (6.2)	3 (2.1)	0 (0.0)
Indiana		6 (3.0)	7 (2.5)	36 (7.0)	49 (7.2)	3 (1.8)	0 (0.0)
Maryland	r	13 (3.6)	26 (6.1)	17 (4.7)	44 (5.9)	0 (0.0)	0 (0.0)
Massachusetts		9 (3.3)	17 (3.8)	28 (3.3)	41 (4.8)	5 (2.3)	0 (0.0)
Michigan		18 (3.3)	25 (3.9)	18 (3.9)	38 (5.2)	1 (1.3)	0 (0.0)
Missouri		5 (2.3)	10 (2.1)	26 (5.3)	58 (5.7)	1 (0.9)	0 (0.0)
North Carolina		3 (2.0)	6 (3.1)	26 (5.2)	64 (6.0)	1 (0.0)	0 (0.0)
Oregon		5 (2.2)	11 (3.5)	25 (3.9)	59 (5.0)	0 (0.0)	0 (0.0)
Pennsylvania		11 (6.2)	15 (2.9)	21 (3.4)	53 (7.0)	1 (0.6)	0 (0.0)
South Carolina		9 (3.6)	13 (4.0)	26 (5.3)	52 (5.7)	0 (0.0)	0 (0.0)
Texas		13 (4.8)	9 (3.0)	28 (5.2)	48 (7.3)	0 (0.0)	2 (1.3)
Districts and Consortia							
Academy School Dist. #20, CO		18 (0.3)	17 (0.3)	22 (0.4)	43 (0.4)	0 (0.0)	0 (0.0)
Chicago Public Schools, IL		0 (0.0)	2 (0.2)	55 (10.7)	41 (10.6)	2 (0.2)	0 (0.0)
Delaware Science Coalition, DE	r	14 (4.9)	24 (6.0)	27 (6.5)	34 (5.5)	0 (0.0)	1 (0.5)
First in the World Consort., IL	r	14 (4.1)	28 (3.7)	18 (4.7)	40 (4.7)	0 (0.0)	0 (0.0)
Fremont/Lincoln/WestSide PS, NE		3 (0.1)	0 (0.0)	33 (7.7)	64 (7.7)	0 (0.0)	0 (0.0)
Guilford County, NC		7 (2.2)	11 (3.7)	18 (5.9)	64 (6.6)	0 (0.0)	0 (0.0)
Jersey City Public Schools, NJ		6 (4.2)	6 (5.1)	42 (4.0)	46 (3.8)	0 (0.0)	0 (0.0)
Miami-Dade County PS, FL	s	7 (4.5)	8 (5.8)	24 (6.8)	58 (11.3)	1 (0.1)	2 (0.3)
Michigan Invitational Group, MI		8 (5.6)	27 (7.1)	8 (2.1)	55 (7.8)	2 (0.1)	0 (0.0)
Montgomery County, MD	s	30 (5.9)	20 (4.0)	14 (4.4)	35 (5.1)	0 (0.0)	0 (0.0)
Naperville Sch. Dist. #203, IL		6 (2.0)	22 (2.5)	6 (1.0)	66 (3.5)	0 (0.0)	0 (0.0)
Project SMART Consortium, OH		18 (5.3)	4 (2.0)	34 (6.9)	42 (6.7)	2 (2.5)	0 (0.2)
Rochester City Sch. Dist., NY		11 (4.2)	7 (2.6)	15 (2.0)	63 (4.5)	4 (1.0)	0 (0.0)
SW Math/Sci. Collaborative, PA		7 (3.4)	23 (4.3)	20 (4.9)	47 (6.1)	3 (0.2)	0 (0.0)
International Avg. (All Countries)		26 (0.5)	24 (0.6)	11 (0.5)	34 (0.6)	4 (0.3)	1 (0.2)

SOURCE: IEA Third International Mathematics and Science Study (TIMSS), 1998-1999.

Background data provided by teachers.

* Categories of topic coverage for fractions and number sense are based on combined responses to questions about the individual mathematics subtopics in the content area described in Exhibit 5.20.

1 For each topic in Exhibit 5.20, teachers were asked if the topic was taught before this year, taught 1-5 periods this year, taught more than 5 periods this year, or not yet taught. Topics taught during this year are included in this category regardless if taught before this year.

States in *italics* did not fully satisfy guidelines for sample participation rates (see Appendix A for details).

( ) Standard errors appear in parentheses. Because results are rounded to the nearest whole number, some totals may appear inconsistent.

A dash (–) indicates data are not available.

An "r" indicates teacher response data available for 70-84% of students. An "s" indicates teacher response data available for 50-69% of students.

Exhibit R2.8 When Measurement Topics Are Taught*

8th Grade Mathematics

	Percentage of Students					
	Taught Topics Before This Year Only		Taught Topics During This Year[1]			Not Yet Taught 50% or More of Topics
	More Than 80% of Topics	More Than 50% Up to and Including 80% of Topics	More Than 50% of Topics Each Taught More Than 5 Periods	More Than 50% of Topics Each Taught at Least 1-5 Periods	50% or Less of Topics Taught	
Countries						
United States	10 (2.2)	11 (1.9)	16 (2.9)	54 (3.6)	3 (0.9)	6 (1.4)
Belgium (Flemish)	33 (3.5)	27 (3.8)	4 (3.4)	19 (3.0)	13 (3.7)	3 (1.4)
Canada r	1 (0.5)	8 (1.6)	21 (2.9)	56 (3.4)	11 (1.4)	2 (0.8)
Chinese Taipei	20 (3.6)	53 (4.4)	3 (1.4)	5 (1.8)	17 (3.3)	2 (1.4)
Czech Republic	50 (5.9)	29 (5.0)	4 (2.0)	14 (3.4)	4 (1.7)	0 (0.0)
England s	8 (2.4)	18 (2.7)	5 (1.3)	58 (3.8)	8 (1.5)	3 (0.9)
Hong Kong, SAR	15 (3.1)	28 (4.2)	5 (1.8)	41 (4.4)	10 (2.8)	1 (1.1)
Italy	29 (3.8)	42 (4.0)	7 (2.3)	15 (2.9)	7 (1.8)	1 (0.6)
Japan	49 (4.6)	26 (4.3)	1 (0.8)	8 (2.1)	5 (2.0)	12 (2.9)
Korea, Rep. of	11 (2.5)	19 (3.3)	8 (2.4)	49 (4.1)	7 (2.0)	6 (1.7)
Netherlands r	6 (3.3)	8 (2.7)	15 (6.2)	51 (6.8)	15 (3.6)	7 (4.7)
Russian Federation	– –	– –	– –	– –	– –	– –
Singapore	39 (4.8)	32 (4.6)	8 (2.5)	19 (3.7)	2 (1.1)	0 (0.0)
States						
Connecticut r	15 (3.7)	17 (5.7)	28 (5.7)	30 (6.2)	6 (2.6)	4 (2.3)
Idaho r	12 (4.6)	4 (2.2)	13 (4.1)	55 (7.1)	3 (1.8)	13 (5.0)
Illinois	12 (4.0)	9 (2.3)	17 (4.4)	58 (5.7)	2 (1.4)	2 (1.5)
Indiana	5 (2.9)	14 (4.5)	15 (3.6)	44 (7.3)	20 (7.2)	2 (1.5)
Maryland r	21 (4.5)	18 (4.9)	9 (3.5)	44 (5.3)	4 (2.2)	4 (2.2)
Massachusetts r	15 (4.9)	17 (4.0)	20 (4.6)	37 (4.2)	6 (2.7)	5 (2.7)
Michigan	19 (4.4)	18 (3.9)	10 (3.8)	45 (6.3)	5 (2.5)	2 (1.3)
Missouri	5 (2.3)	11 (2.7)	12 (3.2)	61 (5.5)	5 (2.4)	5 (3.2)
North Carolina	8 (1.9)	7 (2.5)	12 (3.3)	64 (4.9)	5 (2.3)	5 (2.3)
Oregon	2 (1.6)	15 (4.3)	15 (4.4)	60 (6.8)	6 (3.3)	2 (0.9)
Pennsylvania	15 (6.6)	11 (3.2)	13 (3.6)	47 (4.1)	10 (5.5)	4 (1.7)
South Carolina	12 (4.5)	10 (3.6)	15 (3.8)	62 (5.3)	1 (0.3)	0 (0.0)
Texas	18 (5.2)	5 (2.5)	15 (3.3)	61 (6.3)	1 (0.1)	0 (0.0)
Districts and Consortia						
Academy School Dist. #20, CO	2 (0.1)	20 (0.4)	16 (0.3)	38 (0.3)	14 (0.2)	10 (0.3)
Chicago Public Schools, IL	7 (5.5)	0 (0.0)	35 (7.2)	58 (10.3)	0 (0.0)	0 (0.0)
Delaware Science Coalition, DE r	13 (6.2)	11 (5.2)	17 (6.1)	57 (7.5)	2 (0.1)	1 (0.1)
First in the World Consort., IL r	11 (3.6)	5 (2.6)	16 (7.8)	65 (7.8)	0 (0.0)	3 (0.2)
Fremont/Lincoln/WestSide PS, NE r	13 (1.2)	9 (0.2)	3 (0.1)	54 (6.7)	10 (0.5)	11 (6.7)
Guilford County, NC	15 (5.1)	17 (4.2)	12 (4.4)	46 (6.9)	8 (4.4)	3 (0.1)
Jersey City Public Schools, NJ r	9 (4.2)	0 (0.0)	38 (6.5)	53 (6.8)	0 (0.0)	0 (0.0)
Miami-Dade County PS, FL s	4 (3.6)	3 (2.6)	19 (5.0)	50 (6.9)	13 (8.3)	11 (5.3)
Michigan Invitational Group, MI	14 (5.4)	18 (6.8)	10 (4.6)	50 (10.3)	8 (3.0)	0 (0.0)
Montgomery County, MD s	36 (2.7)	13 (2.2)	10 (5.1)	34 (7.0)	7 (3.3)	0 (0.0)
Naperville Sch. Dist. #203, IL	6 (3.1)	27 (5.1)	8 (0.3)	53 (5.0)	6 (0.2)	0 (0.0)
Project SMART Consortium, OH	7 (3.7)	3 (2.3)	26 (6.4)	63 (6.2)	0 (0.0)	0 (0.0)
Rochester City Sch. Dist., NY	4 (1.8)	30 (5.7)	2 (0.0)	51 (5.0)	6 (2.5)	7 (2.0)
SW Math/Sci. Collaborative, PA	11 (3.5)	16 (4.1)	20 (6.0)	38 (4.8)	10 (4.6)	6 (4.3)
International Avg. (All Countries)	22 (0.6)	23 (0.6)	8 (0.4)	32 (0.7)	8 (0.4)	6 (0.4)

SOURCE: IEA Third International Mathematics and Science Study (TIMSS), 1998-1999.

Background data provided by teachers.

* Categories of topic coverage for measurement are based on combined responses to questions about the individual mathematics subtopics in the content area described in Exhibit 5.21.

1 For each topic in Exhibit 5.21, teachers were asked if the topic was taught before this year, taught 1-5 periods this year, taught more than 5 periods this year, or not yet taught. Topics taught during this year are included in this category regardless if taught before this year.

States in *italics* did not fully satisfy guidelines for sample participation rates (see Appendix A for details).

( ) Standard errors appear in parentheses. Because results are rounded to the nearest whole number, some totals may appear inconsistent.

A dash (–) indicates data are not available.

An "r" indicates teacher response data available for 70-84% of students. An "s" indicates teacher response data available for 50-69% of students.

Exhibit R2.9 When Data Representation, Analysis, and Probability Topics Are Taught*

		Percentage of Students					
		Taught Topics Before This Year Only		Taught Topics During This Year[1]			Not Yet Taught 50% or More of Topics
		More Than 80% of Topics	More Than 50% Up to and Including 80% of Topics	More Than 50% of Topics Each Taught More Than 5 Periods	More Than 50% of Topics Each Taught at Least 1-5 Periods	50% or Less of Topics Taught	
Countries							
United States		6 (1.5)	7 (2.5)	26 (2.4)	53 (3.2)	2 (1.1)	6 (1.3)
Belgium (Flemish)		8 (1.6)	23 (3.0)	0 (0.0)	27 (4.2)	24 (3.0)	18 (4.2)
Canada	r	2 (0.8)	5 (1.6)	27 (3.2)	45 (3.4)	8 (0.8)	13 (3.0)
Chinese Taipei		2 (1.2)	3 (1.4)	1 (0.8)	1 (0.7)	1 (0.0)	92 (2.1)
Czech Republic		2 (1.7)	24 (5.1)	1 (1.0)	7 (2.1)	13 (3.8)	52 (5.3)
England	s	7 (1.7)	15 (3.2)	11 (2.2)	62 (3.9)	3 (1.3)	3 (0.7)
Hong Kong, SAR		3 (1.6)	13 (3.1)	1 (0.9)	7 (2.3)	6 (2.2)	70 (4.2)
Italy		2 (1.1)	17 (2.8)	10 (2.2)	33 (3.9)	4 (1.5)	34 (3.4)
Japan		2 (1.2)	8 (2.7)	1 (0.7)	12 (2.9)	10 (2.6)	68 (4.2)
Korea, Rep. of		3 (1.3)	23 (3.4)	21 (3.2)	38 (4.0)	10 (2.5)	4 (1.6)
Netherlands		0 (0.0)	7 (2.6)	17 (5.8)	48 (6.6)	6 (2.3)	22 (5.7)
Russian Federation		– –	– –	– –	– –	– –	– –
Singapore		2 (1.4)	2 (1.3)	28 (3.7)	54 (3.2)	1 (0.0)	13 (3.3)
States							
Connecticut	s	8 (2.7)	13 (5.3)	37 (6.7)	39 (5.9)	2 (1.5)	1 (0.1)
Idaho	r	6 (2.6)	12 (4.2)	18 (4.9)	53 (8.2)	1 (0.1)	10 (3.6)
Illinois		8 (3.2)	6 (2.5)	26 (5.0)	56 (6.1)	3 (2.0)	2 (1.0)
Indiana		3 (2.0)	6 (3.3)	28 (5.6)	48 (6.1)	5 (2.4)	10 (6.6)
Maryland	r	2 (1.4)	4 (1.7)	44 (5.1)	48 (4.6)	2 (1.7)	0 (0.0)
Massachusetts	r	8 (2.8)	5 (2.4)	34 (5.7)	42 (6.2)	7 (2.2)	5 (2.0)
Michigan	r	13 (4.1)	11 (3.1)	17 (3.8)	53 (4.3)	3 (1.4)	3 (1.5)
Missouri		7 (2.1)	6 (2.4)	19 (5.1)	65 (6.9)	1 (0.0)	3 (2.0)
North Carolina		1 (0.9)	7 (2.6)	21 (4.4)	56 (4.6)	4 (2.9)	10 (3.6)
Oregon		3 (1.8)	4 (2.5)	33 (5.3)	56 (5.3)	1 (0.1)	3 (1.0)
Pennsylvania		10 (3.2)	9 (4.7)	17 (3.7)	53 (7.5)	1 (0.6)	10 (2.8)
South Carolina		5 (2.1)	11 (4.5)	26 (6.2)	56 (7.4)	2 (0.1)	0 (0.0)
Texas		6 (3.0)	5 (3.1)	31 (4.7)	57 (4.7)	0 (0.0)	0 (0.3)
Districts and Consortia							
Academy School Dist. #20, CO		8 (0.3)	4 (0.1)	28 (0.3)	52 (0.4)	3 (0.0)	5 (0.3)
Chicago Public Schools, IL		0 (0.0)	0 (0.0)	36 (10.7)	63 (10.8)	0 (0.0)	1 (1.3)
Delaware Science Coalition, DE	r	8 (4.8)	5 (3.6)	44 (7.3)	37 (6.5)	0 (0.0)	6 (3.2)
First in the World Consort., IL	r	14 (4.5)	11 (5.6)	13 (4.0)	62 (8.2)	0 (0.0)	0 (0.0)
Fremont/Lincoln/WestSide PS, NE		5 (5.2)	12 (6.9)	17 (7.1)	55 (9.8)	5 (5.3)	5 (2.8)
Guilford County, NC		7 (2.5)	10 (5.1)	15 (4.4)	55 (6.0)	0 (0.0)	13 (3.7)
Jersey City Public Schools, NJ		6 (4.2)	0 (0.0)	49 (5.5)	45 (4.9)	0 (0.0)	0 (0.0)
Miami-Dade County PS, FL	s	6 (4.0)	8 (7.1)	20 (5.2)	59 (7.4)	0 (0.0)	6 (4.1)
Michigan Invitational Group, MI		12 (5.2)	5 (3.5)	31 (8.6)	47 (8.5)	5 (5.2)	0 (0.0)
Montgomery County, MD	s	6 (3.4)	12 (3.5)	26 (5.4)	48 (5.1)	7 (4.3)	2 (0.2)
Naperville Sch. Dist. #203, IL		2 (1.9)	0 (0.0)	18 (3.5)	80 (2.6)	0 (0.0)	0 (0.0)
Project SMART Consortium, OH		4 (2.7)	1 (0.7)	23 (5.5)	70 (6.5)	0 (0.0)	3 (2.6)
Rochester City Sch. Dist., NY		6 (3.7)	19 (4.3)	20 (3.3)	40 (4.3)	6 (0.2)	9 (1.8)
SW Math/Sci. Collaborative, PA		14 (5.5)	13 (3.8)	18 (5.9)	43 (6.9)	4 (2.7)	8 (4.5)
International Avg. (All Countries)		5 (0.3)	14 (0.5)	9 (0.4)	30 (0.6)	7 (0.4)	34 (0.6)

SOURCE: IEA Third International Mathematics and Science Study (TIMSS), 1998-1999.

Background data provided by teachers.

* Categories of topic coverage for data representation, analysis, and probability are based on combined responses to questions about the individual mathematics subtopics in the content area described in Exhibit 5.22.

1 For each topic in Exhibit 5.22, teachers were asked if the topic was taught before this year, taught 1-5 periods this year, taught more than 5 periods this year, or not yet taught. Topics taught during this year are included in this category regardless if taught before this year.

States in *italics* did not fully satisfy guidelines for sample participation rates (see Appendix A for details).

( ) Standard errors appear in parentheses. Because results are rounded to the nearest whole number, some totals may appear inconsistent.

A dash (–) indicates data are not available.

An "r" indicates teacher response data available for 70-84% of students. An "s" indicates teacher response data available for 50-69% of students.

		Percentage of Students					
		Taught Topics Before This Year Only		Taught Topics During This Year[1]			Not Yet Taught 50% or More of Topics
		More Than 80% of Topics	More Than 50% Up to and Including 80% of Topics	More Than 50% of Topics Each Taught More Than 5 Periods	More Than 50% of Topics Each Taught at Least 1-5 Periods	50% or Less of Topics Taught	
Countries							
United States		3 (1.0)	7 (1.4)	14 (2.2)	42 (2.9)	10 (2.0)	25 (2.9)
Belgium (Flemish)		0 (0.0)	5 (1.4)	10 (1.9)	47 (3.5)	15 (2.1)	22 (2.4)
Canada	r	2 (0.5)	3 (1.0)	14 (2.9)	52 (3.2)	12 (2.2)	18 (2.6)
Chinese Taipei		1 (0.0)	1 (0.5)	6 (2.1)	18 (3.3)	42 (4.1)	33 (4.1)
Czech Republic		35 (4.6)	23 (4.8)	4 (2.3)	17 (3.1)	17 (3.8)	4 (1.9)
England	s	13 (2.4)	18 (3.1)	2 (0.8)	29 (2.5)	23 (3.4)	15 (2.7)
Hong Kong, SAR		13 (2.7)	21 (3.5)	5 (2.0)	16 (2.7)	30 (4.0)	14 (3.2)
Italy		2 (1.0)	10 (2.8)	9 (2.2)	29 (3.6)	41 (3.9)	9 (2.3)
Japan		2 (1.5)	21 (3.2)	8 (2.4)	35 (4.1)	32 (4.4)	1 (1.0)
Korea, Rep. of		5 (1.8)	6 (1.8)	12 (2.4)	57 (4.4)	19 (3.4)	1 (0.0)
Netherlands		3 (1.3)	17 (4.5)	15 (5.1)	24 (5.1)	25 (4.8)	17 (4.9)
Russian Federation		– –	– –	– –	– –	– –	– –
Singapore		1 (0.0)	1 (0.0)	24 (4.1)	62 (4.4)	5 (2.0)	7 (2.4)
States							
Connecticut	r	1 (1.2)	10 (4.6)	8 (3.4)	34 (6.9)	8 (4.5)	39 (6.5)
Idaho	r	3 (2.2)	6 (2.7)	7 (2.4)	43 (7.6)	8 (4.3)	32 (5.6)
Illinois		6 (2.2)	11 (4.2)	10 (3.1)	49 (6.3)	10 (3.9)	13 (3.7)
Indiana		2 (1.3)	8 (3.4)	8 (3.4)	37 (7.5)	19 (5.1)	27 (5.8)
Maryland	r	4 (1.9)	11 (3.5)	10 (2.3)	31 (6.4)	13 (4.6)	32 (5.8)
Massachusetts	r	2 (1.5)	9 (3.1)	13 (3.8)	31 (6.0)	7 (2.8)	38 (5.7)
Michigan	r	8 (3.7)	17 (4.8)	16 (4.5)	41 (5.4)	5 (2.8)	14 (3.1)
Missouri		4 (1.9)	5 (2.5)	4 (2.0)	62 (6.1)	7 (2.2)	19 (5.3)
North Carolina		1 (1.1)	6 (2.3)	14 (3.2)	64 (4.7)	4 (1.8)	12 (3.5)
Oregon		0 (0.0)	2 (1.4)	14 (4.6)	64 (6.5)	5 (2.7)	15 (4.3)
Pennsylvania		7 (6.0)	7 (2.9)	6 (2.2)	43 (5.1)	9 (2.9)	28 (7.4)
South Carolina		1 (1.0)	8 (3.7)	15 (4.5)	59 (6.9)	6 (2.8)	10 (3.2)
Texas		4 (1.9)	9 (3.3)	11 (2.6)	63 (4.8)	9 (3.9)	4 (2.3)
Districts and Consortia							
Academy School Dist. #20, CO		2 (0.1)	0 (0.0)	21 (0.3)	22 (0.3)	6 (0.1)	49 (0.4)
Chicago Public Schools, IL		2 (2.4)	6 (0.6)	17 (7.3)	55 (8.4)	1 (0.7)	19 (5.1)
Delaware Science Coalition, DE	r	0 (0.0)	10 (5.2)	21 (6.3)	38 (7.4)	11 (2.2)	20 (5.0)
First in the World Consort., IL		3 (1.0)	11 (3.5)	24 (9.2)	36 (9.1)	20 (4.9)	6 (3.0)
Fremont/Lincoln/WestSide PS, NE	r	0 (0.0)	14 (1.4)	22 (1.3)	31 (8.6)	7 (3.6)	26 (9.7)
Guilford County, NC		0 (0.0)	19 (3.4)	18 (5.4)	41 (6.5)	9 (5.4)	13 (4.9)
Jersey City Public Schools, NJ		4 (3.8)	2 (1.9)	36 (6.3)	53 (6.1)	2 (0.1)	3 (0.3)
Miami-Dade County PS, FL	s	0 (0.0)	3 (2.7)	0 (0.0)	41 (7.8)	13 (6.2)	44 (9.6)
Michigan Invitational Group, MI		0 (0.0)	19 (5.5)	19 (6.2)	28 (8.6)	25 (6.7)	9 (3.5)
Montgomery County, MD	s	13 (3.9)	13 (3.7)	15 (3.1)	46 (4.3)	12 (3.7)	0 (0.0)
Naperville Sch. Dist. #203, IL		2 (1.9)	13 (2.7)	10 (0.9)	56 (3.8)	17 (2.5)	3 (2.6)
Project SMART Consortium, OH		1 (0.7)	3 (2.0)	6 (3.5)	69 (7.2)	4 (2.8)	17 (5.6)
Rochester City Sch. Dist., NY		2 (1.8)	8 (3.5)	4 (1.0)	39 (5.6)	17 (3.8)	30 (4.1)
SW Math/Sci. Collaborative, PA		6 (3.5)	5 (2.9)	11 (3.0)	42 (5.8)	18 (6.7)	19 (5.4)
International Avg. (All Countries)		6 (0.3)	10 (0.5)	9 (0.4)	33 (0.6)	20 (0.6)	22 (0.5)

SOURCE: IEA Third International Mathematics and Science Study (TIMSS), 1998-1999.

Background data provided by teachers.

* Categories of topic coverage for geometry are based on combined responses to questions about the individual mathematics subtopics in the content area described in Exhibit 5.23.

1 For each topic in Exhibit 5.23, teachers were asked if the topic was taught before this year, taught 1-5 periods this year, taught more than 5 periods this year, or not yet taught. Topics taught during this year are included in this category regardless if taught before this year.

States in *italics* did not fully satisfy guidelines for sample participation rates (see Appendix A for details).

( ) Standard errors appear in parentheses. Because results are rounded to the nearest whole number, some totals may appear inconsistent.

A dash (–) indicates data are not available.

An "r" indicates teacher response data available for 70-84% of students. An "s" indicates teacher response data available for 50-69% of students.

Exhibit R2.11 When Algebra Topics Are Taught*

		Percentage of Students					
		Taught Topics Before This Year Only		Taught Topics During This Year[1]			Not Yet Taught 50% or More of Topics
		More Than 80% of Topics	More Than 50% Up to and Including 80% of Topics	More Than 50% of Topics Each Taught More Than 5 Periods	More Than 50% of Topics Each Taught at Least 1-5 Periods	50% or Less of Topics Taught	
Countries							
United States		3 (1.2)	0 (0.3)	62 (2.7)	32 (2.6)	0 (0.2)	2 (0.9)
Belgium (Flemish)	r	1 (0.7)	9 (1.9)	20 (2.9)	43 (3.6)	11 (2.1)	16 (3.2)
Canada	r	1 (0.5)	1 (0.4)	54 (3.0)	38 (2.6)	0 (0.0)	6 (2.3)
Chinese Taipei		28 (3.6)	57 (4.0)	4 (1.7)	8 (2.1)	2 (1.1)	1 (0.0)
Czech Republic		2 (1.2)	3 (1.5)	69 (5.0)	20 (4.4)	5 (2.4)	2 (1.7)
England	s	0 (0.0)	8 (2.4)	21 (2.9)	60 (3.3)	4 (1.3)	7 (1.4)
Hong Kong, SAR		4 (1.6)	19 (3.3)	25 (4.0)	43 (3.9)	10 (2.7)	1 (0.0)
Italy		0 (0.0)	1 (0.0)	67 (3.7)	28 (3.3)	0 (0.0)	4 (1.5)
Japan		5 (2.3)	30 (4.2)	38 (3.9)	25 (4.0)	2 (1.1)	0 (0.0)
Korea, Rep. of		5 (1.7)	9 (2.5)	36 (4.0)	48 (4.0)	1 (0.0)	1 (0.7)
Netherlands		1 (0.1)	2 (1.1)	32 (6.4)	34 (6.2)	12 (3.9)	19 (6.0)
Russian Federation		– –	– –	– –	– –	– –	– –
Singapore		2 (1.1)	18 (3.4)	32 (3.9)	48 (4.8)	1 (1.0)	0 (0.0)
States							
Connecticut	r	4 (2.8)	1 (0.1)	76 (6.2)	13 (4.3)	0 (0.0)	6 (2.7)
Idaho	r	0 (0.0)	4 (0.2)	63 (6.3)	21 (5.3)	0 (0.0)	12 (5.3)
Illinois		1 (0.7)	2 (1.4)	69 (5.7)	28 (5.1)	0 (0.0)	1 (0.1)
Indiana		0 (0.0)	0 (0.0)	70 (5.7)	24 (5.3)	0 (0.0)	6 (2.4)
Maryland	r	3 (1.4)	0 (0.0)	56 (5.1)	32 (4.4)	0 (0.0)	9 (3.4)
Massachusetts	r	0 (0.5)	1 (1.3)	64 (6.1)	29 (5.2)	1 (0.9)	4 (2.4)
Michigan		2 (1.3)	4 (2.9)	69 (5.2)	24 (5.8)	0 (0.0)	2 (1.1)
Missouri		1 (1.4)	2 (1.2)	51 (6.5)	45 (6.9)	0 (0.0)	1 (1.0)
North Carolina		0 (0.0)	0 (0.2)	58 (6.4)	42 (6.4)	0 (0.0)	0 (0.0)
Oregon		1 (0.6)	0 (0.0)	58 (5.8)	41 (5.7)	0 (0.0)	1 (0.4)
Pennsylvania		1 (0.1)	5 (4.4)	70 (6.6)	22 (3.2)	0 (0.0)	1 (0.9)
South Carolina		1 (0.6)	1 (1.1)	65 (7.5)	29 (6.7)	0 (0.0)	3 (1.7)
Texas		1 (0.1)	3 (2.6)	51 (6.3)	43 (6.1)	0 (0.0)	2 (1.5)
Districts and Consortia							
Academy School Dist. #20, CO		6 (0.3)	0 (0.0)	72 (0.4)	20 (0.3)	0 (0.0)	2 (0.1)
Chicago Public Schools, IL		0 (0.0)	0 (0.0)	68 (10.4)	32 (10.4)	0 (0.0)	0 (0.0)
Delaware Science Coalition, DE	r	0 (0.0)	0 (0.0)	80 (6.2)	15 (5.3)	0 (0.0)	5 (3.3)
First in the World Consort., IL	r	4 (2.9)	5 (1.0)	75 (6.4)	16 (5.9)	0 (0.0)	0 (0.0)
Fremont/Lincoln/WestSide PS, NE		0 (0.0)	0 (0.0)	73 (8.2)	27 (8.2)	0 (0.0)	0 (0.0)
Guilford County, NC		0 (0.0)	2 (0.6)	63 (4.5)	35 (4.4)	0 (0.0)	0 (0.0)
Jersey City Public Schools, NJ	r	6 (4.2)	0 (0.0)	54 (6.7)	39 (6.1)	0 (0.0)	0 (0.0)
Miami-Dade County PS, FL	s	0 (0.0)	0 (0.0)	61 (10.7)	29 (9.7)	3 (0.5)	7 (4.1)
Michigan Invitational Group, MI		0 (0.0)	0 (0.0)	77 (5.6)	17 (5.1)	4 (0.6)	2 (2.2)
Montgomery County, MD	s	8 (4.3)	1 (0.4)	56 (4.7)	31 (3.3)	0 (0.0)	5 (3.2)
Naperville Sch. Dist. #203, IL		2 (1.9)	2 (0.4)	69 (2.3)	27 (0.9)	0 (0.0)	0 (0.0)
Project SMART Consortium, OH		1 (1.1)	0 (0.0)	60 (6.6)	33 (7.6)	0 (0.0)	6 (4.4)
Rochester City Sch. Dist., NY		0 (0.0)	2 (1.7)	43 (4.9)	48 (4.8)	0 (0.0)	7 (2.0)
SW Math/Sci. Collaborative, PA		0 (0.0)	2 (1.8)	57 (6.9)	38 (7.0)	2 (1.8)	1 (0.1)
International Avg. (All Countries)		4 (0.3)	11 (0.4)	33 (0.7)	40 (0.7)	4 (0.3)	8 (0.4)

SOURCE: IEA Third International Mathematics and Science Study (TIMSS), 1998-1999.

Background data provided by teachers.

* Categories of topic coverage for algebra are based on combined responses to questions about the individual mathematics subtopics in the content area described in Exhibit 5.24.

1 For each topic in Exhibit 5.24, teachers were asked if the topic was taught before this year, taught 1-5 periods this year, taught more than 5 periods this year, or not yet taught. Topics taught during this year are included in this category regardless if taught before this year.

States in *italics* did not fully satisfy guidelines for sample participation rates (see Appendix A for details).

( ) Standard errors appear in parentheses. Because results are rounded to the nearest whole number, some totals may appear inconsistent.

A dash (–) indicates data are not available.

An "r" indicates teacher response data available for 70-84% of students. An "s" indicates teacher response data available for 50-69% of students.

REFERENCE

3 Teachers and Instruction

Teachers' Confidence in Their Preparation to Teach Mathematics Topics

	Percentage of Students Whose Teachers Report Feeling Very Well Prepared to Teach Topic[1]					
	Fractions, decimals and percentages	Ratios and proportions	Measurement – units, instruments, and accuracy	Perimeter, area, and volume	Geometric figures – definitions and properties	Geometric figures – symmetry, motions and transformations, congruence and similarity
Countries						
United States	99 (0.8)	97 (1.1)	84 (2.0)	97 (1.1)	86 (2.7)	75 (2.9)
Belgium (Flemish)	97 (1.4)	93 (1.7)	62 (4.1)	92 (2.0)	93 (2.1)	89 (3.1)
Canada	91 (2.1)	89 (2.4)	r 83 (2.7)	93 (2.2)	77 (2.8)	r 62 (3.4)
Chinese Taipei	80 (3.3)	83 (2.9)	65 (3.8)	77 (3.4)	77 (3.1)	70 (3.5)
Czech Republic	99 (1.3)	98 (1.3)	74 (5.0)	99 (1.3)	96 (2.1)	96 (2.0)
England	– –	– –	– –	– –	– –	– –
Hong Kong, SAR	75 (3.6)	76 (3.7)	67 (4.1)	86 (3.0)	66 (4.0)	53 (4.3)
Italy	79 (3.2)	80 (3.1)	55 (3.6)	86 (2.6)	84 (2.7)	45 (3.9)
Japan	15 (3.2)	20 (3.2)	9 (2.0)	26 (3.7)	23 (3.6)	20 (3.5)
Korea, Rep. of	57 (3.7)	52 (3.9)	38 (3.8)	63 (3.2)	72 (3.2)	63 (3.7)
Netherlands	90 (5.8)	90 (5.9)	69 (5.5)	90 (5.8)	82 (6.3)	79 (6.2)
Russian Federation	– –	– –	– –	– –	– –	– –
Singapore	80 (3.8)	82 (3.4)	76 (4.3)	90 (2.9)	79 (3.6)	69 (4.0)
States						
Connecticut	r 99 (1.4)	r 97 (1.9)	r 86 (5.7)	r 96 (2.2)	r 88 (5.9)	r 80 (5.6)
Idaho	r 97 (1.5)	r 92 (2.8)	r 89 (2.7)	r 95 (2.3)	r 72 (5.2)	r 57 (7.1)
Illinois	100 (0.1)	96 (3.6)	83 (5.1)	97 (2.2)	90 (4.5)	73 (6.1)
Indiana	98 (1.7)	100 (0.0)	86 (4.7)	95 (3.1)	88 (2.6)	72 (6.0)
Maryland	r 98 (1.4)	r 98 (1.4)	r 92 (3.0)	r 100 (0.1)	r 90 (3.5)	r 78 (4.5)
Massachusetts	96 (2.4)	95 (2.6)	88 (4.0)	96 (2.4)	87 (3.2)	74 (4.7)
Michigan	98 (1.9)	99 (0.9)	92 (2.9)	99 (1.0)	90 (3.7)	84 (4.2)
Missouri	98 (1.6)	98 (1.6)	89 (4.0)	98 (1.6)	92 (3.2)	77 (5.3)
North Carolina	98 (1.8)	98 (1.8)	77 (6.1)	98 (1.1)	96 (2.1)	78 (4.4)
Oregon	92 (3.5)	85 (3.8)	84 (4.1)	95 (1.9)	78 (5.2)	73 (6.3)
Pennsylvania	100 (0.0)	95 (4.7)	89 (5.3)	95 (4.7)	91 (5.0)	78 (5.9)
South Carolina	100 (0.0)	97 (2.3)	91 (3.4)	100 (0.0)	99 (1.1)	73 (5.6)
Texas	r 95 (2.7)	r 93 (3.2)	r 90 (3.1)	r 95 (2.7)	r 88 (4.4)	r 88 (3.9)
Districts and Consortia						
Academy School Dist. #20, CO	92 (0.2)	92 (0.2)	92 (0.2)	92 (0.2)	88 (0.3)	92 (0.2)
Chicago Public Schools, IL	99 (0.8)	99 (0.8)	83 (8.5)	97 (2.9)	83 (8.7)	74 (10.4)
Delaware Science Coalition, DE	r 97 (2.3)	r 98 (2.2)	r 94 (2.8)	r 95 (3.4)	r 79 (5.1)	r 70 (5.8)
First in the World Consort., IL	100 (0.0)	95 (5.2)	97 (1.9)	100 (0.0)	93 (5.5)	93 (5.5)
Fremont/Lincoln/WestSide PS, NE	95 (0.2)	95 (0.2)	72 (5.7)	82 (4.9)	76 (4.8)	69 (3.3)
Guilford County, NC	98 (1.7)	98 (1.7)	77 (5.6)	86 (4.5)	87 (4.9)	77 (5.3)
Jersey City Public Schools, NJ	100 (0.0)	100 (0.0)	93 (4.0)	100 (0.0)	94 (2.6)	100 (0.0)
Miami-Dade County PS, FL	s 99 (1.3)	s 95 (3.2)	s 82 (7.2)	s 97 (2.3)	s 86 (6.3)	s 60 (9.3)
Michigan Invitational Group, MI	100 (0.0)	96 (2.7)	87 (6.2)	100 (0.0)	89 (2.7)	70 (6.7)
Montgomery County, MD	s 99 (1.3)	s 96 (3.2)	s 84 (6.3)	s 99 (1.3)	s 90 (5.2)	s 94 (3.7)
Naperville Sch. Dist. #203, IL	100 (0.0)	100 (0.0)	95 (1.9)	98 (1.9)	98 (2.0)	87 (2.1)
Project SMART Consortium, OH	100 (0.0)	96 (1.7)	98 (0.4)	100 (0.0)	87 (4.3)	73 (5.0)
Rochester City Sch. Dist., NY	93 (2.0)	93 (2.0)	76 (4.5)	93 (2.0)	89 (1.5)	82 (3.1)
SW Math/Sci. Collaborative, PA	100 (0.0)	99 (1.3)	95 (2.7)	98 (1.4)	91 (4.0)	81 (5.0)
International Avg. (All Countries)	82 (0.5)	79 (0.5)	65 (0.6)	82 (0.5)	77 (0.6)	65 (0.6)

SOURCE: IEA Third International Mathematics and Science Study (TIMSS), 1998-1999.

Background data provided by teachers

1 Does not include students whose teachers report that they do not teach the topic.

2 Percentage of students averaged across topics.

States in *italics* did not fully satisfy guidelines for sample participation rates (see Appendix A for details).

( ) Standard errors appear in parentheses. Because results are rounded to the nearest whole number, some totals may appear inconsistent.

A dash (–) indicates data are not available.

An "r" indicates teacher response data available for 70-84% of students. An "s" indicates teacher response data available for 50-69% of students.

	Percentage of Students Whose Teachers Report Feeling Very Well Prepared to Teach Topic[1]						
	Coordinate geometry	Algebraic representation	Evaluate and perform operations on algebraic expressions	Solving linear equations and inequalities	Representation and interpretation of data in graphs, charts, and tables	Simple probabilities – understanding and calculations	Average[2]
Countries							
United States	82 (2.6)	94 (1.6)	95 (1.3)	93 (1.5)	94 (1.5)	90 (2.1)	90 (1.2)
Belgium (Flemish)	71 (3.9)	85 (3.1)	86 (2.6)	83 (3.3)	64 (2.9)	30 (4.6)	80 (1.4)
Canada	r 64 (3.6)	r 83 (2.5)	82 (2.5)	r 74 (2.8)	79 (2.9)	r 70 (3.3)	79 (1.7)
Chinese Taipei	81 (3.2)	82 (2.9)	85 (2.9)	84 (3.0)	74 (3.7)	73 (3.7)	78 (2.6)
Czech Republic	84 (4.2)	88 (3.6)	95 (2.4)	97 (1.9)	75 (5.1)	52 (5.6)	88 (1.8)
England	--	--	--	--	--	--	--
Hong Kong, SAR	82 (3.4)	85 (3.2)	87 (3.0)	74 (3.9)	58 (4.4)	58 (3.9)	72 (2.6)
Italy	64 (4.0)	62 (3.8)	79 (2.9)	71 (3.2)	70 (3.1)	53 (4.2)	69 (2.3)
Japan	25 (3.9)	28 (4.4)	33 (4.3)	37 (4.4)	19 (3.3)	19 (3.7)	23 (2.6)
Korea, Rep. of	49 (3.4)	56 (3.9)	74 (3.3)	83 (2.9)	55 (3.8)	67 (3.9)	61 (2.5)
Netherlands	88 (5.9)	87 (6.0)	77 (6.4)	87 (5.9)	85 (5.8)	77 (6.3)	84 (5.3)
Russian Federation	--	--	--	--	--	--	--
Singapore	79 (3.6)	85 (3.3)	86 (3.1)	89 (2.9)	80 (3.6)	46 (5.2)	78 (2.7)
States							
Connecticut	r 82 (6.2)	r 98 (1.7)	r 97 (2.1)	r 96 (2.1)	r 90 (5.4)	r 84 (6.3)	r 91 (3.0)
Idaho	r 70 (6.0)	r 82 (3.5)	r 92 (4.0)	r 88 (4.2)	r 89 (4.9)	r 70 (6.1)	r 83 (2.8)
Illinois	78 (5.6)	89 (5.6)	97 (1.6)	95 (2.1)	93 (3.2)	92 (2.0)	90 (2.0)
Indiana	74 (5.1)	97 (2.0)	99 (1.4)	96 (2.1)	87 (6.6)	80 (7.0)	89 (2.2)
Maryland	r 83 (4.6)	r 97 (1.3)	r 97 (1.3)	r 92 (2.9)	r 91 (3.0)	r 88 (3.9)	r 92 (1.5)
Massachusetts	82 (4.6)	96 (2.3)	96 (2.3)	95 (2.3)	96 (2.2)	78 (4.6)	90 (2.2)
Michigan	89 (3.8)	93 (2.5)	96 (1.9)	95 (2.7)	94 (2.8)	87 (4.3)	93 (1.7)
Missouri	78 (4.9)	93 (2.6)	95 (2.7)	95 (2.2)	97 (1.7)	85 (5.0)	91 (2.0)
North Carolina	86 (3.9)	92 (3.0)	95 (2.4)	91 (3.8)	91 (3.8)	79 (5.4)	90 (2.2)
Oregon	68 (6.1)	83 (4.5)	83 (4.6)	80 (5.0)	93 (3.3)	86 (4.0)	83 (2.8)
Pennsylvania	87 (5.3)	92 (5.1)	94 (4.8)	93 (4.9)	92 (4.9)	88 (5.4)	91 (4.5)
South Carolina	88 (4.2)	97 (1.8)	99 (1.4)	93 (2.1)	97 (2.2)	89 (4.7)	94 (1.4)
Texas	r 82 (5.5)	r 96 (2.0)	r 94 (2.6)	r 95 (2.3)	r 95 (2.1)	r 89 (4.6)	r 91 (2.3)
Districts and Consortia							
Academy School Dist. #20, CO	92 (0.2)	92 (0.2)	92 (0.2)	92 (0.2)	92 (0.2)	92 (0.2)	92 (0.2)
Chicago Public Schools, IL	83 (8.6)	89 (5.9)	93 (4.4)	86 (7.2)	96 (2.8)	97 (2.7)	90 (3.8)
Delaware Science Coalition, DE	r 76 (5.4)	r 85 (5.0)	r 95 (3.6)	r 94 (3.8)	r 90 (2.3)	r 88 (4.6)	r 88 (2.7)
First in the World Consort., IL	91 (5.9)	100 (0.0)	100 (0.0)	98 (1.6)	100 (0.0)	94 (5.3)	97 (2.3)
Fremont/Lincoln/WestSide PS, NE	73 (3.1)	95 (0.2)	95 (0.2)	95 (0.2)	90 (4.9)	84 (1.0)	85 (1.8)
Guilford County, NC	88 (4.7)	95 (2.7)	97 (2.1)	94 (4.0)	87 (5.2)	86 (5.4)	89 (3.1)
Jersey City Public Schools, NJ	74 (3.8)	93 (0.5)	98 (0.1)	78 (3.1)	100 (0.0)	96 (2.2)	94 (1.0)
Miami-Dade County PS, FL	s 69 (6.8)	s 92 (3.5)	s 92 (3.5)	s 90 (4.3)	s 89 (7.1)	s 87 (5.0)	s 86 (2.9)
Michigan Invitational Group, MI	90 (2.3)	96 (2.1)	96 (2.1)	96 (2.1)	90 (5.5)	91 (5.5)	92 (1.5)
Montgomery County, MD	s 88 (4.0)	s 93 (3.0)	s 93 (3.0)	s 91 (3.9)	s 89 (3.7)	s 85 (5.7)	s 91 (2.9)
Naperville Sch. Dist. #203, IL	98 (2.0)	94 (2.1)	100 (0.0)	100 (0.0)	97 (0.1)	97 (0.1)	97 (1.0)
Project SMART Consortium, OH	81 (4.6)	96 (1.9)	97 (2.4)	95 (2.8)	95 (3.7)	93 (4.4)	93 (1.6)
Rochester City Sch. Dist., NY	92 (2.6)	100 (0.0)	100 (0.0)	100 (0.0)	94 (2.0)	100 (0.0)	93 (1.1)
SW Math/Sci. Collaborative, PA	85 (4.4)	99 (0.5)	99 (1.3)	97 (1.6)	95 (2.5)	84 (5.7)	94 (1.4)
International Avg. (All Countries)	66 (0.7)	76 (0.6)	80 (0.6)	81 (0.5)	68 (0.7)	55 (0.7)	73 (0.4)

SOURCE: IEA Third International Mathematics and Science Study (TIMSS), 1998-1999.

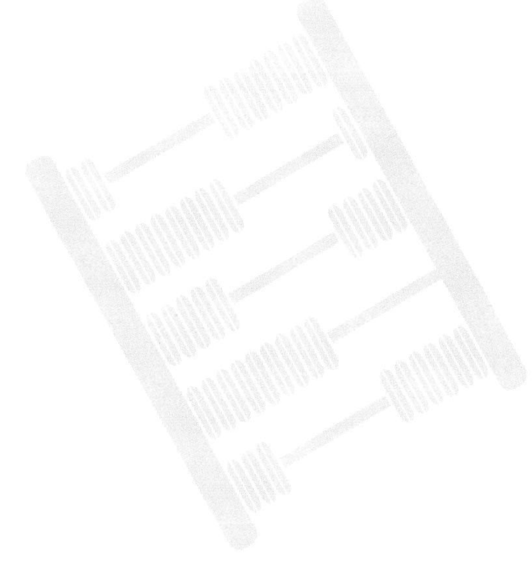

Exhibit R3.2 Shortages of Teachers Qualified to Teach Mathematics Affecting Capacity to Provide Instruction

8th Grade Mathematics

		None		A Little		Some		A Lot	
		Percent of Students	Average Achievement	Percent of Students	Average Achievement	Percent of Students	Average Achievement	Percent of Students	Average Achievement
Countries									
United States	r	62 (4.2)	514 (5.3)	23 (3.5)	497 (6.2)	13 (2.6)	461 (9.3)	3 (1.0)	446 (15.1)
Belgium (Flemish)		84 (3.6)	557 (4.5)	12 (3.3)	554 (22.2)	2 (1.2)	~ ~	1 (0.9)	~ ~
Canada		54 (3.1)	536 (3.1)	27 (3.2)	531 (4.3)	16 (2.4)	514 (7.7)	3 (0.8)	525 (18.6)
Chinese Taipei		43 (4.3)	591 (5.1)	36 (4.4)	584 (7.3)	12 (3.0)	590 (11.4)	9 (2.2)	551 (13.3)
Czech Republic		90 (2.9)	521 (4.7)	6 (2.6)	506 (15.6)	3 (1.8)	527 (26.2)	1 (0.9)	~ ~
England	r	71 (4.1)	508 (5.8)	19 (3.6)	482 (7.3)	8 (2.8)	485 (17.1)	2 (1.2)	~ ~
Hong Kong, SAR		55 (4.5)	596 (6.0)	28 (4.2)	555 (9.6)	10 (2.6)	583 (15.9)	7 (2.1)	602 (15.1)
Italy		38 (4.1)	478 (5.5)	36 (3.6)	483 (7.6)	20 (3.3)	473 (9.0)	6 (1.7)	487 (13.3)
Japan		65 (3.9)	579 (2.1)	17 (2.8)	583 (4.2)	8 (2.4)	577 (6.0)	10 (2.5)	572 (5.5)
Korea, Rep. of		19 (3.2)	589 (5.4)	48 (3.9)	587 (2.5)	22 (3.8)	586 (4.4)	11 (2.7)	591 (4.9)
Netherlands	r	52 (6.5)	546 (11.8)	23 (5.3)	541 (15.4)	19 (6.8)	518 (22.8)	5 (2.0)	560 (22.0)
Russian Federation		51 (3.9)	526 (7.2)	9 (2.3)	523 (24.4)	10 (2.0)	526 (15.2)	30 (3.8)	526 (10.6)
Singapore		58 (4.1)	608 (9.0)	26 (4.1)	599 (10.9)	12 (2.8)	607 (13.9)	4 (1.6)	592 (17.5)
States									
Connecticut	s	68 (9.8)	529 (11.7)	22 (9.1)	526 (19.7)	7 (4.1)	478 (11.4)	3 (2.9)	522 (4.9)
Idaho	r	74 (7.2)	488 (10.1)	18 (5.5)	515 (13.2)	9 (4.7)	523 (8.1)	0 (0.0)	~ ~
Illinois		71 (7.8)	515 (7.0)	17 (6.9)	521 (9.2)	11 (4.5)	511 (16.5)	1 (0.6)	~ ~
Indiana		84 (7.3)	515 (7.7)	10 (4.5)	483 (7.9)	6 (0.3)	550 (3.6)	0 (0.0)	~ ~
Maryland	r	56 (7.5)	490 (6.2)	15 (5.4)	488 (21.2)	19 (6.5)	472 (19.1)	10 (4.9)	524 (22.4)
Massachusetts	s	78 (7.3)	517 (6.1)	18 (6.5)	521 (18.0)	3 (0.3)	557 (6.9)	0 (0.0)	~ ~
Michigan		71 (6.1)	530 (7.3)	19 (5.1)	529 (11.5)	8 (4.1)	512 (20.2)	2 (0.1)	~ ~
Missouri		71 (6.3)	489 (6.5)	16 (5.1)	489 (10.4)	7 (3.6)	510 (17.8)	6 (3.3)	449 (51.5)
North Carolina	r	44 (8.4)	492 (9.9)	37 (7.2)	512 (8.8)	14 (5.9)	474 (8.3)	5 (2.8)	487 (15.0)
Oregon		51 (6.8)	518 (10.5)	39 (6.6)	505 (10.3)	6 (3.5)	542 (12.8)	4 (2.8)	546 (4.2)
Pennsylvania		83 (5.6)	515 (6.6)	12 (4.9)	499 (23.3)	3 (1.9)	509 (34.2)	3 (2.3)	495 (114.3)
South Carolina		34 (7.1)	496 (8.3)	36 (7.7)	505 (17.0)	25 (7.1)	505 (17.9)	5 (3.7)	487 (3.1)
Texas	r	56 (8.5)	539 (7.0)	14 (6.2)	538 (33.3)	20 (6.9)	465 (22.6)	10 (5.5)	488 (12.1)
Districts and Consortia									
Academy School Dist. #20, CO		100 (0.0)	528 (1.8)	0 (0.0)	~ ~	0 (0.0)	~ ~	0 (0.0)	~ ~
Chicago Public Schools, IL	s	58 (11.8)	479 (9.8)	16 (8.9)	444 (10.7)	22 (8.3)	434 (8.8)	4 (4.7)	428 (2.4)
Delaware Science Coalition, DE	r	55 (2.1)	466 (14.2)	24 (1.8)	458 (14.1)	7 (1.8)	447 (24.6)	15 (1.9)	537 (36.7)
First in the World Consort., IL	r	100 (0.0)	557 (7.0)	0 (0.0)	~ ~	0 (0.0)	~ ~	0 (0.0)	~ ~
Fremont/Lincoln/WestSide PS, NE	r	67 (1.5)	481 (10.4)	33 (1.5)	512 (24.7)	0 (0.0)	~ ~	0 (0.0)	~ ~
Guilford County, NC	r	44 (1.2)	519 (9.8)	24 (1.1)	552 (21.4)	23 (0.9)	489 (28.6)	9 (0.4)	451 (24.9)
Jersey City Public Schools, NJ		50 (1.4)	491 (11.0)	15 (1.0)	490 (30.9)	20 (1.0)	450 (6.6)	14 (1.9)	456 (42.5)
Miami-Dade County PS, FL		x x	x x	x x	x x	x x	x x	x x	x x
Michigan Invitational Group, MI		60 (1.6)	535 (9.1)	20 (0.9)	521 (10.3)	15 (1.5)	553 (6.2)	5 (1.2)	497 (12.4)
Montgomery County, MD	s	54 (12.3)	543 (5.5)	13 (9.1)	550 (36.2)	23 (11.9)	524 (6.8)	9 (9.2)	522 (4.9)
Naperville Sch. Dist. #203, IL		79 (1.0)	565 (3.2)	21 (1.0)	583 (5.3)	0 (0.0)	~ ~	0 (0.0)	~ ~
Project SMART Consortium, OH		66 (1.2)	523 (9.6)	15 (1.0)	548 (28.1)	12 (0.7)	483 (7.6)	7 (0.3)	457 (9.8)
Rochester City Sch. Dist., NY	r	55 (1.5)	448 (8.1)	16 (0.5)	436 (18.0)	10 (0.9)	482 (35.2)	19 (1.3)	411 (13.2)
SW Math/Sci. Collaborative, PA		83 (6.0)	527 (6.7)	9 (5.1)	475 (14.8)	6 (3.7)	480 (19.6)	1 (0.1)	~ ~
International Avg. (All Countries)		47 (0.6)	493 (1.4)	20 (0.5)	484 (2.6)	14 (0.5)	478 (2.4)	19 (0.5)	477 (3.0)

SOURCE: IEA Third International Mathematics and Science Study (TIMSS), 1998-1999.

Background data provided by schools.

States in *italics* did not fully satisfy guidelines for sample participation rates (see Appendix A for details).

( ) Standard errors appear in parentheses. Because results are rounded to the nearest whole number, some totals may appear inconsistent.

A tilde (~) indicates insufficient data to report achievement.

An "r" indicates school response data available for 70-84% of students. An "s" indicates school response data available for 50-69% of students. An "x" indicates school response data available for <50% of students.

Exhibit R3.3 **Percentage of Students Whose Mathematics Teachers Agree or Strongly Agree with Statements About the Nature of Mathematics and Mathematics Teaching**

8th Grade Mathematics

TIMSS 1999 Benchmarking
Boston College

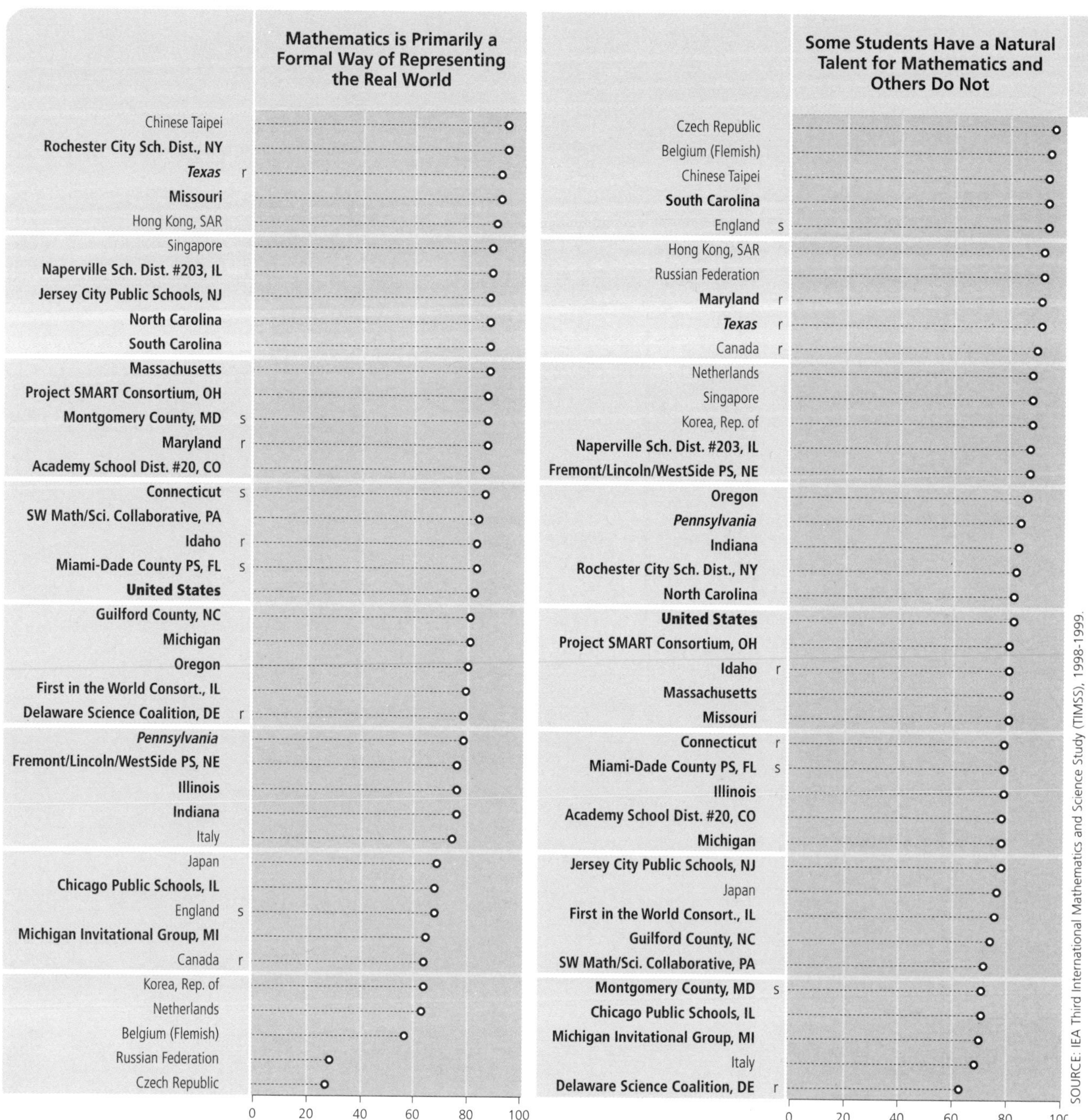

SOURCE: IEA Third International Mathematics and Science Study (TIMSS), 1998-1999.

Background data provided by teachers.

States in *italics* did not fully satisfy guidelines for sample participation rates (see Appendix A for details).

An "r" indicates teacher response data available for 70-84% of students. An "s" indicates teacher response data available for 50-69% of students.

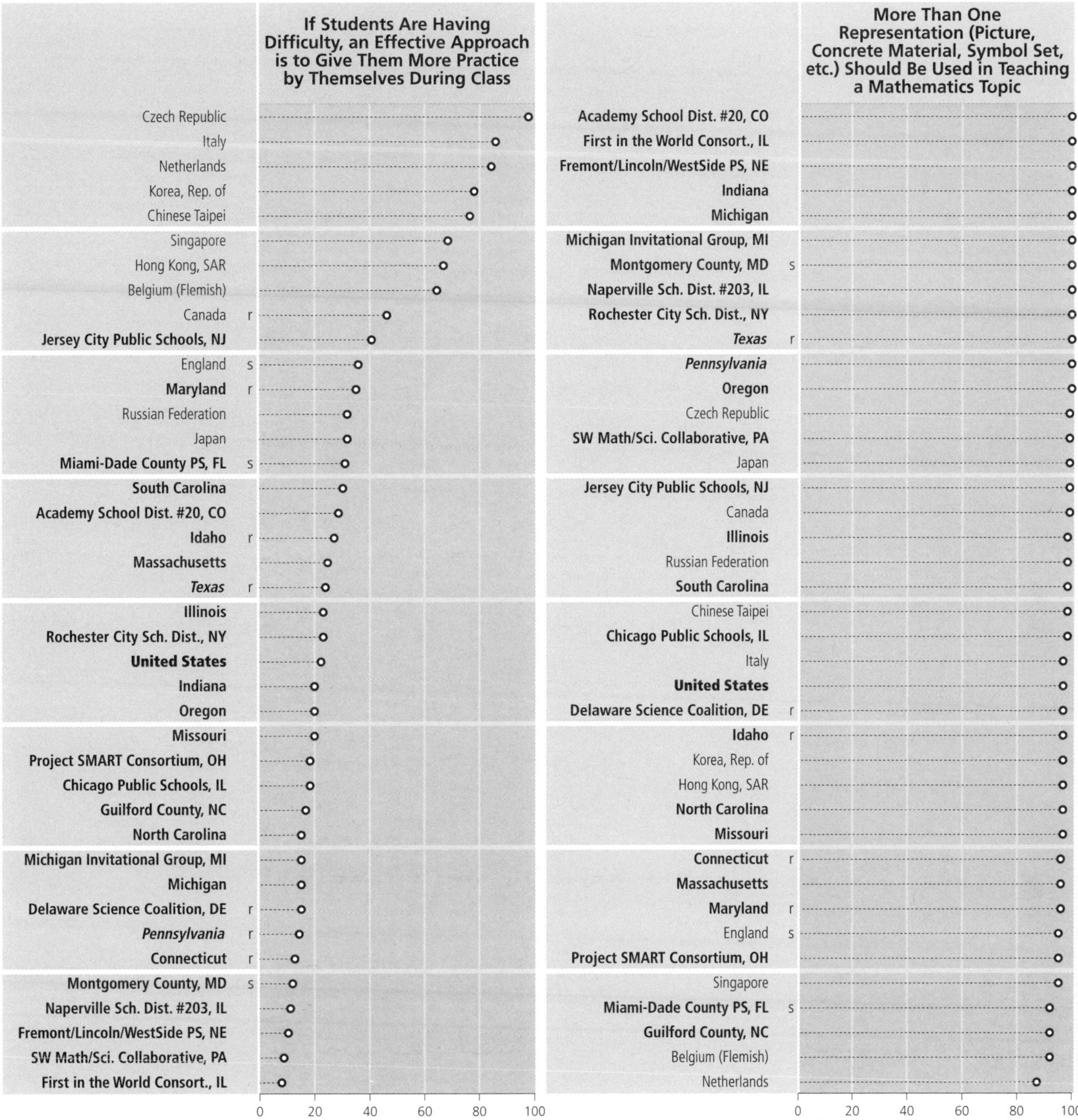

If Students Are Having Difficulty, an Effective Approach is to Give Them More Practice by Themselves During Class		More Than One Representation (Picture, Concrete Material, Symbol Set, etc.) Should Be Used in Teaching a Mathematics Topic	
Czech Republic		**Academy School Dist. #20, CO**	
Italy		**First in the World Consort., IL**	
Netherlands		**Fremont/Lincoln/WestSide PS, NE**	
Korea, Rep. of		**Indiana**	
Chinese Taipei		**Michigan**	
Singapore		**Michigan Invitational Group, MI**	
Hong Kong, SAR		**Montgomery County, MD**	s
Belgium (Flemish)		**Naperville Sch. Dist. #203, IL**	
Canada	r	**Rochester City Sch. Dist., NY**	
Jersey City Public Schools, NJ		***Texas***	r
England	s	***Pennsylvania***	
Maryland	r	**Oregon**	
Russian Federation		Czech Republic	
Japan		**SW Math/Sci. Collaborative, PA**	
Miami-Dade County PS, FL	s	Japan	
South Carolina		**Jersey City Public Schools, NJ**	
Academy School Dist. #20, CO		Canada	
Idaho	r	**Illinois**	
Massachusetts		Russian Federation	
Texas	r	**South Carolina**	
Illinois		Chinese Taipei	
Rochester City Sch. Dist., NY		**Chicago Public Schools, IL**	
United States		Italy	
Indiana		**United States**	
Oregon		**Delaware Science Coalition, DE**	r
Missouri		**Idaho**	r
Project SMART Consortium, OH		Korea, Rep. of	
Chicago Public Schools, IL		Hong Kong, SAR	
Guilford County, NC		**North Carolina**	
North Carolina		**Missouri**	
Michigan Invitational Group, MI		**Connecticut**	r
Michigan		**Massachusetts**	
Delaware Science Coalition, DE	r	**Maryland**	r
Pennsylvania	r	England	s
Connecticut	r	**Project SMART Consortium, OH**	
Montgomery County, MD	s	Singapore	
Naperville Sch. Dist. #203, IL		**Miami-Dade County PS, FL**	s
Fremont/Lincoln/WestSide PS, NE		**Guilford County, NC**	
SW Math/Sci. Collaborative, PA		Belgium (Flemish)	
First in the World Consort., IL		Netherlands	

Axis: 0 20 40 60 80 100 (both panels)

SOURCE: IEA Third International Mathematics and Science Study (TIMSS), 1998-1999.

Exhibit R3.4 Percentage of Students Whose Mathematics Teachers Think Particular Abilities Are Very Important for Students' Success in Mathematics in School

8th Grade Mathematics

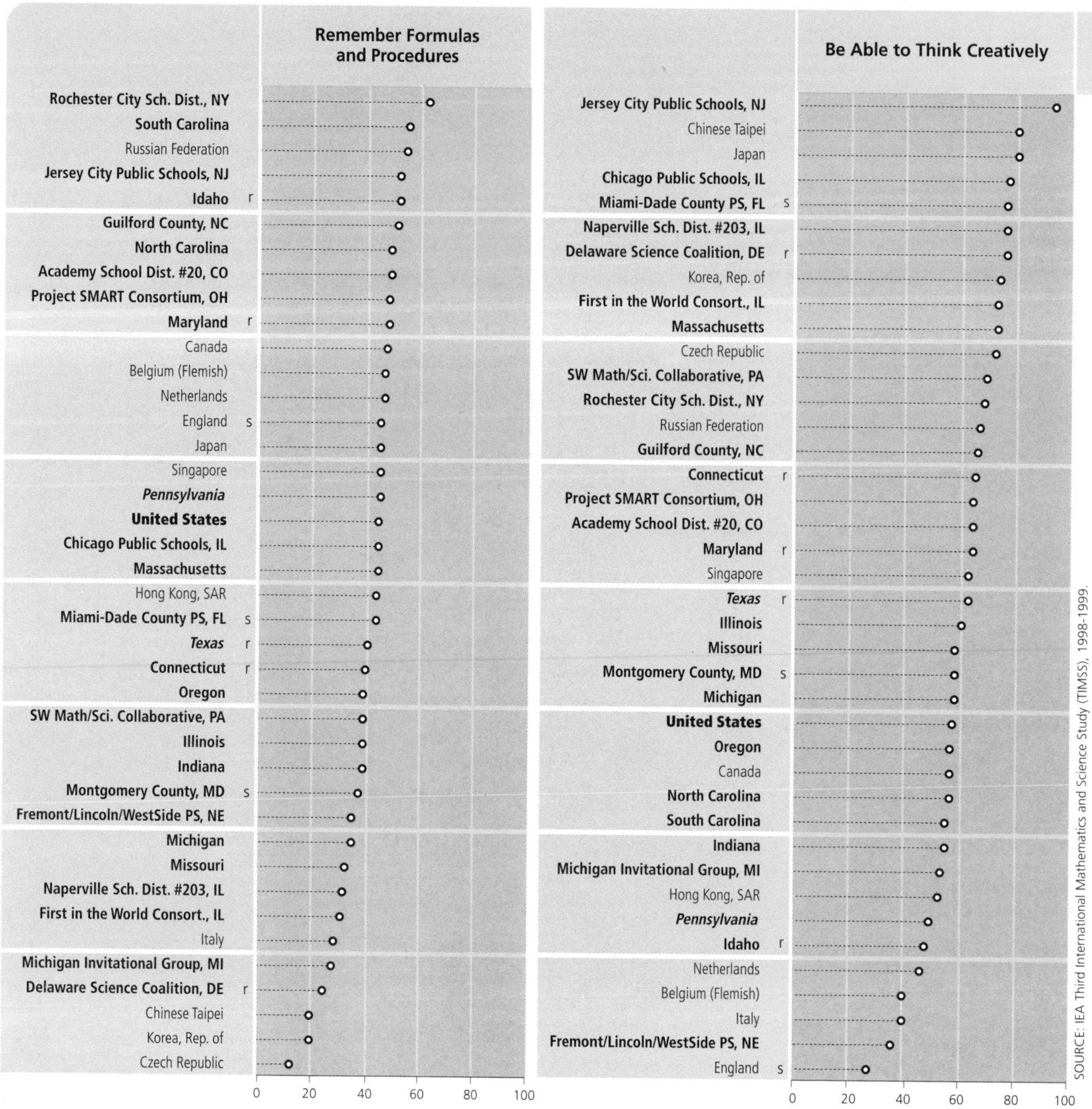

Remember Formulas and Procedures

Rochester City Sch. Dist., NY
South Carolina
Russian Federation
Jersey City Public Schools, NJ
Idaho r
Guilford County, NC
North Carolina
Academy School Dist. #20, CO
Project SMART Consortium, OH
Maryland r
Canada
Belgium (Flemish)
Netherlands
England s
Japan
Singapore
Pennsylvania
United States
Chicago Public Schools, IL
Massachusetts
Hong Kong, SAR
Miami-Dade County PS, FL s
Texas r
Connecticut r
Oregon
SW Math/Sci. Collaborative, PA
Illinois
Indiana
Montgomery County, MD s
Fremont/Lincoln/WestSide PS, NE
Michigan
Missouri
Naperville Sch. Dist. #203, IL
First in the World Consort., IL
Italy
Michigan Invitational Group, MI
Delaware Science Coalition, DE r
Chinese Taipei
Korea, Rep. of
Czech Republic

0 20 40 60 80 100

Be Able to Think Creatively

Jersey City Public Schools, NJ
Chinese Taipei
Japan
Chicago Public Schools, IL
Miami-Dade County PS, FL s
Naperville Sch. Dist. #203, IL
Delaware Science Coalition, DE r
Korea, Rep. of
First in the World Consort., IL
Massachusetts
Czech Republic
SW Math/Sci. Collaborative, PA
Rochester City Sch. Dist., NY
Russian Federation
Guilford County, NC
Connecticut r
Project SMART Consortium, OH
Academy School Dist. #20, CO
Maryland r
Singapore
Texas r
Illinois
Missouri
Montgomery County, MD s
Michigan
United States
Oregon
Canada
North Carolina
South Carolina
Indiana
Michigan Invitational Group, MI
Hong Kong, SAR
Pennsylvania
Idaho r
Netherlands
Belgium (Flemish)
Italy
Fremont/Lincoln/WestSide PS, NE
England s

0 20 40 60 80 100

SOURCE: IEA Third International Mathematics and Science Study (TIMSS), 1998-1999.

Background data provided by teachers.

States in *italics* did not fully satisfy guidelines for sample participation rates (see Appendix A for details).

An "r" indicates teacher response data available for 70-84% of students. An "s" indicates teacher response data available for 50-69% of students.

Percentage of Students Whose Mathematics Teachers Think Particular Abilities Are Very Important for Students' Success in Mathematics in School

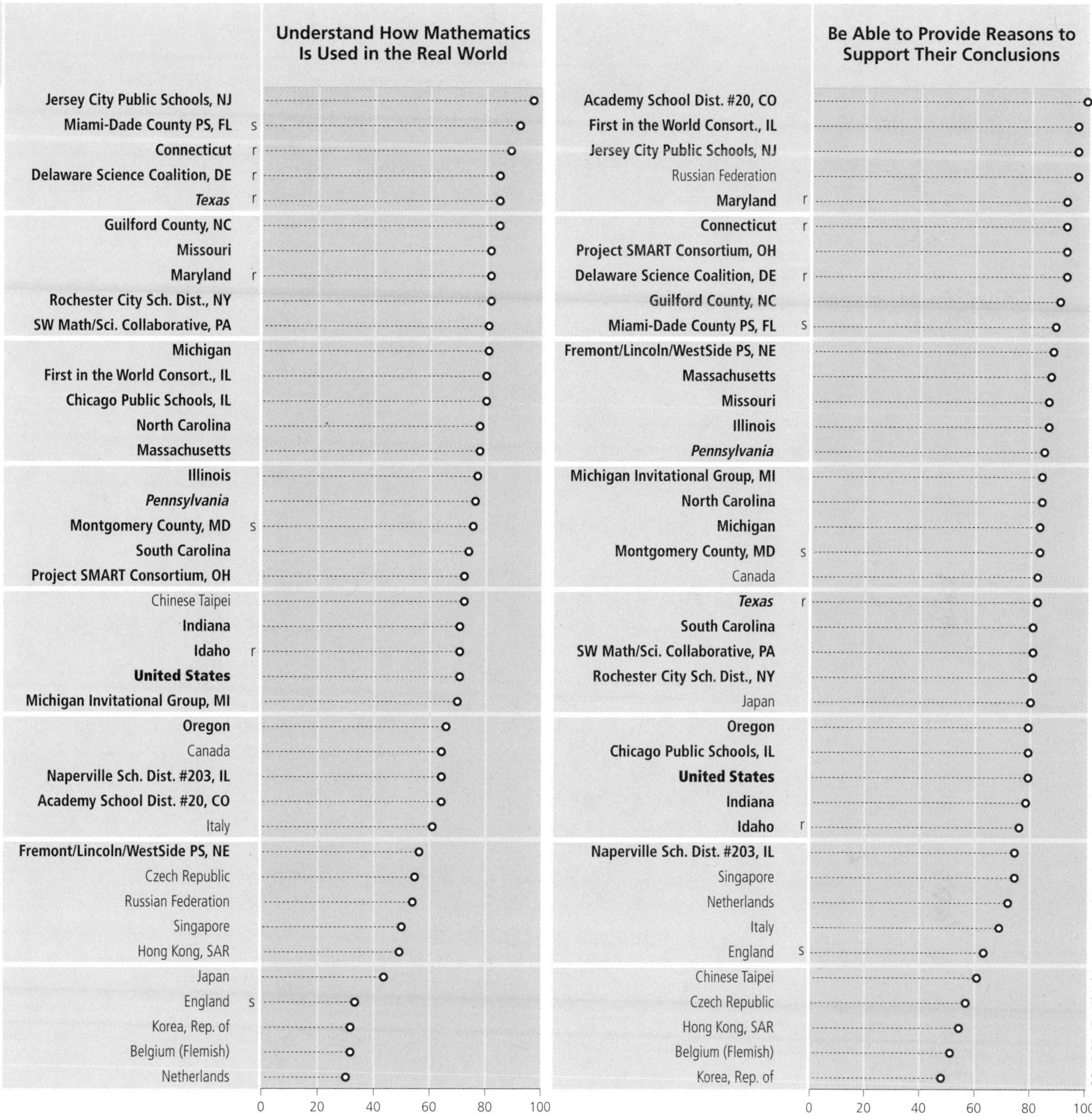

8th Grade Mathematics

Understand How Mathematics Is Used in the Real World

Jersey City Public Schools, NJ	
Miami-Dade County PS, FL	s
Connecticut	r
Delaware Science Coalition, DE	r
Texas	r
Guilford County, NC	
Missouri	
Maryland	r
Rochester City Sch. Dist., NY	
SW Math/Sci. Collaborative, PA	
Michigan	
First in the World Consort., IL	
Chicago Public Schools, IL	
North Carolina	
Massachusetts	
Illinois	
Pennsylvania	
Montgomery County, MD	s
South Carolina	
Project SMART Consortium, OH	
Chinese Taipei	
Indiana	
Idaho	r
United States	
Michigan Invitational Group, MI	
Oregon	
Canada	
Naperville Sch. Dist. #203, IL	
Academy School Dist. #20, CO	
Italy	
Fremont/Lincoln/WestSide PS, NE	
Czech Republic	
Russian Federation	
Singapore	
Hong Kong, SAR	
Japan	
England	s
Korea, Rep. of	
Belgium (Flemish)	
Netherlands	

Scale: 0 20 40 60 80 100

Be Able to Provide Reasons to Support Their Conclusions

Academy School Dist. #20, CO	
First in the World Consort., IL	
Jersey City Public Schools, NJ	
Russian Federation	
Maryland	r
Connecticut	r
Project SMART Consortium, OH	
Delaware Science Coalition, DE	r
Guilford County, NC	
Miami-Dade County PS, FL	s
Fremont/Lincoln/WestSide PS, NE	
Massachusetts	
Missouri	
Illinois	
Pennsylvania	
Michigan Invitational Group, MI	
North Carolina	
Michigan	
Montgomery County, MD	s
Canada	
Texas	r
South Carolina	
SW Math/Sci. Collaborative, PA	
Rochester City Sch. Dist., NY	
Japan	
Oregon	
Chicago Public Schools, IL	
United States	
Indiana	
Idaho	r
Naperville Sch. Dist. #203, IL	
Singapore	
Netherlands	
Italy	
England	s
Chinese Taipei	
Czech Republic	
Hong Kong, SAR	
Belgium (Flemish)	
Korea, Rep. of	

Scale: 0 20 40 60 80 100

SOURCE: IEA Third International Mathematics and Science Study (TIMSS), 1998-1999.

Exhibit R3.5 How Teachers Spend Their Formally Scheduled School Time

TIMSS 1999 Benchmarking
Boston College

8th Grade Mathematics

	Percentage of Formally Scheduled School Time Averaged Across Students				
	Teaching Mathematics, Science, and Other Subjects	Teaching Mathematics	Curriculum Planning[1]	Administrative Duties	Other Activities[2]
Countries					
United States	75 (1.3)	65 (1.7)	13 (0.8)	2 (0.4)	11 (0.8)
Belgium (Flemish)	r 85 (1.1)	r 70 (1.7)	r 3 (0.5)	r 1 (0.3)	r 10 (0.8)
Canada	78 (1.0)	43 (1.3)	7 (0.4)	2 (0.5)	14 (1.0)
Chinese Taipei	55 (2.0)	55 (2.0)	9 (0.9)	4 (0.9)	32 (1.6)
Czech Republic	70 (2.1)	46 (2.1)	12 (1.6)	3 (0.5)	15 (1.0)
England	s 87 (0.9)	s 80 (1.3)	s 1 (0.3)	s 2 (0.3)	s 10 (0.8)
Hong Kong, SAR	x x	x x	x x	x x	x x
Italy	87 (1.1)	52 (0.8)	7 (0.8)	0 (0.1)	6 (0.6)
Japan	68 (1.8)	63 (1.9)	7 (0.8)	4 (0.4)	21 (1.4)
Korea, Rep. of	54 (1.3)	54 (1.3)	11 (0.6)	14 (0.8)	21 (0.9)
Netherlands [3]	s 88 (0.9)	s 75 (2.6)	– –	– –	s 12 (0.9)
Russian Federation [4]	– –	– –	– –	– –	– –
Singapore	73 (0.8)	55 (1.8)	– –	3 (0.3)	24 (0.7)
States					
Connecticut	r 68 (1.6)	r 61 (1.8)	r 15 (1.1)	r 3 (0.8)	r 14 (1.6)
Idaho	r 78 (1.9)	r 67 (3.4)	r 12 (1.5)	r 1 (0.6)	r 9 (1.2)
Illinois	70 (1.9)	57 (2.3)	16 (1.2)	2 (0.7)	12 (1.4)
Indiana	72 (3.0)	64 (2.8)	16 (1.2)	0 (0.3)	12 (2.8)
Maryland	r 72 (1.7)	r 70 (1.8)	r 19 (1.7)	r 2 (0.4)	r 7 (0.8)
Massachusetts	71 (1.7)	67 (1.9)	14 (1.3)	3 (0.8)	12 (1.2)
Michigan	75 (1.7)	60 (2.9)	14 (1.3)	2 (1.2)	9 (1.5)
Missouri	75 (1.7)	69 (2.3)	15 (1.2)	1 (0.5)	9 (1.3)
North Carolina	63 (1.7)	54 (2.6)	21 (1.0)	2 (0.4)	15 (1.9)
Oregon	81 (1.3)	64 (1.7)	12 (1.0)	0 (0.1)	8 (1.2)
Pennsylvania	71 (2.0)	61 (1.7)	11 (1.0)	2 (0.5)	17 (1.7)
South Carolina	68 (2.4)	65 (2.5)	19 (1.7)	r 1 (0.4)	12 (1.5)
Texas	r 66 (2.8)	r 59 (2.5)	r 16 (1.4)	r 3 (0.9)	r 16 (2.3)
Districts and Consortia					
Academy School Dist. #20, CO	70 (0.1)	50 (0.1)	22 (0.1)	1 (0.0)	8 (0.1)
Chicago Public Schools, IL	81 (1.6)	50 (3.6)	r 12 (1.5)	3 (1.0)	5 (1.2)
Delaware Science Coalition, DE	r 72 (1.7)	r 63 (2.0)	r 19 (1.0)	r 1 (0.1)	r 8 (1.0)
First in the World Consort., IL	68 (1.3)	62 (3.0)	21 (1.1)	2 (0.4)	9 (1.1)
Fremont/Lincoln/WestSide PS, NE	72 (0.9)	71 (1.3)	24 (1.1)	0 (0.2)	4 (0.3)
Guilford County, NC	58 (1.9)	48 (2.1)	23 (1.4)	3 (0.6)	16 (1.4)
Jersey City Public Schools, NJ	r 81 (1.0)	r 69 (1.6)	r 12 (0.8)	r 0 (0.0)	r 7 (0.5)
Miami-Dade County PS, FL	s 77 (3.3)	s 74 (3.2)	s 12 (2.2)	s 1 (0.4)	s 11 (2.3)
Michigan Invitational Group, MI	78 (1.4)	67 (2.0)	11 (1.4)	2 (0.7)	10 (1.3)
Montgomery County, MD	s 63 (2.9)	s 61 (3.0)	s 19 (3.1)	s 7 (1.0)	s 11 (1.4)
Naperville Sch. Dist. #203, IL	64 (0.8)	50 (1.1)	15 (0.8)	2 (0.4)	19 (0.4)
Project SMART Consortium, OH	70 (1.5)	67 (1.7)	15 (1.4)	0 (0.3)	15 (1.1)
Rochester City Sch. Dist., NY	66 (0.9)	64 (1.2)	13 (0.8)	9 (0.6)	12 (0.8)
SW Math/Sci. Collaborative, PA	69 (2.5)	63 (2.8)	8 (1.3)	3 (0.8)	20 (1.8)
International Avg. (All Countries)	71 (0.3)	60 (0.3)	9 (0.2)	4 (0.1)	16 (0.2)

SOURCE: IEA Third International Mathematics and Science Study (TIMSS), 1998-1999.

Background data provided by teachers.

1 Includes individual curriculum planning and cooperative curriculum planning.

2 Includes student supervision (other than teaching), student counseling/appraisal, other non-student contact time, and other activities.

3 Netherlands: Data in other activities category reflects the total reported for curriculum planning, administrative duties, and other activities.

4 Russian Federation: Formally scheduled school time is for instruction only; teachers are not formally scheduled for other activities.

States in *italics* did not fully satisfy guidelines for sample participation rates (see Appendix A for details).

( ) Standard errors appear in parentheses. Because results are rounded to the nearest whole number, some totals may appear inconsistent.

A dash (–) indicates data are not available.

An "r" indicates teacher response data available for 70-84% of students. An "s" indicates teacher response data available for 50-69% of students. An "x" indicates teacher response data available for <50% of students.

Average Number of Instructional Days in the School Year[1]

Korea, Rep. of		225 (0.7)
Japan		223 (0.6)
Chinese Taipei		221 (0.4)
Italy		210 (0.0)
Czech Republic		197 (0.8)
Russian Federation		195 (1.2)
Netherlands	r	191 (2.0)
England	r	190 (0.3)
Canada		188 (0.3)
First in the World Consort., IL	s	185 (0.3)
Montgomery County, MD	s	184 (0.6)
Michigan Invitational Group, MI		183 (0.1)
Rochester City Sch. Dist., NY	r	182 (0.1)
Michigan		182 (0.3)
Connecticut	s	181 (0.3)
Pennsylvania		181 (0.4)
Maryland	r	181 (0.6)
Indiana		181 (0.2)
SW Math/Sci. Collaborative, PA		181 (0.5)
South Carolina		181 (0.4)
Chicago Public Schools, IL	r	180 (1.1)
Jersey City Public Schools, NJ		180 (0.0)
Massachusetts	s	180 (0.2)
Project SMART Consortium, OH	r	180 (0.1)
Texas	s	180 (0.9)
Singapore		180 (0.0)
Guilford County, NC	r	180 (0.0)
North Carolina	r	180 (0.0)
United States	r	180 (0.4)
Delaware Science Coalition, DE	r	179 (0.0)
Idaho	r	179 (0.5)
Fremont/Lincoln/WestSide PS, NE	r	179 (0.1)
Illinois		179 (0.4)
Naperville Sch. Dist. #203, IL		178 (0.1)
Oregon		177 (3.2)
Missouri	r	176 (0.4)
Hong Kong, SAR	r	176 (2.7)
Belgium (Flemish)		175 (0.0)
Academy School Dist. #20, CO		172 (0.0)
Miami-Dade County PS, FL		x x

International Avg. (All Countries)	193 (0.2)

SOURCE: IEA Third International Mathematics and Science Study (TIMSS), 1998-1999.

Background data provided by schools.

1 Days reported averaged across students.

States in *italics* did not satisfy guidelines for sample participation rates (see Appendix A for details).

( ) Standard errors appear in parentheses. Because results are rounded to the nearest whole number, some totals may appear inconsistent.

An "r" indicates school response data available for 70-84% of students. An "s" indicates school response data available for 50-69% of students. An "x" indicates school response data available for <50% of students.

Asking Students to Do Problem-Solving Activities or Computation During Mathematics Lessons

8th Grade Mathematics

	Percentage of Students Whose Teachers Report Most or Every Lesson				
	Explain Reasoning Behind an Idea	Represent and Analyze Relationships Using Tables, Charts, or Graphs	Work on Problems for Which There Is No Immediately Obvious Method of Solution	Write Equations to Represent Relationships	Practice Computational Skills
Countries					
United States	72 (3.1)	24 (2.6)	19 (2.6)	54 (2.8)	66 (3.7)
Belgium (Flemish)	67 (3.5)	2 (0.8)	5 (3.1)	12 (2.0)	76 (2.7)
Canada	76 (2.7)	20 (2.6)	18 (2.8)	44 (2.8)	64 (3.3)
Chinese Taipei	54 (4.1)	39 (4.4)	11 (2.1)	57 (4.2)	54 (4.5)
Czech Republic	94 (2.6)	21 (4.0)	22 (4.2)	69 (5.4)	81 (4.4)
England	s 70 (3.0)	s 13 (2.7)	s 6 (2.1)	s 22 (3.4)	52 (3.6)
Hong Kong, SAR	33 (4.1)	17 (3.5)	18 (3.5)	60 (3.5)	78 (3.1)
Italy	84 (3.0)	44 (3.8)	45 (3.6)	30 (3.4)	65 (3.4)
Japan	82 (3.2)	62 (4.1)	41 (4.4)	80 (3.0)	62 (4.5)
Korea, Rep. of	65 (3.2)	38 (4.0)	28 (3.5)	65 (3.9)	55 (4.3)
Netherlands	65 (5.6)	28 (5.6)	33 (6.4)	23 (5.4)	40 (4.8)
Russian Federation	84 (3.3)	11 (2.5)	13 (2.9)	40 (3.7)	88 (2.3)
Singapore	44 (4.9)	14 (3.2)	15 (3.4)	29 (3.8)	51 (4.4)
States					
Connecticut	r 83 (6.5)	r 24 (6.4)	r 18 (4.7)	r 59 (7.3)	r 50 (7.6)
Idaho	r 56 (6.9)	r 14 (5.0)	r 23 (6.7)	r 54 (6.2)	r 84 (4.3)
Illinois	67 (6.3)	14 (3.4)	13 (2.8)	38 (5.2)	58 (6.2)
Indiana	73 (5.8)	16 (3.6)	16 (4.1)	58 (6.6)	75 (5.4)
Maryland	r 72 (4.7)	r 32 (5.6)	r 25 (4.9)	r 46 (6.2)	r 59 (5.5)
Massachusetts	79 (5.2)	25 (5.2)	23 (4.7)	52 (6.0)	64 (5.4)
Michigan	70 (5.7)	26 (5.5)	29 (6.7)	56 (5.0)	68 (4.7)
Missouri	50 (6.1)	14 (4.0)	14 (4.2)	45 (6.1)	85 (4.7)
North Carolina	76 (5.1)	19 (4.3)	22 (5.8)	37 (6.2)	63 (5.7)
Oregon	58 (6.1)	16 (4.5)	11 (3.3)	40 (6.4)	63 (5.9)
Pennsylvania	80 (4.0)	15 (5.3)	11 (2.8)	57 (6.2)	64 (3.5)
South Carolina	73 (4.7)	23 (4.7)	20 (4.0)	49 (6.3)	81 (5.2)
Texas	71 (6.0)	33 (7.2)	28 (6.9)	52 (7.8)	77 (5.7)
Districts and Consortia					
Academy School Dist. #20, CO	81 (0.3)	20 (0.3)	21 (0.4)	71 (0.4)	62 (0.4)
Chicago Public Schools, IL	79 (8.3)	20 (8.1)	10 (5.1)	25 (10.1)	62 (11.7)
Delaware Science Coalition, DE	r 84 (5.9)	r 35 (5.9)	r 9 (3.4)	r 41 (7.7)	r 35 (6.4)
First in the World Consort., IL	87 (3.2)	40 (5.4)	44 (8.5)	66 (9.6)	37 (7.4)
Fremont/Lincoln/WestSide PS, NE	r 74 (8.7)	r 14 (1.2)	r 8 (7.0)	r 50 (7.4)	r 86 (7.2)
Guilford County, NC	81 (5.4)	29 (6.5)	25 (5.5)	49 (6.7)	61 (6.8)
Jersey City Public Schools, NJ	r 96 (2.5)	r 62 (6.7)	r 51 (6.8)	r 60 (6.1)	r 57 (6.4)
Miami-Dade County PS, FL	s 85 (5.8)	s 36 (10.8)	s 21 (7.8)	s 41 (8.1)	73 (8.8)
Michigan Invitational Group, MI	91 (2.8)	49 (8.3)	48 (6.8)	64 (8.7)	51 (7.8)
Montgomery County, MD	s 79 (5.0)	s 18 (6.0)	s 19 (8.3)	s 52 (4.1)	63 (8.0)
Naperville Sch. Dist. #203, IL	76 (2.5)	35 (5.2)	19 (4.1)	90 (1.9)	51 (4.3)
Project SMART Consortium, OH	86 (4.3)	8 (3.8)	15 (4.4)	48 (5.1)	73 (6.0)
Rochester City Sch. Dist., NY	69 (4.1)	8 (2.3)	8 (2.7)	57 (4.3)	47 (5.0)
SW Math/Sci. Collaborative, PA	77 (5.2)	27 (6.9)	10 (4.0)	52 (5.9)	54 (7.7)
International Avg. (All Countries)	70 (0.6)	26 (0.6)	21 (0.6)	43 (0.6)	73 (0.6)

SOURCE: IEA Third International Mathematics and Science Study (TIMSS), 1998-1999.

Background data provided by teachers.

States in *italics* did not fully satisfy guidelines for sample participation rates (see Appendix A for details).

( ) Standard errors appear in parentheses. Because results are rounded to the nearest whole number, some totals may appear inconsistent.

An "r" indicates teacher response data available for 70-84% of students. An "s" indicates teacher response data available for 50-69% of students.

Exhibit R3.8 Students Using Things from Everyday Life in Solving Mathematics Problems

8th Grade Mathematics

	Almost Always		Pretty Often		Once in a While		Never	
	Percent of Students	Average Achievement	Percent of Students	Average Achievement	Percent of Students	Average Achievement	Percent of Students	Average Achievement
Countries								
United States	23 (0.9)	489 (5.1)	31 (0.8)	509 (3.8)	34 (0.7)	515 (4.1)	12 (0.7)	493 (6.8)
Belgium (Flemish)	7 (0.7)	531 (13.8)	20 (0.8)	560 (4.4)	47 (1.1)	567 (4.0)	27 (1.1)	552 (4.8)
Canada	19 (0.7)	518 (3.6)	33 (0.8)	534 (4.3)	36 (0.9)	536 (2.7)	11 (0.7)	531 (4.6)
Chinese Taipei	11 (0.5)	596 (6.2)	31 (0.8)	600 (4.0)	43 (0.8)	590 (4.4)	15 (0.7)	540 (6.5)
Czech Republic	11 (0.9)	522 (8.1)	36 (1.4)	530 (4.2)	37 (1.5)	519 (5.0)	16 (1.2)	499 (8.3)
England	14 (0.8)	486 (6.3)	41 (1.3)	499 (5.2)	34 (1.0)	505 (4.9)	11 (0.8)	489 (8.2)
Hong Kong, SAR	6 (0.3)	573 (8.0)	24 (0.8)	583 (6.5)	56 (0.9)	587 (4.0)	15 (0.7)	570 (6.1)
Italy	10 (0.6)	456 (7.7)	19 (0.8)	494 (6.2)	38 (1.0)	489 (4.0)	33 (1.2)	471 (4.5)
Japan	2 (0.2)	~ ~	17 (0.7)	590 (3.5)	55 (0.8)	583 (2.1)	27 (1.0)	564 (3.3)
Korea, Rep. of	3 (0.3)	580 (7.5)	12 (0.6)	602 (3.2)	47 (0.8)	595 (2.3)	37 (0.8)	573 (2.7)
Netherlands	7 (0.6)	544 (10.2)	20 (0.9)	549 (8.5)	49 (1.6)	542 (8.1)	23 (1.5)	530 (8.4)
Russian Federation	14 (0.8)	506 (6.7)	23 (1.2)	538 (8.3)	46 (1.4)	533 (6.0)	17 (0.9)	521 (6.3)
Singapore	16 (0.8)	578 (7.8)	34 (0.9)	606 (6.9)	36 (1.1)	617 (6.3)	14 (0.8)	599 (6.1)
States								
Connecticut	24 (1.3)	497 (7.9)	33 (1.4)	518 (10.1)	32 (1.1)	525 (9.6)	11 (1.7)	508 (15.3)
Idaho	24 (1.4)	480 (9.7)	29 (1.3)	498 (8.1)	36 (1.9)	508 (6.7)	12 (1.2)	481 (8.7)
Illinois	22 (1.5)	487 (8.9)	33 (1.2)	513 (7.4)	35 (1.5)	525 (6.9)	10 (1.1)	500 (8.0)
Indiana	23 (1.7)	497 (8.2)	29 (2.0)	518 (8.5)	36 (1.9)	528 (7.5)	12 (1.4)	515 (11.9)
Maryland	26 (1.0)	484 (6.1)	36 (1.3)	507 (6.3)	27 (1.0)	506 (6.9)	11 (0.8)	480 (8.9)
Massachusetts	20 (1.2)	497 (8.0)	34 (1.4)	518 (6.5)	35 (1.3)	519 (6.0)	12 (1.0)	517 (10.0)
Michigan	23 (1.3)	508 (8.5)	33 (1.6)	527 (7.2)	33 (1.4)	527 (7.4)	11 (1.0)	496 (10.2)
Missouri	26 (1.5)	480 (6.4)	30 (1.8)	491 (6.3)	32 (1.7)	500 (5.3)	11 (1.0)	483 (12.2)
North Carolina	29 (1.3)	477 (6.2)	33 (1.5)	501 (9.4)	29 (1.2)	511 (7.9)	9 (1.0)	492 (7.7)
Oregon	20 (1.1)	500 (7.6)	32 (1.3)	521 (6.2)	37 (1.4)	526 (5.9)	11 (1.1)	496 (10.1)
Pennsylvania	20 (0.9)	484 (7.1)	32 (1.1)	514 (5.7)	36 (1.3)	521 (7.0)	13 (0.8)	499 (9.5)
South Carolina	23 (1.5)	478 (7.2)	32 (1.3)	505 (8.8)	33 (1.5)	519 (7.8)	12 (1.4)	500 (11.5)
Texas	21 (1.5)	497 (13.4)	31 (1.1)	527 (9.4)	36 (1.5)	536 (7.7)	13 (1.1)	503 (13.8)
Districts and Consortia								
Academy School Dist. #20, CO	25 (0.9)	520 (4.1)	36 (1.3)	531 (3.0)	32 (1.3)	544 (3.9)	8 (0.8)	493 (9.4)
Chicago Public Schools, IL	21 (1.9)	454 (8.9)	30 (1.5)	465 (7.3)	36 (2.3)	472 (6.8)	14 (1.8)	452 (7.4)
Delaware Science Coalition, DE	23 (1.4)	471 (10.8)	32 (2.1)	489 (10.5)	34 (1.3)	496 (10.0)	12 (1.9)	468 (12.3)
First in the World Consort., IL	17 (1.7)	526 (11.4)	31 (1.3)	551 (6.0)	42 (1.5)	576 (6.9)	10 (1.1)	572 (14.2)
Fremont/Lincoln/WestSide PS, NE	26 (2.5)	465 (10.7)	36 (2.2)	493 (9.0)	28 (1.9)	514 (9.6)	10 (1.4)	474 (20.3)
Guilford County, NC	19 (1.6)	481 (8.8)	30 (1.5)	515 (8.6)	40 (1.9)	531 (8.4)	11 (1.1)	507 (12.2)
Jersey City Public Schools, NJ	39 (1.5)	467 (10.9)	35 (1.8)	482 (10.8)	22 (1.8)	491 (7.7)	4 (0.5)	449 (21.4)
Miami-Dade County PS, FL	27 (1.8)	401 (13.8)	26 (1.9)	427 (9.5)	29 (2.0)	442 (11.2)	18 (2.2)	434 (7.6)
Michigan Invitational Group, MI	23 (1.8)	525 (7.8)	38 (2.2)	534 (7.2)	31 (1.9)	539 (8.5)	8 (0.8)	526 (13.1)
Montgomery County, MD	19 (1.2)	514 (8.6)	35 (1.7)	541 (6.0)	37 (1.5)	550 (4.9)	9 (1.2)	527 (11.9)
Naperville Sch. Dist. #203, IL	28 (1.8)	555 (4.4)	32 (1.5)	570 (3.5)	33 (1.6)	583 (5.3)	7 (0.7)	567 (10.2)
Project SMART Consortium, OH	21 (1.7)	507 (8.3)	32 (1.6)	518 (7.3)	35 (2.2)	538 (9.4)	12 (1.4)	507 (12.1)
Rochester City Sch. Dist., NY r	30 (1.8)	439 (7.6)	27 (2.1)	460 (9.2)	27 (1.6)	474 (9.3)	17 (1.5)	441 (12.7)
SW Math/Sci. Collaborative, PA	19 (1.3)	503 (6.2)	32 (1.8)	523 (9.0)	37 (2.2)	526 (8.3)	13 (1.5)	502 (11.8)
International Avg. (All Countries)	15 (0.1)	474 (1.4)	26 (0.2)	493 (0.9)	39 (0.2)	497 (0.9)	19 (0.2)	478 (1.0)

SOURCE: IEA Third International Mathematics and Science Study (TIMSS), 1998-1999.

Background data provided by students.

States in *italics* did not fully satisfy guidelines for sample participation rates (see Appendix A for details).

( ) Standard errors appear in parentheses. Because results are rounded to the nearest whole number, some totals may appear inconsistent.

A tilde (~) indicates insufficient data to report achievement.

An "r" indicates a 70-84% student response rate.

	Almost Always		Pretty Often		Once in a While		Never	
	Percent of Students	Average Achievement	Percent of Students	Average Achievement	Percent of Students	Average Achievement	Percent of Students	Average Achievement
Countries								
United States	42 (2.3)	516 (5.1)	28 (1.1)	506 (4.6)	22 (1.5)	493 (4.8)	8 (1.2)	471 (7.4)
Belgium (Flemish)	16 (1.3)	553 (6.4)	30 (1.6)	567 (6.0)	49 (2.1)	560 (4.5)	5 (1.2)	529 (11.0)
Canada	44 (1.9)	532 (3.2)	35 (1.2)	535 (3.0)	17 (1.5)	523 (4.1)	3 (0.9)	523 (14.8)
Chinese Taipei	2 (0.2)	~ ~	7 (0.4)	543 (7.9)	45 (1.0)	591 (4.0)	46 (1.1)	591 (5.0)
Czech Republic	14 (2.1)	520 (7.5)	27 (1.7)	530 (6.8)	48 (2.5)	518 (4.6)	12 (2.0)	507 (8.0)
England	30 (1.6)	518 (6.5)	53 (1.4)	500 (4.3)	16 (1.1)	460 (5.9)	1 (0.3)	~ ~
Hong Kong, SAR	26 (1.4)	591 (5.0)	51 (1.0)	582 (4.5)	21 (1.4)	575 (5.2)	2 (0.5)	~ ~
Italy	35 (1.9)	485 (4.4)	22 (0.9)	483 (5.3)	26 (1.5)	477 (5.6)	16 (1.7)	473 (7.6)
Japan	0 (0.1)	~ ~	3 (0.4)	545 (8.2)	28 (2.1)	580 (3.7)	68 (2.3)	580 (2.4)
Korea, Rep. of	0 (0.1)	~ ~	1 (0.2)	~ ~	13 (0.5)	574 (5.1)	86 (0.7)	590 (2.3)
Netherlands	67 (2.0)	547 (6.5)	29 (1.6)	533 (9.5)	4 (0.6)	494 (16.4)	0 (0.1)	~ ~
Russian Federation	16 (1.2)	508 (8.7)	19 (1.0)	528 (8.4)	47 (1.3)	533 (6.3)	18 (1.7)	532 (7.6)
Singapore	36 (1.5)	610 (7.4)	50 (1.0)	610 (6.1)	14 (1.2)	572 (6.5)	1 (0.2)	~ ~
States								
Connecticut	43 (3.1)	525 (9.9)	34 (2.3)	516 (8.5)	18 (3.0)	497 (13.7)	5 (1.0)	463 (23.5)
Idaho	38 (5.3)	496 (8.2)	28 (1.8)	501 (9.6)	25 (3.2)	496 (8.5)	9 (3.0)	471 (11.8)
Illinois	48 (3.8)	526 (7.0)	25 (1.9)	517 (6.3)	20 (2.2)	482 (7.3)	7 (2.8)	451 (14.5)
Indiana	33 (3.8)	522 (9.9)	29 (1.9)	523 (7.8)	30 (3.2)	512 (8.0)	8 (1.9)	478 (12.2)
Maryland	42 (3.6)	510 (5.2)	32 (1.6)	501 (7.3)	21 (2.6)	482 (9.3)	5 (1.3)	447 (16.9)
Massachusetts	37 (4.0)	519 (8.9)	32 (1.7)	512 (5.1)	25 (3.0)	515 (6.8)	6 (1.5)	491 (21.6)
Michigan	51 (3.1)	527 (5.9)	29 (2.3)	518 (9.1)	16 (2.0)	509 (10.5)	5 (1.3)	475 (15.3)
Missouri	55 (3.2)	500 (6.3)	24 (1.8)	487 (5.4)	13 (2.3)	473 (12.9)	8 (2.0)	468 (20.4)
North Carolina	54 (3.5)	494 (7.0)	31 (1.8)	498 (8.2)	11 (1.8)	493 (12.8)	4 (2.0)	516 (28.8)
Oregon	60 (2.8)	525 (6.3)	30 (1.9)	509 (6.1)	9 (1.4)	489 (10.7)	2 (0.4)	~ ~
Pennsylvania	38 (2.5)	516 (7.8)	31 (3.6)	513 (8.2)	22 (2.2)	498 (8.2)	9 (4.0)	491 (6.0)
South Carolina	25 (2.7)	518 (13.4)	30 (2.1)	506 (8.7)	36 (2.6)	502 (8.6)	9 (2.1)	452 (17.4)
Texas	23 (2.8)	545 (10.2)	24 (2.2)	535 (12.0)	39 (2.5)	508 (9.4)	14 (2.4)	492 (15.3)
Districts and Consortia								
Academy School Dist. #20, CO	68 (1.3)	542 (2.0)	22 (1.2)	520 (4.2)	6 (0.7)	466 (8.5)	4 (0.5)	447 (10.9)
Chicago Public Schools, IL	13 (2.9)	458 (11.6)	25 (3.0)	471 (7.0)	52 (2.8)	467 (7.2)	10 (3.0)	432 (7.3)
Delaware Science Coalition, DE	32 (3.0)	482 (13.8)	35 (1.9)	491 (7.9)	29 (2.2)	486 (12.3)	5 (1.4)	462 (22.4)
First in the World Consort., IL	56 (2.7)	568 (7.1)	32 (2.5)	555 (7.6)	10 (1.5)	538 (10.2)	2 (0.8)	~ ~
Fremont/Lincoln/WestSide PS, NE	46 (2.8)	491 (12.3)	36 (2.2)	494 (8.5)	14 (2.1)	480 (7.7)	4 (1.7)	448 (15.8)
Guilford County, NC	46 (3.4)	509 (9.3)	34 (1.9)	518 (10.4)	17 (2.7)	528 (12.0)	3 (1.3)	479 (14.1)
Jersey City Public Schools, NJ	68 (2.5)	484 (9.8)	25 (1.9)	470 (8.4)	6 (1.0)	438 (12.1)	1 (0.3)	~ ~
Miami-Dade County PS, FL	23 (3.5)	416 (16.0)	26 (2.6)	422 (11.7)	36 (2.8)	434 (8.9)	16 (3.8)	427 (12.6)
Michigan Invitational Group, MI	62 (3.0)	539 (5.6)	28 (2.2)	526 (8.2)	7 (1.6)	526 (17.4)	3 (1.3)	485 (13.0)
Montgomery County, MD	54 (3.4)	550 (5.5)	36 (2.8)	530 (5.9)	7 (1.9)	517 (12.1)	2 (0.5)	~ ~
Naperville Sch. Dist. #203, IL	71 (1.3)	571 (3.5)	21 (1.4)	573 (3.8)	7 (0.6)	555 (15.6)	1 (0.3)	~ ~
Project SMART Consortium, OH	26 (2.2)	541 (13.4)	32 (2.1)	529 (9.3)	29 (3.0)	512 (8.7)	13 (2.0)	485 (9.8)
Rochester City Sch. Dist., NY r	21 (3.3)	444 (16.2)	20 (1.9)	456 (12.6)	35 (2.4)	472 (7.4)	24 (2.4)	440 (9.4)
SW Math/Sci. Collaborative, PA	45 (5.0)	525 (6.9)	29 (3.2)	518 (9.3)	23 (3.9)	506 (11.5)	3 (1.3)	488 (23.8)
International Avg. (All Countries)	19 (0.2)	469 (2.0)	20 (0.2)	482 (1.1)	29 (0.2)	488 (1.2)	32 (0.3)	476 (2.2)

SOURCE: IEA Third International Mathematics and Science Study (TIMSS), 1998-1999.

Background data provided by students.

* The use of calculators on TIMSS was not allowed in 1995 or in 1999.

States in *italics* did not fully satisfy guidelines for sample participation rates (see Appendix A for details).

( ) Standard errors appear in parentheses. Because results are rounded to the nearest whole number, some totals may appear inconsistent.

A tilde (~) indicates insufficient data to report achievement.

An "r" indicates a 70-84% student response rate.

Exhibit R3.10 Ways in Which Calculators Are Used*

	Percentage of Students					
	Never or Hardly Ever Use Calculators	Ways in Which Students Use Calculators At Least Once or Twice a Week				
		Checking Answers	Tests and Exams	Routine Computations	Solving Complex Problems	Exploring Number Concepts
Countries						
United States	5 (1.2)	69 (3.5)	45 (3.2)	62 (3.9)	75 (3.2)	59 (3.2)
Belgium (Flemish)	7 (2.5)	55 (4.6)	34 (3.5)	45 (4.7)	53 (4.5)	27 (4.0)
Canada	3 (1.0)	82 (2.3)	58 (2.9)	r 80 (2.8)	91 (1.8)	r 69 (3.0)
Chinese Taipei	55 (4.6)	5 (1.7)	2 (1.0)	9 (2.5)	9 (2.5)	8 (2.3)
Czech Republic	5 (2.3)	60 (4.9)	18 (3.6)	49 (4.9)	60 (5.0)	22 (4.6)
England	0 (0.4)	s 74 (3.6)	s 32 (3.4)	s 76 (3.0)	s 72 (3.8)	s 41 (4.1)
Hong Kong, SAR	1 (0.9)	85 (2.8)	58 (4.2)	88 (2.8)	74 (3.6)	39 (3.9)
Italy	13 (2.1)	66 (2.8)	14 (2.8)	58 (3.2)	68 (3.4)	21 (3.3)
Japan	81 (3.3)	0 (0.0)	1 (0.7)	5 (2.0)	4 (1.7)	1 (1.0)
Korea, Rep. of	72 (3.4)	1 (0.9)	1 (0.9)	2 (1.2)	5 (1.7)	2 (1.2)
Netherlands	1 (0.1)	91 (2.9)	68 (5.4)	97 (1.8)	90 (3.0)	65 (6.3)
Russian Federation	14 (2.7)	58 (3.9)	3 (1.3)	47 (4.4)	46 (3.8)	15 (3.0)
Singapore	0 (0.0)	88 (3.0)	61 (4.5)	85 (3.5)	93 (2.5)	67 (3.9)
States						
Connecticut	5 (1.8)	r 77 (5.3)	r 64 (6.4)	r 70 (6.2)	r 86 (4.7)	r 75 (5.8)
Idaho	8 (4.0)	r 66 (7.3)	r 32 (5.0)	r 49 (6.9)	r 74 (6.3)	r 51 (6.6)
Illinois	7 (4.0)	80 (5.3)	43 (6.6)	72 (5.1)	81 (5.2)	68 (5.8)
Indiana	5 (2.5)	69 (5.2)	22 (5.5)	57 (6.1)	69 (5.9)	52 (7.5)
Maryland	1 (0.8)	r 67 (5.8)	r 55 (5.3)	r 70 (6.4)	r 77 (5.6)	r 67 (5.9)
Massachusetts	4 (1.9)	70 (5.5)	52 (6.1)	62 (6.2)	72 (5.7)	65 (6.5)
Michigan	1 (0.9)	74 (4.8)	55 (6.5)	80 (4.6)	89 (3.1)	77 (4.4)
Missouri	5 (3.0)	77 (5.2)	49 (5.7)	74 (4.7)	78 (4.3)	69 (4.9)
North Carolina	1 (0.7)	69 (6.2)	63 (5.3)	72 (4.8)	85 (3.2)	78 (3.8)
Oregon	0 (0.3)	91 (2.8)	52 (6.9)	85 (2.8)	87 (3.4)	71 (5.2)
Pennsylvania	12 (5.6)	70 (6.8)	48 (8.0)	64 (7.2)	69 (7.4)	53 (7.5)
South Carolina	16 (4.7)	48 (6.2)	27 (6.1)	45 (6.1)	56 (6.4)	45 (6.9)
Texas	11 (4.7)	43 (6.1)	28 (4.9)	37 (6.1)	51 (6.2)	40 (5.7)
Districts and Consortia						
Academy School Dist. #20, CO	1 (0.2)	92 (0.2)	77 (0.3)	81 (0.3)	95 (0.3)	83 (0.3)
Chicago Public Schools, IL	14 (7.8)	57 (9.4)	13 (7.5)	47 (8.3)	62 (8.6)	45 (11.4)
Delaware Science Coalition, DE	4 (2.9)	r 63 (6.4)	r 44 (4.9)	r 55 (6.4)	r 72 (5.2)	r 57 (6.1)
First in the World Consort., IL	0 (0.0)	95 (3.2)	69 (9.0)	76 (3.6)	94 (3.9)	93 (3.9)
Fremont/Lincoln/WestSide PS, NE	0 (0.0)	63 (7.4)	44 (5.7)	77 (6.9)	90 (3.3)	66 (1.3)
Guilford County, NC	2 (0.6)	62 (7.4)	51 (7.2)	55 (6.9)	80 (6.3)	59 (8.0)
Jersey City Public Schools, NJ	0 (0.0)	96 (0.5)	89 (1.9)	93 (1.9)	96 (0.5)	97 (2.9)
Miami-Dade County PS, FL	7 (4.8)	s 71 (9.4)	s 39 (10.9)	s 59 (10.0)	s 77 (9.0)	s 60 (12.5)
Michigan Invitational Group, MI	2 (0.0)	89 (4.9)	80 (5.7)	83 (4.6)	96 (2.9)	90 (4.0)
Montgomery County, MD	0 (0.0)	s 86 (3.0)	s 75 (5.7)	s 91 (4.0)	s 96 (3.1)	s 87 (5.0)
Naperville Sch. Dist. #203, IL	0 (0.0)	100 (0.0)	93 (0.8)	90 (2.5)	99 (0.6)	90 (3.2)
Project SMART Consortium, OH	14 (4.3)	60 (4.4)	33 (4.7)	47 (4.1)	60 (4.5)	49 (4.8)
Rochester City Sch. Dist., NY	20 (4.0)	34 (4.4)	13 (4.5)	39 (4.8)	34 (4.5)	26 (4.3)
SW Math/Sci. Collaborative, PA	1 (1.0)	78 (4.9)	66 (7.2)	71 (6.7)	82 (5.3)	64 (7.9)
International Avg. (All Countries)	28 (0.5)	44 (0.6)	20 (0.4)	43 (0.6)	43 (0.6)	26 (0.5)

SOURCE: IEA Third International Mathematics and Science Study (TIMSS), 1998-1999.

Background data provided by teachers.

* The use of calculators on TIMSS was not allowed in 1995 or in 1999.

States in *italics* did not fully satisfy guidelines for sample participation rates (see Appendix A for details).

( ) Standard errors appear in parentheses. Because results are rounded to the nearest whole number, some totals may appear inconsistent.

An "r" indicates teacher response data available for 70-84% of students. An "s" indicates teacher response data available for 50-69% of students.

Exhibit R3.11 Amount of Mathematics Homework

	Percentage of Students Taught by Teachers						
	Assigning Homework Three Times a Week or More Often		Assigning Homework Once or Twice a Week		Assigning Homework Less Than Once a Week		Never Assigning Homework
	30 Minutes or Less	More Than 30 Minutes	30 Minutes or Less	More Than 30 Minutes	30 Minutes or Less	More Than 30 Minutes	
Countries							
United States	63 (2.8)	27 (2.4)	7 (1.6)	1 (0.6)	1 (0.7)	0 (0.0)	1 (0.7)
Belgium (Flemish)	15 (2.7)	2 (0.9)	48 (4.9)	9 (2.0)	18 (3.3)	3 (1.2)	5 (3.2)
Canada	58 (3.5)	16 (2.6)	22 (2.8)	3 (1.5)	1 (0.7)	0 (0.0)	0 (0.0)
Chinese Taipei	21 (3.1)	25 (3.6)	27 (3.4)	23 (3.4)	2 (1.1)	2 (1.1)	0 (0.0)
Czech Republic	15 (3.7)	0 (0.2)	69 (5.1)	2 (1.3)	13 (3.8)	0 (0.0)	0 (0.0)
England	3 (1.4)	1 (0.6)	48 (3.9)	46 (4.1)	2 (0.8)	1 (0.3)	0 (0.0)
Hong Kong, SAR	30 (4.0)	19 (3.1)	26 (3.1)	23 (3.7)	2 (1.2)	1 (0.0)	0 (0.0)
Italy	15 (2.6)	68 (3.8)	5 (1.7)	12 (2.5)	0 (0.0)	0 (0.0)	0 (0.0)
Japan	14 (2.7)	3 (1.4)	27 (4.0)	8 (2.1)	34 (4.3)	6 (2.0)	9 (2.3)
Korea, Rep. of	24 (3.3)	9 (2.3)	29 (3.4)	15 (2.7)	14 (2.6)	6 (2.0)	2 (0.7)
Netherlands	73 (4.4)	9 (2.6)	13 (3.3)	3 (1.3)	1 (0.6)	0 (0.0)	1 (0.6)
Russian Federation	42 (4.5)	57 (4.6)	0 (0.5)	0 (0.0)	0 (0.0)	0 (0.0)	0 (0.0)
Singapore	26 (4.2)	54 (4.3)	8 (2.1)	12 (2.4)	0 (0.0)	0 (0.0)	0 (0.0)
States							
Connecticut	69 (6.9)	30 (6.5)	1 (1.4)	0 (0.0)	0 (0.0)	0 (0.0)	0 (0.0)
Idaho	59 (5.4)	18 (4.1)	12 (4.5)	0 (0.0)	4 (1.3)	0 (0.0)	7 (4.3)
Illinois	72 (5.6)	27 (5.5)	0 (0.4)	0 (0.4)	0 (0.0)	0 (0.0)	0 (0.3)
Indiana	78 (5.7)	18 (4.9)	4 (1.9)	1 (0.6)	0 (0.0)	0 (0.0)	0 (0.0)
Maryland	73 (4.2)	16 (3.3)	8 (2.0)	1 (1.0)	2 (1.8)	0 (0.0)	0 (0.1)
Massachusetts	57 (7.2)	41 (7.0)	1 (0.1)	0 (0.0)	0 (0.0)	0 (0.0)	2 (1.9)
Michigan	52 (5.0)	31 (4.5)	9 (3.3)	5 (2.1)	0 (0.0)	0 (0.0)	3 (2.7)
Missouri	59 (6.1)	26 (5.9)	12 (3.9)	0 (0.0)	1 (1.1)	0 (0.0)	1 (0.7)
North Carolina	57 (6.4)	23 (5.4)	16 (3.4)	1 (0.5)	4 (2.4)	0 (0.0)	0 (0.0)
Oregon	64 (5.5)	22 (4.8)	9 (2.3)	0 (0.0)	3 (2.1)	0 (0.0)	2 (1.5)
Pennsylvania	69 (5.7)	26 (5.7)	5 (2.4)	0 (0.0)	0 (0.0)	0 (0.0)	0 (0.1)
South Carolina	54 (7.6)	32 (6.3)	14 (3.6)	0 (0.0)	0 (0.0)	0 (0.0)	0 (0.1)
Texas	41 (6.7)	34 (6.8)	14 (4.3)	7 (4.0)	2 (1.8)	0 (0.0)	2 (1.5)
Districts and Consortia							
Academy School Dist. #20, CO	20 (0.3)	77 (0.3)	1 (0.0)	0 (0.0)	2 (0.1)	0 (0.0)	0 (0.0)
Chicago Public Schools, IL	56 (9.1)	39 (9.1)	3 (0.2)	2 (2.2)	0 (0.0)	0 (0.0)	0 (0.0)
Delaware Science Coalition, DE	75 (5.8)	17 (5.1)	8 (2.8)	0 (0.0)	0 (0.0)	0 (0.0)	0 (0.0)
First in the World Consort., IL	61 (5.2)	39 (5.2)	0 (0.0)	0 (0.0)	0 (0.0)	0 (0.0)	0 (0.0)
Fremont/Lincoln/WestSide PS, NE	72 (3.2)	23 (3.2)	5 (0.2)	0 (0.0)	0 (0.0)	0 (0.0)	0 (0.0)
Guilford County, NC	59 (8.1)	28 (6.3)	10 (5.0)	2 (0.1)	2 (0.1)	0 (0.0)	0 (0.0)
Jersey City Public Schools, NJ	51 (5.7)	46 (6.4)	2 (2.5)	0 (0.0)	0 (0.0)	0 (0.0)	0 (0.0)
Miami-Dade County PS, FL	53 (9.1)	24 (5.7)	21 (7.3)	2 (2.0)	0 (0.0)	0 (0.0)	0 (0.0)
Michigan Invitational Group, MI	61 (9.8)	29 (7.0)	10 (6.9)	0 (0.0)	0 (0.0)	0 (0.0)	0 (0.0)
Montgomery County, MD	49 (4.5)	42 (4.9)	3 (1.4)	4 (4.4)	1 (0.2)	0 (0.0)	1 (0.6)
Naperville Sch. Dist. #203, IL	68 (2.4)	30 (2.3)	0 (0.0)	0 (0.0)	2 (0.1)	0 (0.0)	0 (0.0)
Project SMART Consortium, OH	68 (6.6)	23 (6.4)	6 (3.2)	3 (2.5)	0 (0.0)	0 (0.0)	0 (0.0)
Rochester City Sch. Dist., NY	60 (5.1)	21 (5.4)	15 (3.4)	0 (0.0)	0 (0.0)	5 (0.4)	0 (0.0)
SW Math/Sci. Collaborative, PA	59 (6.2)	35 (5.5)	5 (2.4)	0 (0.0)	1 (0.9)	0 (0.0)	0 (0.0)
International Avg. (All Countries)	41 (0.6)	26 (0.5)	16 (0.5)	10 (0.4)	4 (0.2)	2 (0.2)	1 (0.1)

SOURCE: IEA Third International Mathematics and Science Study (TIMSS), 1998-1999.

Background data provided by teachers.

States in *italics* did not fully satisfy guidelines for sample participation rates (see Appendix A for details).

( ) Standard errors appear in parentheses. Because results are rounded to the nearest whole number, some totals may appear inconsistent.

Exhibit R3.12 Assigning Mathematics Homework Based on Projects and Investigations*

8th Grade Mathematics

		Sometimes or Always		Never or Rarely	
		Percent of Students	Average Achievement	Percent of Students	Average Achievement
Countries					
United States		24 (2.8)	499 (8.3)	76 (2.8)	504 (4.9)
Belgium (Flemish)		3 (1.1)	567 (19.0)	97 (1.1)	567 (3.9)
Canada	r	24 (2.9)	531 (4.9)	76 (2.9)	535 (3.4)
Chinese Taipei		4 (1.6)	577 (22.9)	96 (1.6)	585 (4.1)
Czech Republic		3 (1.7)	537 (9.7)	97 (1.7)	520 (4.6)
England	s	14 (2.5)	518 (9.1)	86 (2.5)	509 (5.6)
Hong Kong, SAR		3 (1.4)	636 (14.9)	97 (1.4)	581 (4.4)
Italy		30 (3.5)	468 (6.9)	70 (3.5)	484 (4.7)
Japan		1 (0.7)	~ ~	99 (0.7)	579 (1.8)
Korea, Rep. of		16 (2.9)	586 (5.6)	84 (2.9)	588 (2.3)
Netherlands		10 (5.3)	510 (44.2)	90 (5.3)	540 (6.9)
Russian Federation		34 (3.7)	539 (8.4)	66 (3.7)	520 (6.9)
Singapore		20 (3.6)	616 (14.5)	80 (3.6)	602 (6.8)
States					
Connecticut	r	38 (6.7)	507 (9.7)	62 (6.7)	529 (13.8)
Idaho	r	14 (3.9)	497 (14.8)	86 (3.9)	498 (9.5)
Illinois		22 (5.3)	503 (19.9)	78 (5.3)	513 (7.2)
Indiana		18 (4.9)	497 (14.3)	82 (4.9)	522 (7.5)
Maryland	r	30 (5.8)	488 (12.9)	70 (5.8)	482 (7.6)
Massachusetts		35 (4.5)	519 (10.1)	65 (4.5)	508 (7.0)
Michigan		14 (4.4)	537 (13.1)	86 (4.4)	522 (7.7)
Missouri		24 (6.0)	490 (8.1)	76 (6.0)	492 (6.6)
North Carolina		21 (4.4)	484 (13.8)	79 (4.4)	494 (6.6)
Oregon		33 (7.9)	528 (10.6)	67 (7.9)	508 (6.4)
Pennsylvania		21 (5.4)	513 (17.7)	79 (5.4)	510 (6.2)
South Carolina		32 (7.2)	513 (14.5)	68 (7.2)	499 (8.3)
Texas	r	25 (6.3)	515 (26.0)	75 (6.3)	526 (8.3)
Districts and Consortia					
Academy School Dist. #20, CO		19 (0.2)	570 (2.7)	81 (0.2)	522 (2.0)
Chicago Public Schools, IL		29 (10.4)	463 (15.9)	71 (10.4)	468 (6.0)
Delaware Science Coalition, DE	r	28 (5.8)	499 (22.0)	72 (5.8)	473 (12.1)
First in the World Consort., IL	r	6 (3.9)	536 (33.4)	94 (3.9)	557 (7.1)
Fremont/Lincoln/WestSide PS, NE		19 (5.9)	454 (18.6)	81 (5.9)	495 (9.7)
Guilford County, NC		24 (5.2)	529 (19.2)	76 (5.2)	504 (9.2)
Jersey City Public Schools, NJ		94 (3.3)	482 (10.5)	6 (3.3)	425 (36.0)
Miami-Dade County PS, FL	s	32 (7.7)	438 (11.3)	68 (7.7)	413 (17.9)
Michigan Invitational Group, MI		29 (7.5)	532 (6.2)	71 (7.5)	534 (6.9)
Montgomery County, MD	s	40 (8.8)	541 (9.4)	60 (8.8)	530 (6.7)
Naperville Sch. Dist. #203, IL		7 (3.5)	644 (3.3)	93 (3.5)	562 (3.3)
Project SMART Consortium, OH		33 (6.7)	508 (11.9)	67 (6.7)	527 (9.5)
Rochester City Sch. Dist., NY		3 (2.0)	433 (29.0)	97 (2.0)	442 (7.1)
SW Math/Sci. Collaborative, PA		14 (5.5)	528 (13.1)	86 (5.5)	517 (8.4)
International Avg. (All Countries)		18 (0.5)	491 (2.2)	82 (0.5)	487 (0.8)

SOURCE: IEA Third International Mathematics and Science Study (TIMSS), 1998-1999.

Background data provided by teachers.

* Based on average response to questions about assigning homework based on small investigation(s) or gathering data, working individually on long term projects or experiments, and working as a small group on long term projects or experiments.

States in *italics* did not fully satisfy guidelines for sample participation rates (see Appendix A for details).

( ) Standard errors appear in parentheses. Because results are rounded to the nearest whole number, some totals may appear inconsistent.

A tilde (~) indicates insufficient data to report achievement.

An "r" indicates teacher response data available for 70-84% of students. An "s" indicates teacher response data available for 50-69% of students.

Frequency of Having a Quiz or Test in Mathematics Class

	Almost Always		Pretty Often		Once in a While or Never	
	Percent of Students	Average Achievement	Percent of Students	Average Achievement	Percent of Students	Average Achievement
Countries						
United States	40 (1.6)	491 (3.7)	46 (1.3)	520 (4.3)	14 (0.9)	493 (6.1)
Belgium (Flemish)	19 (0.9)	534 (4.1)	56 (1.7)	568 (4.1)	25 (1.6)	558 (8.8)
Canada	25 (1.6)	516 (3.9)	50 (1.4)	534 (2.9)	25 (1.1)	541 (4.1)
Chinese Taipei	27 (1.2)	589 (5.1)	46 (1.0)	590 (4.2)	27 (1.5)	576 (6.5)
Czech Republic	7 (1.2)	494 (10.9)	33 (1.7)	522 (4.9)	60 (2.3)	522 (4.5)
England	14 (0.9)	463 (5.6)	46 (1.6)	500 (4.8)	40 (1.9)	507 (5.4)
Hong Kong, SAR	9 (0.7)	569 (7.3)	37 (1.6)	579 (4.9)	54 (2.2)	587 (5.2)
Italy	9 (0.8)	452 (8.7)	19 (0.9)	476 (6.3)	72 (1.3)	485 (3.7)
Japan	12 (1.4)	571 (6.1)	30 (1.5)	582 (3.2)	58 (2.1)	579 (2.6)
Korea, Rep. of	7 (0.5)	587 (6.9)	18 (0.8)	601 (4.5)	75 (1.2)	584 (1.9)
Netherlands	14 (1.5)	510 (10.5)	47 (1.9)	538 (6.7)	39 (1.7)	555 (8.2)
Russian Federation	26 (1.1)	517 (6.6)	50 (1.3)	541 (6.5)	24 (1.3)	510 (6.5)
Singapore	19 (0.9)	597 (8.3)	45 (1.0)	609 (6.5)	36 (1.4)	602 (6.4)
States						
Connecticut	41 (1.9)	500 (9.2)	48 (1.7)	528 (9.7)	11 (1.1)	501 (11.3)
Idaho	41 (2.9)	485 (6.9)	45 (1.9)	509 (8.3)	14 (1.7)	482 (11.3)
Illinois	42 (1.8)	501 (6.9)	47 (1.6)	521 (7.9)	11 (1.3)	494 (6.8)
Indiana	35 (1.8)	508 (7.6)	50 (1.6)	522 (6.9)	15 (1.8)	514 (13.5)
Maryland	39 (1.6)	483 (6.7)	46 (1.6)	513 (6.5)	15 (1.6)	488 (9.2)
Massachusetts	36 (2.1)	499 (5.8)	50 (1.7)	527 (6.7)	13 (1.3)	502 (8.5)
Michigan	37 (2.0)	508 (8.5)	50 (1.7)	529 (6.8)	13 (1.6)	512 (8.2)
Missouri	34 (2.9)	479 (5.0)	45 (2.3)	502 (5.4)	21 (2.6)	486 (11.5)
North Carolina	45 (2.1)	488 (7.0)	45 (1.5)	507 (8.0)	10 (1.2)	480 (12.2)
Oregon	32 (2.1)	504 (7.7)	49 (1.8)	528 (6.5)	19 (2.0)	504 (8.3)
Pennsylvania	36 (1.3)	494 (8.5)	52 (1.5)	519 (5.8)	12 (0.9)	506 (7.5)
South Carolina	46 (2.3)	484 (8.4)	44 (1.9)	523 (6.7)	9 (1.4)	497 (11.5)
Texas	34 (2.1)	518 (9.1)	46 (1.6)	537 (8.3)	20 (2.0)	489 (13.3)
Districts and Consortia						
Academy School Dist. #20, CO	42 (1.5)	519 (3.3)	47 (1.4)	545 (3.1)	11 (0.8)	500 (6.8)
Chicago Public Schools, IL	35 (3.8)	454 (6.4)	44 (1.6)	471 (6.8)	21 (3.6)	463 (12.6)
Delaware Science Coalition, DE	39 (1.5)	474 (8.7)	48 (1.7)	499 (10.1)	13 (1.2)	471 (16.4)
First in the World Consort., IL	40 (1.8)	553 (8.6)	53 (2.2)	568 (5.6)	7 (1.8)	535 (8.1)
Fremont/Lincoln/WestSide PS, NE	40 (1.6)	473 (8.7)	48 (2.2)	507 (9.9)	12 (1.4)	473 (12.5)
Guilford County, NC	41 (2.2)	501 (8.7)	50 (1.8)	528 (8.6)	9 (0.9)	495 (16.0)
Jersey City Public Schools, NJ	50 (1.6)	471 (8.7)	40 (1.3)	485 (10.7)	11 (1.3)	478 (12.3)
Miami-Dade County PS, FL	45 (2.8)	417 (9.8)	37 (2.9)	443 (9.9)	18 (4.0)	410 (15.6)
Michigan Invitational Group, MI	37 (2.9)	534 (9.3)	52 (3.1)	538 (4.7)	12 (2.7)	507 (8.8)
Montgomery County, MD	37 (2.1)	523 (5.3)	52 (1.9)	550 (4.5)	11 (0.7)	529 (11.2)
Naperville Sch. Dist. #203, IL	42 (1.4)	556 (3.6)	52 (1.3)	582 (3.6)	6 (0.7)	559 (8.7)
Project SMART Consortium, OH	37 (2.9)	509 (7.3)	48 (2.1)	535 (8.9)	16 (1.7)	513 (10.0)
Rochester City Sch. Dist., NY r	53 (2.4)	441 (7.3)	37 (2.5)	469 (8.7)	10 (1.3)	465 (15.2)
SW Math/Sci. Collaborative, PA	30 (2.1)	503 (7.5)	56 (1.9)	528 (7.7)	14 (1.8)	504 (9.1)
International Avg. (All Countries)	21 (0.2)	473 (1.2)	36 (0.2)	493 (0.9)	43 (0.2)	490 (0.9)

SOURCE: IEA Third International Mathematics and Science Study (TIMSS), 1998-1999.

Background data provided by students.

States in *italics* did not fully satisfy guidelines for sample participation rates (see Appendix A for details).

( ) Standard errors appear in parentheses. Because results are rounded to the nearest whole number, some totals may appear inconsistent.

An "r" indicates a 70-84% student response rate.

REFERENCE

4 School Contexts for Learning and Instruction

Shortages or Inadequacies in General Facilities and Materials That Affect Schools' Capacity to Provide Mathematics Instruction Some or A Lot

	Percentage of Students Affected by Shortage or Inadequacy				
	Instuctional Materials	Budget for Supplies	School Buildings/ Grounds	Heating/Cooling and Lighting	Instructional Space
Countries					
United States	r 22 (2.9)	r 27 (4.1)	r 33 (3.4)	r 17 (3.5)	r 33 (3.4)
Belgium (Flemish)	6 (2.2)	5 (2.1)	20 (3.3)	4 (1.8)	20 (4.2)
Canada	45 (2.8)	43 (2.8)	29 (2.8)	11 (1.9)	25 (2.4)
Chinese Taipei	45 (4.4)	45 (4.0)	59 (4.1)	41 (3.9)	51 (4.0)
Czech Republic	22 (5.0)	52 (5.5)	15 (3.3)	5 (1.4)	11 (3.3)
England	r 37 (4.9)	r 31 (4.5)	r 42 (5.3)	r 17 (3.6)	r 38 (5.1)
Hong Kong, SAR	35 (3.9)	21 (3.9)	57 (4.8)	24 (3.4)	57 (4.6)
Italy	28 (3.5)	28 (3.6)	31 (3.7)	15 (2.7)	35 (3.4)
Japan	17 (2.9)	14 (3.0)	29 (3.8)	31 (3.5)	34 (3.5)
Korea, Rep. of	37 (3.9)	29 (4.0)	51 (4.5)	52 (4.2)	55 (4.2)
Netherlands	r 10 (4.0)	r 19 (6.4)	r 45 (7.0)	r 9 (2.8)	r 26 (5.3)
Russian Federation	92 (2.4)	81 (3.1)	73 (3.6)	63 (4.4)	69 (3.2)
Singapore	10 (2.2)	7 (2.0)	23 (2.6)	11 (2.4)	26 (3.3)
States					
Connecticut	s 15 (5.4)	s 18 (7.2)	s 23 (8.0)	s 9 (5.0)	s 28 (8.3)
Idaho	r 15 (6.5)	r 34 (9.4)	r 24 (5.3)	r 23 (4.7)	r 29 (8.1)
Illinois	15 (5.1)	19 (5.0)	16 (5.4)	12 (4.6)	31 (7.2)
Indiana	17 (6.8)	17 (5.2)	20 (6.2)	15 (5.2)	33 (7.8)
Maryland	r 36 (6.8)	r 27 (6.8)	r 29 (6.9)	r 27 (5.9)	r 41 (5.8)
Massachusetts	s 26 (6.6)	s 18 (6.4)	s 38 (6.9)	s 17 (6.0)	s 26 (7.5)
Michigan	14 (4.1)	20 (4.9)	27 (7.5)	10 (4.3)	36 (7.4)
Missouri	23 (7.3)	24 (7.3)	26 (6.2)	18 (5.0)	40 (6.7)
North Carolina	r 35 (7.3)	r 37 (7.6)	r 40 (7.8)	r 23 (7.5)	r 48 (8.3)
Oregon	56 (6.9)	57 (7.3)	42 (6.9)	22 (7.6)	56 (7.5)
Pennsylvania	7 (2.8)	8 (3.5)	24 (8.3)	15 (5.3)	27 (7.7)
South Carolina	14 (6.0)	r 26 (7.1)	39 (8.4)	21 (7.2)	50 (9.3)
Texas	r 10 (5.9)	r 18 (7.0)	r 29 (7.4)	r 16 (7.1)	r 33 (8.6)
Districts and Consortia					
Academy School Dist. #20, CO	0 (0.0)	25 (0.3)	0 (0.0)	0 (0.0)	0 (0.0)
Chicago Public Schools, IL	s 13 (7.5)	s 27 (12.5)	s 34 (11.5)	s 32 (11.8)	s 29 (12.5)
Delaware Science Coalition, DE	r 31 (2.1)	r 36 (2.1)	r 27 (1.7)	r 26 (1.9)	r 28 (2.3)
First in the World Consort., IL	r 13 (0.4)	r 11 (0.4)	r 19 (0.8)	r 0 (0.0)	r 28 (0.9)
Fremont/Lincoln/WestSide PS, NE	r 33 (1.5)	r 49 (1.6)	r 39 (1.5)	r 49 (1.6)	r 57 (1.7)
Guilford County, NC	s 0 (0.0)	s 0 (0.0)	s 30 (1.0)	r 43 (1.1)	r 36 (1.0)
Jersey City Public Schools, NJ	11 (0.7)	9 (0.7)	43 (1.4)	16 (0.8)	r 65 (1.7)
Miami-Dade County PS, FL	s 17 (9.6)	s 25 (12.0)	s 53 (14.8)	s 3 (2.9)	s 59 (11.5)
Michigan Invitational Group, MI	30 (1.3)	30 (1.3)	38 (1.7)	22 (1.0)	49 (1.6)
Montgomery County, MD	s 18 (10.2)	s 5 (5.4)	s 25 (12.0)	s 23 (9.1)	s 53 (8.9)
Naperville Sch. Dist. #203, IL	24 (1.5)	24 (1.5)	24 (1.5)	0 (0.0)	45 (1.5)
Project SMART Consortium, OH	22 (0.8)	r 24 (1.4)	r 29 (1.4)	22 (0.9)	29 (1.3)
Rochester City Sch. Dist., NY	r 41 (1.5)	r 41 (1.5)	r 0 (0.0)	r 16 (0.5)	r 16 (0.5)
SW Math/Sci. Collaborative, PA	18 (7.4)	13 (5.8)	18 (7.3)	17 (7.1)	23 (7.3)
International Avg. (All Countries)	45 (0.6)	47 (0.6)	50 (0.7)	36 (0.6)	47 (0.6)

SOURCE: IEA Third International Mathematics and Science Study (TIMSS), 1998-1999.

Background data provided by schools.

States in *italics* did not fully satisfy guidelines for sample participation rates (see Appendix A for details).

( ) Standard errors appear in parentheses. Because results are rounded to the nearest whole number, some totals may appear inconsistent.

An "r" indicates school response data available for 70-84% of students. An "s" indicates school response data available for 50-69% of students.

Shortages or Inadequacies in Equipment and Materials for Mathematics Instruction That Affect Schools' Capacity to Provide Mathematics Instruction Some or A Lot

8th Grade Mathematics

	Percentage of Students Affected by Shortage or Inadequacy				
	Computers for Mathematics Instruction	Computer Software for Mathematics Instruction	Calculators for Mathematics Instruction	Library Materials Relevant to Mathematics Instruction	Audio-Visual Resources for Mathematics Instruction
Countries					
United States	r 47 (4.2)	r 48 (4.1)	r 16 (3.4)	r 29 (4.0)	r 25 (3.2)
Belgium (Flemish)	30 (3.3)	25 (3.4)	2 (1.1)	15 (3.0)	19 (3.0)
Canada	47 (2.9)	59 (2.7)	26 (3.0)	34 (3.7)	34 (3.4)
Chinese Taipei	57 (4.3)	66 (4.2)	47 (4.7)	56 (4.4)	63 (4.2)
Czech Republic	37 (4.0)	33 (4.0)	8 (3.0)	16 (3.8)	14 (3.5)
England	r 51 (5.2)	r 53 (5.0)	r 23 (3.8)	r 29 (4.5)	r 30 (4.5)
Hong Kong, SAR	61 (4.8)	70 (4.2)	16 (3.1)	32 (3.4)	45 (4.5)
Italy	38 (3.8)	50 (3.7)	18 (3.3)	32 (3.8)	42 (3.6)
Japan	33 (4.2)	42 (4.2)	7 (2.2)	12 (2.8)	25 (3.9)
Korea, Rep. of	64 (4.0)	71 (3.8)	42 (3.9)	58 (4.5)	70 (4.0)
Netherlands	r 46 (6.8)	r 50 (6.6)	r 2 (1.2)	r 18 (5.0)	r 18 (4.3)
Russian Federation	89 (2.2)	87 (2.9)	69 (4.3)	71 (3.8)	82 (3.5)
Singapore	30 (4.2)	44 (4.8)	5 (1.8)	13 (2.8)	17 (3.4)
States					
Connecticut	s 35 (8.6)	s 31 (9.6)	s 7 (5.4)	s 24 (9.0)	s 18 (8.5)
Idaho	r 51 (9.2)	r 53 (9.2)	r 18 (5.3)	r 25 (7.9)	r 22 (8.1)
Illinois	34 (6.8)	33 (7.1)	14 (6.4)	29 (5.6)	22 (5.2)
Indiana	33 (8.6)	34 (8.2)	10 (6.7)	23 (7.7)	20 (6.8)
Maryland	r 49 (7.5)	r 52 (8.0)	r 36 (7.7)	r 37 (7.5)	r 37 (7.3)
Massachusetts	s 40 (8.0)	s 40 (8.0)	s 12 (5.9)	s 18 (6.4)	s 23 (7.3)
Michigan	45 (7.4)	45 (6.9)	9 (3.7)	17 (5.7)	27 (6.1)
Missouri	63 (7.7)	67 (6.9)	29 (7.4)	31 (7.0)	38 (7.9)
North Carolina	r 57 (8.7)	r 70 (7.7)	r 18 (6.0)	r 46 (6.7)	r 44 (6.2)
Oregon	49 (8.0)	49 (8.0)	23 (7.0)	24 (6.4)	25 (7.3)
Pennsylvania	42 (6.1)	47 (6.8)	14 (5.3)	33 (7.0)	28 (7.3)
South Carolina	48 (8.3)	48 (9.0)	25 (6.9)	33 (8.5)	21 (6.7)
Texas	r 43 (8.0)	r 40 (7.5)	r 20 (5.9)	r 22 (6.7)	r 19 (6.3)
Districts and Consortia					
Academy School Dist. #20, CO	0 (0.0)	17 (0.4)	0 (0.0)	0 (0.0)	0 (0.0)
Chicago Public Schools, IL	s 43 (10.7)	s 47 (10.9)	s 15 (8.0)	s 43 (14.5)	s 51 (13.0)
Delaware Science Coalition, DE	r 65 (1.9)	r 67 (2.1)	r 40 (2.1)	r 28 (2.2)	r 37 (2.1)
First in the World Consort., IL	r 0 (0.0)	r 0 (0.0)	r 0 (0.0)	s 9 (1.2)	r 8 (1.1)
Fremont/Lincoln/WestSide PS, NE	r 35 (1.5)	r 49 (1.6)	r 22 (1.4)	r 49 (1.6)	r 49 (1.6)
Guilford County, NC	r 34 (1.1)	r 34 (1.1)	r 0 (0.0)	r 20 (0.7)	r 20 (0.7)
Jersey City Public Schools, NJ	41 (1.6)	45 (1.4)	9 (2.0)	r 37 (1.4)	27 (1.1)
Miami-Dade County PS, FL	s 33 (14.3)	s 33 (14.3)	s 30 (11.9)	s 20 (10.7)	s 33 (12.5)
Michigan Invitational Group, MI	17 (1.4)	37 (1.5)	26 (1.7)	12 (1.2)	15 (1.4)
Montgomery County, MD	s 13 (9.3)	s 13 (9.3)	s 28 (14.2)	s 5 (5.4)	s 0 (0.0)
Naperville Sch. Dist. #203, IL	0 (0.0)	0 (0.0)	0 (0.0)	0 (0.0)	0 (0.0)
Project SMART Consortium, OH	42 (1.3)	44 (1.4)	23 (1.2)	21 (1.1)	15 (1.0)
Rochester City Sch. Dist., NY	r 35 (1.4)	r 35 (1.4)	r 35 (1.4)	r 35 (1.4)	r 35 (1.4)
SW Math/Sci. Collaborative, PA	30 (8.0)	39 (8.3)	17 (7.2)	26 (6.5)	29 (6.9)
International Avg. (All Countries)	57 (0.7)	59 (0.7)	35 (0.6)	46 (0.6)	50 (0.6)

SOURCE: IEA Third International Mathematics and Science Study (TIMSS), 1998-1999.

Background data provided by schools.

States in *italics* did not fully satisfy guidelines for sample participation rates (see Appendix A for details).

( ) Standard errors appear in parentheses. Because results are rounded to the nearest whole number, some totals may appear inconsistent.

An "r" indicates school response data available for 70-84% of students. An "s" indicates school response data available for 50-69% of students.

Exhibit R4.3 Availability of Computers for Instructional Purposes

8th Grade Mathematics

		Percentage of Students by Number of Students per Computer[1]				Percentage of Students in Schools Without Any Computers
		Fewer than 15 Students per Computer	15-30 Students per Computer	31-50 Students per Computer	More than 50 Students per Computer	
Countries						
United States	s	97 (1.8)	3 (1.8)	0 (0.0)	0 (0.0)	0 (0.0)
Belgium (Flemish)	r	83 (3.0)	9 (2.2)	1 (0.8)	4 (1.7)	4 (1.6)
Canada		100 (0.0)	0 (0.0)	0 (0.0)	0 (0.0)	0 (0.0)
Chinese Taipei		90 (2.5)	9 (2.6)	1 (0.8)	0 (0.0)	0 (0.0)
Czech Republic		89 (3.0)	2 (1.4)	5 (2.4)	0 (0.0)	3 (1.2)
England	r	100 (0.0)	0 (0.0)	0 (0.0)	0 (0.0)	0 (0.0)
Hong Kong, SAR	r	86 (3.3)	3 (1.5)	4 (1.8)	3 (1.3)	5 (2.2)
Italy		64 (3.4)	19 (2.9)	7 (2.2)	3 (1.3)	6 (1.6)
Japan		92 (2.7)	5 (1.8)	0 (0.0)	0 (0.0)	3 (1.9)
Korea, Rep. of		75 (3.6)	14 (3.2)	6 (1.8)	5 (1.8)	1 (0.0)
Netherlands	s	99 (1.0)	1 (0.1)	0 (0.0)	0 (0.0)	0 (0.0)
Russian Federation		37 (4.9)	6 (2.0)	1 (0.0)	3 (1.5)	53 (4.8)
Singapore		98 (1.3)	2 (1.3)	0 (0.0)	0 (0.0)	0 (0.0)
States						
Connecticut		x x	x x	x x	x x	x x
Idaho	s	100 (0.0)	0 (0.0)	0 (0.0)	0 (0.0)	0 (0.0)
Illinois	r	100 (0.0)	0 (0.0)	0 (0.0)	0 (0.0)	0 (0.0)
Indiana		100 (0.0)	0 (0.0)	0 (0.0)	0 (0.0)	0 (0.0)
Maryland	s	97 (3.2)	3 (3.2)	0 (0.0)	0 (0.0)	0 (0.0)
Massachusetts	s	94 (3.9)	6 (3.9)	0 (0.0)	0 (0.0)	0 (0.0)
Michigan	r	100 (0.0)	0 (0.0)	0 (0.0)	0 (0.0)	0 (0.0)
Missouri	r	100 (0.0)	0 (0.0)	0 (0.0)	0 (0.0)	0 (0.0)
North Carolina	r	97 (2.6)	3 (0.2)	0 (0.0)	0 (0.0)	0 (0.0)
Oregon	r	98 (2.0)	0 (0.0)	0 (0.0)	2 (2.0)	0 (0.0)
Pennsylvania	r	91 (5.6)	2 (0.3)	0 (0.3)	0 (0.0)	6 (5.2)
South Carolina	r	97 (2.6)	3 (2.6)	0 (0.0)	0 (0.0)	0 (0.0)
Texas	s	100 (0.0)	0 (0.0)	0 (0.0)	0 (0.0)	0 (0.0)
Districts and Consortia						
Academy School Dist. #20, CO	r	100 (0.0)	0 (0.0)	0 (0.0)	0 (0.0)	0 (0.0)
Chicago Public Schools, IL		x x	x x	x x	x x	x x
Delaware Science Coalition, DE	s	97 (0.2)	3 (0.2)	0 (0.0)	0 (0.0)	0 (0.0)
First in the World Consort., IL	r	100 (0.0)	0 (0.0)	0 (0.0)	0 (0.0)	0 (0.0)
Fremont/Lincoln/WestSide PS, NE	s	100 (0.0)	0 (0.0)	0 (0.0)	0 (0.0)	0 (0.0)
Guilford County, NC	s	100 (0.0)	0 (0.0)	0 (0.0)	0 (0.0)	0 (0.0)
Jersey City Public Schools, NJ	r	90 (0.3)	10 (0.3)	0 (0.0)	0 (0.0)	0 (0.0)
Miami-Dade County PS, FL		x x	x x	x x	x x	x x
Michigan Invitational Group, MI	r	100 (0.0)	0 (0.0)	0 (0.0)	0 (0.0)	0 (0.0)
Montgomery County, MD		x x	x x	x x	x x	x x
Naperville Sch. Dist. #203, IL		100 (0.0)	0 (0.0)	0 (0.0)	0 (0.0)	0 (0.0)
Project SMART Consortium, OH	r	100 (0.0)	0 (0.0)	0 (0.0)	0 (0.0)	0 (0.0)
Rochester City Sch. Dist., NY	r	100 (0.0)	0 (0.0)	0 (0.0)	0 (0.0)	0 (0.0)
SW Math/Sci. Collaborative, PA	r	99 (1.5)	0 (0.0)	1 (0.2)	0 (0.0)	0 (0.0)
International Avg. (All Countries)		60 (0.4)	6 (0.3)	3 (0.2)	6 (0.3)	25 (0.4)

SOURCE: IEA Third International Mathematics and Science Study (TIMSS), 1998-1999.

Background data provided by schools.

1 Based on ratio of grade 8 enrollment to total computers for instructional use by grade 8 teachers and students.

States in *italics* did not fully satisfy guidelines for sample participation rates (see Appendix A for details).

( ) Standard errors appear in parentheses. Because results are rounded to the nearest whole number, some totals may appear inconsistent.

An "r" indicates school response data available for 70-84% of students. An "s" indicates school response data available for 50-69% of students. An "x" indicates school response data available for <50% of students.

		Percentage of Students by Level of Access			
		Access to World Wide Web (with or without e-mail)	Access to E-mail Only	No Internet Access but Planning to Get Internet Access by 2001	No Access and No Immediate Plans to Obtain Access
Countries					
United States	r	91 (3.1)	0 (0.0)	9 (2.8)	0 (0.0)
Belgium (Flemish)		73 (4.0)	1 (0.7)	24 (3.9)	2 (1.2)
Canada		96 (1.2)	1 (0.5)	3 (1.0)	0 (0.3)
Chinese Taipei		89 (2.8)	5 (1.9)	6 (2.0)	0 (0.0)
Czech Republic		34 (5.1)	2 (1.7)	45 (5.4)	19 (3.8)
England	r	86 (3.4)	1 (0.1)	13 (3.3)	0 (0.0)
Hong Kong, SAR	r	85 (3.7)	0 (0.0)	15 (3.7)	0 (0.0)
Italy		41 (4.2)	4 (1.6)	54 (4.2)	2 (1.2)
Japan		29 (3.9)	2 (1.1)	29 (4.0)	41 (4.2)
Korea, Rep. of		48 (4.4)	0 (0.0)	46 (4.3)	6 (1.9)
Netherlands	r	81 (7.1)	3 (1.9)	15 (7.0)	1 (0.7)
Russian Federation		5 (1.4)	0 (0.0)	16 (2.8)	79 (2.4)
Singapore		89 (3.0)	1 (0.9)	10 (2.8)	0 (0.0)
States					
Connecticut	s	99 (1.5)	0 (0.0)	1 (0.2)	0 (0.0)
Idaho	r	100 (0.0)	0 (0.0)	0 (0.0)	0 (0.0)
Illinois		92 (2.6)	0 (0.0)	8 (2.6)	0 (0.0)
Indiana		86 (6.6)	0 (0.0)	8 (3.0)	6 (0.3)
Maryland	r	95 (3.5)	0 (0.0)	5 (3.5)	0 (0.0)
Massachusetts	s	90 (5.7)	0 (0.0)	10 (5.7)	0 (0.0)
Michigan		91 (3.1)	0 (0.0)	9 (3.1)	0 (0.0)
Missouri		87 (5.6)	0 (0.0)	11 (5.1)	2 (2.2)
North Carolina	r	95 (3.5)	2 (0.2)	3 (2.6)	0 (0.0)
Oregon		91 (4.7)	0 (0.0)	9 (4.7)	0 (0.0)
Pennsylvania	r	83 (3.5)	0 (0.0)	15 (3.2)	1 (1.3)
South Carolina	r	100 (0.0)	0 (0.0)	0 (0.0)	0 (0.0)
Texas	r	100 (0.0)	0 (0.0)	0 (0.0)	0 (0.0)
Districts and Consortia					
Academy School Dist. #20, CO		100 (0.0)	0 (0.0)	0 (0.0)	0 (0.0)
Chicago Public Schools, IL	r	44 (13.8)	0 (0.0)	56 (13.8)	0 (0.0)
Delaware Science Coalition, DE	s	100 (0.0)	0 (0.0)	0 (0.0)	0 (0.0)
First in the World Consort., IL	r	100 (0.0)	0 (0.0)	0 (0.0)	0 (0.0)
Fremont/Lincoln/WestSide PS, NE	r	100 (0.0)	0 (0.0)	0 (0.0)	0 (0.0)
Guilford County, NC	r	100 (0.0)	0 (0.0)	0 (0.0)	0 (0.0)
Jersey City Public Schools, NJ	r	90 (1.1)	10 (1.1)	0 (0.0)	0 (0.0)
Miami-Dade County PS, FL		x x	x x	x x	x x
Michigan Invitational Group, MI		94 (1.3)	0 (0.0)	6 (1.3)	0 (0.0)
Montgomery County, MD	s	100 (0.0)	0 (0.0)	0 (0.0)	0 (0.0)
Naperville Sch. Dist. #203, IL		100 (0.0)	0 (0.0)	0 (0.0)	0 (0.0)
Project SMART Consortium, OH		91 (0.3)	0 (0.0)	9 (0.3)	0 (0.0)
Rochester City Sch. Dist., NY	s	69 (1.6)	0 (0.0)	31 (1.6)	0 (0.0)
SW Math/Sci. Collaborative, PA	r	95 (3.5)	0 (0.0)	5 (3.5)	0 (0.0)
International Avg. (All Countries)		41 (0.5)	1 (0.2)	29 (0.6)	29 (0.5)

SOURCE: IEA Third International Mathematics and Science Study (TIMSS), 1998-1999.

Background data provided by schools.

States in *italics* did not fully satisfy guidelines for sample participation rates (see Appendix A for details).

( ) Standard errors appear in parentheses. Because results are rounded to the nearest whole number, some totals may appear inconsistent.

An "r" indicates school response data available for 70-84% of students. An "s" indicates school response data available for 50-69% of students. An "x" indicates school response data available for <50% of students.

APPENDIX A

Overview of TIMSS Benchmarking Procedures: Mathematics Achievement

A

History

TIMSS 1999 represents the continuation of a long series of studies conducted by the International Association for the Evaluation of Educational Achievement (IEA). Since its inception in 1959, the IEA has conducted more than 15 studies of cross-national achievement in the curricular areas of mathematics, science, language, civics, and reading. The Third International Mathematics and Science Study (TIMSS), conducted in 1994-1995, was the largest and most complex IEA study, and included both mathematics and science at third and fourth grades, seventh and eighth grades, and the final year of secondary school. In 1999, TIMSS again assessed eighth-grade students in both mathematics and science to measure trends in student achievement since 1995. TIMSS 1999 was also known as TIMSS-Repeat, or TIMSS-R.[1]

To provide U.S. states and school districts with an opportunity to benchmark the performance of their students against that of students in the high-performing TIMSS countries, the International Study Center at Boston College, with the support of the National Center for Education Statistics and the National Science Foundation, established the TIMSS 1999 Benchmarking Study. Through this project, the TIMSS mathematics and science achievement tests and questionnaires were administered to representative samples of students in participating states and school districts in the spring of 1999, at the same time the tests and questionnaires were administered in the TIMSS countries. Participation in TIMSS Benchmarking was intended to help states and districts understand their comparative educational standing, assess the rigor and effectiveness of their own mathematics and science programs in an international context, and improve the teaching and learning of mathematics and science.

Participants in TIMSS Benchmarking

Thirteen states availed of the opportunity to participate in the Benchmarking Study. Eight public school districts and six consortia also participated, for a total of fourteen districts and consortia. They are listed in Exhibit 1 of the Introduction, together with the 38 countries that took part in TIMSS 1999.

1 The TIMSS 1999 results for mathematics and science, respectively, are reported in Mullis, I.V.S., Martin, M.O., Gonzalez, E.J., Gregory, K.D., Garden, R.A., O'Connor, K.M., Chrostowski, S.J., and Smith, T.A. (2000), *TIMSS 1999 International Mathematics Report: Findings from IEA's Repeat of the Third International Mathematics and Science Study at the Eighth Grade*, Chestnut Hill, MA: Boston College, and in Martin, M.O., Mullis, I.V.S., Gonzalez, E.J., Gregory, K.D., Smith, T.A., Chrostowski, S.J., Garden, R.A., and O'Connor, K.M. (2000), *TIMSS 1999 International Science Report: Findings from IEA's Repeat of the Third International Mathematics and Science Study at the Eighth Grade*, Chestnut Hill, MA: Boston College.

Developing the TIMSS 1999 Mathematics Test

The TIMSS curriculum framework underlying the mathematics tests was developed for TIMSS in 1995 by groups of mathematics educators with input from the TIMSS National Research Coordinators (NRCs). As shown in Exhibit A.1, the mathematics curriculum framework contains three dimensions or aspects. The *content* aspect represents the subject matter content of school mathematics. The *performance expectations* aspect describes, in a non-hierarchical way, the many kinds of performances or behaviors that might be expected of students in school mathematics. The *perspectives* aspect focuses on the development of students' attitudes, interest, and motivation in mathematics. Because the frameworks were developed to include content, performance expectations, and perspectives for the entire span of curricula from the beginning of schooling through the completion of secondary school, some aspects may not be reflected in the eighth-grade TIMSS assessment.[2] Working within the framework, mathematics test specifications for TIMSS in 1995 were developed that included items representing a wide range of mathematics topics and eliciting a range of skills from the students. The 1995 tests were developed through an international consensus involving input from experts in mathematics and measurement specialists, ensuring they reflected current thinking and priorities in mathematics.

About one-third of the items in the 1995 assessment were kept secure to measure trends over time; the remaining items were released for public use. An essential part of the development of the 1999 assessment, therefore, was to replace the released items with items of similar content, format, and difficulty. With the assistance of the Science and Mathematics Item Replacement Committee, a group of internationally prominent mathematics and science educators nominated by participating countries to advise on subject-matter issues in the assessment, over 300 mathematics and science items were developed as potential replacements. After an extensive process of review and field testing, 114 items were selected for use as replacements in the 1999 mathematics assessment.

Exhibit A.2 presents the five content areas included in the 1999 mathematics test and the numbers of items and score points in each area. Distributions are also included for the five performance categories derived from the performance expectations aspect of the curriculum framework. About one-fourth of the items were in the free-response format, requiring students to generate and write their own answers. Designed to take about one-third of students' test time, some free-response questions asked for short answers while others required

2 The complete TIMSS curriculum frameworks can be found in Robitaille, D.F., et al. (1993), *TIMSS Monograph No. 1: Curriculum Frameworks for Mathematics and Science*, Vancouver, BC: Pacific Educational Press.

extended responses with students showing their work or providing explanations for their answers. The remaining questions used a multiple-choice format. In scoring the tests, correct answers to most questions were worth one point. Consistent with the approach of allotting students longer response time for the constructed-response questions than for multiple-choice questions, however, responses to some of these questions (particularly those requiring extended responses) were evaluated for partial credit, with a fully correct answer being awarded two points (see later section on scoring). The total number of score points available for analysis thus somewhat exceeds the number of items.

Every effort was made to help ensure that the tests represented the curricula of the participating countries and that the items exhibited no bias towards or against particular countries. The final forms of the tests were endorsed by the NRCs of the participating countries.[3]

3 For a full discussion of the TIMSS 1999 test development effort, please see Garden, R.A. and Smith, T.A. (2000), "TIMSS Test Development" in M.O. Martin, K.D. Gregory, K.M. O'Connor, and S.E. Stemler (eds.), *TIMSS 1999 Benchmarking Technical Report*, Chestnut Hill, MA: Boston College.

Content	Performance Expectations	Perspectives
Numbers	Knowing	Attitudes
Measurement	Using Routine Procedures	Careers
Geometry	Investigating and Problem Solving	Participation
Proportionality	Mathematical Reasoning	Increasing Interest
Functions, Relations, and Equations	Communicating	Habits of Mind
Data Representation		
Probability and Statistics		
Elementary Analysis, Validation, and Structure		

SOURCE: IEA Third International Mathematics and Science Study (TIMSS), 1998-1999.

Distribution of Mathematics Items by Content Reporting Category and Performance Category

8th Grade Mathematics

Content Category	Percentage of Items	Total Number of Items	Number of Multiple-Choice Items	Number of Free-Response Items[1]	Number of Score Points[2]
Fractions and Number Sense	38	61	47	14	62
Measurement	15	24	15	9	26
Data Representation, Analysis and Probability	13	21	19	2	22
Geometry	13	21	20	1	21
Algebra	22	35	24	11	38
Total	100	162	125	37	169

Performance Category	Percentage of Items	Total Number of Items	Number of Multiple-Choice Items	Number of Free-Response Items[1]	Number of Score Points[2]
Knowing	19	30	28	2	30
Using Routine Procedures	23	38	28	10	39
Using Complex Procedures	24	39	34	5	40
Investigating and Solving Problems	31	51	34	17	53
Communicating and Reasoning	2	4	1	3	7
Total	100	162	125	37	169

SOURCE: IEA Third International Mathematics and Science Study (TIMSS), 1998-1999.

1 Free-response items include both short-answer and extended-response types.

2 In scoring the tests, correct answers to most items were worth one point. However, responses to some free-response items were evaluated for partial credit with a fully correct answer awarded up to two points. Thus, the number of score points exceeds the number of items in the test.

TIMSS Test Design

Not all of the students in the TIMSS assessment responded to all of the mathematics items. To ensure broad subject-matter coverage without overburdening individual students, TIMSS used a rotated design that included both the mathematics and science items. Thus, the same students participated in both the mathematics and science testing. As in 1995, the 1999 assessment consisted of eight booklets, each requiring 90 minutes of response time. Each participating student was assigned one booklet only. In accordance with the design, the mathematics and science items were assembled into 26 clusters (labeled A through Z). The secure trend items were in clusters A through H, and items replacing the released 1995 items in clusters I through Z. Eight of the clusters were designed to take 12 minutes to complete; 10 of the clusters, 22 minutes; and 8 clusters, 10 minutes. In all, the design provided 396 testing minutes, 198 for mathematics and 198 for science. Cluster A was a core cluster assigned to all booklets. The remaining clusters were assigned to the booklets in accordance with the rotated design so that representative samples of students responded to each cluster.[4]

Background Questionnaires

TIMSS in 1999 administered a broad array of questionnaires to collect data on the educational context for student achievement and to measure trends since 1995. *National Research Coordinators*, with the assistance of their curriculum experts, provided detailed information on the organization, emphases, and content coverage of the mathematics and science curriculum. The *students* who were tested answered questions pertaining to their attitudes towards mathematics and science, their academic self-concept, classroom activities, home background, and out-of-school activities. The mathematics and science *teachers* of sampled students responded to questions about teaching emphasis on the topics in the curriculum frameworks, instructional practices, professional training and education, and their views on mathematics and science. The heads of *schools* responded to questions about school staffing and resources, mathematics and science course offerings, and teacher support.

4 The 1999 TIMSS test design is identical to the design for 1995, which is fully documented in Adams, R. and Gonzalez, E. (1996), "TIMSS Test Design" in M.O. Martin and D.L. Kelly (eds.), *Third International Mathematics and Science Study Technical Report, Volume I*, Chestnut Hill, MA: Boston College.

Translation and Verification

The TIMSS instruments were prepared in English and translated into 33 languages, with 10 of the 38 countries collecting data in two languages. In addition, it sometimes was necessary to modify the international versions for cultural reasons, even in the nine countries that tested in English. This process represented an enormous effort for the national centers, with many checks along the way. The translation effort included (1) developing explicit guidelines for translation and cultural adaptation; (2) translation of the instruments by the national centers in accordance with the guidelines, using two or more independent translations; (3) consultation with subject-matter experts on cultural adaptations to ensure that the meaning and difficulty of items did not change; (4) verification of translation quality by professional translators from an independent translation company; (5) corrections by the national centers in accordance with the suggestions made; (6) verification by the International Study Center that corrections were made; and (7) a series of statistical checks after the testing to detect items that did not perform comparably across countries.[5]

Population Definition and Sampling

TIMSS in 1995 had as its target population students enrolled in the two adjacent grades that contained the largest proportion of 13-year-old students at the time of testing, which were seventh- and eighth-grade students in most countries. TIMSS in 1999 used the same definition to identify the target grades, but assessed students in the upper of the two grades only, which was the eighth grade in most countries, including the United States.[6] The eighth grade was the target population for all of the Benchmarking participants.

The selection of valid and efficient samples was essential to the success of TIMSS and of the Benchmarking Study. For TIMSS internationally, NRCs, including Westat, the sampling and data collection coordinator for TIMSS in the United States, received training in how to select the school and student samples and in the use of the sampling software, and worked in close consultation with Statistics Canada, the TIMSS sampling consultants, on all phases of sampling. As well as conducting the sampling and data collection for the U.S. national TIMSS sample, Westat was also responsible for sampling and data collection in each of the Benchmarking states, districts, and consortia.

5 More details about the translation verification procedures can be found in O'Connor, K., and Malak, B. (2000), "Translation and Cultural Adaptation of the TIMSS Instruments" in M.O. Martin, K.D. Gregory, K.M. O'Connor, and S.E. Stemler (eds.), *TIMSS 1999 Benchmarking Technical Report*, Chestnut Hill, MA: Boston College.

6 The sample design for TIMSS is described in detail in Foy, P., and Joncas, M. (2000), "TIMSS Sample Design" in M.O. Martin, K.D. Gregory and S.E. Stemler (eds.), *TIMSS 1999 Technical Report*, Chestnut Hill, MA: Boston College. Sampling for the Benchmarking project is described in Fowler, J., Rizzo, L., and Rust, K. (2001), "TIMSS Benchmarking Sampling Design and Implementation" in M.O. Martin, K.D. Gregory, K.M. O'Connor, and S.E. Stemler (eds.), *TIMSS 1999 Benchmarking Technical Report*, Chestnut Hill, MA: Boston College.

To document the quality of the school and student samples in each of the TIMSS countries, staff from Statistics Canada and the International Study Center worked with the TIMSS sampling referee (Keith Rust, Westat) to review sampling plans, sampling frames, and sampling implementation. Particular attention was paid to coverage of the target population and to participation by the sampled schools and students. The data from the few countries that did not fully meet all of the sampling guidelines are annotated in the TIMSS international reports, and are also annotated in this report. The TIMSS samples for the Benchmarking participants were also carefully reviewed in light of the TIMSS sampling guidelines, and the results annotated where appropriate. Since Westat was the sampling contractor for the Benchmarking project, the role of sampling referee for the Benchmarking review was filled by Pierre Foy, of Statistics Canada.

Although all countries and Benchmarking participants were expected to draw samples representative of the entire internationally desired population (all students in the upper of the two adjacent grades with the greatest proportion of 13-year-olds), the few countries where this was not possible were permitted to define a national desired population that excluded part of the internationally desired population. Exhibit A.3 shows any differences in coverage between the international and national desired populations. Almost all TIMSS countries achieved 100 percent coverage (36 out of 38), with Lithuania and Latvia the exceptions. Consequently, the results for Lithuania are annotated, and because coverage fell below 65 percent for Latvia, the Latvian results are labeled "Latvia (LSS)," for Latvian-Speaking Schools. Additionally, because of scheduling difficulties, Lithuania was unable to test its eighth-grade students in May 1999 as planned. Instead, the students were tested in September 1999, when they had moved into the ninth grade. The results for Lithuania are annotated to reflect this as well. Exhibit A.3 also shows that the sampling plans for the Benchmarking participants all incorporated 100 percent coverage of the desired population. Four of the 13 states (Idaho, Indiana, Michigan, and Pennsylvania) as well as the Southwest Pennsylvania Math and Science Collaborative included private schools as well as public schools.

In operationalizing their desired eighth-grade population, countries and Benchmarking participants could define a population to be sampled that excluded a small percentage (less than 10 percent) of certain kinds of schools or students that would be very difficult or resource-intensive to test (e.g., schools for students with special needs or schools that were very small or located in extremely rural areas). Exhibit A.3 also shows that the degree of such exclusions was small. Among countries, only Israel reached the 10 percent limit, and among Benchmarking participants, only Guilford County and Montgomery County did so. All three are annotated as such in the achievement chapters of this report.

Within countries, TIMSS used a two-stage sample design, in which the first stage involved selecting about 150 public and private schools in each country. Within each school, countries were to use random procedures to select one mathematics class at the eighth grade. All of the students in that class were to participate in the TIMSS testing. This approach was designed to yield a representative sample of about 3,750 students per country. Typically, between 450 and 3,750 students responded to each achievement item in each country, depending on the booklets in which the items appeared.

States participating in the Benchmarking study were required to sample at least 50 schools and approximately 2,000 eighth-grade students. School districts and consortia were required to sample at least 25 schools and at least 1,000 students. Where there were fewer than 25 schools in a district or consortium, all schools were to be included, and the within-school sample increased to yield the total of 1,000 students.

Exhibits A.4 and A.5 present achieved sample sizes for schools and students, respectively, for the TIMSS countries and for the Benchmarking participants. Where a district or consortium was part of a state that also participated, the state sample was augmented by the district or consortium sample, properly weighted in accordance with its size. Schools in a state that were sampled as part of the U.S. national TIMSS sample were also used to augment the state sample. For example, the Illinois sample consists of 90 schools, 41 from the state Benchmarking sample (including five schools from the national TIMSS sample), 27 from the Chicago Public Schools, 17 from the First in the World Consortium, and five from the Naperville School District.

Exhibit A.6 shows the participation rates for schools, students, and overall, both with and without the use of replacement schools, for TIMSS countries and Benchmarking participants. All of the countries met the guideline for sampling participation – 85 percent of both the schools and students, or a combined rate (the product of school and student participation) of 75 percent – although Belgium (Flemish), England, Hong Kong, and the Netherlands did so only after including replacement schools, and are annotated accordingly in the achievement chapters.

With the exception of Pennsylvania and Texas, all the Benchmarking participants met the sampling guidelines, although Indiana did so only after including replacement schools. Indiana is annotated to reflect this in the achievement chapters, and Pennsylvania and Texas are italicized in all exhibits in this report.

	International Desired Population		National Desired Population		
	Coverage	Notes on Coverage	School-Level Exclusions	Within-Sample Exclusions	Overall Exclusions
United States	100%		0%	4%	4%
Australia	100%		1%	1%	2%
Belgium (Flemish)	100%		1%	0%	1%
Bulgaria	100%		5%	0%	5%
Canada	100%		4%	2%	6%
Chile	100%		3%	0%	3%
Chinese Taipei	100%		1%	1%	2%
Cyprus	100%		0%	1%	1%
Czech Republic	100%		5%	0%	5%
England	100%		2%	3%	5%
Finland	100%		3%	0%	4%
Hong Kong, SAR	100%		1%	0%	1%
Hungary	100%		4%	0%	4%
Indonesia	100%		0%	0%	0%
Iran, Islamic Rep. of	100%		4%	0%	4%
Israel	100%		8%	8%	16%
Italy	100%		4%	2%	7%
Japan	100%		1%	0%	1%
Jordan	100%		2%	1%	3%
Korea, Rep. of	100%		2%	2%	4%
Latvia (LSS)	61%	Latvian-speaking students only	4%	0%	4%
Lithuania	87%	Lithuanian-speaking students only	5%	0%	5%
Macedonia, Rep. of	100%		1%	0%	1%
Malaysia	100%		5%	0%	5%
Moldova	100%		2%	0%	2%
Morocco	100%		1%	0%	1%
Netherlands	100%		1%	0%	1%
New Zealand	100%		2%	1%	2%
Philippines	100%		3%	0%	3%
Romania	100%		4%	0%	4%
Russian Federation	100%		1%	1%	2%
Singapore	100%		0%	0%	0%
Slovak Republic	100%		7%	0%	7%
Slovenia	100%		3%	0%	3%
South Africa	100%		2%	0%	2%
Thailand	100%		3%	0%	3%
Tunisia	100%		0%	0%	0%
Turkey	100%		2%	0%	2%

SOURCE: IEA Third International Mathematics and Science Study (TIMSS), 1998-1999.

	International Desired Population		National Desired Population		
	Coverage	Notes on Coverage	School-Level Exclusions	Within-Sample Exclusions	Overall Exclusions
States					
Connecticut	100%		0%	5%	5%
Idaho	100%	Included private schools	0%	2%	2%
Illinois	100%		0%	4%	4%
Indiana	100%	Included private schools	0%	6%	6%
Maryland	100%		0%	6%	6%
Massachusetts	100%		0%	5%	5%
Michigan	100%	Included private schools	0%	2%	2%
Missouri	100%		0%	4%	4%
North Carolina	100%		0%	4%	4%
Oregon	100%		0%	5%	5%
Pennsylvania	100%	Included private schools	0%	6%	6%
South Carolina	100%		0%	2%	2%
Texas	100%		0%	4%	4%
Districts and Consortia					
Academy School Dist. #20, CO	100%		NA	2%	2%
Chicago Public Schools, IL	100%		NA	4%	4%
Delaware Science Coalition, DE	100%		NA	5%	5%
First in the World Consort., IL	100%		NA	2%	2%
Fremont/Lincoln/WestSide PS, NE	100%		NA	2%	2%
Guilford County, NC	100%		NA	10%	10%
Jersey City Public Schools, NJ	100%		NA	6%	6%
Miami-Dade County PS, FL	100%		NA	7%	7%
Michigan Invitational Group, MI	100%		NA	2%	2%
Montgomery County, MD	100%		NA	17%	17%
Naperville Sch. Dist. #203, IL	100%		NA	7%	7%
Project SMART Consortium, OH	100%		NA	2%	2%
Rochester City Sch. Dist., NY	100%		NA	1%	1%
SW Math/Sci. Collaborative, PA	100%	Included private schools	NA	4%	4%

SOURCE: IEA Third International Mathematics and Science Study (TIMSS), 1998-1999.

8th Grade Mathematics

	Number of Schools in Original Sample	Number of Eligible Schools in Original Sample	Number of Schools in Original Sample That Participated	Number of Replacement Schools That Participated	Total Number of Schools That Participated
United States	250	246	202	19	221
Australia	184	182	152	18	170
Belgium (Flemish)	150	150	106	29	135
Bulgaria	172	169	163	0	163
Canada	410	398	376	9	385
Chile	186	185	181	4	185
Chinese Taipei	150	150	150	0	150
Cyprus	61	61	61	0	61
Czech Republic	150	142	136	6	142
England	150	150	76	52	128
Finland	160	160	155	4	159
Hong Kong, SAR	180	180	135	2	137
Hungary	150	150	147	0	147
Indonesia	150	150	132	18	150
Iran, Islamic Rep. of	170	170	164	6	170
Israel	150	139	137	2	139
Italy	180	180	170	10	180
Japan	150	150	140	0	140
Jordan	150	147	146	1	147
Korea, Rep. of	150	150	150	0	150
Latvia (LSS)	150	148	143	2	145
Lithuania	150	150	150	0	150
Macedonia, Rep. of	150	150	149	0	149
Malaysia	150	150	148	2	150
Moldova	150	150	145	5	150
Morocco	174	174	172	1	173
Netherlands	150	148	86	40	126
New Zealand	156	156	145	7	152
Philippines	150	150	148	2	150
Romania	150	150	147	0	147
Russian Federation	190	190	186	3	189
Singapore	145	145	145	0	145
Slovak Republic	150	150	143	2	145
Slovenia	150	150	147	2	149
South Africa	225	219	183	11	194
Thailand	150	150	143	7	150
Tunisia	150	149	126	23	149
Turkey	204	204	202	2	204

SOURCE: IEA Third International Mathematics and Science Study (TIMSS), 1998-1999.

8th Grade Mathematics

	Number of Schools in Original Sample	Number of Eligible Schools in Original Sample	Number of Schools in Original Sample That Participated	Number of Replacement Schools That Participated	Total Number of Schools That Participated
States					
Connecticut	54	54	52	0	52
Idaho	54	54	47	0	47
Illinois	90	90	85	0	85
Indiana	61	61	39	13	52
Maryland	79	77	73	0	73
Massachusetts	59	58	57	0	57
Michigan	66	62	55	2	57
Missouri	57	55	43	8	51
North Carolina	71	68	67	0	67
Oregon	51	51	45	0	45
Pennsylvania	116	113	80	0	80
South Carolina	53	53	49	0	49
Texas	71	70	51	1	52
Districts and Consortia					
Academy School Dist. #20, CO	4	4	4	0	4
Chicago Public Schools, IL	27	27	26	0	26
Delaware Science Coalition, DE	25	25	25	0	25
First in the World Consort., IL	17	17	15	0	15
Fremont/Lincoln/WestSide PS, NE	12	12	12	0	12
Guilford County, NC	17	17	17	0	17
Jersey City Public Schools, NJ	25	25	24	0	24
Miami-Dade County PS, FL	25	25	25	0	25
Michigan Invitational Group, MI	21	21	21	0	21
Montgomery County, MD	25	25	25	0	25
Naperville Sch. Dist. #203, IL	5	5	5	0	5
Project SMART Consortium, OH	24	24	24	0	24
Rochester City Sch. Dist., NY	7	7	7	0	7
SW Math/Sci. Collaborative, PA	50	49	39	0	39

SOURCE: IEA Third International Mathematics and Science Study (TIMSS), 1998-1999.

	Within-School Student Participation (Weighted Percentage)	Number of Sampled Students in Participating Schools	Number of Students Withdrawn from Class/School	Number of Students Excluded	Number of Eligible Students	Number of Students Absent	Number of Students Assessed
United States	94%	9981	115	142	9724	652	9072
Australia	90%	4600	96	53	4451	419	4032
Belgium (Flemish)	97%	5387	12	0	5375	116	5259
Bulgaria	96%	3461	63	0	3398	126	3272
Canada	96%	9490	84	245	9161	391	8770
Chile	96%	6283	119	18	6146	239	5907
Chinese Taipei	99%	5889	30	42	5817	45	5772
Cyprus	97%	3296	38	32	3226	110	3116
Czech Republic	96%	3640	24	0	3616	163	3453
England	90%	3400	27	115	3258	298	2960
Finland	96%	3060	17	13	3030	110	2920
Hong Kong, SAR	98%	5310	18	1	5291	112	5179
Hungary	95%	3350	0	0	3350	167	3183
Indonesia	97%	6162	106	1	6055	207	5848
Iran, Islamic Rep. of	98%	5497	104	0	5393	92	5301
Israel	94%	4670	29	187	4454	259	4195
Italy	97%	3531	23	86	3422	94	3328
Japan	95%	4996	15	12	4969	224	4745
Jordan	99%	5300	130	42	5128	76	5052
Korea, Rep. of	100%	6285	29	128	6128	14	6114
Latvia (LSS)	93%	3128	16	4	3108	235	2873
Lithuania	89%	2668	0	0	2668	307	2361
Macedonia, Rep. of	98%	4096	0	0	4096	73	4023
Malaysia	99%	5713	98	0	5615	38	5577
Moldova	98%	3824	23	0	3801	90	3711
Morocco	92%	5841	42	0	5799	397	5402
Netherlands	95%	3099	12	0	3087	125	2962
New Zealand	94%	3966	96	22	3848	235	3613
Philippines	92%	7591	461	0	7130	529	6601
Romania	98%	3514	36	0	3478	53	3425
Russian Federation	97%	4557	48	34	4475	143	4332
Singapore	98%	5100	37	0	5063	97	4966
Slovak Republic	98%	3695	149	0	3546	49	3497
Slovenia	95%	3287	0	4	3283	174	3109
South Africa	93%	9071	256	0	8815	669	8146
Thailand	99%	5831	59	0	5772	40	5732
Tunisia	98%	5189	45	0	5144	93	5051
Turkey	99%	7972	49	0	7923	82	7841

SOURCE: IEA Third International Mathematics and Science Study (TIMSS), 1998-1999.

	Within-School Student Participation (Weighted Percentage)	Number of Sampled Students in Participating Schools	Number of Students Withdrawn from Class/School	Number of Students Excluded	Number of Eligible Students	Number of Students Absent	Number of Students Assessed
States							
Connecticut	94%	2190	6	43	2141	124	2023
Idaho	95%	1968	17	27	1924	94	1847
Illinois	96%	5144	30	136	4978	227	4781
Indiana	95%	2175	9	27	2139	102	2046
Maryland	94%	3877	21	339	3517	221	3317
Massachusetts	95%	2538	18	54	2466	131	2353
Michigan	96%	2811	7	44	2760	143	2623
Missouri	94%	2147	27	40	2080	128	1979
North Carolina	94%	3502	34	191	3277	214	3097
Oregon	93%	2044	24	29	1991	126	1889
Pennsylvania	95%	3463	18	60	3385	167	3236
South Carolina	94%	2177	18	36	2123	130	2011
Texas	93%	2189	18	44	2127	149	1996
Districts and Consortia							
Academy School Dist. #20, CO	94%	1329	0	15	1314	81	1233
Chicago Public Schools, IL	94%	1227	13	21	1193	74	1132
Delaware Science Coalition, DE	92%	1389	16	18	1355	103	1268
First in the World Consort., IL	96%	782	1	2	779	30	750
Fremont/Lincoln/WestSide PS, NE	95%	1178	20	25	1133	60	1093
Guilford County, NC	92%	1215	17	121	1077	76	1018
Jersey City Public Schools, NJ	94%	1116	5	47	1064	65	1004
Miami-Dade County PS, FL	91%	1356	23	10	1323	117	1229
Michigan Invitational Group, MI	91%	994	0	11	983	80	903
Montgomery County, MD	94%	1481	13	254	1214	72	1155
Naperville Sch. Dist. #203, IL	96%	1343	9	84	1250	47	1212
Project SMART Consortium, OH	94%	1188	11	18	1159	74	1096
Rochester City Sch. Dist., NY	84%	1165	8	9	1148	190	966
SW Math/Sci. Collaborative, PA	95%	1638	14	21	1603	79	1538

SOURCE: IEA Third International Mathematics and Science Study (TIMSS), 1998-1999.

	School Participation		Student Participation	Overall Participation	
	Before Replacement	After Replacement		Before Replacement	After Replacement
United States	83%	90%	94%	78%	85%
Australia	83%	93%	90%	75%	84%
Belgium (Flemish)	72%	89%	97%	70%	87%
Bulgaria	97%	97%	96%	93%	93%
Canada	92%	95%	96%	88%	92%
Chile	98%	100%	96%	94%	96%
Chinese Taipei	100%	100%	99%	99%	99%
Cyprus	100%	100%	97%	97%	97%
Czech Republic	94%	100%	96%	90%	96%
England	49%	85%	90%	45%	77%
Finland	97%	100%	96%	93%	96%
Hong Kong, SAR	75%	76%	98%	74%	75%
Hungary	98%	98%	95%	93%	93%
Indonesia	84%	100%	97%	81%	97%
Iran, Islamic Rep. of	96%	100%	98%	95%	98%
Israel	98%	100%	94%	93%	94%
Italy	94%	100%	97%	91%	97%
Japan	93%	93%	95%	89%	89%
Jordan	99%	100%	99%	98%	99%
Korea, Rep. of	100%	100%	100%	100%	100%
Latvia (LSS)	96%	98%	93%	89%	91%
Lithuania	100%	100%	89%	89%	89%
Macedonia, Rep. of	99%	99%	98%	98%	98%
Malaysia	99%	100%	99%	98%	99%
Moldova	96%	100%	98%	94%	98%
Morocco	99%	99%	92%	91%	92%
Netherlands	62%	85%	95%	59%	81%
New Zealand	93%	97%	94%	87%	91%
Philippines	98%	100%	92%	91%	92%
Romania	98%	98%	98%	97%	97%
Russian Federation	98%	100%	97%	95%	97%
Singapore	100%	100%	98%	98%	98%
Slovak Republic	95%	96%	98%	93%	94%
Slovenia	98%	99%	95%	93%	94%
South Africa	85%	91%	93%	79%	84%
Thailand	93%	100%	99%	93%	99%
Tunisia	84%	100%	98%	82%	98%
Turkey	99%	100%	99%	98%	99%

SOURCE: IEA Third International Mathematics and Science Study (TIMSS), 1998-1999.

	School Participation		Student Participation	Overall Participation	
	Before Replacement	After Replacement		Before Replacement	After Replacement
States					
Connecticut	96%	96%	94%	90%	90%
Idaho	88%	88%	95%	83%	83%
Illinois	95%	95%	96%	91%	91%
Indiana	61%	83%	95%	58%	79%
Maryland	94%	94%	94%	88%	88%
Massachusetts	98%	98%	95%	93%	93%
Michigan	89%	92%	96%	85%	88%
Missouri	79%	94%	94%	75%	88%
North Carolina	98%	98%	94%	92%	92%
Oregon	89%	89%	93%	83%	83%
Pennsylvania	66%	66%	95%	63%	63%
South Carolina	92%	92%	94%	86%	86%
Texas	73%	74%	93%	67%	69%
Districts and Consortia					
Academy School Dist. #20, CO	100%	100%	94%	94%	94%
Chicago Public Schools, IL	95%	95%	94%	90%	90%
Delaware Science Coalition, DE	100%	100%	92%	92%	92%
First in the World Consort., IL	93%	93%	96%	90%	90%
Fremont/Lincoln/WestSide PS, NE	100%	100%	95%	95%	95%
Guilford County, NC	100%	100%	92%	92%	92%
Jersey City Public Schools, NJ	97%	97%	94%	91%	91%
Miami-Dade County PS, FL	100%	100%	91%	91%	91%
Michigan Invitational Group, MI	100%	100%	91%	91%	91%
Montgomery County, MD	100%	100%	94%	94%	94%
Naperville Sch. Dist. #203, IL	100%	100%	96%	96%	96%
Project SMART Consortium, OH	100%	100%	94%	94%	94%
Rochester City Sch. Dist., NY	100%	100%	84%	84%	84%
SW Math/Sci. Collaborative, PA	78%	78%	95%	75%	75%

SOURCE: IEA Third International Mathematics and Science Study (TIMSS), 1998-1999.

Data Collection

Each participating country was responsible for carrying out all aspects of the data collection, using standardized procedures developed for the study. Training manuals were created for school coordinators and test administrators that explained procedures for receipt and distribution of materials as well as for the activities related to the testing sessions. These manuals covered procedures for test security, standardized scripts to regulate directions and timing, rules for answering students' questions, and steps to ensure that identification on the test booklets and questionnaires corresponded to the information on the forms used to track students. As the data collection contractor for the U.S. national TIMSS, Westat was fully acquainted with the TIMSS procedures, and applied them in each of the Benchmarking jurisdictions in the same way as in the national data collection.

Each country was responsible for conducting quality control procedures and describing this effort in the NRC's report documenting procedures used in the study. In addition, the International Study Center considered it essential to monitor compliance with standardized procedures through an international program of quality control site visits. NRCs were asked to nominate one or more persons unconnected with their national center, such as retired school teachers, to serve as quality control monitors for their countries. The International Study Center developed manuals for the monitors and briefed them in two-day training sessions about TIMSS, the responsibilities of the national centers in conducting the study, and their own roles and responsibilities. In all, 71 international quality control monitors participated in this training.

The international quality control monitors interviewed the NRCs about data collection plans and procedures. They also visited a sample of 15 schools where they observed testing sessions and interviewed school coordinators.[7] Quality control monitors interviewed school coordinators in all 38 countries, and observed a total of 550 testing sessions. The results of the interviews conducted by the international quality control monitors indicated that, in general, NRCs had prepared well for data collection and, despite the heavy demands of the schedule and shortages of resources, were able to conduct the data collection efficiently and professionally. Similarly, the TIMSS tests appeared to have been administered in compliance with international procedures, including the activities before the testing session, those during testing, and the school-level activities related to receiving, distributing, and returning material from the national centers.

7 Steps taken to ensure high-quality data collection in TIMSS internationally are described in detail in O'Connor, K., and Stemler, S. (2000), "Quality Control in the TIMSS Data Collection" in M.O. Martin, K.D. Gregory and S.E. Stemler (eds.), *TIMSS 1999 Technical Report*, Chestnut Hill, MA: Boston College.

As a parallel quality control effort for the Benchmarking project, the International Study Center recruited and trained a team of 18 quality control observers, and sent them to observe the data collection activities of the Westat test administrators in a sample of about 10 percent of the schools in the study (98 schools in all).[8] In line with the experience internationally, the observers reported that the data collection was conducted successfully according to the prescribed procedures, and that no serious problems were encountered.

Scoring the Free-Response Items

Because about one-third of the written test time was devoted to free-response items, TIMSS needed to develop procedures for reliably evaluating student responses within and across countries. Scoring used two-digit codes with rubrics specific to each item. The first digit designates the correctness level of the response. The second digit, combined with the first, represents a diagnostic code identifying specific types of approaches, strategies, or common errors and misconceptions. Although not used in this report, analyses of responses based on the second digit should provide insight into ways to help students better understand mathematics concepts and problem-solving approaches.

To ensure reliable scoring procedures based on the TIMSS rubrics, the International Study Center prepared detailed guides containing the rubrics and explanations of how to implement them, together with example student responses for the various rubric categories. These guides, along with training packets containing extensive examples of student responses for practice in applying the rubrics, were used as a basis for intensive training in scoring the free-response items. The training sessions were designed to help representatives of national centers who would then be responsible for training personnel in their countries to apply the two-digit codes reliably. In the United States, the scoring was conducted by National Computer Systems (NCS) under contract to Westat. To ensure that student responses from the Benchmarking participants were scored in the same way as those from the U.S. national sample, NCS had both sets of data scored at the same time and by the same scoring staff.

To gather and document empirical information about the within-country agreement among scorers, TIMSS arranged to have systematic subsamples of at least 100 students' responses to each item coded independently by two readers. Exhibit A.7 shows the average and range of the within-country percent of exact agreement between scorers on the

8 Quality control measures for the Benchmarking project are described in O'Connor, K. and Stemler, S. (2001), "Quality Control in the TIMSS Benchmarking Data Collection" in M.O. Martin, K.D. Gregory, K.M. O'Connor, and S.E. Stemler (eds.), *TIMSS 1999 Benchmarking Technical Report*, Chestnut Hill, MA: Boston College.

free-response items in the mathematics test for 37 of the 38 countries. A high percentage of exact agreement was observed, with an overall average of 99 percent across the 37 countries. The TIMSS data from the reliability studies indicate that scoring procedures were robust for the mathematics items, especially for the correctness score used for the analyses in this report. In the United States, the average percent exact agreement was 99 percent for the correctness score and 96 percent for the diagnostic score. Since the Benchmarking data were combined with the U.S. national TIMSS sample for scoring purposes, this high level of scoring reliability applies to the Benchmarking data also.

Exhibit A.7 TIMSS 1999 Within-Country Free-Response Scoring Reliability Data for Mathematics Items

8th Grade Mathematics

	Correctness Score Agreement			Diagnostic Score Agreement		
	Average of Exact Percent Agreement Across Items	Range of Exact Percent Agreement		Average of Exact Percent Agreement Across Items	Range of Exact Percent Agreement	
		Min	Max		Min	Max
United States	99	96	100	97	89	100
Australia	98	94	100	95	80	100
Belgium (Flemish)	99	92	100	98	91	100
Bulgaria	99	94	100	96	73	100
Canada	98	88	100	94	80	99
Chile	99	94	100	97	88	100
Chinese Taipei	100	98	100	99	93	100
Cyprus	–	–	–	–	–	–
Czech Republic	97	81	100	92	63	99
England	99	96	100	97	87	100
Finland	99	97	100	97	90	100
Hong Kong, SAR	98	84	100	95	80	100
Hungary	98	87	100	96	76	100
Indonesia	99	92	100	94	79	100
Iran, Islamic Rep.	99	93	100	94	74	100
Israel	98	92	100	95	81	100
Italy	99	95	100	97	89	100
Japan	99	90	100	96	88	100
Jordan	99	96	100	96	89	100
Korea, Rep. of	98	88	100	96	73	100
Latvia (LSS)	99	96	100	96	79	100
Lithuania	99	90	100	98	88	100
Macedonia, Rep. of	99	97	100	98	95	100
Malaysia	100	98	100	99	97	100
Moldova	97	92	100	94	86	99
Morocco	97	84	100	88	65	99
Netherlands	99	85	100	94	79	100
New Zealand	99	95	100	95	88	100
Philippines	99	97	100	95	84	100
Romania	99	96	100	97	92	100
Russian Federation	100	98	100	98	92	100
Singapore	99	94	100	97	87	100
Slovak Republic	99	97	100	99	96	100
Slovenia	100	99	100	96	83	100
South Africa	99	93	100	96	85	99
Thailand	100	100	100	100	100	100
Tunisia	98	92	100	96	88	100
Turkey	100	97	100	99	97	100
International Avg.	99	93	100	96	85	100

SOURCE: IEA Third International Mathematics and Science Study (TIMSS), 1998-1999.

A dash (–) indicates data are not available.

Test Reliability

Exhibit A.8 displays the mathematics test reliability coefficient for each country and Benchmarking participant. This coefficient is the median KR-20 reliability across the eight test booklets. Among countries, median reliabilities ranged from 0.76 in the Philippines to 0.94 in Chinese Taipei. The international median, 0.89, is the median of the reliability coefficients for all countries. Reliability coefficients among Benchmarking participants were generally close to the international median, ranging from 0.88 to 0.91 across states, and from 0.84 to 0.91 across districts and consortia.

Countries	Reliability Coefficient[1]
United States	0.90
Australia	0.90
Belgium (Flemish)	0.89
Bulgaria	0.90
Canada	0.88
Chile	0.83
Chinese Taipei	0.94
Cyprus	0.87
Czech Republic	0.90
England	0.90
Finland	0.86
Hong Kong, SAR	0.89
Hungary	0.91
Indonesia	0.87
Iran, Islamic Rep.	0.83
Israel	0.90
Italy	0.89
Japan	0.91
Jordan	0.89
Korea, Rep. of	0.91
Latvia (LSS)	0.89
Lithuania	0.89
Macedonia, Rep. of	0.88
Malaysia	0.90
Moldova	0.88
Morocco	0.69
Netherlands	0.89
New Zealand	0.91
Philippines	0.76
Romania	0.90
Russian Federation	0.91
Singapore	0.90
Slovak Republic	0.89
Slovenia	0.90
South Africa	0.77
Thailand	0.87
Tunisia	0.79
Turkey	0.86
International Median	0.89

States	Reliability Coefficient[1]
Connecticut	0.90
Idaho	0.89
Illinois	0.89
Indiana	0.89
Maryland	0.91
Massachusetts	0.90
Michigan	0.90
Missouri	0.88
North Carolina	0.90
Oregon	0.90
Pennsylvania	0.90
South Carolina	0.91
Texas	0.91
Districts and Consortia	
Academy School Dist. #20, CO	0.89
Chicago Public Schools, IL	0.84
Delaware Science Coalition, DE	0.89
First in the World Consort., IL	0.91
Fremont/Lincoln/WestSide PS, NE	0.90
Guilford County, NC	0.90
Jersey City Public Schools, NJ	0.89
Miami-Dade County PS, FL	0.86
Michigan Invitational Group, MI	0.87
Montgomery County, MD	0.90
Naperville Sch. Dist. #203, IL	0.88
Project SMART Consortium, OH	0.89
Rochester City Sch. Dist., NY	0.86
SW Math/Sci. Collaborative, PA	0.89

SOURCE: IEA Third International Mathematics and Science Study (TIMSS), 1998-1999.

1 For each country and jurisdiction, the reliability coefficient is the median KR-20 reliability across the eight test booklets.

Data Processing

To ensure the availability of comparable, high-quality data for analysis, TIMSS took rigorous quality control steps to create the international database.[9] TIMSS prepared manuals and software for countries to use in entering their data, so that the information would be in a standardized international format before being forwarded to the IEA Data Processing Center in Hamburg for creation of the international database. Upon arrival at the Data Processing Center, the data underwent an exhaustive cleaning process. This involved several iterative steps and procedures designed to identify, document, and correct deviations from the international instruments, file structures, and coding schemes. The process also emphasized consistency of information within national data sets and appropriate linking among the many student, teacher, and school data files. In the United States, the creation of the data files for both the Benchmarking participants and the U.S. national TIMSS effort was the responsibility of Westat, working closely with NCS. After the data files were checked carefully by Westat, they were sent to the IEA Data Processing Center, where they underwent further validity checks before being forwarded to the International Study Center.

IRT Scaling and Data Analysis

The general approach to reporting the TIMSS achievement data was based primarily on item response theory (IRT) scaling methods.[10] The mathematics results were summarized using a family of 2-parameter and 3-parameter IRT models for dichotomously-scored items (right or wrong), and generalized partial credit models for items with 0, 1, or 2 available score points. The IRT scaling method produces a score by averaging the responses of each student to the items that he or she took in a way that takes into account the difficulty and discriminating power of each item. The methodology used in TIMSS includes refinements that enable reliable scores to be produced even though individual students responded to relatively small subsets of the total mathematics item pool. Achievement scales were produced for each of the five mathematics content areas (fractions and number sense, measurement, data representation, analysis, and probability, geometry, and algebra), as well as for mathematics overall.

The IRT methodology was preferred for developing comparable estimates of performance for all students, since students answered different test items depending upon which of the eight test booklets they received. The IRT analysis provides a common scale on which performance can be

9 These steps are detailed in Hastedt, D., and Gonzalez, E. (2000), "Data Management and Database Construction" in M.O. Martin, K.D. Gregory, K.M. O'Connor, and S.E. Stemler (eds.), *TIMSS 1999 Benchmarking Technical Report*, Chestnut Hill, MA: Boston College.

10 For a detailed description of the TIMSS scaling, see Yamamoto, K., and Kulick, E. (2000), "Scaling Methods and Procedures for the TIMSS Mathematics and Science Scales" in M.O. Martin, K.D. Gregory, K.M. O'Connor, and S.E. Stemler (eds.), *TIMSS 1999 Benchmarking Technical Report*, Chestnut Hill, MA: Boston College.

compared across countries. In addition to providing a basis for estimating mean achievement, scale scores permit estimates of how students within countries vary and provide information on percentiles of performance. To provide a reliable measure of student achievement in both 1999 and 1995, the overall mathematics scale was calibrated using students from the countries that participated in both years. When all countries participating in 1995 at the eighth grade are treated equally, the TIMSS scale average over those countries is 500 and the standard deviation is 100. Since the countries varied in size, each country was weighted to contribute equally to the mean and standard deviation of the scale. The average and standard deviation of the scale scores are arbitrary and do not affect scale interpretation. When the metric of the scale had been established, students from the countries that tested in 1999 but not 1995 were assigned scores on the basis of the new scale. IRT scales were also created for each of the five mathematics content areas for the 1999 data. Students from the Benchmarking samples were assigned scores on the overall mathematics scale as well as in each of the five mathematics content areas using the same item parameters and estimation procedures as for TIMSS internationally.

To allow more accurate estimation of summary statistics for student subpopulations, the TIMSS scaling made use of plausible-value technology, whereby five separate estimates of each student's score were generated on each scale, based on the student's responses to the items in the student's booklet and the student's background characteristics. The five score estimates are known as "plausible values," and the variability between them encapsulates the uncertainty inherent in the score estimation process.

Estimating Sampling Error

Because the statistics presented in this report are estimates of performance based on samples of students, rather than the values that could be calculated if every student in every country or Benchmarking jurisdiction had answered every question, it is important to have measures of the degree of uncertainty of the estimates. The jackknife procedure was used to estimate the standard error associated with each statistic presented in this report.[11] The jackknife standard errors also include an error component due to variation between the five plausible values generated for each student. The use of confidence intervals, based on the standard errors, provides a way to make inferences about the popu-

[11] Procedures for computing jackknifed standard errors are presented in Gonzalez, E. and Foy, P. (2000), "Estimation of Sampling Variance" in M.O. Martin, K.D. Gregory, K.M. O'Connor, and S.E. Stemler (eds.), *TIMSS 1999 Benchmarking Technical Report*, Chestnut Hill, MA: Boston College.

lation means and proportions in a manner that reflects the uncertainty associated with the sample estimates. An estimated sample statistic plus or minus two standard errors represents a 95 percent confidence interval for the corresponding population result.

Making Multiple Comparisons

This report makes extensive use of statistical hypothesis-testing to provide a basis for evaluating the significance of differences in percentages and in average achievement scores. Each separate test follows the usual convention of holding to 0.05 the probability that reported differences could be due to sampling variability alone. However, in exhibits where statistical significance tests are reported, the results of many tests are reported simultaneously, usually at least one for each country and Benchmarking participant in the exhibit. The significance tests in these exhibits are based on a Bonferroni procedure for multiple comparisons that hold to 0.05 the probability of erroneously declaring a statistic (mean or percentage) for one entity to be different from that for another entity. In the multiple comparison charts (Exhibit 1.2 and those in Appendix B), the Bonferroni procedure adjusts for the number of entities in the chart, minus one. In exhibits where a country or Benchmarking participant statistic is compared to the international average, the adjustment is for the number of entities.[12]

Setting International Benchmarks of Student Achievement

International benchmarks of student achievement were computed at each grade level for both mathematics and science. The benchmarks are points in the weighted international distribution of achievement scores that separate the 10 percent of students located on top of the distribution, the top 25 percent of students, the top 50 percent, and the bottom 25 percent. The percentage of students in each country and Benchmarking jurisdiction meeting or exceeding the international benchmarks is reported. The benchmarks correspond to the 90th, 75th, 50th, and 25th percentiles of the international distribution of achievement. When computing these percentiles, each country contributed as many students to the distribution as there were students in the target population in the country. That is, each country's contribution to setting the international benchmarks was proportional to the estimated population enrolled at the eighth grade.

In order to interpret the TIMSS scale scores and analyze achievement at the international benchmarks, TIMSS conducted a scale anchoring analysis to describe achievement of students at those four points on the scale. Scale anchoring is a way of describing students' performance at different

[12] The application of the Bonferroni procedures is described in Gonzalez, E., and Gregory, K. (2000), "Reporting Student Achievement in Mathematics and Science" in M.O. Martin, K.D. Gregory, K.M. O'Connor, and S.E. Stemler (eds.), *TIMSS 1999 Benchmarking Technical Report*, Chestnut Hill, MA: Boston College.

points on a scale in terms of what they know and can do. It involves a statistical component, in which items that discriminate between successive points on the scale are identified, and a judgmental component in which subject-matter experts examine the items and generalize to students' knowledge and understandings.[13]

Mathematics Curriculum Questionnaire

In an effort to collect information about the content of the intended curriculum in mathematics, TIMSS asked National Research Coordinators and Coordinators from the Benchmarking jurisdictions to complete a questionnaire about the structure, organization, and content coverage of their curricula. Coordinators reviewed 56 mathematics topics and reported the percentage of their eighth-grade students for which each topic was intended in their curriculum. Although most topic descriptions were used without modification, there were occasions when Coordinators found it necessary to expand on or qualify the topic description to describe their situation accurately. The country-specific adaptations to the mathematics curriculum questionnaire are presented in Exhibit A.9. No adaptations to the list of topics were necessary for the U.S. national version, nor were any adaptations made by any Benchmarking participants.

13 The scale anchoring procedure is described fully in Gregory, K., and Mullis, I. (2000), "Describing International Benchmarks of Student Achievement" in M.O. Martin, K.D. Gregory, K.M. O'Connor, and S.E. Stemler (eds.), *TIMSS 1999 Benchmarking Technical Report*, Chestnut Hill, MA: Boston College. An application of the procedure to the 1995 TIMSS data may be found in Kelly, D.L., Mullis, I.V.S., and Martin, M.O. (2000), *Profiles of Student Achievement in Mathematics at the TIMSS International Benchmarks: U.S. Performance and Standards in an International Context*, Chestnut Hill, MA: Boston College.

	Topic	Response	Comments
Bulgaria	Geometry: Congruence and similarity	All or almost all of the students (at least 90%)	Similarity not included in curriculum through grade 8.
Czech Republic	Measurement: Volume of other solids (e.g., pyramids, cylinders, cones, spheres)	All or almost all of the students (at least 90%)	Volume of pyramids, cones, & spheres not included in curriculum through grade 8.
	Geometry: Congruence and similarity	All or almost all of the students (at least 90%)	Similarity not included in curriculum through grade 8.
Finland	Fractions and Number Sense: Concepts of ratio and proportion; ratio and proportion problems	Not included in curriculum through grade 8	Concepts of ratio and proportion included in curriculum through grade 8.
	Geometry: Symmetry and transformations (reflection and rotation)	Not included in curriculum through grade 8	Symmetry included in curriculum through grade 8.
	Algebra: Representing situations algebraically; formulas	All or almost all of the students (at least 90%)	Formulas not included in curriculum through grade 8.
Israel	Fractions and Number Sense: Whole numbers–including place values, factorization and operations (+, -, x, ÷)	All or almost all of the students (at least 90%)	Factorization not included in curriculum through grade 8.
	Fractions and Number Sense: Computations with common fractions	All or almost all of the students (at least 90%)	Division with common fractions not included in curriculum through grade 8.
	Fractions and Number Sense: Computations with decimal fractions	All or almost all of the students (at least 90%)	Division with decimal fractions not included in curriculum through grade 8.
	Measurement: Estimates of measurement; accuracy of measurement	Only the most advanced students (10% or less)	Accuracy of measurement not included in curriculum through grade 8.
	Geometry: Simple two dimensional geometry – angles on a straight line, parallel lines, triangles and quadrilaterals	About half of the students	Quadrilaterals not included in curriculum through grade 8.
	Geometry: Congruence and similarity	All or almost all of the students (at least 90%)	Similarity not included in curriculum through grade 8.
Japan	Fractions and Number Sense: Prime factors, highest common factor, lowest common multiple, rules for divisibility	Not included in curriculum through grade 8	Highest common factor and lowest common multiple included in curriculum through grade 8.
Korea, Rep. of	Fractions and Number Sense: Number lines	All or almost all of the students (at least 90%)	Whole number and integer number lines included in curriculum through grade 8. The real number line is taught in grade 9.
	Geometry: Cartesian coordinates of points in a plane	Not included in curriculum through grade 8	Linear function and its graph included in curriculum through grade 8.
Morocco	Geometry: Symmetry and transformations (reflection and rotation)	All or almost all of the students (at least 90%)	Transformations (reflection & rotation) not included in curriculum through grade 8.
Netherlands	Geometry: Congruence and similarity	Not included in curriculum through grade 8	Symmetry taught to all or almost all of the students.
New Zealand	Fractions and Number Sense: Computations with common fractions	All or almost all of the students (at least 90%)	Division with common fractions not included in curriculum through grade 8.
	Fractions and Number Sense: Square roots (of perfect squares less than 144), small integer exponents	All or almost all of the students (at least 90%)	Small integer exponents taught to about half of the students.
	Algebra: Representing situations algebraically; formulas	About half of the students	Formulas not included in curriculum through grade 8.
	Algebra: Using the graph of a relationship to interpolate/extrapolate	All or almost all of the students (at least 90%)	Using the graph of a relationship to extrapolate not included in curriculum through grade 8.
Russian Federation	Measurement: Perimeter and area of simple shapes – triangles, rectangles, and circles	About half of the students	Perimeter and area of rectangles and circles included in curriculum through grade 8.
	Geometry: Congruence and similarity	About half of the students	Congruence included in curriculum through grade 8.
South Africa	Measurement: Volume of other solids (e.g., pyramids, cylinders, cones, spheres)	All or almost all of the students (at least 90%)	Volume of pyramids, cones, & spheres not included in curriculum through grade 8.
Tunisia	Geometry: Symmetry and transformations (reflection and rotation)	All or almost all of the students (at least 90%)	Rotation not included in curriculum through grade 8.

SOURCE: IEA Third International Mathematics and Science Study (TIMSS), 1998-1999.

APPENDIX

B Multiple Comparisons of Average Achievement in Mathematics Content Areas

Instructions: Read across the row for a participant to compare performance with the participants listed along the top of the chart. The symbols indicate whether the average achievement of the participant in the row is significantly lower than that of the comparison participant, significantly higher than that of the comparison participant, or if there is no statistically significant difference between the average achievement of the two participants.

	Singapore	Hong Kong, SAR	Chinese Taipei	Korea, Rep. of	Japan	**Naperville Sch. Dist. #203, IL**	**First in the World Consort., IL**	Belgium (Flemish)	Netherlands	**Montgomery County, MD**	**Michigan Invitational Group, MI**	**Academy School Dist. #20, CO**	Canada	Malaysia	Finland	**Project SMART Consortium, OH**	Slovenia	***Texas***	**Indiana**	Hungary	Slovak Republic	**Michigan**	**SW Math/Sci. Collaborative, PA**	**Connecticut**	**Massachusetts**	**Oregon**	Australia	***Pennsylvania***	**Illinois**	Russian Federation	**Guilford County, NC**	**United States**
Singapore		▲	▲	▲	▲	▲	▲	▲	▲	▲	▲	▲	▲	▲	▲	▲	▲	▲	▲	▲	▲	▲	▲	▲	▲	▲	▲	▲	▲	▲	▲	▲
Hong Kong, SAR	▼		●	●	●	●	●	▲	▲	▲	▲	▲	▲	▲	▲	▲	▲	▲	▲	▲	▲	▲	▲	▲	▲	▲	▲	▲	▲	▲	▲	▲
Chinese Taipei	▼	●		●	●	●	●	▲	▲	▲	▲	▲	▲	▲	▲	▲	▲	▲	▲	▲	▲	▲	▲	▲	▲	▲	▲	▲	▲	▲	▲	▲
Korea, Rep. of	▼	●	●		●	●	●	●	▲	▲	▲	▲	▲	▲	▲	▲	▲	▲	▲	▲	▲	▲	▲	▲	▲	▲	▲	▲	▲	▲	▲	▲
Japan	▼	●	●	●		●	●	●	▲	▲	▲	▲	▲	▲	▲	▲	▲	▲	▲	▲	▲	▲	▲	▲	▲	▲	▲	▲	▲	▲	▲	▲
Naperville Sch. Dist. #203, IL	▼	●	●	●	●		●	●	●	▲	▲	▲	▲	▲	▲	▲	▲	▲	▲	▲	▲	▲	▲	▲	▲	▲	▲	▲	▲	▲	▲	▲
First in the World Consort., IL	▼	●	●	●	●	●		●	●	●	▲	▲	▲	▲	▲	▲	▲	▲	▲	▲	▲	▲	▲	▲	▲	▲	▲	▲	▲	▲	▲	▲
Belgium (Flemish)	▼	▼	▼	●	●	●	●		●	●	▲	▲	▲	▲	▲	▲	▲	●	▲	▲	▲	▲	▲	▲	▲	▲	▲	▲	▲	▲	▲	▲
Netherlands	▼	▼	▼	▼	▼	●	●	●		●	●	●	●	●	●	●	●	●	●	●	●	●	●	●	●	●	●	●	●	▲	●	▲
Montgomery County, MD	▼	▼	▼	▼	▼	▼	●	●	●		●	●	●	●	●	●	●	●	●	●	●	●	●	●	●	●	●	●	●	▲	●	▲
Michigan Invitational Group, MI	▼	▼	▼	▼	▼	▼	▼	▼	●	●		●	●	●	●	●	●	●	●	●	●	●	●	●	●	●	●	●	●	●	●	▲
Academy School Dist. #20, CO	▼	▼	▼	▼	▼	▼	▼	▼	●	●	●		●	●	●	●	●	●	●	●	●	●	●	●	●	●	●	●	●	●	●	▲
Canada	▼	▼	▼	▼	▼	▼	▼	▼	●	●	●	●		●	●	●	●	●	●	●	●	●	●	●	●	●	●	●	●	●	●	▲
Malaysia	▼	▼	▼	▼	▼	▼	▼	▼	●	●	●	●	●		●	●	●	●	●	●	●	●	●	●	●	●	●	●	●	●	●	▲
Finland	▼	▼	▼	▼	▼	▼	▼	▼	●	●	●	●	●	●		●	●	●	●	●	●	●	●	●	●	●	●	●	●	●	●	▲
Project SMART Consortium, OH	▼	▼	▼	▼	▼	▼	▼	▼	●	●	●	●	●	●	●		●	●	●	●	●	●	●	●	●	●	●	●	●	●	●	●
Slovenia	▼	▼	▼	▼	▼	▼	▼	▼	●	●	●	●	●	●	●	●		●	●	●	●	●	●	●	●	●	●	●	●	●	●	●
Texas	▼	▼	▼	▼	▼	▼	▼	●	●	●	●	●	●	●	●	●	●		●	●	●	●	●	●	●	●	●	●	●	●	●	●
Indiana	▼	▼	▼	▼	▼	▼	▼	▼	●	●	●	●	●	●	●	●	●	●		●	●	●	●	●	●	●	●	●	●	●	●	●
Hungary	▼	▼	▼	▼	▼	▼	▼	▼	●	●	●	●	●	●	●	●	●	●	●		●	●	●	●	●	●	●	●	●	●	●	●
Slovak Republic	▼	▼	▼	▼	▼	▼	▼	▼	●	●	●	●	●	●	●	●	●	●	●	●		●	●	●	●	●	●	●	●	●	●	●
Michigan	▼	▼	▼	▼	▼	▼	▼	▼	●	●	●	●	●	●	●	●	●	●	●	●	●		●	●	●	●	●	●	●	●	●	●
SW Math/Sci. Collaborative, PA	▼	▼	▼	▼	▼	▼	▼	▼	●	●	●	●	●	●	●	●	●	●	●	●	●	●		●	●	●	●	●	●	●	●	●
Connecticut	▼	▼	▼	▼	▼	▼	▼	▼	●	●	●	●	●	●	●	●	●	●	●	●	●	●	●		●	●	●	●	●	●	●	●
Massachusetts	▼	▼	▼	▼	▼	▼	▼	▼	●	●	●	●	●	●	●	●	●	●	●	●	●	●	●	●		●	●	●	●	●	●	●
Oregon	▼	▼	▼	▼	▼	▼	▼	▼	●	●	●	●	●	●	●	●	●	●	●	●	●	●	●	●	●		●	●	●	●	●	●
Australia	▼	▼	▼	▼	▼	▼	▼	▼	●	●	●	●	●	●	●	●	●	●	●	●	●	●	●	●	●	●		●	●	●	●	●
Pennsylvania	▼	▼	▼	▼	▼	▼	▼	▼	●	●	●	●	●	●	●	●	●	●	●	●	●	●	●	●	●	●	●		●	●	●	●
Illinois	▼	▼	▼	▼	▼	▼	▼	▼	●	●	●	●	●	●	●	●	●	●	●	●	●	●	●	●	●	●	●	●		●	●	●
Russian Federation	▼	▼	▼	▼	▼	▼	▼	▼	▼	▼	●	●	●	●	●	●	●	●	●	●	●	●	●	●	●	●	●	●	●		●	●
Guilford County, NC	▼	▼	▼	▼	▼	▼	▼	▼	●	●	●	●	●	●	●	●	●	●	●	●	●	●	●	●	●	●	●	●	●	●		●
United States	▼	▼	▼	▼	▼	▼	▼	▼	▼	▼	▼	▼	▼	▼	▼	●	●	●	●	●	●	●	●	●	●	●	●	●	●	●	●	
South Carolina	▼	▼	▼	▼	▼	▼	▼	▼	▼	▼	●	▼	▼	●	●	●	●	●	●	●	●	●	●	●	●	●	●	●	●	●	●	●
Czech Republic	▼	▼	▼	▼	▼	▼	▼	▼	▼	▼	▼	▼	▼	▼	▼	●	▼	●	●	●	●	●	●	●	●	●	●	●	●	●	●	●
Idaho	▼	▼	▼	▼	▼	▼	▼	▼	▼	▼	▼	▼	▼	●	▼	●	●	●	●	●	●	●	●	●	●	●	●	●	●	●	●	●
Bulgaria	▼	▼	▼	▼	▼	▼	▼	▼	▼	▼	▼	▼	▼	▼	▼	●	●	●	●	●	●	●	●	●	●	●	●	●	●	●	●	●
Maryland	▼	▼	▼	▼	▼	▼	▼	▼	▼	▼	▼	▼	▼	▼	▼	●	▼	●	●	▼	●	●	●	●	●	●	●	●	●	●	●	●
Fremont/Lincoln/WestSide PS, NE	▼	▼	▼	▼	▼	▼	▼	▼	▼	▼	▼	▼	▼	▼	▼	●	▼	●	●	▼	▼	●	●	●	●	●	●	●	●	●	●	●
England	▼	▼	▼	▼	▼	▼	▼	▼	▼	▼	▼	▼	▼	▼	▼	▼	▼	●	▼	▼	▼	▼	▼	●	▼	▼	▼	●	●	●	●	●
North Carolina	▼	▼	▼	▼	▼	▼	▼	▼	▼	▼	▼	▼	▼	▼	▼	●	▼	●	●	▼	▼	●	●	●	●	●	●	●	●	●	●	●
Missouri	▼	▼	▼	▼	▼	▼	▼	▼	▼	▼	▼	▼	▼	▼	▼	▼	▼	●	▼	▼	▼	▼	▼	●	●	●	▼	●	●	●	●	●
Latvia (LSS)	▼	▼	▼	▼	▼	▼	▼	▼	▼	▼	▼	▼	▼	▼	▼	▼	▼	●	▼	▼	▼	▼	▼	●	▼	▼	▼	▼	●	●	●	●
New Zealand	▼	▼	▼	▼	▼	▼	▼	▼	▼	▼	▼	▼	▼	▼	▼	▼	▼	▼	▼	▼	▼	▼	▼	●	▼	▼	▼	▼	●	●	●	●
Delaware Science Coalition, DE	▼	▼	▼	▼	▼	▼	▼	▼	▼	▼	▼	▼	▼	▼	▼	▼	▼	▼	▼	▼	▼	▼	▼	●	▼	▼	▼	●	●	●	●	●
Jersey City Public Schools, NJ	▼	▼	▼	▼	▼	▼	▼	▼	▼	▼	▼	▼	▼	▼	▼	▼	▼	▼	▼	▼	▼	▼	▼	▼	▼	▼	▼	▼	▼	●	●	●
Cyprus	▼	▼	▼	▼	▼	▼	▼	▼	▼	▼	▼	▼	▼	▼	▼	▼	▼	▼	▼	▼	▼	▼	▼	▼	▼	▼	▼	▼	▼	▼	▼	▼
Lithuania	▼	▼	▼	▼	▼	▼	▼	▼	▼	▼	▼	▼	▼	▼	▼	▼	▼	▼	▼	▼	▼	▼	▼	▼	▼	▼	▼	▼	▼	▼	▼	▼
Chicago Public Schools, IL	▼	▼	▼	▼	▼	▼	▼	▼	▼	▼	▼	▼	▼	▼	▼	▼	▼	▼	▼	▼	▼	▼	▼	▼	▼	▼	▼	▼	▼	▼	▼	▼
Israel	▼	▼	▼	▼	▼	▼	▼	▼	▼	▼	▼	▼	▼	▼	▼	▼	▼	▼	▼	▼	▼	▼	▼	▼	▼	▼	▼	▼	▼	▼	▼	▼
Thailand	▼	▼	▼	▼	▼	▼	▼	▼	▼	▼	▼	▼	▼	▼	▼	▼	▼	▼	▼	▼	▼	▼	▼	▼	▼	▼	▼	▼	▼	▼	▼	▼
Italy	▼	▼	▼	▼	▼	▼	▼	▼	▼	▼	▼	▼	▼	▼	▼	▼	▼	▼	▼	▼	▼	▼	▼	▼	▼	▼	▼	▼	▼	▼	▼	▼
Moldova	▼	▼	▼	▼	▼	▼	▼	▼	▼	▼	▼	▼	▼	▼	▼	▼	▼	▼	▼	▼	▼	▼	▼	▼	▼	▼	▼	▼	▼	▼	▼	▼
Romania	▼	▼	▼	▼	▼	▼	▼	▼	▼	▼	▼	▼	▼	▼	▼	▼	▼	▼	▼	▼	▼	▼	▼	▼	▼	▼	▼	▼	▼	▼	▼	▼
Rochester City Sch. Dist., NY	▼	▼	▼	▼	▼	▼	▼	▼	▼	▼	▼	▼	▼	▼	▼	▼	▼	▼	▼	▼	▼	▼	▼	▼	▼	▼	▼	▼	▼	▼	▼	▼
Tunisia	▼	▼	▼	▼	▼	▼	▼	▼	▼	▼	▼	▼	▼	▼	▼	▼	▼	▼	▼	▼	▼	▼	▼	▼	▼	▼	▼	▼	▼	▼	▼	▼
Iran, Islamic Rep.	▼	▼	▼	▼	▼	▼	▼	▼	▼	▼	▼	▼	▼	▼	▼	▼	▼	▼	▼	▼	▼	▼	▼	▼	▼	▼	▼	▼	▼	▼	▼	▼
Macedonia, Rep. of	▼	▼	▼	▼	▼	▼	▼	▼	▼	▼	▼	▼	▼	▼	▼	▼	▼	▼	▼	▼	▼	▼	▼	▼	▼	▼	▼	▼	▼	▼	▼	▼
Miami-Dade County PS, FL	▼	▼	▼	▼	▼	▼	▼	▼	▼	▼	▼	▼	▼	▼	▼	▼	▼	▼	▼	▼	▼	▼	▼	▼	▼	▼	▼	▼	▼	▼	▼	▼
Jordan	▼	▼	▼	▼	▼	▼	▼	▼	▼	▼	▼	▼	▼	▼	▼	▼	▼	▼	▼	▼	▼	▼	▼	▼	▼	▼	▼	▼	▼	▼	▼	▼
Turkey	▼	▼	▼	▼	▼	▼	▼	▼	▼	▼	▼	▼	▼	▼	▼	▼	▼	▼	▼	▼	▼	▼	▼	▼	▼	▼	▼	▼	▼	▼	▼	▼
Indonesia	▼	▼	▼	▼	▼	▼	▼	▼	▼	▼	▼	▼	▼	▼	▼	▼	▼	▼	▼	▼	▼	▼	▼	▼	▼	▼	▼	▼	▼	▼	▼	▼
Chile	▼	▼	▼	▼	▼	▼	▼	▼	▼	▼	▼	▼	▼	▼	▼	▼	▼	▼	▼	▼	▼	▼	▼	▼	▼	▼	▼	▼	▼	▼	▼	▼
Philippines	▼	▼	▼	▼	▼	▼	▼	▼	▼	▼	▼	▼	▼	▼	▼	▼	▼	▼	▼	▼	▼	▼	▼	▼	▼	▼	▼	▼	▼	▼	▼	▼
Morocco	▼	▼	▼	▼	▼	▼	▼	▼	▼	▼	▼	▼	▼	▼	▼	▼	▼	▼	▼	▼	▼	▼	▼	▼	▼	▼	▼	▼	▼	▼	▼	▼
South Africa	▼	▼	▼	▼	▼	▼	▼	▼	▼	▼	▼	▼	▼	▼	▼	▼	▼	▼	▼	▼	▼	▼	▼	▼	▼	▼	▼	▼	▼	▼	▼	▼

States in *italics* did not fully satisfy guidelines for sample participation rates (see Appendix A for details).

	South Carolina	Czech Republic	**Idaho**	Bulgaria	**Maryland**	**Fremont/Lincoln/WestSide PS, NE**	England	**North Carolina**	**Missouri**	Latvia (LSS)	New Zealand	**Delaware Science Coalition, DE**	**Jersey City Public Schools, NJ**	Cyprus	Lithuania	**Chicago Public Schools, IL**	Israel	Thailand	Italy	Moldova	Romania	**Rochester City Sch. Dist., NY**	Tunisia	Iran, Islamic Rep.	Macedonia, Rep. of	**Miami-Dade County PS, FL**	Jordan	Turkey	Indonesia	Chile	Philippines	Morocco	South Africa
Singapore	▲	▲	▲	▲	▲	▲	▲	▲	▲	▲	▲	▲	▲	▲	▲	▲	▲	▲	▲	▲	▲	▲	▲	▲	▲	▲	▲	▲	▲	▲	▲	▲	▲
Hong Kong, SAR	▲	▲	▲	▲	▲	▲	▲	▲	▲	▲	▲	▲	▲	▲	▲	▲	▲	▲	▲	▲	▲	▲	▲	▲	▲	▲	▲	▲	▲	▲	▲	▲	▲
Chinese Taipei	▲	▲	▲	▲	▲	▲	▲	▲	▲	▲	▲	▲	▲	▲	▲	▲	▲	▲	▲	▲	▲	▲	▲	▲	▲	▲	▲	▲	▲	▲	▲	▲	▲
Korea, Rep. of	▲	▲	▲	▲	▲	▲	▲	▲	▲	▲	▲	▲	▲	▲	▲	▲	▲	▲	▲	▲	▲	▲	▲	▲	▲	▲	▲	▲	▲	▲	▲	▲	▲
Japan	▲	▲	▲	▲	▲	▲	▲	▲	▲	▲	▲	▲	▲	▲	▲	▲	▲	▲	▲	▲	▲	▲	▲	▲	▲	▲	▲	▲	▲	▲	▲	▲	▲
Naperville Sch. Dist. #203, IL	▲	▲	▲	▲	▲	▲	▲	▲	▲	▲	▲	▲	▲	▲	▲	▲	▲	▲	▲	▲	▲	▲	▲	▲	▲	▲	▲	▲	▲	▲	▲	▲	▲
First in the World Consort., IL	▲	▲	▲	▲	▲	▲	▲	▲	▲	▲	▲	▲	▲	▲	▲	▲	▲	▲	▲	▲	▲	▲	▲	▲	▲	▲	▲	▲	▲	▲	▲	▲	▲
Belgium (Flemish)	▲	▲	▲	▲	▲	▲	▲	▲	▲	▲	▲	▲	▲	▲	▲	▲	▲	▲	▲	▲	▲	▲	▲	▲	▲	▲	▲	▲	▲	▲	▲	▲	▲
Netherlands	▲	▲	▲	▲	▲	▲	▲	▲	▲	▲	▲	▲	▲	▲	▲	▲	▲	▲	▲	▲	▲	▲	▲	▲	▲	▲	▲	▲	▲	▲	▲	▲	▲
Montgomery County, MD	▲	▲	▲	▲	▲	▲	▲	▲	▲	▲	▲	▲	▲	▲	▲	▲	▲	▲	▲	▲	▲	▲	▲	▲	▲	▲	▲	▲	▲	▲	▲	▲	▲
Michigan Invitational Group, MI	●	▲	▲	▲	▲	▲	▲	▲	▲	▲	▲	▲	▲	▲	▲	▲	▲	▲	▲	▲	▲	▲	▲	▲	▲	▲	▲	▲	▲	▲	▲	▲	▲
Academy School Dist. #20, CO	▲	▲	▲	▲	▲	▲	▲	▲	▲	▲	▲	▲	▲	▲	▲	▲	▲	▲	▲	▲	▲	▲	▲	▲	▲	▲	▲	▲	▲	▲	▲	▲	▲
Canada	▲	▲	▲	▲	▲	▲	▲	▲	▲	▲	▲	▲	▲	▲	▲	▲	▲	▲	▲	▲	▲	▲	▲	▲	▲	▲	▲	▲	▲	▲	▲	▲	▲
Malaysia	●	▲	●	▲	▲	▲	▲	▲	▲	▲	▲	▲	▲	▲	▲	▲	▲	▲	▲	▲	▲	▲	▲	▲	▲	▲	▲	▲	▲	▲	▲	▲	▲
Finland	●	▲	▲	▲	▲	▲	▲	▲	▲	▲	▲	▲	▲	▲	▲	▲	▲	▲	▲	▲	▲	▲	▲	▲	▲	▲	▲	▲	▲	▲	▲	▲	▲
Project SMART Consortium, OH	●	●	●	●	●	●	▲	●	▲	▲	▲	▲	▲	▲	▲	▲	▲	▲	▲	▲	▲	▲	▲	▲	▲	▲	▲	▲	▲	▲	▲	▲	▲
Slovenia	●	▲	●	●	▲	▲	▲	▲	▲	▲	▲	▲	▲	▲	▲	▲	▲	▲	▲	▲	▲	▲	▲	▲	▲	▲	▲	▲	▲	▲	▲	▲	▲
Texas	●	●	●	●	●	●	●	●	●	●	▲	▲	▲	▲	▲	▲	▲	▲	▲	▲	▲	▲	▲	▲	▲	▲	▲	▲	▲	▲	▲	▲	▲
Indiana	●	●	●	●	●	●	▲	●	▲	▲	▲	▲	▲	▲	▲	▲	▲	▲	▲	▲	▲	▲	▲	▲	▲	▲	▲	▲	▲	▲	▲	▲	▲
Hungary	●	●	●	●	▲	▲	▲	▲	▲	▲	▲	▲	▲	▲	▲	▲	▲	▲	▲	▲	▲	▲	▲	▲	▲	▲	▲	▲	▲	▲	▲	▲	▲
Slovak Republic	●	●	●	●	●	▲	▲	▲	▲	▲	▲	▲	▲	▲	▲	▲	▲	▲	▲	▲	▲	▲	▲	▲	▲	▲	▲	▲	▲	▲	▲	▲	▲
Michigan	●	●	●	●	●	●	▲	●	▲	▲	▲	▲	▲	▲	▲	▲	▲	▲	▲	▲	▲	▲	▲	▲	▲	▲	▲	▲	▲	▲	▲	▲	▲
SW Math/Sci. Collaborative, PA	●	●	●	●	●	●	▲	●	▲	▲	▲	▲	▲	▲	▲	▲	▲	▲	▲	▲	▲	▲	▲	▲	▲	▲	▲	▲	▲	▲	▲	▲	▲
Connecticut	●	●	●	●	●	●	●	●	●	●	●	●	▲	▲	▲	▲	▲	▲	▲	▲	▲	▲	▲	▲	▲	▲	▲	▲	▲	▲	▲	▲	▲
Massachusetts	●	●	●	●	●	●	▲	●	●	▲	▲	▲	▲	▲	▲	▲	▲	▲	▲	▲	▲	▲	▲	▲	▲	▲	▲	▲	▲	▲	▲	▲	▲
Oregon	●	●	●	●	●	●	▲	●	●	▲	▲	▲	▲	▲	▲	▲	▲	▲	▲	▲	▲	▲	▲	▲	▲	▲	▲	▲	▲	▲	▲	▲	▲
Australia	●	●	●	●	●	●	▲	●	▲	▲	▲	▲	▲	▲	▲	▲	▲	▲	▲	▲	▲	▲	▲	▲	▲	▲	▲	▲	▲	▲	▲	▲	▲
Pennsylvania	●	●	●	●	●	●	●	●	●	▲	▲	●	▲	▲	▲	▲	▲	▲	▲	▲	▲	▲	▲	▲	▲	▲	▲	▲	▲	▲	▲	▲	▲
Illinois	●	●	●	●	●	●	●	●	●	●	●	●	▲	▲	▲	▲	▲	▲	▲	▲	▲	▲	▲	▲	▲	▲	▲	▲	▲	▲	▲	▲	▲
Russian Federation	●	●	●	●	●	●	●	●	●	●	●	●	●	▲	▲	▲	▲	▲	▲	▲	▲	▲	▲	▲	▲	▲	▲	▲	▲	▲	▲	▲	▲
Guilford County, NC	●	●	●	●	●	●	●	●	●	●	●	●	●	▲	▲	▲	▲	▲	▲	▲	▲	▲	▲	▲	▲	▲	▲	▲	▲	▲	▲	▲	▲
United States	●	●	●	●	●	●	●	●	●	●	●	●	●	▲	▲	▲	▲	▲	▲	▲	▲	▲	▲	▲	▲	▲	▲	▲	▲	▲	▲	▲	▲
South Carolina		●	●	●	●	●	●	●	●	●	●	●	●	▲	▲	▲	▲	▲	▲	▲	▲	▲	▲	▲	▲	▲	▲	▲	▲	▲	▲	▲	▲
Czech Republic	●		●	●	●	●	●	●	●	●	●	●	●	▲	▲	▲	▲	▲	▲	▲	▲	▲	▲	▲	▲	▲	▲	▲	▲	▲	▲	▲	▲
Idaho	●	●		●	●	●	●	●	●	●	●	●	●	▲	▲	▲	▲	▲	▲	▲	▲	▲	▲	▲	▲	▲	▲	▲	▲	▲	▲	▲	▲
Bulgaria	●	●	●		●	●	●	●	●	●	●	●	●	●	●	▲	▲	▲	▲	▲	▲	▲	▲	▲	▲	▲	▲	▲	▲	▲	▲	▲	▲
Maryland	●	●	●	●		●	●	●	●	●	●	●	●	●	●	●	▲	▲	▲	▲	▲	▲	▲	▲	▲	▲	▲	▲	▲	▲	▲	▲	▲
Fremont/Lincoln/WestSide PS, NE	●	●	●	●	●		●	●	●	●	●	●	●	●	●	●	▲	▲	▲	▲	▲	▲	▲	▲	▲	▲	▲	▲	▲	▲	▲	▲	▲
England	●	●	●	●	●	●		●	●	●	●	●	●	▲	▲	▲	▲	▲	▲	▲	▲	▲	▲	▲	▲	▲	▲	▲	▲	▲	▲	▲	▲
North Carolina	●	●	●	●	●	●	●		●	●	●	●	●	●	●	●	●	●	●	▲	▲	▲	▲	▲	▲	▲	▲	▲	▲	▲	▲	▲	▲
Missouri	●	●	●	●	●	●	●	●		●	●	●	●	●	●	●	▲	▲	▲	▲	▲	▲	▲	▲	▲	▲	▲	▲	▲	▲	▲	▲	▲
Latvia (LSS)	●	●	●	●	●	●	●	●	●		●	●	●	▲	●	●	▲	▲	▲	▲	▲	▲	▲	▲	▲	▲	▲	▲	▲	▲	▲	▲	▲
New Zealand	●	●	●	●	●	●	●	●	●	●		●	●	●	●	●	●	●	●	▲	▲	▲	▲	▲	▲	▲	▲	▲	▲	▲	▲	▲	▲
Delaware Science Coalition, DE	●	●	●	●	●	●	●	●	●	●	●		●	●	●	●	●	●	●	●	●	●	▲	▲	▲	▲	▲	▲	▲	▲	▲	▲	▲
Jersey City Public Schools, NJ	●	●	●	●	●	●	●	●	●	●	●	●		●	●	●	●	●	●	●	●	●	▲	▲	▲	▲	▲	▲	▲	▲	▲	▲	▲
Cyprus	▼	▼	▼	●	●	●	▼	●	●	▼	●	●	●		●	●	●	●	●	●	▲	▲	▲	▲	▲	▲	▲	▲	▲	▲	▲	▲	▲
Lithuania	▼	▼	▼	●	●	●	▼	●	●	●	●	●	●	●		●	●	●	●	●	●	●	▲	▲	▲	▲	▲	▲	▲	▲	▲	▲	▲
Chicago Public Schools, IL	▼	▼	▼	▼	●	●	▼	●	●	●	●	●	●	●	●		●	●	●	●	●	●	▲	▲	▲	▲	▲	▲	▲	▲	▲	▲	▲
Israel	▼	▼	▼	▼	▼	▼	▼	●	▼	▼	●	●	●	●	●	●		●	●	●	●	●	▲	▲	▲	▲	▲	▲	▲	▲	▲	▲	▲
Thailand	▼	▼	▼	▼	▼	▼	▼	●	▼	▼	●	●	●	●	●	●	●		●	●	●	●	▲	▲	▲	▲	▲	▲	▲	▲	▲	▲	▲
Italy	▼	▼	▼	▼	▼	▼	▼	●	▼	▼	●	●	●	●	●	●	●	●		●	●	●	▲	▲	▲	▲	▲	▲	▲	▲	▲	▲	▲
Moldova	▼	▼	▼	▼	▼	▼	▼	▼	▼	▼	▼	●	●	●	●	●	●	●	●		●	●	▲	▲	▲	●	▲	▲	▲	▲	▲	▲	▲
Romania	▼	▼	▼	▼	▼	▼	▼	▼	▼	▼	▼	●	●	▼	●	●	●	●	●	●		●	●	●	●	●	▲	▲	▲	▲	▲	▲	▲
Rochester City Sch. Dist., NY	▼	▼	▼	▼	▼	▼	▼	▼	▼	▼	▼	●	●	▼	●	●	●	●	●	●	●		●	●	●	●	▲	▲	▲	▲	▲	▲	▲
Tunisia	▼	▼	▼	▼	▼	▼	▼	▼	▼	▼	▼	▼	▼	▼	▼	▼	▼	▼	▼	▼	●	●		●	●	●	●	●	▲	▲	▲	▲	▲
Iran, Islamic Rep.	▼	▼	▼	▼	▼	▼	▼	▼	▼	▼	▼	▼	▼	▼	▼	▼	▼	▼	▼	▼	●	●	●		●	●	●	●	▲	▲	▲	▲	▲
Macedonia, Rep. of	▼	▼	▼	▼	▼	▼	▼	▼	▼	▼	▼	▼	▼	▼	▼	▼	▼	▼	▼	▼	●	●	●	●		●	●	●	▲	▲	▲	▲	▲
Miami-Dade County PS, FL	▼	▼	▼	▼	▼	▼	▼	▼	▼	▼	▼	▼	▼	▼	▼	▼	▼	▼	▼	●	●	●	●	●	●		●	●	●	●	▲	▲	▲
Jordan	▼	▼	▼	▼	▼	▼	▼	▼	▼	▼	▼	▼	▼	▼	▼	▼	▼	▼	▼	▼	▼	▼	●	●	●	●		●	▲	▲	▲	▲	▲
Turkey	▼	▼	▼	▼	▼	▼	▼	▼	▼	▼	▼	▼	▼	▼	▼	▼	▼	▼	▼	▼	▼	▼	●	●	●	●	●		▲	▲	▲	▲	▲
Indonesia	▼	▼	▼	▼	▼	▼	▼	▼	▼	▼	▼	▼	▼	▼	▼	▼	▼	▼	▼	▼	▼	▼	▼	▼	▼	●	▼	▼		●	▲	▲	▲
Chile	▼	▼	▼	▼	▼	▼	▼	▼	▼	▼	▼	▼	▼	▼	▼	▼	▼	▼	▼	▼	▼	▼	▼	▼	▼	●	▼	▼	●		●	▲	▲
Philippines	▼	▼	▼	▼	▼	▼	▼	▼	▼	▼	▼	▼	▼	▼	▼	▼	▼	▼	▼	▼	▼	▼	▼	▼	▼	▼	▼	▼	▼	●		▲	▲
Morocco	▼	▼	▼	▼	▼	▼	▼	▼	▼	▼	▼	▼	▼	▼	▼	▼	▼	▼	▼	▼	▼	▼	▼	▼	▼	▼	▼	▼	▼	▼	▼		▲
South Africa	▼	▼	▼	▼	▼	▼	▼	▼	▼	▼	▼	▼	▼	▼	▼	▼	▼	▼	▼	▼	▼	▼	▼	▼	▼	▼	▼	▼	▼	▼	▼	▼	

▲ Average achievement significantly higher than comparison participant

● No statistically significant difference from comparison participant

▼ Average achievement significantly lower than comparison participant

Significance tests adjusted for multiple comparisons

SOURCE: IEA Third International Mathematics and Science Study (TIMSS), 1998-1999.

Instructions: Read across the row for a participant to compare performance with the participants listed along the top of the chart. The symbols indicate whether the average achievement of the participant in the row is significantly lower than that of the comparison participant, significantly higher than that of the comparison participant, or if there is no statistically significant difference between the average achievement of the two participants.

	Singapore	Korea, Rep. of	Hong Kong, SAR	Chinese Taipei	Japan	Belgium (Flemish)	**Naperville Sch. Dist. #203, IL**	Hungary	Netherlands	Slovak Republic	**First in the World Consort., IL**	Czech Republic	Australia	Russian Federation	Slovenia	Canada	Finland	**Michigan Invitational Group, MI**	**Montgomery County, MD**	Malaysia	England	**Academy School Dist. #20, CO**	Latvia (LSS)	Italy	**Oregon**	**Project SMART Consortium, OH**	Bulgaria	New Zealand	**SW Math/Sci. Collaborative, PA**	**Michigan**	**Connecticut**	**Massachusetts**
Singapore		▲	▲	▲	▲	▲	▲	▲	▲	▲	▲	▲	▲	▲	▲	▲	▲	▲	▲	▲	▲	▲	▲	▲	▲	▲	▲	▲	▲	▲	▲	▲
Korea, Rep. of	▼		●	●	▲	▲	▲	▲	▲	▲	▲	▲	▲	▲	▲	▲	▲	▲	▲	▲	▲	▲	▲	▲	▲	▲	▲	▲	▲	▲	▲	▲
Hong Kong, SAR	▼	●		●	●	●	●	▲	▲	▲	▲	▲	▲	▲	▲	▲	▲	▲	▲	▲	▲	▲	▲	▲	▲	▲	▲	▲	▲	▲	▲	▲
Chinese Taipei	▼	●	●		●	▲	▲	▲	▲	▲	▲	▲	▲	▲	▲	▲	▲	▲	▲	▲	▲	▲	▲	▲	▲	▲	▲	▲	▲	▲	▲	▲
Japan	▼	▼	●	●		●	●	▲	●	▲	▲	▲	▲	▲	▲	▲	▲	▲	▲	▲	▲	▲	▲	▲	▲	▲	▲	▲	▲	▲	▲	▲
Belgium (Flemish)	▼	▼	●	▼	●		●	●	●	●	●	●	●	●	▲	▲	▲	▲	▲	▲	▲	▲	▲	▲	▲	▲	▲	▲	▲	▲	▲	▲
Naperville Sch. Dist. #203, IL	▼	▼	●	▼	●	●		●	●	●	●	●	▲	▲	▲	▲	▲	▲	▲	▲	▲	▲	▲	▲	▲	▲	▲	▲	▲	▲	▲	▲
Hungary	▼	▼	▼	▼	▼	●	●		●	●	●	●	●	●	●	▲	●	▲	▲	▲	▲	▲	▲	▲	▲	▲	▲	▲	▲	▲	▲	▲
Netherlands	▼	▼	▼	▼	●	●	●	●		●	●	●	●	●	●	●	●	●	●	●	▲	▲	▲	▲	▲	▲	▲	▲	▲	▲	▲	▲
Slovak Republic	▼	▼	▼	▼	▼	●	●	●	●		●	●	●	●	●	▲	●	●	▲	▲	▲	▲	▲	▲	▲	▲	▲	▲	▲	▲	▲	▲
First in the World Consort., IL	▼	▼	▼	▼	▼	●	●	●	●	●		●	●	●	●	●	●	●	●	●	▲	▲	▲	▲	▲	▲	▲	▲	▲	▲	▲	▲
Czech Republic	▼	▼	▼	▼	▼	●	●	●	●	●	●		●	●	●	●	●	●	●	●	▲	▲	▲	▲	▲	▲	▲	▲	▲	▲	▲	▲
Australia	▼	▼	▼	▼	▼	●	▼	●	●	●	●	●		●	●	●	●	●	●	●	▲	▲	▲	▲	▲	▲	▲	▲	▲	▲	▲	▲
Russian Federation	▼	▼	▼	▼	▼	●	▼	●	●	●	●	●	●		●	●	●	●	●	●	●	●	●	▲	●	●	▲	▲	▲	▲	▲	▲
Slovenia	▼	▼	▼	▼	▼	▼	▼	●	●	●	●	●	●	●		●	●	●	●	●	●	●	▲	▲	▲	●	▲	▲	▲	▲	▲	▲
Canada	▼	▼	▼	▼	▼	▼	▼	▼	●	▼	●	●	●	●	●		●	●	●	●	●	▲	▲	▲	●	●	▲	▲	▲	▲	●	▲
Finland	▼	▼	▼	▼	▼	▼	▼	●	●	●	●	●	●	●	●	●		●	●	●	●	●	●	●	●	●	●	▲	●	●	●	▲
Michigan Invitational Group, MI	▼	▼	▼	▼	▼	▼	▼	▼	●	●	●	●	●	●	●	●	●		●	●	●	●	●	●	●	●	●	●	●	●	●	●
Montgomery County, MD	▼	▼	▼	▼	▼	▼	▼	▼	●	▼	●	●	●	●	●	●	●	●		●	●	●	●	●	●	●	●	●	●	●	●	●
Malaysia	▼	▼	▼	▼	▼	▼	▼	▼	●	▼	●	●	●	●	●	●	●	●	●		●	●	●	●	●	●	●	●	●	●	●	●
England	▼	▼	▼	▼	▼	▼	▼	▼	▼	▼	▼	▼	▼	●	●	●	●	●	●	●		●	●	●	●	●	●	●	●	●	●	●
Academy School Dist. #20, CO	▼	▼	▼	▼	▼	▼	▼	▼	▼	▼	▼	▼	▼	●	●	▼	●	●	●	●	●		●	●	●	●	●	●	●	●	●	●
Latvia (LSS)	▼	▼	▼	▼	▼	▼	▼	▼	▼	▼	▼	▼	▼	●	▼	▼	●	●	●	●	●	●		●	●	●	●	●	●	●	●	●
Italy	▼	▼	▼	▼	▼	▼	▼	▼	▼	▼	▼	▼	▼	▼	▼	▼	●	●	●	●	●	●	●		●	●	●	●	●	●	●	●
Oregon	▼	▼	▼	▼	▼	▼	▼	▼	▼	▼	▼	▼	▼	●	▼	●	●	●	●	●	●	●	●	●		●	●	●	●	●	●	●
Project SMART Consortium, OH	▼	▼	▼	▼	▼	▼	▼	▼	▼	▼	▼	▼	▼	●	●	●	●	●	●	●	●	●	●	●	●		●	●	●	●	●	●
Bulgaria	▼	▼	▼	▼	▼	▼	▼	▼	▼	▼	▼	▼	▼	▼	▼	▼	●	●	●	●	●	●	●	●	●	●		●	●	●	●	●
New Zealand	▼	▼	▼	▼	▼	▼	▼	▼	▼	▼	▼	▼	▼	▼	▼	▼	▼	●	●	●	●	●	●	●	●	●	●		●	●	●	●
SW Math/Sci. Collaborative, PA	▼	▼	▼	▼	▼	▼	▼	▼	▼	▼	▼	▼	▼	▼	▼	▼	●	●	●	●	●	●	●	●	●	●	●	●		●	●	●
Michigan	▼	▼	▼	▼	▼	▼	▼	▼	▼	▼	▼	▼	▼	▼	▼	▼	●	●	●	●	●	●	●	●	●	●	●	●	●		●	●
Connecticut	▼	▼	▼	▼	▼	▼	▼	▼	▼	▼	▼	▼	▼	▼	▼	●	●	●	●	●	●	●	●	●	●	●	●	●	●	●		●
Massachusetts	▼	▼	▼	▼	▼	▼	▼	▼	▼	▼	▼	▼	▼	▼	▼	▼	▼	●	●	●	●	●	●	●	●	●	●	●	●	●	●	
Illinois	▼	▼	▼	▼	▼	▼	▼	▼	▼	▼	▼	▼	▼	▼	▼	▼	▼	●	▼	●	●	●	●	●	●	●	●	●	●	●	●	●
Romania	▼	▼	▼	▼	▼	▼	▼	▼	▼	▼	▼	▼	▼	▼	▼	▼	▼	▼	▼	▼	●	●	●	●	●	●	●	●	●	●	●	●
Pennsylvania	▼	▼	▼	▼	▼	▼	▼	▼	▼	▼	▼	▼	▼	▼	▼	▼	▼	▼	▼	▼	●	●	●	●	●	●	●	●	●	●	●	●
Indiana	▼	▼	▼	▼	▼	▼	▼	▼	▼	▼	▼	▼	▼	▼	▼	▼	▼	●	▼	●	●	●	●	●	●	●	●	●	●	●	●	●
Texas	▼	▼	▼	▼	▼	▼	▼	▼	▼	▼	▼	▼	▼	▼	▼	▼	●	●	●	●	●	●	●	●	●	●	●	●	●	●	●	●
Guilford County, NC	▼	▼	▼	▼	▼	▼	▼	▼	▼	▼	▼	▼	▼	▼	▼	▼	▼	●	▼	▼	●	●	●	●	●	●	●	●	●	●	●	●
United States	▼	▼	▼	▼	▼	▼	▼	▼	▼	▼	▼	▼	▼	▼	▼	▼	▼	▼	▼	▼	▼	▼	▼	●	●	●	●	●	●	●	●	●
Idaho	▼	▼	▼	▼	▼	▼	▼	▼	▼	▼	▼	▼	▼	▼	▼	▼	▼	▼	▼	▼	●	●	●	●	●	●	●	●	●	●	●	●
Maryland	▼	▼	▼	▼	▼	▼	▼	▼	▼	▼	▼	▼	▼	▼	▼	▼	▼	▼	▼	▼	▼	▼	▼	●	●	●	●	●	●	●	●	●
Moldova	▼	▼	▼	▼	▼	▼	▼	▼	▼	▼	▼	▼	▼	▼	▼	▼	▼	▼	▼	▼	▼	▼	▼	●	●	●	●	●	●	●	●	●
South Carolina	▼	▼	▼	▼	▼	▼	▼	▼	▼	▼	▼	▼	▼	▼	▼	▼	▼	▼	▼	▼	▼	▼	▼	●	●	●	●	●	●	●	●	●
Missouri	▼	▼	▼	▼	▼	▼	▼	▼	▼	▼	▼	▼	▼	▼	▼	▼	▼	▼	▼	▼	▼	▼	▼	▼	●	●	●	●	●	●	●	●
Fremont/Lincoln/WestSide PS, NE	▼	▼	▼	▼	▼	▼	▼	▼	▼	▼	▼	▼	▼	▼	▼	▼	▼	▼	▼	▼	▼	▼	▼	●	●	●	●	●	●	●	●	●
North Carolina	▼	▼	▼	▼	▼	▼	▼	▼	▼	▼	▼	▼	▼	▼	▼	▼	▼	▼	▼	▼	▼	▼	▼	▼	●	●	●	●	●	●	●	●
Cyprus	▼	▼	▼	▼	▼	▼	▼	▼	▼	▼	▼	▼	▼	▼	▼	▼	▼	▼	▼	▼	▼	▼	▼	▼	▼	●	▼	▼	●	●	●	●
Lithuania	▼	▼	▼	▼	▼	▼	▼	▼	▼	▼	▼	▼	▼	▼	▼	▼	▼	▼	▼	▼	▼	▼	▼	▼	▼	▼	▼	▼	▼	●	●	●
Thailand	▼	▼	▼	▼	▼	▼	▼	▼	▼	▼	▼	▼	▼	▼	▼	▼	▼	▼	▼	▼	▼	▼	▼	▼	▼	▼	▼	▼	▼	▼	●	●
Delaware Science Coalition, DE	▼	▼	▼	▼	▼	▼	▼	▼	▼	▼	▼	▼	▼	▼	▼	▼	▼	▼	▼	▼	▼	▼	▼	▼	▼	▼	▼	▼	▼	●	●	●
Israel	▼	▼	▼	▼	▼	▼	▼	▼	▼	▼	▼	▼	▼	▼	▼	▼	▼	▼	▼	▼	▼	▼	▼	▼	▼	▼	▼	▼	▼	▼	▼	▼
Macedonia, Rep. of	▼	▼	▼	▼	▼	▼	▼	▼	▼	▼	▼	▼	▼	▼	▼	▼	▼	▼	▼	▼	▼	▼	▼	▼	▼	▼	▼	▼	▼	▼	▼	▼
Jersey City Public Schools, NJ	▼	▼	▼	▼	▼	▼	▼	▼	▼	▼	▼	▼	▼	▼	▼	▼	▼	▼	▼	▼	▼	▼	▼	▼	▼	▼	▼	▼	▼	▼	▼	▼
Tunisia	▼	▼	▼	▼	▼	▼	▼	▼	▼	▼	▼	▼	▼	▼	▼	▼	▼	▼	▼	▼	▼	▼	▼	▼	▼	▼	▼	▼	▼	▼	▼	▼
Chicago Public Schools, IL	▼	▼	▼	▼	▼	▼	▼	▼	▼	▼	▼	▼	▼	▼	▼	▼	▼	▼	▼	▼	▼	▼	▼	▼	▼	▼	▼	▼	▼	▼	▼	▼
Jordan	▼	▼	▼	▼	▼	▼	▼	▼	▼	▼	▼	▼	▼	▼	▼	▼	▼	▼	▼	▼	▼	▼	▼	▼	▼	▼	▼	▼	▼	▼	▼	▼
Turkey	▼	▼	▼	▼	▼	▼	▼	▼	▼	▼	▼	▼	▼	▼	▼	▼	▼	▼	▼	▼	▼	▼	▼	▼	▼	▼	▼	▼	▼	▼	▼	▼
Rochester City Sch. Dist., NY	▼	▼	▼	▼	▼	▼	▼	▼	▼	▼	▼	▼	▼	▼	▼	▼	▼	▼	▼	▼	▼	▼	▼	▼	▼	▼	▼	▼	▼	▼	▼	▼
Chile	▼	▼	▼	▼	▼	▼	▼	▼	▼	▼	▼	▼	▼	▼	▼	▼	▼	▼	▼	▼	▼	▼	▼	▼	▼	▼	▼	▼	▼	▼	▼	▼
Miami-Dade County PS, FL	▼	▼	▼	▼	▼	▼	▼	▼	▼	▼	▼	▼	▼	▼	▼	▼	▼	▼	▼	▼	▼	▼	▼	▼	▼	▼	▼	▼	▼	▼	▼	▼
Iran, Islamic Rep.	▼	▼	▼	▼	▼	▼	▼	▼	▼	▼	▼	▼	▼	▼	▼	▼	▼	▼	▼	▼	▼	▼	▼	▼	▼	▼	▼	▼	▼	▼	▼	▼
Indonesia	▼	▼	▼	▼	▼	▼	▼	▼	▼	▼	▼	▼	▼	▼	▼	▼	▼	▼	▼	▼	▼	▼	▼	▼	▼	▼	▼	▼	▼	▼	▼	▼
Philippines	▼	▼	▼	▼	▼	▼	▼	▼	▼	▼	▼	▼	▼	▼	▼	▼	▼	▼	▼	▼	▼	▼	▼	▼	▼	▼	▼	▼	▼	▼	▼	▼
Morocco	▼	▼	▼	▼	▼	▼	▼	▼	▼	▼	▼	▼	▼	▼	▼	▼	▼	▼	▼	▼	▼	▼	▼	▼	▼	▼	▼	▼	▼	▼	▼	▼
South Africa	▼	▼	▼	▼	▼	▼	▼	▼	▼	▼	▼	▼	▼	▼	▼	▼	▼	▼	▼	▼	▼	▼	▼	▼	▼	▼	▼	▼	▼	▼	▼	▼

States in *italics* did not fully satisfy guidelines for sample participation rates (see Appendix A for details).

	Illinois	Romania	Pennsylvania	Indiana	Texas	Guilford County, NC	United States	Idaho	Maryland	Moldova	South Carolina	Missouri	Fremont/Lincoln/WestSide PS, NE	North Carolina	Cyprus	Lithuania	Thailand	Delaware Science Coalition, DE	Israel	Macedonia, Rep. of	Jersey City Public Schools, NJ	Tunisia	Chicago Public Schools, IL	Jordan	Turkey	Rochester City Sch. Dist., NY	Chile	Miami-Dade County PS, FL	Iran, Islamic Rep.	Indonesia	Philippines	Morocco	South Africa
Singapore	▲	▲	▲	▲	▲	▲	▲	▲	▲	▲	▲	▲	▲	▲	▲	▲	▲	▲	▲	▲	▲	▲	▲	▲	▲	▲	▲	▲	▲	▲	▲	▲	▲
Korea, Rep. of	▲	▲	▲	▲	▲	▲	▲	▲	▲	▲	▲	▲	▲	▲	▲	▲	▲	▲	▲	▲	▲	▲	▲	▲	▲	▲	▲	▲	▲	▲	▲	▲	▲
Hong Kong, SAR	▲	▲	▲	▲	▲	▲	▲	▲	▲	▲	▲	▲	▲	▲	▲	▲	▲	▲	▲	▲	▲	▲	▲	▲	▲	▲	▲	▲	▲	▲	▲	▲	▲
Chinese Taipei	▲	▲	▲	▲	▲	▲	▲	▲	▲	▲	▲	▲	▲	▲	▲	▲	▲	▲	▲	▲	▲	▲	▲	▲	▲	▲	▲	▲	▲	▲	▲	▲	▲
Japan	▲	▲	▲	▲	▲	▲	▲	▲	▲	▲	▲	▲	▲	▲	▲	▲	▲	▲	▲	▲	▲	▲	▲	▲	▲	▲	▲	▲	▲	▲	▲	▲	▲
Belgium (Flemish)	▲	▲	▲	▲	▲	▲	▲	▲	▲	▲	▲	▲	▲	▲	▲	▲	▲	▲	▲	▲	▲	▲	▲	▲	▲	▲	▲	▲	▲	▲	▲	▲	▲
Naperville Sch. Dist. #203, IL	▲	▲	▲	▲	▲	▲	▲	▲	▲	▲	▲	▲	▲	▲	▲	▲	▲	▲	▲	▲	▲	▲	▲	▲	▲	▲	▲	▲	▲	▲	▲	▲	▲
Hungary	▲	▲	▲	▲	▲	▲	▲	▲	▲	▲	▲	▲	▲	▲	▲	▲	▲	▲	▲	▲	▲	▲	▲	▲	▲	▲	▲	▲	▲	▲	▲	▲	▲
Netherlands	▲	▲	▲	▲	▲	▲	▲	▲	▲	▲	▲	▲	▲	▲	▲	▲	▲	▲	▲	▲	▲	▲	▲	▲	▲	▲	▲	▲	▲	▲	▲	▲	▲
Slovak Republic	▲	▲	▲	▲	▲	▲	▲	▲	▲	▲	▲	▲	▲	▲	▲	▲	▲	▲	▲	▲	▲	▲	▲	▲	▲	▲	▲	▲	▲	▲	▲	▲	▲
First in the World Consort., IL	▲	▲	▲	▲	▲	▲	▲	▲	▲	▲	▲	▲	▲	▲	▲	▲	▲	▲	▲	▲	▲	▲	▲	▲	▲	▲	▲	▲	▲	▲	▲	▲	▲
Czech Republic	▲	▲	▲	▲	▲	▲	▲	▲	▲	▲	▲	▲	▲	▲	▲	▲	▲	▲	▲	▲	▲	▲	▲	▲	▲	▲	▲	▲	▲	▲	▲	▲	▲
Australia	▲	▲	▲	▲	▲	▲	▲	▲	▲	▲	▲	▲	▲	▲	▲	▲	▲	▲	▲	▲	▲	▲	▲	▲	▲	▲	▲	▲	▲	▲	▲	▲	▲
Russian Federation	▲	▲	▲	▲	▲	▲	▲	▲	▲	▲	▲	▲	▲	▲	▲	▲	▲	▲	▲	▲	▲	▲	▲	▲	▲	▲	▲	▲	▲	▲	▲	▲	▲
Slovenia	▲	▲	▲	▲	▲	▲	▲	▲	▲	▲	▲	▲	▲	▲	▲	▲	▲	▲	▲	▲	▲	▲	▲	▲	▲	▲	▲	▲	▲	▲	▲	▲	▲
Canada	▲	▲	▲	▲	▲	▲	▲	▲	▲	▲	▲	▲	▲	▲	▲	▲	▲	▲	▲	▲	▲	▲	▲	▲	▲	▲	▲	▲	▲	▲	▲	▲	▲
Finland	▲	▲	▲	▲	●	▲	▲	▲	▲	▲	▲	▲	▲	▲	▲	▲	▲	▲	▲	▲	▲	▲	▲	▲	▲	▲	▲	▲	▲	▲	▲	▲	▲
Michigan Invitational Group, MI	●	▲	▲	●	●	●	▲	▲	▲	▲	▲	▲	▲	▲	▲	▲	▲	▲	▲	▲	▲	▲	▲	▲	▲	▲	▲	▲	▲	▲	▲	▲	▲
Montgomery County, MD	▲	▲	▲	▲	●	▲	▲	▲	▲	▲	▲	▲	▲	▲	▲	▲	▲	▲	▲	▲	▲	▲	▲	▲	▲	▲	▲	▲	▲	▲	▲	▲	▲
Malaysia	●	▲	▲	●	●	▲	▲	▲	▲	▲	▲	▲	▲	▲	▲	▲	▲	▲	▲	▲	▲	▲	▲	▲	▲	▲	▲	▲	▲	▲	▲	▲	▲
England	●	●	●	●	●	●	▲	●	▲	▲	▲	▲	▲	▲	▲	▲	▲	▲	▲	▲	▲	▲	▲	▲	▲	▲	▲	▲	▲	▲	▲	▲	▲
Academy School Dist. #20, CO	●	●	●	●	●	●	▲	●	▲	▲	▲	▲	▲	▲	▲	▲	▲	▲	▲	▲	▲	▲	▲	▲	▲	▲	▲	▲	▲	▲	▲	▲	▲
Latvia (LSS)	●	●	●	●	●	●	▲	●	▲	▲	▲	▲	▲	▲	▲	▲	▲	▲	▲	▲	▲	▲	▲	▲	▲	▲	▲	▲	▲	▲	▲	▲	▲
Italy	●	●	●	●	●	●	●	●	●	●	●	▲	●	▲	▲	▲	▲	▲	▲	▲	▲	▲	▲	▲	▲	▲	▲	▲	▲	▲	▲	▲	▲
Oregon	●	●	●	●	●	●	●	●	●	●	●	●	●	●	▲	▲	▲	▲	▲	▲	▲	▲	▲	▲	▲	▲	▲	▲	▲	▲	▲	▲	▲
Project SMART Consortium, OH	●	●	●	●	●	●	●	●	●	●	●	●	●	●	●	▲	▲	▲	▲	▲	▲	▲	▲	▲	▲	▲	▲	▲	▲	▲	▲	▲	▲
Bulgaria	●	●	●	●	●	●	●	●	●	●	●	●	●	●	▲	▲	▲	▲	▲	▲	▲	▲	▲	▲	▲	▲	▲	▲	▲	▲	▲	▲	▲
New Zealand	●	●	●	●	●	●	●	●	●	●	●	●	●	●	▲	▲	▲	▲	▲	▲	▲	▲	▲	▲	▲	▲	▲	▲	▲	▲	▲	▲	▲
SW Math/Sci. Collaborative, PA	●	●	●	●	●	●	●	●	●	●	●	●	●	●	●	▲	▲	▲	▲	▲	▲	▲	▲	▲	▲	▲	▲	▲	▲	▲	▲	▲	▲
Michigan	●	●	●	●	●	●	●	●	●	●	●	●	●	●	●	●	▲	●	▲	▲	▲	▲	▲	▲	▲	▲	▲	▲	▲	▲	▲	▲	▲
Connecticut	●	●	●	●	●	●	●	●	●	●	●	●	●	●	●	●	●	●	▲	▲	▲	▲	▲	▲	▲	▲	▲	▲	▲	▲	▲	▲	▲
Massachusetts	●	●	●	●	●	●	●	●	●	●	●	●	●	●	●	●	●	●	▲	▲	▲	▲	▲	▲	▲	▲	▲	▲	▲	▲	▲	▲	▲
Illinois		●	●	●	●	●	●	●	●	●	●	●	●	●	●	●	▲	●	▲	▲	▲	▲	▲	▲	▲	▲	▲	▲	▲	▲	▲	▲	▲
Romania	●		●	●	●	●	●	●	●	●	●	●	●	●	●	▲	▲	▲	▲	▲	▲	▲	▲	▲	▲	▲	▲	▲	▲	▲	▲	▲	▲
Pennsylvania	●	●		●	●	●	●	●	●	●	●	●	●	●	●	●	●	●	▲	▲	▲	▲	▲	▲	▲	▲	▲	▲	▲	▲	▲	▲	▲
Indiana	●	●	●		●	●	●	●	●	●	●	●	●	●	●	●	●	●	▲	▲	▲	▲	▲	▲	▲	▲	▲	▲	▲	▲	▲	▲	▲
Texas	●	●	●	●		●	●	●	●	●	●	●	●	●	●	●	●	●	●	▲	●	▲	▲	▲	▲	▲	▲	▲	▲	▲	▲	▲	▲
Guilford County, NC	●	●	●	●	●		●	●	●	●	●	●	●	●	●	●	●	●	▲	▲	▲	▲	▲	▲	▲	▲	▲	▲	▲	▲	▲	▲	▲
United States	●	●	●	●	●	●		●	●	●	●	●	●	●	●	●	●	●	▲	▲	▲	▲	▲	▲	▲	▲	▲	▲	▲	▲	▲	▲	▲
Idaho	●	●	●	●	●	●	●		●	●	●	●	●	●	●	●	●	●	●	●	●	▲	▲	▲	▲	▲	▲	▲	▲	▲	▲	▲	▲
Maryland	●	●	●	●	●	●	●	●		●	●	●	●	●	●	●	●	●	●	▲	●	▲	▲	▲	▲	▲	▲	▲	▲	▲	▲	▲	▲
Moldova	●	●	●	●	●	●	●	●	●		●	●	●	●	●	●	●	●	●	▲	●	▲	▲	▲	▲	▲	▲	▲	▲	▲	▲	▲	▲
South Carolina	●	●	●	●	●	●	●	●	●	●		●	●	●	●	●	●	●	●	●	●	▲	▲	▲	▲	▲	▲	▲	▲	▲	▲	▲	▲
Missouri	●	●	●	●	●	●	●	●	●	●	●		●	●	●	●	●	●	●	●	●	▲	▲	▲	▲	▲	▲	▲	▲	▲	▲	▲	▲
Fremont/Lincoln/WestSide PS, NE	●	●	●	●	●	●	●	●	●	●	●	●		●	●	●	●	●	●	●	●	▲	●	▲	▲	▲	▲	▲	▲	▲	▲	▲	▲
North Carolina	●	●	●	●	●	●	●	●	●	●	●	●	●		●	●	●	●	●	●	●	▲	●	▲	▲	▲	▲	▲	▲	▲	▲	▲	▲
Cyprus	●	●	●	●	●	●	●	●	●	●	●	●	●	●		●	●	●	●	●	●	▲	▲	▲	▲	▲	▲	▲	▲	▲	▲	▲	▲
Lithuania	●	▼	●	●	●	●	●	●	●	●	●	●	●	●	●		●	●	●	●	●	▲	●	▲	▲	▲	▲	▲	▲	▲	▲	▲	▲
Thailand	▼	▼	●	●	●	●	●	●	●	●	●	●	●	●	●	●		●	●	●	●	●	●	▲	●	▲	▲	▲	▲	▲	▲	▲	▲
Delaware Science Coalition, DE	●	▼	●	●	●	●	●	●	●	●	●	●	●	●	●	●	●		●	●	●	●	●	●	●	▲	▲	▲	▲	▲	▲	▲	▲
Israel	▼	▼	▼	▼	●	▼	▼	●	●	●	●	●	●	●	●	●	●	●		●	●	●	●	●	●	▲	▲	▲	▲	▲	▲	▲	▲
Macedonia, Rep. of	▼	▼	▼	▼	▼	▼	▼	●	▼	▼	●	●	●	●	●	●	●	●	●		●	●	●	●	●	▲	▲	▲	▲	▲	▲	▲	▲
Jersey City Public Schools, NJ	▼	▼	▼	▼	●	▼	▼	●	●	●	●	●	●	●	●	●	●	●	●	●		●	●	●	●	●	▲	▲	▲	▲	▲	▲	▲
Tunisia	▼	▼	▼	▼	▼	▼	▼	▼	▼	▼	▼	▼	▼	▼	▼	▼	●	●	●	●	●		●	●	●	▲	▲	▲	▲	▲	▲	▲	▲
Chicago Public Schools, IL	▼	▼	▼	▼	▼	▼	▼	▼	▼	▼	▼	▼	●	●	▼	●	●	●	●	●	●	●		●	●	●	●	●	▲	▲	▲	▲	▲
Jordan	▼	▼	▼	▼	▼	▼	▼	▼	▼	▼	▼	▼	▼	▼	▼	▼	▼	●	●	●	●	●	●		●	●	▲	●	▲	▲	▲	▲	▲
Turkey	▼	▼	▼	▼	▼	▼	▼	▼	▼	▼	▼	▼	▼	▼	▼	▼	●	●	●	●	●	●	●	●		●	●	●	▲	▲	▲	▲	▲
Rochester City Sch. Dist., NY	▼	▼	▼	▼	▼	▼	▼	▼	▼	▼	▼	▼	▼	▼	▼	▼	▼	▼	▼	▼	●	▼	●	●	●		●	●	●	●	▲	▲	▲
Chile	▼	▼	▼	▼	▼	▼	▼	▼	▼	▼	▼	▼	▼	▼	▼	▼	▼	▼	▼	▼	▼	▼	●	▼	●	●		●	●	●	▲	▲	▲
Miami-Dade County PS, FL	▼	▼	▼	▼	▼	▼	▼	▼	▼	▼	▼	▼	▼	▼	▼	▼	▼	▼	▼	▼	▼	▼	●	●	●	●	●		●	●	▲	▲	▲
Iran, Islamic Rep.	▼	▼	▼	▼	▼	▼	▼	▼	▼	▼	▼	▼	▼	▼	▼	▼	▼	▼	▼	▼	▼	▼	▼	▼	▼	●	●	●		●	▲	▲	▲
Indonesia	▼	▼	▼	▼	▼	▼	▼	▼	▼	▼	▼	▼	▼	▼	▼	▼	▼	▼	▼	▼	▼	▼	▼	▼	▼	●	●	●	●		▲	▲	▲
Philippines	▼	▼	▼	▼	▼	▼	▼	▼	▼	▼	▼	▼	▼	▼	▼	▼	▼	▼	▼	▼	▼	▼	▼	▼	▼	▼	▼	▼	▼	▼		●	▲
Morocco	▼	▼	▼	▼	▼	▼	▼	▼	▼	▼	▼	▼	▼	▼	▼	▼	▼	▼	▼	▼	▼	▼	▼	▼	▼	▼	▼	▼	▼	▼	●		▲
South Africa	▼	▼	▼	▼	▼	▼	▼	▼	▼	▼	▼	▼	▼	▼	▼	▼	▼	▼	▼	▼	▼	▼	▼	▼	▼	▼	▼	▼	▼	▼	▼	▼	

▲ Average achievement significantly higher than comparison participant

● No statistically significant difference from comparison participant

▼ Average achievement significantly lower than comparison participant

Significance tests adjusted for multiple comparisons

SOURCE: IEA Third International Mathematics and Science Study (TIMSS), 1998-1999.

Exhibit B.3 Multiple Comparisons of Average Achievement in Data Representation, Analysis, and Probability

8th Grade Mathematics

Instructions: Read across the row for a participant to compare performance with the participants listed along the top of the chart. The symbols indicate whether the average achievement of the participant in the row is significantly lower than that of the comparison participant, significantly higher than that of the comparison participant, or if there is no statistically significant difference between the average achievement of the two participants.

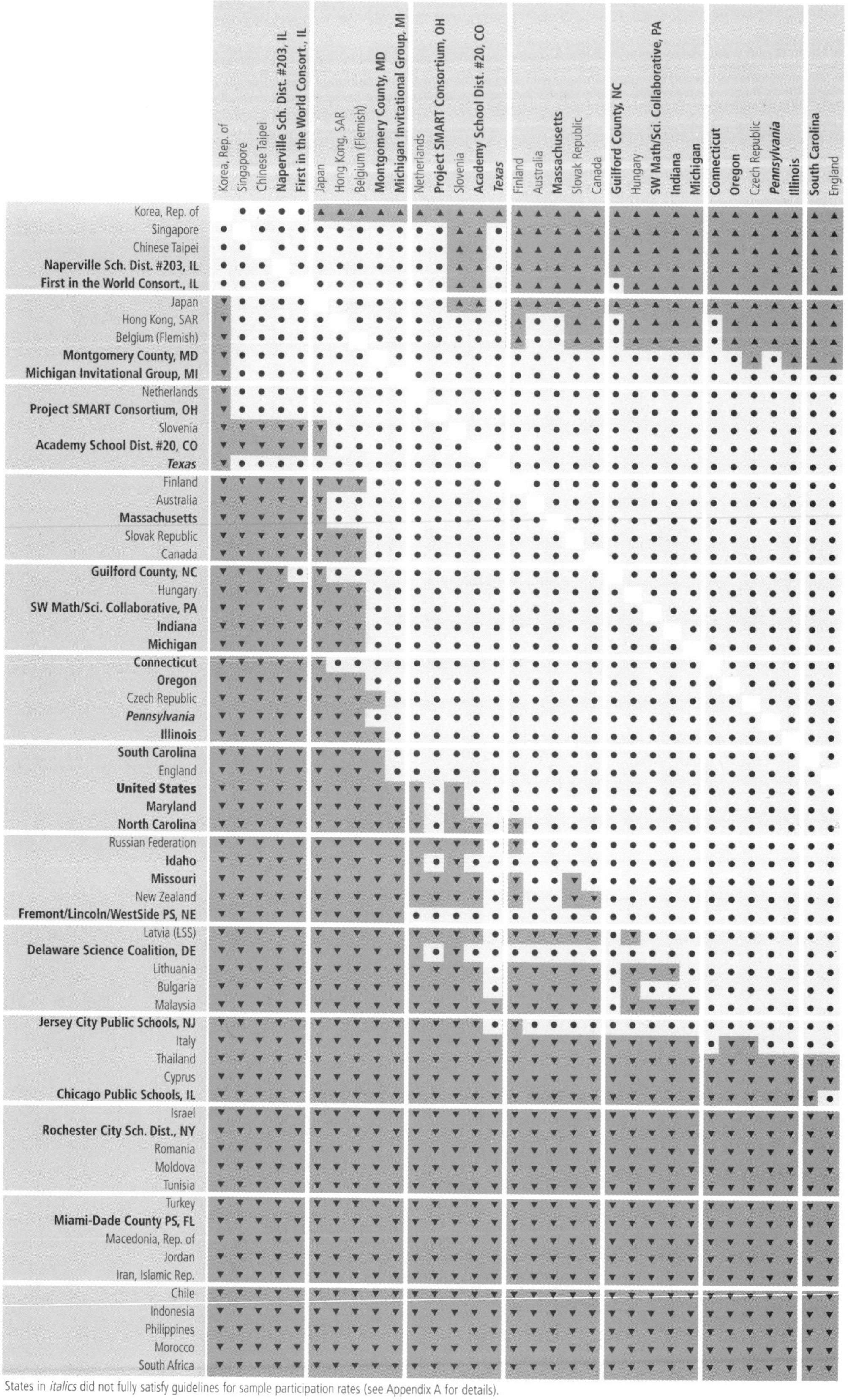

	Korea, Rep. of	Singapore	Chinese Taipei	**Naperville Sch. Dist. #203, IL**	**First in the World Consort., IL**	Japan	Hong Kong, SAR	Belgium (Flemish)	**Montgomery County, MD**	**Michigan Invitational Group, MI**	Netherlands	**Project SMART Consortium, OH**	Slovenia	**Academy School Dist. #20, CO**	***Texas***	Finland	Australia	**Massachusetts**	Slovak Republic	Canada	**Guilford County, NC**	Hungary	**SW Math/Sci. Collaborative, PA**	**Indiana**	**Michigan**	**Connecticut**	**Oregon**	Czech Republic	***Pennsylvania***	**Illinois**	**South Carolina**	England
Korea, Rep. of		●	●	●	●	▲	▲	▲	▲	▲	▲	▲	▲	▲	▲	▲	▲	▲	▲	▲	▲	▲	▲	▲	▲	▲	▲	▲	▲	▲	▲	▲
Singapore	●		●	●	●	●	●	●	●	●	●	●	▲	▲	●	▲	▲	▲	▲	▲	▲	▲	▲	▲	▲	▲	▲	▲	▲	▲	▲	▲
Chinese Taipei	●	●		●	●	●	●	●	●	●	●	●	▲	▲	●	▲	▲	▲	▲	▲	▲	▲	▲	▲	▲	▲	▲	▲	▲	▲	▲	▲
Naperville Sch. Dist. #203, IL	●	●	●		●	●	●	●	●	●	●	●	▲	▲	●	▲	▲	▲	▲	▲	▲	▲	▲	▲	▲	▲	▲	▲	▲	▲	▲	▲
First in the World Consort., IL	●	●	●	●		●	●	●	●	●	●	●	▲	▲	●	▲	▲	▲	▲	▲	●	▲	▲	▲	▲	▲	▲	▲	▲	▲	▲	▲
Japan	▼	●	●	●	●		●	●	●	●	●	●	▲	▲	●	▲	▲	▲	▲	▲	▲	▲	▲	▲	▲	▲	▲	▲	▲	▲	▲	▲
Hong Kong, SAR	▼	●	●	●	●	●		●	●	●	●	●	●	●	●	▲	●	●	▲	▲	●	▲	▲	▲	▲	●	▲	▲	▲	▲	▲	▲
Belgium (Flemish)	▼	●	●	●	●	●	●		●	●	●	●	●	●	●	▲	●	●	▲	▲	●	▲	▲	▲	▲	●	▲	▲	▲	▲	▲	▲
Montgomery County, MD	▼	●	●	●	●	●	●	●		●	●	●	●	●	●	●	●	●	●	●	●	●	●	●	●	●	●	▲	●	▲	▲	▲
Michigan Invitational Group, MI	▼	●	●	●	●	●	●	●	●		●	●	●	●	●	●	●	●	●	●	●	●	●	●	●	●	●	●	●	●	●	●
Netherlands	▼	●	●	●	●	●	●	●	●	●		●	●	●	●	●	●	●	●	●	●	●	●	●	●	●	●	●	●	●	●	●
Project SMART Consortium, OH	▼	●	●	●	●	●	●	●	●	●	●		●	●	●	●	●	●	●	●	●	●	●	●	●	●	●	●	●	●	●	●
Slovenia	▼	▼	▼	▼	▼	▼	●	●	●	●	●	●		●	●	●	●	●	●	●	●	●	●	●	●	●	●	●	●	●	●	●
Academy School Dist. #20, CO	▼	▼	▼	▼	▼	▼	●	●	●	●	●	●	●		●	●	●	●	●	●	●	●	●	●	●	●	●	●	●	●	●	●
Texas	▼	●	●	●	●	●	●	●	●	●	●	●	●	●		●	●	●	●	●	●	●	●	●	●	●	●	●	●	●	●	●
Finland	▼	▼	▼	▼	▼	▼	▼	▼	●	●	●	●	●	●	●		●	●	●	●	●	●	●	●	●	●	●	●	●	●	●	●
Australia	▼	▼	▼	▼	▼	▼	●	●	●	●	●	●	●	●	●	●		●	●	●	●	●	●	●	●	●	●	●	●	●	●	●
Massachusetts	▼	▼	▼	▼	▼	▼	●	●	●	●	●	●	●	●	●	●	●		●	●	●	●	●	●	●	●	●	●	●	●	●	●
Slovak Republic	▼	▼	▼	▼	▼	▼	▼	▼	●	●	●	●	●	●	●	●	●	●		●	●	●	●	●	●	●	●	●	●	●	●	●
Canada	▼	▼	▼	▼	▼	▼	▼	▼	●	●	●	●	●	●	●	●	●	●	●		●	●	●	●	●	●	●	●	●	●	●	●
Guilford County, NC	▼	▼	▼	▼	●	▼	●	●	●	●	●	●	●	●	●	●	●	●	●	●		●	●	●	●	●	●	●	●	●	●	●
Hungary	▼	▼	▼	▼	▼	▼	▼	▼	●	●	●	●	●	●	●	●	●	●	●	●	●		●	●	●	●	●	●	●	●	●	●
SW Math/Sci. Collaborative, PA	▼	▼	▼	▼	▼	▼	▼	▼	●	●	●	●	●	●	●	●	●	●	●	●	●	●		●	●	●	●	●	●	●	●	●
Indiana	▼	▼	▼	▼	▼	▼	▼	▼	●	●	●	●	●	●	●	●	●	●	●	●	●	●	●		●	●	●	●	●	●	●	●
Michigan	▼	▼	▼	▼	▼	▼	▼	▼	●	●	●	●	●	●	●	●	●	●	●	●	●	●	●	●		●	●	●	●	●	●	●
Connecticut	▼	▼	▼	▼	▼	▼	●	●	●	●	●	●	●	●	●	●	●	●	●	●	●	●	●	●	●		●	●	●	●	●	●
Oregon	▼	▼	▼	▼	▼	▼	▼	▼	●	●	●	●	●	●	●	●	●	●	●	●	●	●	●	●	●	●		●	●	●	●	●
Czech Republic	▼	▼	▼	▼	▼	▼	▼	▼	▼	●	●	●	●	●	●	●	●	●	●	●	●	●	●	●	●	●	●		●	●	●	●
Pennsylvania	▼	▼	▼	▼	▼	▼	▼	▼	●	●	●	●	●	●	●	●	●	●	●	●	●	●	●	●	●	●	●	●		●	●	●
Illinois	▼	▼	▼	▼	▼	▼	▼	▼	▼	●	●	●	●	●	●	●	●	●	●	●	●	●	●	●	●	●	●	●	●		●	●
South Carolina	▼	▼	▼	▼	▼	▼	▼	▼	▼	●	●	●	●	●	●	●	●	●	●	●	●	●	●	●	●	●	●	●	●	●		●
England	▼	▼	▼	▼	▼	▼	▼	▼	▼	●	●	●	●	●	●	●	●	●	●	●	●	●	●	●	●	●	●	●	●	●	●	
United States	▼	▼	▼	▼	▼	▼	▼	▼	▼	▼	▼	●	▼	●	●	●	●	●	●	●	●	●	●	●	●	●	●	●	●	●	●	●
Maryland	▼	▼	▼	▼	▼	▼	▼	▼	▼	▼	▼	●	▼	●	●	●	●	●	●	●	●	●	●	●	●	●	●	●	●	●	●	●
North Carolina	▼	▼	▼	▼	▼	▼	▼	▼	▼	▼	▼	●	▼	▼	●	▼	●	●	●	●	●	●	●	●	●	●	●	●	●	●	●	●
Russian Federation	▼	▼	▼	▼	▼	▼	▼	▼	▼	▼	▼	▼	▼	▼	●	▼	●	●	●	●	●	●	●	●	●	●	●	●	●	●	●	●
Idaho	▼	▼	▼	▼	▼	▼	▼	▼	▼	▼	▼	●	▼	●	●	●	●	●	●	●	●	●	●	●	●	●	●	●	●	●	●	●
Missouri	▼	▼	▼	▼	▼	▼	▼	▼	▼	▼	▼	▼	▼	▼	●	▼	●	●	▼	●	●	●	●	●	●	●	●	●	●	●	●	●
New Zealand	▼	▼	▼	▼	▼	▼	▼	▼	▼	▼	▼	▼	▼	▼	●	▼	●	●	▼	▼	●	●	●	●	●	●	●	●	●	●	●	●
Fremont/Lincoln/WestSide PS, NE	▼	▼	▼	▼	▼	▼	▼	▼	▼	▼	●	●	●	●	●	●	●	●	●	●	●	●	●	●	●	●	●	●	●	●	●	●
Latvia (LSS)	▼	▼	▼	▼	▼	▼	▼	▼	▼	▼	▼	▼	▼	▼	●	▼	▼	▼	▼	▼	●	▼	●	●	●	●	●	●	●	●	●	●
Delaware Science Coalition, DE	▼	▼	▼	▼	▼	▼	▼	▼	▼	▼	▼	●	▼	●	●	●	●	●	●	●	●	●	●	●	●	●	●	●	●	●	●	●
Lithuania	▼	▼	▼	▼	▼	▼	▼	▼	▼	▼	▼	▼	▼	▼	●	▼	▼	▼	▼	▼	●	▼	▼	▼	●	●	●	●	●	●	●	●
Bulgaria	▼	▼	▼	▼	▼	▼	▼	▼	▼	▼	▼	▼	▼	▼	●	▼	▼	▼	▼	▼	●	▼	●	●	●	●	●	●	●	●	●	●
Malaysia	▼	▼	▼	▼	▼	▼	▼	▼	▼	▼	▼	▼	▼	▼	▼	▼	▼	▼	▼	▼	●	▼	▼	▼	▼	●	●	●	●	●	●	●
Jersey City Public Schools, NJ	▼	▼	▼	▼	▼	▼	▼	▼	▼	▼	▼	▼	▼	▼	●	▼	●	●	●	●	●	●	●	●	●	●	●	●	●	●	●	●
Italy	▼	▼	▼	▼	▼	▼	▼	▼	▼	▼	▼	▼	▼	▼	▼	▼	▼	▼	▼	▼	▼	▼	▼	▼	▼	●	▼	▼	●	●	●	●
Thailand	▼	▼	▼	▼	▼	▼	▼	▼	▼	▼	▼	▼	▼	▼	▼	▼	▼	▼	▼	▼	▼	▼	▼	▼	▼	▼	▼	▼	▼	▼	▼	▼
Cyprus	▼	▼	▼	▼	▼	▼	▼	▼	▼	▼	▼	▼	▼	▼	▼	▼	▼	▼	▼	▼	▼	▼	▼	▼	▼	▼	▼	▼	▼	▼	▼	▼
Chicago Public Schools, IL	▼	▼	▼	▼	▼	▼	▼	▼	▼	▼	▼	▼	▼	▼	▼	▼	▼	▼	▼	▼	▼	▼	▼	▼	▼	▼	▼	▼	▼	▼	▼	●
Israel	▼	▼	▼	▼	▼	▼	▼	▼	▼	▼	▼	▼	▼	▼	▼	▼	▼	▼	▼	▼	▼	▼	▼	▼	▼	▼	▼	▼	▼	▼	▼	▼
Rochester City Sch. Dist., NY	▼	▼	▼	▼	▼	▼	▼	▼	▼	▼	▼	▼	▼	▼	▼	▼	▼	▼	▼	▼	▼	▼	▼	▼	▼	▼	▼	▼	▼	▼	▼	▼
Romania	▼	▼	▼	▼	▼	▼	▼	▼	▼	▼	▼	▼	▼	▼	▼	▼	▼	▼	▼	▼	▼	▼	▼	▼	▼	▼	▼	▼	▼	▼	▼	▼
Moldova	▼	▼	▼	▼	▼	▼	▼	▼	▼	▼	▼	▼	▼	▼	▼	▼	▼	▼	▼	▼	▼	▼	▼	▼	▼	▼	▼	▼	▼	▼	▼	▼
Tunisia	▼	▼	▼	▼	▼	▼	▼	▼	▼	▼	▼	▼	▼	▼	▼	▼	▼	▼	▼	▼	▼	▼	▼	▼	▼	▼	▼	▼	▼	▼	▼	▼
Turkey	▼	▼	▼	▼	▼	▼	▼	▼	▼	▼	▼	▼	▼	▼	▼	▼	▼	▼	▼	▼	▼	▼	▼	▼	▼	▼	▼	▼	▼	▼	▼	▼
Miami-Dade County PS, FL	▼	▼	▼	▼	▼	▼	▼	▼	▼	▼	▼	▼	▼	▼	▼	▼	▼	▼	▼	▼	▼	▼	▼	▼	▼	▼	▼	▼	▼	▼	▼	▼
Macedonia, Rep. of	▼	▼	▼	▼	▼	▼	▼	▼	▼	▼	▼	▼	▼	▼	▼	▼	▼	▼	▼	▼	▼	▼	▼	▼	▼	▼	▼	▼	▼	▼	▼	▼
Jordan	▼	▼	▼	▼	▼	▼	▼	▼	▼	▼	▼	▼	▼	▼	▼	▼	▼	▼	▼	▼	▼	▼	▼	▼	▼	▼	▼	▼	▼	▼	▼	▼
Iran, Islamic Rep.	▼	▼	▼	▼	▼	▼	▼	▼	▼	▼	▼	▼	▼	▼	▼	▼	▼	▼	▼	▼	▼	▼	▼	▼	▼	▼	▼	▼	▼	▼	▼	▼
Chile	▼	▼	▼	▼	▼	▼	▼	▼	▼	▼	▼	▼	▼	▼	▼	▼	▼	▼	▼	▼	▼	▼	▼	▼	▼	▼	▼	▼	▼	▼	▼	▼
Indonesia	▼	▼	▼	▼	▼	▼	▼	▼	▼	▼	▼	▼	▼	▼	▼	▼	▼	▼	▼	▼	▼	▼	▼	▼	▼	▼	▼	▼	▼	▼	▼	▼
Philippines	▼	▼	▼	▼	▼	▼	▼	▼	▼	▼	▼	▼	▼	▼	▼	▼	▼	▼	▼	▼	▼	▼	▼	▼	▼	▼	▼	▼	▼	▼	▼	▼
Morocco	▼	▼	▼	▼	▼	▼	▼	▼	▼	▼	▼	▼	▼	▼	▼	▼	▼	▼	▼	▼	▼	▼	▼	▼	▼	▼	▼	▼	▼	▼	▼	▼
South Africa	▼	▼	▼	▼	▼	▼	▼	▼	▼	▼	▼	▼	▼	▼	▼	▼	▼	▼	▼	▼	▼	▼	▼	▼	▼	▼	▼	▼	▼	▼	▼	▼

States in *italics* did not fully satisfy guidelines for sample participation rates (see Appendix A for details).

	United States	Maryland	North Carolina	Russian Federation	Idaho	Missouri	New Zealand	Fremont/Lincoln/WestSide PS, NE	Latvia (LSS)	Delaware Science Coalition, DE	Lithuania	Bulgaria	Malaysia	Jersey City Public Schools, NJ	Italy	Thailand	Cyprus	Chicago Public Schools, IL	Israel	Rochester City Sch. Dist., NY	Romania	Moldova	Tunisia	Turkey	Miami-Dade County PS, FL	Macedonia, Rep. of	Jordan	Iran, Islamic Rep.	Chile	Indonesia	Philippines	Morocco	South Africa
Korea, Rep. of	▲	▲	▲	▲	▲	▲	▲	▲	▲	▲	▲	▲	▲	▲	▲	▲	▲	▲	▲	▲	▲	▲	▲	▲	▲	▲	▲	▲	▲	▲	▲	▲	▲
Singapore	▲	▲	▲	▲	▲	▲	▲	▲	▲	▲	▲	▲	▲	▲	▲	▲	▲	▲	▲	▲	▲	▲	▲	▲	▲	▲	▲	▲	▲	▲	▲	▲	▲
Chinese Taipei	▲	▲	▲	▲	▲	▲	▲	▲	▲	▲	▲	▲	▲	▲	▲	▲	▲	▲	▲	▲	▲	▲	▲	▲	▲	▲	▲	▲	▲	▲	▲	▲	▲
Naperville Sch. Dist. #203, IL	▲	▲	▲	▲	▲	▲	▲	▲	▲	▲	▲	▲	▲	▲	▲	▲	▲	▲	▲	▲	▲	▲	▲	▲	▲	▲	▲	▲	▲	▲	▲	▲	▲
First in the World Consort., IL	▲	▲	▲	▲	▲	▲	▲	▲	▲	▲	▲	▲	▲	▲	▲	▲	▲	▲	▲	▲	▲	▲	▲	▲	▲	▲	▲	▲	▲	▲	▲	▲	▲
Japan	▲	▲	▲	▲	▲	▲	▲	▲	▲	▲	▲	▲	▲	▲	▲	▲	▲	▲	▲	▲	▲	▲	▲	▲	▲	▲	▲	▲	▲	▲	▲	▲	▲
Hong Kong, SAR	▲	▲	▲	▲	▲	▲	▲	▲	▲	▲	▲	▲	▲	▲	▲	▲	▲	▲	▲	▲	▲	▲	▲	▲	▲	▲	▲	▲	▲	▲	▲	▲	▲
Belgium (Flemish)	▲	▲	▲	▲	▲	▲	▲	▲	▲	▲	▲	▲	▲	▲	▲	▲	▲	▲	▲	▲	▲	▲	▲	▲	▲	▲	▲	▲	▲	▲	▲	▲	▲
Montgomery County, MD	▲	▲	▲	▲	▲	▲	▲	▲	▲	▲	▲	▲	▲	▲	▲	▲	▲	▲	▲	▲	▲	▲	▲	▲	▲	▲	▲	▲	▲	▲	▲	▲	▲
Michigan Invitational Group, MI	▲	▲	▲	▲	▲	▲	▲	▲	▲	▲	▲	▲	▲	▲	▲	▲	▲	▲	▲	▲	▲	▲	▲	▲	▲	▲	▲	▲	▲	▲	▲	▲	▲
Netherlands	▲	▲	▲	▲	▲	▲	▲	●	▲	▲	▲	▲	▲	▲	▲	▲	▲	▲	▲	▲	▲	▲	▲	▲	▲	▲	▲	▲	▲	▲	▲	▲	▲
Project SMART Consortium, OH	●	●	●	▲	●	▲	▲	●	▲	●	▲	▲	▲	▲	▲	▲	▲	▲	▲	▲	▲	▲	▲	▲	▲	▲	▲	▲	▲	▲	▲	▲	▲
Slovenia	▲	▲	▲	▲	▲	▲	▲	●	▲	▲	▲	▲	▲	▲	▲	▲	▲	▲	▲	▲	▲	▲	▲	▲	▲	▲	▲	▲	▲	▲	▲	▲	▲
Academy School Dist. #20, CO	●	●	▲	▲	●	▲	▲	●	▲	●	▲	▲	▲	▲	▲	▲	▲	▲	▲	▲	▲	▲	▲	▲	▲	▲	▲	▲	▲	▲	▲	▲	▲
Texas	●	●	●	●	●	●	●	●	●	●	●	●	▲	●	▲	▲	▲	▲	▲	▲	▲	▲	▲	▲	▲	▲	▲	▲	▲	▲	▲	▲	▲
Finland	●	●	▲	▲	●	▲	▲	●	▲	●	▲	▲	▲	▲	▲	▲	▲	▲	▲	▲	▲	▲	▲	▲	▲	▲	▲	▲	▲	▲	▲	▲	▲
Australia	●	●	●	●	●	●	●	●	▲	●	▲	▲	▲	●	▲	▲	▲	▲	▲	▲	▲	▲	▲	▲	▲	▲	▲	▲	▲	▲	▲	▲	▲
Massachusetts	●	●	●	●	●	●	●	●	▲	●	▲	▲	▲	●	▲	▲	▲	▲	▲	▲	▲	▲	▲	▲	▲	▲	▲	▲	▲	▲	▲	▲	▲
Slovak Republic	●	●	●	●	●	▲	▲	●	▲	●	▲	▲	▲	●	▲	▲	▲	▲	▲	▲	▲	▲	▲	▲	▲	▲	▲	▲	▲	▲	▲	▲	▲
Canada	●	●	●	●	●	●	▲	●	▲	●	▲	▲	▲	●	▲	▲	▲	▲	▲	▲	▲	▲	▲	▲	▲	▲	▲	▲	▲	▲	▲	▲	▲
Guilford County, NC	●	●	●	●	●	●	●	●	●	●	●	●	●	●	▲	▲	▲	▲	▲	▲	▲	▲	▲	▲	▲	▲	▲	▲	▲	▲	▲	▲	▲
Hungary	●	●	●	●	●	●	●	●	▲	●	▲	▲	▲	●	▲	▲	▲	▲	▲	▲	▲	▲	▲	▲	▲	▲	▲	▲	▲	▲	▲	▲	▲
SW Math/Sci. Collaborative, PA	●	●	●	●	●	●	●	●	●	●	▲	●	▲	●	▲	▲	▲	▲	▲	▲	▲	▲	▲	▲	▲	▲	▲	▲	▲	▲	▲	▲	▲
Indiana	●	●	●	●	●	●	●	●	●	●	▲	●	▲	●	▲	▲	▲	▲	▲	▲	▲	▲	▲	▲	▲	▲	▲	▲	▲	▲	▲	▲	▲
Michigan	●	●	●	●	●	●	●	●	●	●	●	●	▲	●	▲	▲	▲	▲	▲	▲	▲	▲	▲	▲	▲	▲	▲	▲	▲	▲	▲	▲	▲
Connecticut	●	●	●	●	●	●	●	●	●	●	●	●	●	●	●	▲	▲	▲	▲	▲	▲	▲	▲	▲	▲	▲	▲	▲	▲	▲	▲	▲	▲
Oregon	●	●	●	●	●	●	●	●	●	●	●	●	●	●	▲	▲	▲	▲	▲	▲	▲	▲	▲	▲	▲	▲	▲	▲	▲	▲	▲	▲	▲
Czech Republic	●	●	●	●	●	●	●	●	●	●	●	●	●	●	▲	▲	▲	▲	▲	▲	▲	▲	▲	▲	▲	▲	▲	▲	▲	▲	▲	▲	▲
Pennsylvania	●	●	●	●	●	●	●	●	●	●	●	●	●	●	●	▲	▲	▲	▲	▲	▲	▲	▲	▲	▲	▲	▲	▲	▲	▲	▲	▲	▲
Illinois	●	●	●	●	●	●	●	●	●	●	●	●	●	●	●	▲	▲	▲	▲	▲	▲	▲	▲	▲	▲	▲	▲	▲	▲	▲	▲	▲	▲
South Carolina	●	●	●	●	●	●	●	●	●	●	●	●	●	●	●	▲	▲	▲	▲	▲	▲	▲	▲	▲	▲	▲	▲	▲	▲	▲	▲	▲	▲
England	●	●	●	●	●	●	●	●	●	●	●	●	●	●	●	▲	▲	●	▲	▲	▲	▲	▲	▲	▲	▲	▲	▲	▲	▲	▲	▲	▲
United States		●	●	●	●	●	●	●	●	●	●	●	●	●	●	▲	▲	▲	▲	▲	▲	▲	▲	▲	▲	▲	▲	▲	▲	▲	▲	▲	▲
Maryland	●		●	●	●	●	●	●	●	●	●	●	●	●	●	▲	▲	▲	▲	▲	▲	▲	▲	▲	▲	▲	▲	▲	▲	▲	▲	▲	▲
North Carolina	●	●		●	●	●	●	●	●	●	●	●	●	●	●	▲	▲	▲	▲	▲	▲	▲	▲	▲	▲	▲	▲	▲	▲	▲	▲	▲	▲
Russian Federation	●	●	●		●	●	●	●	●	●	●	●	●	●	●	▲	▲	▲	▲	▲	▲	▲	▲	▲	▲	▲	▲	▲	▲	▲	▲	▲	▲
Idaho	●	●	●	●		●	●	●	●	●	●	●	●	●	●	●	▲	●	▲	▲	▲	▲	▲	▲	▲	▲	▲	▲	▲	▲	▲	▲	▲
Missouri	●	●	●	●	●		●	●	●	●	●	●	●	●	●	▲	▲	●	▲	▲	▲	▲	▲	▲	▲	▲	▲	▲	▲	▲	▲	▲	▲
New Zealand	●	●	●	●	●	●		●	●	●	●	●	●	●	●	▲	▲	●	▲	▲	▲	▲	▲	▲	▲	▲	▲	▲	▲	▲	▲	▲	▲
Fremont/Lincoln/WestSide PS, NE	●	●	●	●	●	●	●		●	●	●	●	●	●	●	●	●	●	●	●	▲	▲	▲	▲	▲	▲	▲	▲	▲	▲	▲	▲	▲
Latvia (LSS)	●	●	●	●	●	●	●	●		●	●	●	●	●	●	●	▲	●	▲	▲	▲	▲	▲	▲	▲	▲	▲	▲	▲	▲	▲	▲	▲
Delaware Science Coalition, DE	●	●	●	●	●	●	●	●	●		●	●	●	●	●	●	●	●	●	●	▲	▲	▲	▲	▲	▲	▲	▲	▲	▲	▲	▲	▲
Lithuania	●	●	●	●	●	●	●	●	●	●		●	●	●	●	●	▲	●	▲	▲	▲	▲	▲	▲	▲	▲	▲	▲	▲	▲	▲	▲	▲
Bulgaria	●	●	●	●	●	●	●	●	●	●	●		●	●	●	●	●	●	●	●	▲	▲	▲	▲	▲	▲	▲	▲	▲	▲	▲	▲	▲
Malaysia	●	●	●	●	●	●	●	●	●	●	●	●		●	●	●	●	●	▲	▲	▲	▲	▲	▲	▲	▲	▲	▲	▲	▲	▲	▲	▲
Jersey City Public Schools, NJ	●	●	●	●	●	●	●	●	●	●	●	●	●		●	●	●	●	●	●	▲	▲	▲	▲	▲	▲	▲	▲	▲	▲	▲	▲	▲
Italy	●	●	●	●	●	●	●	●	●	●	●	●	●	●		●	●	●	●	●	▲	▲	▲	▲	▲	▲	▲	▲	▲	▲	▲	▲	▲
Thailand	▼	▼	▼	▼	●	▼	▼	●	●	●	●	●	●	●	●		●	●	●	●	▲	▲	▲	▲	●	▲	▲	▲	▲	▲	▲	▲	▲
Cyprus	▼	▼	▼	▼	▼	▼	▼	●	▼	●	▼	●	●	●	●	●		●	●	●	●	●	▲	▲	●	▲	▲	▲	▲	▲	▲	▲	▲
Chicago Public Schools, IL	▼	▼	▼	▼	●	●	●	●	●	●	●	●	●	●	●	●	●		●	●	●	●	●	▲	●	●	▲	▲	▲	▲	▲	▲	▲
Israel	▼	▼	▼	▼	▼	▼	▼	●	▼	●	▼	●	▼	●	●	●	●	●		●	●	●	●	▲	●	▲	▲	▲	▲	▲	▲	▲	▲
Rochester City Sch. Dist., NY	▼	▼	▼	▼	▼	▼	▼	●	▼	●	▼	●	▼	●	●	●	●	●	●		●	●	●	●	●	●	●	▲	▲	▲	▲	▲	▲
Romania	▼	▼	▼	▼	▼	▼	▼	▼	▼	▼	▼	▼	▼	▼	▼	▼	●	●	●	●		●	●	●	●	●	●	●	▲	▲	▲	▲	▲
Moldova	▼	▼	▼	▼	▼	▼	▼	▼	▼	▼	▼	▼	▼	▼	▼	▼	●	●	●	●	●		●	●	●	●	●	●	●	▲	▲	▲	▲
Tunisia	▼	▼	▼	▼	▼	▼	▼	▼	▼	▼	▼	▼	▼	▼	▼	▼	▼	●	●	●	●	●		●	●	●	●	●	●	▲	▲	▲	▲
Turkey	▼	▼	▼	▼	▼	▼	▼	▼	▼	▼	▼	▼	▼	▼	▼	▼	▼	▼	▼	●	●	●	●		●	●	●	●	▲	▲	▲	▲	▲
Miami-Dade County PS, FL	▼	▼	▼	▼	▼	▼	▼	▼	▼	▼	▼	▼	▼	▼	▼	●	●	●	●	●	●	●	●	●		●	●	●	●	●	▲	▲	▲
Macedonia, Rep. of	▼	▼	▼	▼	▼	▼	▼	▼	▼	▼	▼	▼	▼	▼	▼	▼	▼	●	▼	●	●	●	●	●	●		●	●	●	●	▲	▲	▲
Jordan	▼	▼	▼	▼	▼	▼	▼	▼	▼	▼	▼	▼	▼	▼	▼	▼	▼	▼	▼	●	●	●	●	●	●	●		●	●	●	▲	▲	▲
Iran, Islamic Rep.	▼	▼	▼	▼	▼	▼	▼	▼	▼	▼	▼	▼	▼	▼	▼	▼	▼	▼	▼	▼	●	●	●	●	●	●	●		●	●	▲	▲	▲
Chile	▼	▼	▼	▼	▼	▼	▼	▼	▼	▼	▼	▼	▼	▼	▼	▼	▼	▼	▼	▼	▼	●	●	▼	●	●	●	●		●	▲	▲	▲
Indonesia	▼	▼	▼	▼	▼	▼	▼	▼	▼	▼	▼	▼	▼	▼	▼	▼	▼	▼	▼	▼	▼	▼	▼	▼	●	●	●	●	●		●	▲	▲
Philippines	▼	▼	▼	▼	▼	▼	▼	▼	▼	▼	▼	▼	▼	▼	▼	▼	▼	▼	▼	▼	▼	▼	▼	▼	▼	▼	▼	▼	▼	●		▲	▲
Morocco	▼	▼	▼	▼	▼	▼	▼	▼	▼	▼	▼	▼	▼	▼	▼	▼	▼	▼	▼	▼	▼	▼	▼	▼	▼	▼	▼	▼	▼	▼	▼		▲
South Africa	▼	▼	▼	▼	▼	▼	▼	▼	▼	▼	▼	▼	▼	▼	▼	▼	▼	▼	▼	▼	▼	▼	▼	▼	▼	▼	▼	▼	▼	▼	▼	▼	

▲ Average achievement significantly higher than comparison participant

● No statistically significant difference from comparison participant

▼ Average achievement significantly lower than comparison participant

Significance tests adjusted for multiple comparisons

SOURCE: IEA Third International Mathematics and Science Study (TIMSS), 1998-1999.

Exhibit B.4 Multiple Comparisons of Average Achievement in Geometry

Instructions: Read across the row for a participant to compare performance with the participants listed along the top of the chart. The symbols indicate whether the average achievement of the participant in the row is significantly lower than that of the comparison participant, significantly higher than that of the comparison participant, or if there is no statistically significant difference between the average achievement of the two participants.

	Japan	Korea, Rep. of	Singapore	Chinese Taipei	Hong Kong, SAR	Belgium (Flemish)	**Naperville Sch. Dist. #203, IL**	Slovak Republic	Bulgaria	Latvia (LSS)	Russian Federation	**First in the World Consort., IL**	Netherlands	Czech Republic	Canada	Slovenia	**Montgomery County, MD**	**Academy School Dist. #20, CO**	Australia	Malaysia	Lithuania	**Michigan Invitational Group, MI**	Finland	**Guilford County, NC**	Hungary	Romania	***Texas***	**Michigan**	**Oregon**	Thailand	Cyprus	Tunisia
Japan		•	•	•	•	▲	▲	▲	▲	▲	▲	▲	▲	▲	▲	▲	▲	▲	▲	▲	▲	▲	▲	▲	▲	▲	▲	▲	▲	▲	▲	▲
Korea, Rep. of	•		•	•	•	▲	▲	▲	▲	▲	▲	▲	▲	▲	▲	▲	▲	▲	▲	▲	▲	▲	▲	▲	▲	▲	▲	▲	▲	▲	▲	▲
Singapore	•	•		•	•	▲	▲	▲	▲	▲	▲	▲	▲	▲	▲	▲	▲	▲	▲	▲	▲	▲	▲	▲	▲	▲	▲	▲	▲	▲	▲	▲
Chinese Taipei	•	•	•		•	▲	▲	▲	▲	▲	▲	▲	▲	▲	▲	▲	▲	▲	▲	▲	▲	▲	▲	▲	▲	▲	▲	▲	▲	▲	▲	▲
Hong Kong, SAR	•	•	•	•		▲	▲	▲	▲	▲	▲	▲	▲	▲	▲	▲	▲	▲	▲	▲	▲	▲	▲	▲	▲	▲	▲	▲	▲	▲	▲	▲
Belgium (Flemish)	▼	▼	▼	▼	▼		•	•	•	•	•	•	•	•	▲	▲	▲	▲	▲	▲	▲	▲	▲	▲	▲	▲	▲	▲	▲	▲	▲	▲
Naperville Sch. Dist. #203, IL	▼	▼	▼	▼	▼	•		•	•	•	•	•	•	•	▲	•	▲	▲	▲	▲	▲	▲	▲	▲	▲	▲	▲	▲	▲	▲	▲	▲
Slovak Republic	▼	▼	▼	▼	▼	•	•		•	•	•	•	•	•	•	•	•	•	▲	▲	▲	•	▲	▲	▲	▲	▲	▲	▲	▲	▲	▲
Bulgaria	▼	▼	▼	▼	▼	•	•	•		•	•	•	•	•	•	•	•	•	▲	▲	▲	•	▲	▲	▲	▲	▲	▲	▲	▲	▲	▲
Latvia (LSS)	▼	▼	▼	▼	▼	•	•	•	•		•	•	•	•	•	•	•	•	•	▲	▲	•	▲	▲	▲	▲	▲	▲	▲	▲	▲	▲
Russian Federation	▼	▼	▼	▼	▼	•	•	•	•	•		•	•	•	•	•	•	•	•	▲	•	•	▲	▲	▲	▲	▲	▲	▲	▲	▲	▲
First in the World Consort., IL	▼	▼	▼	▼	▼	•	•	•	•	•	•		•	•	•	•	•	•	•	•	•	•	•	•	•	•	•	•	•	▲	▲	▲
Netherlands	▼	▼	▼	▼	▼	•	•	•	•	•	•	•		•	•	•	•	•	•	•	•	•	•	•	▲	▲	•	•	▲	▲	▲	▲
Czech Republic	▼	▼	▼	▼	▼	•	•	•	•	•	•	•	•		•	•	•	•	•	•	•	•	•	•	▲	•	•	•	•	▲	▲	▲
Canada	▼	▼	▼	▼	▼	▼	▼	•	•	•	•	•	•	•		•	•	•	•	•	•	•	•	•	•	•	•	•	•	▲	▲	▲
Slovenia	▼	▼	▼	▼	▼	▼	•	•	•	•	•	•	•	•	•		•	•	•	•	•	•	•	•	•	•	•	•	•	•	•	•
Montgomery County, MD	▼	▼	▼	▼	▼	▼	▼	•	•	•	•	•	•	•	•	•		•	•	•	•	•	•	•	•	•	•	•	•	•	•	•
Academy School Dist. #20, CO	▼	▼	▼	▼	▼	▼	▼	•	•	•	•	•	•	•	•	•	•		•	•	•	•	•	•	•	•	•	•	•	•	•	•
Australia	▼	▼	▼	▼	▼	▼	▼	▼	▼	•	•	•	•	•	•	•	•	•		•	•	•	•	•	•	•	•	•	•	•	•	•
Malaysia	▼	▼	▼	▼	▼	▼	▼	▼	▼	▼	▼	•	•	•	•	•	•	•	•		•	•	•	•	•	•	•	•	•	•	•	•
Lithuania	▼	▼	▼	▼	▼	▼	▼	▼	▼	▼	•	•	•	•	•	•	•	•	•	•		•	•	•	•	•	•	•	•	•	•	•
Michigan Invitational Group, MI	▼	▼	▼	▼	▼	▼	▼	•	•	•	•	•	•	•	•	•	•	•	•	•	•		•	•	•	•	•	•	•	•	•	•
Finland	▼	▼	▼	▼	▼	▼	▼	▼	▼	▼	▼	•	•	•	•	•	•	•	•	•	•	•		•	•	•	•	•	•	•	•	•
Guilford County, NC	▼	▼	▼	▼	▼	▼	▼	▼	▼	▼	▼	•	•	•	•	•	•	•	•	•	•	•	•		•	•	•	•	•	•	•	•
Hungary	▼	▼	▼	▼	▼	▼	▼	▼	▼	▼	▼	•	▼	▼	•	•	•	•	•	•	•	•	•	•		•	•	•	•	•	•	•
Romania	▼	▼	▼	▼	▼	▼	▼	▼	▼	▼	▼	•	▼	•	•	•	•	•	•	•	•	•	•	•	•		•	•	•	•	•	•
Texas	▼	▼	▼	▼	▼	▼	▼	▼	▼	▼	▼	•	•	•	•	•	•	•	•	•	•	•	•	•	•	•		•	•	•	•	•
Michigan	▼	▼	▼	▼	▼	▼	▼	▼	▼	▼	▼	•	•	•	•	•	•	•	•	•	•	•	•	•	•	•	•		•	•	•	•
Oregon	▼	▼	▼	▼	▼	▼	▼	▼	▼	▼	▼	•	▼	•	•	•	•	•	•	•	•	•	•	•	•	•	•	•		•	•	•
Thailand	▼	▼	▼	▼	▼	▼	▼	▼	▼	▼	▼	▼	▼	▼	▼	•	•	•	•	•	•	•	•	•	•	•	•	•	•		•	•
Cyprus	▼	▼	▼	▼	▼	▼	▼	▼	▼	▼	▼	▼	▼	▼	▼	•	•	•	•	•	•	•	•	•	•	•	•	•	•	•		•
Tunisia	▼	▼	▼	▼	▼	▼	▼	▼	▼	▼	▼	▼	▼	▼	▼	•	•	•	•	•	•	•	•	•	•	•	•	•	•	•	•	
Illinois	▼	▼	▼	▼	▼	▼	▼	▼	▼	▼	▼	•	▼	▼	•	•	•	•	•	•	•	•	•	•	•	•	•	•	•	•	•	•
SW Math/Sci. Collaborative, PA	▼	▼	▼	▼	▼	▼	▼	▼	▼	▼	▼	•	•	•	•	•	•	•	•	•	•	•	•	•	•	•	•	•	•	•	•	•
Italy	▼	▼	▼	▼	▼	▼	▼	▼	▼	▼	▼	▼	▼	▼	▼	•	•	•	•	•	•	•	•	•	•	•	•	•	•	•	•	•
Moldova	▼	▼	▼	▼	▼	▼	▼	▼	▼	▼	▼	▼	▼	▼	▼	•	•	•	•	•	•	•	•	•	•	•	•	•	•	•	•	•
New Zealand	▼	▼	▼	▼	▼	▼	▼	▼	▼	▼	▼	▼	▼	▼	▼	▼	▼	▼	•	•	•	•	•	•	•	•	•	•	•	•	•	•
Project SMART Consortium, OH	▼	▼	▼	▼	▼	▼	▼	▼	▼	▼	▼	▼	▼	▼	•	•	•	•	•	•	•	•	•	•	•	•	•	•	•	•	•	•
Massachusetts	▼	▼	▼	▼	▼	▼	▼	▼	▼	▼	▼	▼	▼	▼	▼	▼	•	•	•	•	•	•	•	•	•	•	•	•	•	•	•	•
South Carolina	▼	▼	▼	▼	▼	▼	▼	▼	▼	▼	▼	▼	▼	▼	▼	•	•	•	•	•	•	•	•	•	•	•	•	•	•	•	•	•
Indiana	▼	▼	▼	▼	▼	▼	▼	▼	▼	▼	▼	▼	▼	▼	▼	•	•	•	•	•	•	•	•	•	•	•	•	•	•	•	•	•
North Carolina	▼	▼	▼	▼	▼	▼	▼	▼	▼	▼	▼	▼	▼	▼	▼	▼	▼	▼	•	•	•	•	•	•	•	•	•	•	•	•	•	•
United States	▼	▼	▼	▼	▼	▼	▼	▼	▼	▼	▼	▼	▼	▼	▼	▼	▼	▼	▼	▼	•	•	•	•	•	•	•	•	•	•	•	•
Pennsylvania	▼	▼	▼	▼	▼	▼	▼	▼	▼	▼	▼	▼	▼	▼	▼	▼	▼	▼	▼	▼	•	•	•	•	•	•	•	•	•	•	•	•
England	▼	▼	▼	▼	▼	▼	▼	▼	▼	▼	▼	▼	▼	▼	▼	▼	▼	▼	▼	▼	▼	•	▼	•	•	•	•	•	•	•	•	•
Connecticut	▼	▼	▼	▼	▼	▼	▼	▼	▼	▼	▼	▼	▼	▼	▼	▼	▼	•	•	•	•	•	•	•	•	•	•	•	•	•	•	•
Fremont/Lincoln/WestSide PS, NE	▼	▼	▼	▼	▼	▼	▼	▼	▼	▼	▼	▼	▼	▼	▼	▼	▼	▼	▼	▼	▼	•	▼	•	•	•	•	•	•	•	•	•
Maryland	▼	▼	▼	▼	▼	▼	▼	▼	▼	▼	▼	▼	▼	▼	▼	▼	▼	▼	▼	▼	▼	•	▼	•	•	•	•	•	•	•	•	•
Missouri	▼	▼	▼	▼	▼	▼	▼	▼	▼	▼	▼	▼	▼	▼	▼	▼	▼	▼	▼	▼	▼	•	▼	•	▼	•	•	•	•	•	•	•
Idaho	▼	▼	▼	▼	▼	▼	▼	▼	▼	▼	▼	▼	▼	▼	▼	▼	▼	▼	▼	▼	▼	•	▼	•	•	•	•	•	•	•	•	•
Israel	▼	▼	▼	▼	▼	▼	▼	▼	▼	▼	▼	▼	▼	▼	▼	▼	▼	▼	▼	▼	▼	▼	▼	•	▼	•	•	•	•	▼	•	▼
Macedonia, Rep. of	▼	▼	▼	▼	▼	▼	▼	▼	▼	▼	▼	▼	▼	▼	▼	▼	▼	▼	▼	▼	▼	▼	▼	•	▼	•	•	•	•	▼	•	▼
Jersey City Public Schools, NJ	▼	▼	▼	▼	▼	▼	▼	▼	▼	▼	▼	▼	▼	▼	▼	▼	▼	▼	▼	▼	▼	▼	▼	•	▼	•	•	•	•	•	•	•
Delaware Science Coalition, DE	▼	▼	▼	▼	▼	▼	▼	▼	▼	▼	▼	▼	▼	▼	▼	▼	▼	▼	▼	▼	▼	▼	▼	▼	▼	▼	•	•	•	▼	▼	▼
Chicago Public Schools, IL	▼	▼	▼	▼	▼	▼	▼	▼	▼	▼	▼	▼	▼	▼	▼	▼	▼	▼	▼	▼	▼	▼	▼	▼	▼	▼	•	•	•	▼	▼	▼
Jordan	▼	▼	▼	▼	▼	▼	▼	▼	▼	▼	▼	▼	▼	▼	▼	▼	▼	▼	▼	▼	▼	▼	▼	▼	▼	▼	▼	▼	▼	▼	▼	▼
Iran, Islamic Rep.	▼	▼	▼	▼	▼	▼	▼	▼	▼	▼	▼	▼	▼	▼	▼	▼	▼	▼	▼	▼	▼	▼	▼	▼	▼	▼	▼	▼	▼	▼	▼	▼
Indonesia	▼	▼	▼	▼	▼	▼	▼	▼	▼	▼	▼	▼	▼	▼	▼	▼	▼	▼	▼	▼	▼	▼	▼	▼	▼	▼	▼	▼	▼	▼	▼	▼
Rochester City Sch. Dist., NY	▼	▼	▼	▼	▼	▼	▼	▼	▼	▼	▼	▼	▼	▼	▼	▼	▼	▼	▼	▼	▼	▼	▼	▼	▼	▼	▼	▼	▼	▼	▼	▼
Turkey	▼	▼	▼	▼	▼	▼	▼	▼	▼	▼	▼	▼	▼	▼	▼	▼	▼	▼	▼	▼	▼	▼	▼	▼	▼	▼	▼	▼	▼	▼	▼	▼
Miami-Dade County PS, FL	▼	▼	▼	▼	▼	▼	▼	▼	▼	▼	▼	▼	▼	▼	▼	▼	▼	▼	▼	▼	▼	▼	▼	▼	▼	▼	▼	▼	▼	▼	▼	▼
Chile	▼	▼	▼	▼	▼	▼	▼	▼	▼	▼	▼	▼	▼	▼	▼	▼	▼	▼	▼	▼	▼	▼	▼	▼	▼	▼	▼	▼	▼	▼	▼	▼
Morocco	▼	▼	▼	▼	▼	▼	▼	▼	▼	▼	▼	▼	▼	▼	▼	▼	▼	▼	▼	▼	▼	▼	▼	▼	▼	▼	▼	▼	▼	▼	▼	▼
Philippines	▼	▼	▼	▼	▼	▼	▼	▼	▼	▼	▼	▼	▼	▼	▼	▼	▼	▼	▼	▼	▼	▼	▼	▼	▼	▼	▼	▼	▼	▼	▼	▼
South Africa	▼	▼	▼	▼	▼	▼	▼	▼	▼	▼	▼	▼	▼	▼	▼	▼	▼	▼	▼	▼	▼	▼	▼	▼	▼	▼	▼	▼	▼	▼	▼	▼

States in *italics* did not fully satisfy guidelines for sample participation rates (see Appendix A for details).

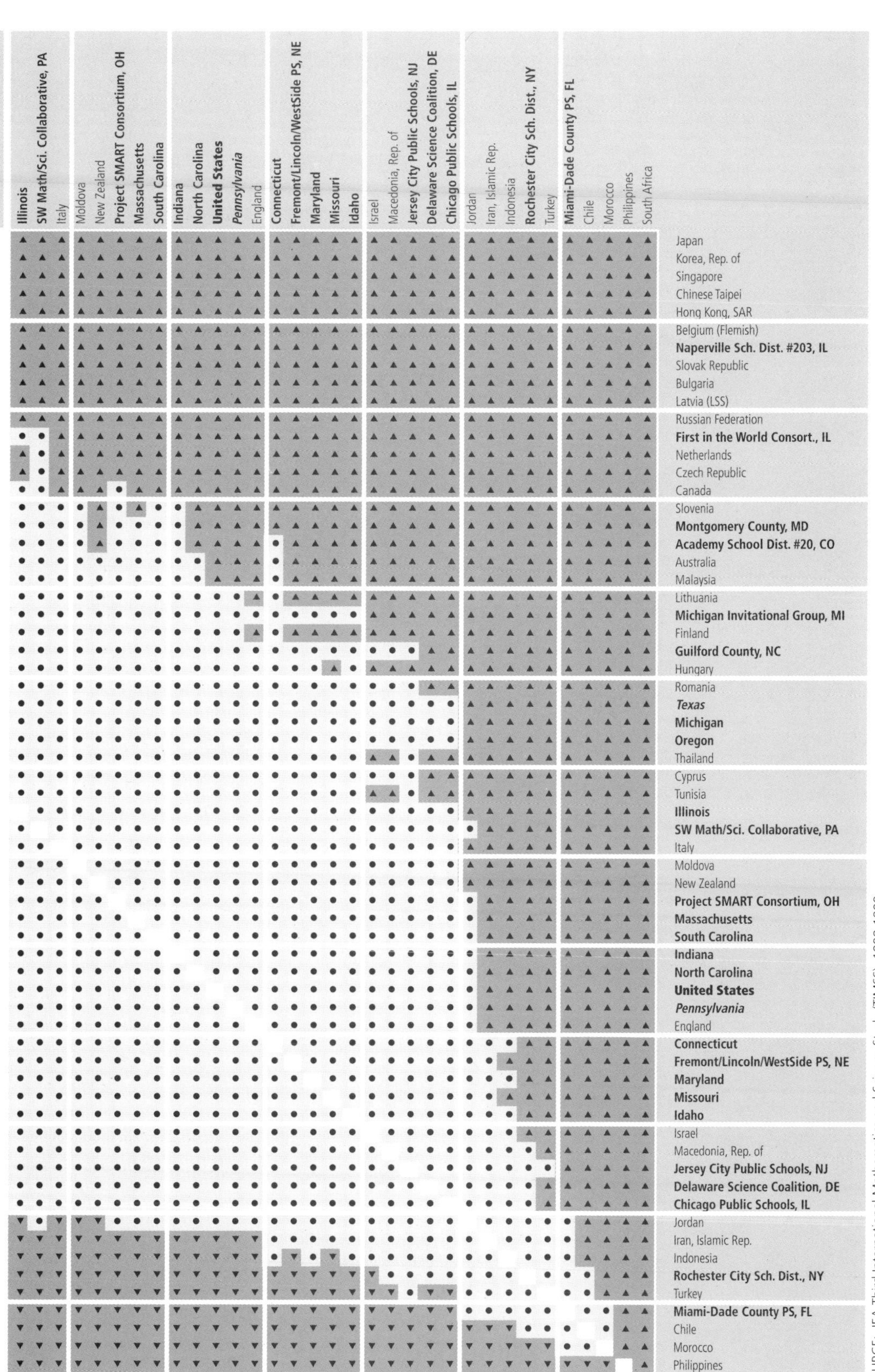

▲ Average achievement significantly higher than comparison participant

● No statistically significant difference from comparison participant

▼ Average achievement significantly lower than comparison participant

Significance tests adjusted for multiple comparisons

SOURCE: IEA Third International Mathematics and Science Study (TIMSS), 1998-1999.

Exhibit B.5 Multiple Comparisons of Average Achievement in Algebra

8th Grade Mathematics

Instructions: Read across the row for a participant to compare performance with the participants listed along the top of the chart. The symbols indicate whether the average achievement of the participant in the row is significantly lower than that of the comparison participant, significantly higher than that of the comparison participant, or if there is no statistically significant difference between the average achievement of the two participants.

	Chinese Taipei	Korea, Rep. of	Singapore	Japan	Hong Kong, SAR	**Naperville Sch. Dist. #203, IL**	**First in the World Consort., IL**	Belgium (Flemish)	**Montgomery County, MD**	Hungary	**Michigan Invitational Group, MI**	**Academy School Dist. #20, CO**	Russian Federation	Slovak Republic	Slovenia	Canada	**Guilford County, NC**	Netherlands	**Massachusetts**	**Project SMART Consortium, OH**	**Michigan**	Australia	**SW Math/Sci. Collaborative, PA**	**Oregon**	**Indiana**	Czech Republic	***Texas***	**Connecticut**	**Illinois**	Bulgaria	**South Carolina**	***Pennsylvania***
Chinese Taipei		●	●	●	●	▲	▲	▲	▲	▲	▲	▲	▲	▲	▲	▲	▲	▲	▲	▲	▲	▲	▲	▲	▲	▲	▲	▲	▲	▲	▲	▲
Korea, Rep. of	●		●	▲	●	▲	▲	▲	▲	▲	▲	▲	▲	▲	▲	▲	▲	▲	▲	▲	▲	▲	▲	▲	▲	▲	▲	▲	▲	▲	▲	▲
Singapore	●	●		●	●	●	●	▲	▲	▲	▲	▲	▲	▲	▲	▲	▲	▲	▲	▲	▲	▲	▲	▲	▲	▲	▲	▲	▲	▲	▲	▲
Japan	●	▼	●		●	●	●	▲	▲	▲	▲	▲	▲	▲	▲	▲	▲	▲	▲	▲	▲	▲	▲	▲	▲	▲	▲	▲	▲	▲	▲	▲
Hong Kong, SAR	●	●	●	●		●	●	▲	▲	▲	▲	▲	▲	▲	▲	▲	▲	▲	▲	▲	▲	▲	▲	▲	▲	▲	▲	▲	▲	▲	▲	▲
Naperville Sch. Dist. #203, IL	▼	▼	●	●	●		●	▲	▲	▲	▲	▲	▲	▲	▲	▲	▲	▲	▲	▲	▲	▲	▲	▲	▲	▲	▲	▲	▲	▲	▲	▲
First in the World Consort., IL	▼	▼	●	●	●	●		●	●	▲	●	▲	▲	▲	▲	▲	▲	▲	▲	▲	▲	▲	▲	▲	▲	▲	▲	▲	▲	▲	▲	▲
Belgium (Flemish)	▼	▼	▼	▼	▼	▼	●		●	●	●	●	●	●	●	●	●	●	●	●	●	●	●	●	●	▲	●	●	▲	▲	▲	▲
Montgomery County, MD	▼	▼	▼	▼	▼	▼	●	●		●	●	●	●	●	●	●	●	●	●	●	●	●	●	●	●	▲	●	●	▲	▲	▲	▲
Hungary	▼	▼	▼	▼	▼	▼	▼	●	●		●	●	●	●	●	●	●	●	●	●	●	●	●	●	●	▲	●	●	▲	▲	▲	▲
Michigan Invitational Group, MI	▼	▼	▼	▼	▼	▼	●	●	●	●		●	●	●	●	●	●	●	●	●	●	●	●	●	●	●	●	●	●	●	●	●
Academy School Dist. #20, CO	▼	▼	▼	▼	▼	▼	▼	●	●	●	●		●	●	●	●	●	●	●	●	●	●	●	●	●	▲	●	●	●	▲	●	●
Russian Federation	▼	▼	▼	▼	▼	▼	▼	●	●	●	●	●		●	●	●	●	●	●	●	●	●	●	●	●	●	●	●	●	●	●	●
Slovak Republic	▼	▼	▼	▼	▼	▼	▼	●	●	●	●	●	●		●	●	●	●	●	●	●	●	●	●	●	●	●	●	●	●	●	●
Slovenia	▼	▼	▼	▼	▼	▼	▼	●	●	●	●	●	●	●		●	●	●	●	●	●	●	●	●	●	●	●	●	●	●	●	●
Canada	▼	▼	▼	▼	▼	▼	▼	●	●	●	●	●	●	●	●		●	●	●	●	●	●	●	●	●	●	●	●	●	●	●	●
Guilford County, NC	▼	▼	▼	▼	▼	▼	▼	●	●	●	●	●	●	●	●	●		●	●	●	●	●	●	●	●	●	●	●	●	●	●	●
Netherlands	▼	▼	▼	▼	▼	▼	▼	●	●	●	●	●	●	●	●	●	●		●	●	●	●	●	●	●	●	●	●	●	●	●	●
Massachusetts	▼	▼	▼	▼	▼	▼	▼	●	●	●	●	●	●	●	●	●	●	●		●	●	●	●	●	●	●	●	●	●	●	●	●
Project SMART Consortium, OH	▼	▼	▼	▼	▼	▼	▼	●	●	●	●	●	●	●	●	●	●	●	●		●	●	●	●	●	●	●	●	●	●	●	●
Michigan	▼	▼	▼	▼	▼	▼	▼	●	●	●	●	●	●	●	●	●	●	●	●	●		●	●	●	●	●	●	●	●	●	●	●
Australia	▼	▼	▼	▼	▼	▼	▼	●	●	●	●	●	●	●	●	●	●	●	●	●	●		●	●	●	●	●	●	●	●	●	●
SW Math/Sci. Collaborative, PA	▼	▼	▼	▼	▼	▼	▼	●	●	●	●	●	●	●	●	●	●	●	●	●	●	●		●	●	●	●	●	●	●	●	●
Oregon	▼	▼	▼	▼	▼	▼	▼	●	●	●	●	●	●	●	●	●	●	●	●	●	●	●	●		●	●	●	●	●	●	●	●
Indiana	▼	▼	▼	▼	▼	▼	▼	●	●	●	●	●	●	●	●	●	●	●	●	●	●	●	●	●		●	●	●	●	●	●	●
Czech Republic	▼	▼	▼	▼	▼	▼	▼	▼	▼	▼	●	▼	●	●	●	●	●	●	●	●	●	●	●	●	●		●	●	●	●	●	●
Texas	▼	▼	▼	▼	▼	▼	▼	●	●	●	●	●	●	●	●	●	●	●	●	●	●	●	●	●	●	●		●	●	●	●	●
Connecticut	▼	▼	▼	▼	▼	▼	▼	●	●	●	●	●	●	●	●	●	●	●	●	●	●	●	●	●	●	●	●		●	●	●	●
Illinois	▼	▼	▼	▼	▼	▼	▼	▼	▼	▼	●	●	●	●	●	●	●	●	●	●	●	●	●	●	●	●	●	●		●	●	●
Bulgaria	▼	▼	▼	▼	▼	▼	▼	▼	▼	▼	●	▼	●	●	●	●	●	●	●	●	●	●	●	●	●	●	●	●	●		●	●
South Carolina	▼	▼	▼	▼	▼	▼	▼	▼	▼	▼	●	●	●	●	●	●	●	●	●	●	●	●	●	●	●	●	●	●	●	●		●
Pennsylvania	▼	▼	▼	▼	▼	▼	▼	▼	▼	▼	●	●	●	●	●	●	●	●	●	●	●	●	●	●	●	●	●	●	●	●	●	
North Carolina	▼	▼	▼	▼	▼	▼	▼	▼	▼	▼	●	▼	●	●	●	●	●	●	●	●	●	●	●	●	●	●	●	●	●	●	●	●
United States	▼	▼	▼	▼	▼	▼	▼	▼	▼	▼	▼	▼	▼	●	▼	▼	●	●	●	●	●	●	●	●	●	●	●	●	●	●	●	●
Malaysia	▼	▼	▼	▼	▼	▼	▼	▼	▼	▼	▼	▼	▼	●	▼	▼	●	●	●	●	●	●	●	●	●	●	●	●	●	●	●	●
Idaho	▼	▼	▼	▼	▼	▼	▼	▼	▼	▼	▼	▼	▼	●	●	▼	●	●	●	●	●	●	●	●	●	●	●	●	●	●	●	●
Maryland	▼	▼	▼	▼	▼	▼	▼	▼	▼	▼	▼	▼	▼	▼	▼	▼	●	●	●	●	●	●	●	●	●	●	●	●	●	●	●	●
Latvia (LSS)	▼	▼	▼	▼	▼	▼	▼	▼	▼	▼	▼	▼	▼	▼	▼	▼	▼	●	▼	●	●	●	●	●	●	●	●	●	●	●	●	●
England	▼	▼	▼	▼	▼	▼	▼	▼	▼	▼	▼	▼	▼	▼	▼	▼	▼	●	●	●	●	●	●	●	●	●	●	●	●	●	●	●
Finland	▼	▼	▼	▼	▼	▼	▼	▼	▼	▼	▼	▼	▼	▼	▼	▼	▼	●	▼	●	▼	▼	●	●	●	▼	●	●	●	●	●	●
Delaware Science Coalition, DE	▼	▼	▼	▼	▼	▼	▼	▼	▼	▼	▼	▼	▼	●	●	▼	●	●	●	●	●	●	●	●	●	●	●	●	●	●	●	●
New Zealand	▼	▼	▼	▼	▼	▼	▼	▼	▼	▼	▼	▼	▼	▼	▼	▼	▼	●	▼	●	●	▼	●	●	●	●	●	●	●	●	●	●
Jersey City Public Schools, NJ	▼	▼	▼	▼	▼	▼	▼	▼	▼	▼	▼	▼	▼	▼	▼	▼	●	●	●	●	●	●	●	●	●	●	●	●	●	●	●	●
Fremont/Lincoln/WestSide PS, NE	▼	▼	▼	▼	▼	▼	▼	▼	▼	▼	▼	▼	▼	▼	▼	▼	●	●	●	●	●	●	●	●	●	●	●	●	●	●	●	●
Missouri	▼	▼	▼	▼	▼	▼	▼	▼	▼	▼	▼	▼	▼	▼	▼	▼	▼	●	▼	●	▼	▼	●	●	●	●	●	●	●	●	●	●
Lithuania	▼	▼	▼	▼	▼	▼	▼	▼	▼	▼	▼	▼	▼	▼	▼	▼	▼	▼	▼	▼	▼	▼	▼	▼	▼	▼	●	●	▼	▼	▼	▼
Italy	▼	▼	▼	▼	▼	▼	▼	▼	▼	▼	▼	▼	▼	▼	▼	▼	▼	▼	▼	▼	▼	▼	▼	▼	▼	▼	▼	▼	▼	▼	▼	▼
Romania	▼	▼	▼	▼	▼	▼	▼	▼	▼	▼	▼	▼	▼	▼	▼	▼	▼	▼	▼	▼	▼	▼	▼	▼	▼	▼	▼	▼	▼	▼	▼	▼
Israel	▼	▼	▼	▼	▼	▼	▼	▼	▼	▼	▼	▼	▼	▼	▼	▼	▼	▼	▼	▼	▼	▼	▼	▼	▼	▼	▼	▼	▼	▼	▼	▼
Cyprus	▼	▼	▼	▼	▼	▼	▼	▼	▼	▼	▼	▼	▼	▼	▼	▼	▼	▼	▼	▼	▼	▼	▼	▼	▼	▼	▼	▼	▼	▼	▼	▼
Moldova	▼	▼	▼	▼	▼	▼	▼	▼	▼	▼	▼	▼	▼	▼	▼	▼	▼	▼	▼	▼	▼	▼	▼	▼	▼	▼	▼	▼	▼	▼	▼	▼
Chicago Public Schools, IL	▼	▼	▼	▼	▼	▼	▼	▼	▼	▼	▼	▼	▼	▼	▼	▼	▼	▼	▼	▼	▼	▼	▼	▼	▼	▼	▼	▼	▼	▼	▼	▼
Rochester City Sch. Dist., NY	▼	▼	▼	▼	▼	▼	▼	▼	▼	▼	▼	▼	▼	▼	▼	▼	▼	▼	▼	▼	▼	▼	▼	▼	▼	▼	▼	▼	▼	▼	▼	▼
Macedonia, Rep. of	▼	▼	▼	▼	▼	▼	▼	▼	▼	▼	▼	▼	▼	▼	▼	▼	▼	▼	▼	▼	▼	▼	▼	▼	▼	▼	▼	▼	▼	▼	▼	▼
Thailand	▼	▼	▼	▼	▼	▼	▼	▼	▼	▼	▼	▼	▼	▼	▼	▼	▼	▼	▼	▼	▼	▼	▼	▼	▼	▼	▼	▼	▼	▼	▼	▼
Tunisia	▼	▼	▼	▼	▼	▼	▼	▼	▼	▼	▼	▼	▼	▼	▼	▼	▼	▼	▼	▼	▼	▼	▼	▼	▼	▼	▼	▼	▼	▼	▼	▼
Miami-Dade County PS, FL	▼	▼	▼	▼	▼	▼	▼	▼	▼	▼	▼	▼	▼	▼	▼	▼	▼	▼	▼	▼	▼	▼	▼	▼	▼	▼	▼	▼	▼	▼	▼	▼
Jordan	▼	▼	▼	▼	▼	▼	▼	▼	▼	▼	▼	▼	▼	▼	▼	▼	▼	▼	▼	▼	▼	▼	▼	▼	▼	▼	▼	▼	▼	▼	▼	▼
Iran, Islamic Rep.	▼	▼	▼	▼	▼	▼	▼	▼	▼	▼	▼	▼	▼	▼	▼	▼	▼	▼	▼	▼	▼	▼	▼	▼	▼	▼	▼	▼	▼	▼	▼	▼
Turkey	▼	▼	▼	▼	▼	▼	▼	▼	▼	▼	▼	▼	▼	▼	▼	▼	▼	▼	▼	▼	▼	▼	▼	▼	▼	▼	▼	▼	▼	▼	▼	▼
Indonesia	▼	▼	▼	▼	▼	▼	▼	▼	▼	▼	▼	▼	▼	▼	▼	▼	▼	▼	▼	▼	▼	▼	▼	▼	▼	▼	▼	▼	▼	▼	▼	▼
Chile	▼	▼	▼	▼	▼	▼	▼	▼	▼	▼	▼	▼	▼	▼	▼	▼	▼	▼	▼	▼	▼	▼	▼	▼	▼	▼	▼	▼	▼	▼	▼	▼
Morocco	▼	▼	▼	▼	▼	▼	▼	▼	▼	▼	▼	▼	▼	▼	▼	▼	▼	▼	▼	▼	▼	▼	▼	▼	▼	▼	▼	▼	▼	▼	▼	▼
Philippines	▼	▼	▼	▼	▼	▼	▼	▼	▼	▼	▼	▼	▼	▼	▼	▼	▼	▼	▼	▼	▼	▼	▼	▼	▼	▼	▼	▼	▼	▼	▼	▼
South Africa	▼	▼	▼	▼	▼	▼	▼	▼	▼	▼	▼	▼	▼	▼	▼	▼	▼	▼	▼	▼	▼	▼	▼	▼	▼	▼	▼	▼	▼	▼	▼	▼

States in *italics* did not fully satisfy guidelines for sample participation rates (see Appendix A for details).

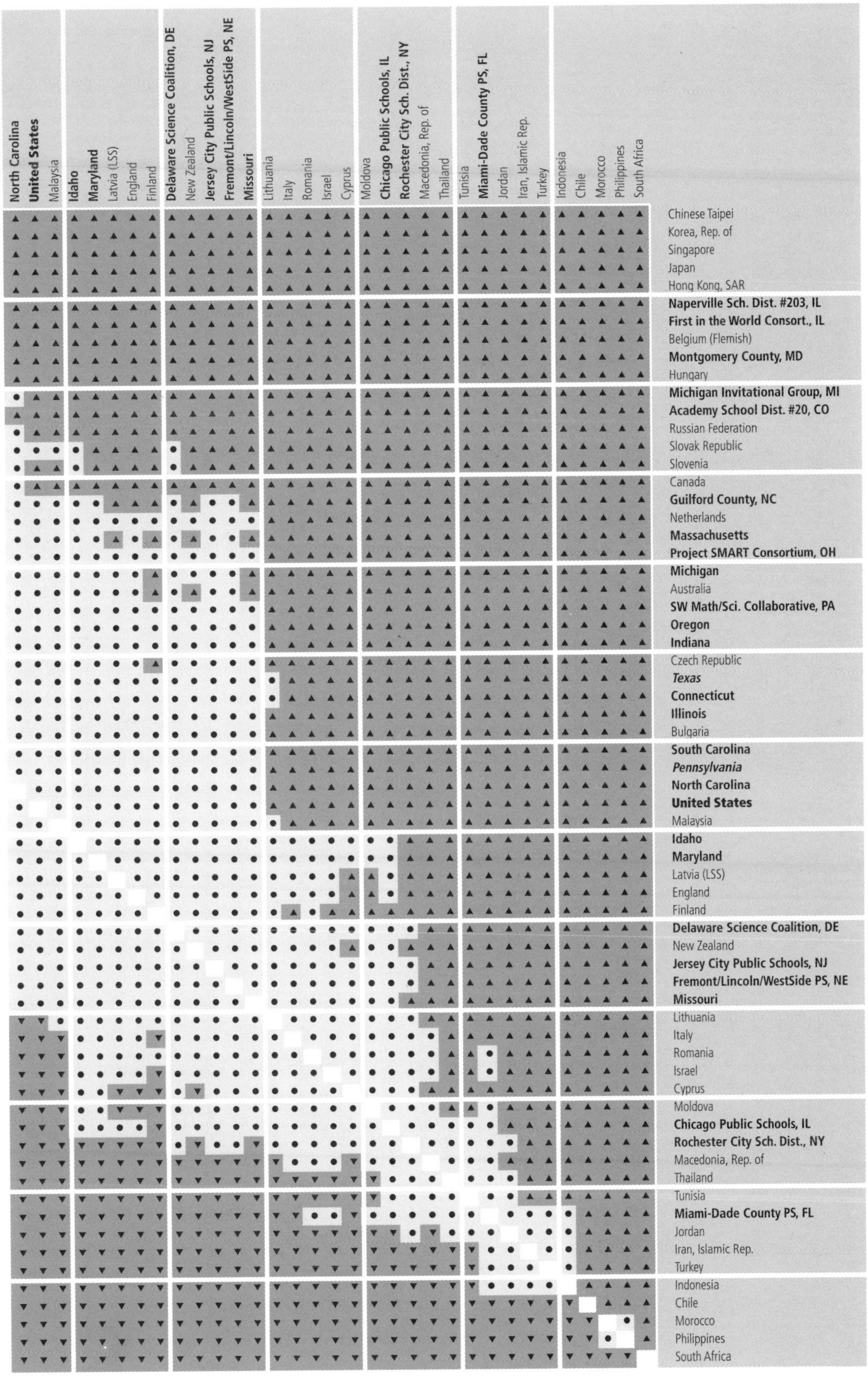

	North Carolina	**United States**	Malaysia	**Idaho**	**Maryland**	Latvia (LSS)	England	Finland	**Delaware Science Coalition, DE**	New Zealand	**Jersey City Public Schools, NJ**	**Fremont/Lincoln/WestSide PS, NE**	**Missouri**	Lithuania	Italy	Romania	Israel	Cyprus	Moldova	**Chicago Public Schools, IL**	**Rochester City Sch. Dist., NY**	Macedonia, Rep. of	Thailand	Tunisia	**Miami-Dade County PS, FL**	Jordan	Iran, Islamic Rep.	Turkey	Indonesia	Chile	Morocco	Philippines	South Africa
Chinese Taipei	▲	▲	▲	▲	▲	▲	▲	▲	▲	▲	▲	▲	▲	▲	▲	▲	▲	▲	▲	▲	▲	▲	▲	▲	▲	▲	▲	▲	▲	▲	▲	▲	▲
Korea, Rep. of	▲	▲	▲	▲	▲	▲	▲	▲	▲	▲	▲	▲	▲	▲	▲	▲	▲	▲	▲	▲	▲	▲	▲	▲	▲	▲	▲	▲	▲	▲	▲	▲	▲
Singapore	▲	▲	▲	▲	▲	▲	▲	▲	▲	▲	▲	▲	▲	▲	▲	▲	▲	▲	▲	▲	▲	▲	▲	▲	▲	▲	▲	▲	▲	▲	▲	▲	▲
Japan	▲	▲	▲	▲	▲	▲	▲	▲	▲	▲	▲	▲	▲	▲	▲	▲	▲	▲	▲	▲	▲	▲	▲	▲	▲	▲	▲	▲	▲	▲	▲	▲	▲
Hong Kong, SAR	▲	▲	▲	▲	▲	▲	▲	▲	▲	▲	▲	▲	▲	▲	▲	▲	▲	▲	▲	▲	▲	▲	▲	▲	▲	▲	▲	▲	▲	▲	▲	▲	▲
Naperville Sch. Dist. #203, IL	▲	▲	▲	▲	▲	▲	▲	▲	▲	▲	▲	▲	▲	▲	▲	▲	▲	▲	▲	▲	▲	▲	▲	▲	▲	▲	▲	▲	▲	▲	▲	▲	▲
First in the World Consort., IL	▲	▲	▲	▲	▲	▲	▲	▲	▲	▲	▲	▲	▲	▲	▲	▲	▲	▲	▲	▲	▲	▲	▲	▲	▲	▲	▲	▲	▲	▲	▲	▲	▲
Belgium (Flemish)	▲	▲	▲	▲	▲	▲	▲	▲	▲	▲	▲	▲	▲	▲	▲	▲	▲	▲	▲	▲	▲	▲	▲	▲	▲	▲	▲	▲	▲	▲	▲	▲	▲
Montgomery County, MD	▲	▲	▲	▲	▲	▲	▲	▲	▲	▲	▲	▲	▲	▲	▲	▲	▲	▲	▲	▲	▲	▲	▲	▲	▲	▲	▲	▲	▲	▲	▲	▲	▲
Hungary	▲	▲	▲	▲	▲	▲	▲	▲	▲	▲	▲	▲	▲	▲	▲	▲	▲	▲	▲	▲	▲	▲	▲	▲	▲	▲	▲	▲	▲	▲	▲	▲	▲
Michigan Invitational Group, MI	●	▲	▲	▲	▲	▲	▲	▲	▲	▲	▲	▲	▲	▲	▲	▲	▲	▲	▲	▲	▲	▲	▲	▲	▲	▲	▲	▲	▲	▲	▲	▲	▲
Academy School Dist. #20, CO	▲	▲	▲	▲	▲	▲	▲	▲	▲	▲	▲	▲	▲	▲	▲	▲	▲	▲	▲	▲	▲	▲	▲	▲	▲	▲	▲	▲	▲	▲	▲	▲	▲
Russian Federation	●	▲	▲	▲	▲	▲	▲	▲	▲	▲	▲	▲	▲	▲	▲	▲	▲	▲	▲	▲	▲	▲	▲	▲	▲	▲	▲	▲	▲	▲	▲	▲	▲
Slovak Republic	●	●	●	●	▲	▲	▲	▲	●	▲	▲	▲	▲	▲	▲	▲	▲	▲	▲	▲	▲	▲	▲	▲	▲	▲	▲	▲	▲	▲	▲	▲	▲
Slovenia	●	▲	▲	●	▲	▲	▲	▲	●	▲	▲	▲	▲	▲	▲	▲	▲	▲	▲	▲	▲	▲	▲	▲	▲	▲	▲	▲	▲	▲	▲	▲	▲
Canada	●	▲	▲	▲	▲	▲	▲	▲	▲	▲	▲	▲	▲	▲	▲	▲	▲	▲	▲	▲	▲	▲	▲	▲	▲	▲	▲	▲	▲	▲	▲	▲	▲
Guilford County, NC	●	●	●	●	●	▲	▲	▲	●	▲	●	●	▲	▲	▲	▲	▲	▲	▲	▲	▲	▲	▲	▲	▲	▲	▲	▲	▲	▲	▲	▲	▲
Netherlands	●	●	●	●	●	●	●	●	●	●	●	●	●	▲	▲	▲	▲	▲	▲	▲	▲	▲	▲	▲	▲	▲	▲	▲	▲	▲	▲	▲	▲
Massachusetts	●	●	●	●	●	▲	●	▲	●	▲	●	●	▲	▲	▲	▲	▲	▲	▲	▲	▲	▲	▲	▲	▲	▲	▲	▲	▲	▲	▲	▲	▲
Project SMART Consortium, OH	●	●	●	●	●	●	●	●	●	●	●	●	●	▲	▲	▲	▲	▲	▲	▲	▲	▲	▲	▲	▲	▲	▲	▲	▲	▲	▲	▲	▲
Michigan	●	●	●	●	●	●	●	▲	●	●	●	●	▲	▲	▲	▲	▲	▲	▲	▲	▲	▲	▲	▲	▲	▲	▲	▲	▲	▲	▲	▲	▲
Australia	●	●	●	●	●	●	●	▲	●	▲	●	●	▲	▲	▲	▲	▲	▲	▲	▲	▲	▲	▲	▲	▲	▲	▲	▲	▲	▲	▲	▲	▲
SW Math/Sci. Collaborative, PA	●	●	●	●	●	●	●	●	●	●	●	●	●	▲	▲	▲	▲	▲	▲	▲	▲	▲	▲	▲	▲	▲	▲	▲	▲	▲	▲	▲	▲
Oregon	●	●	●	●	●	●	●	●	●	●	●	●	●	▲	▲	▲	▲	▲	▲	▲	▲	▲	▲	▲	▲	▲	▲	▲	▲	▲	▲	▲	▲
Indiana	●	●	●	●	●	●	●	●	●	●	●	●	●	▲	▲	▲	▲	▲	▲	▲	▲	▲	▲	▲	▲	▲	▲	▲	▲	▲	▲	▲	▲
Czech Republic	●	●	●	●	●	●	●	▲	●	●	●	●	●	▲	▲	▲	▲	▲	▲	▲	▲	▲	▲	▲	▲	▲	▲	▲	▲	▲	▲	▲	▲
Texas	●	●	●	●	●	●	●	●	●	●	●	●	●	●	▲	▲	▲	▲	▲	▲	▲	▲	▲	▲	▲	▲	▲	▲	▲	▲	▲	▲	▲
Connecticut	●	●	●	●	●	●	●	●	●	●	●	●	●	●	▲	▲	▲	▲	▲	▲	▲	▲	▲	▲	▲	▲	▲	▲	▲	▲	▲	▲	▲
Illinois	●	●	●	●	●	●	●	●	●	●	●	●	●	▲	▲	▲	▲	▲	▲	▲	▲	▲	▲	▲	▲	▲	▲	▲	▲	▲	▲	▲	▲
Bulgaria	●	●	●	●	●	●	●	●	●	●	●	●	●	▲	▲	▲	▲	▲	▲	▲	▲	▲	▲	▲	▲	▲	▲	▲	▲	▲	▲	▲	▲
South Carolina	●	●	●	●	●	●	●	●	●	●	●	●	●	▲	▲	▲	▲	▲	▲	▲	▲	▲	▲	▲	▲	▲	▲	▲	▲	▲	▲	▲	▲
Pennsylvania	●	●	●	●	●	●	●	●	●	●	●	●	●	▲	▲	▲	▲	▲	▲	▲	▲	▲	▲	▲	▲	▲	▲	▲	▲	▲	▲	▲	▲
North Carolina		●	●	●	●	●	●	●	●	●	●	●	●	▲	▲	▲	▲	▲	▲	▲	▲	▲	▲	▲	▲	▲	▲	▲	▲	▲	▲	▲	▲
United States	●		●	●	●	●	●	●	●	●	●	●	●	▲	▲	▲	▲	▲	▲	▲	▲	▲	▲	▲	▲	▲	▲	▲	▲	▲	▲	▲	▲
Malaysia	●	●		●	●	●	●	●	●	●	●	●	●	●	▲	▲	▲	▲	▲	▲	▲	▲	▲	▲	▲	▲	▲	▲	▲	▲	▲	▲	▲
Idaho	●	●	●		●	●	●	●	●	●	●	●	●	●	●	●	●	●	●	●	▲	▲	▲	▲	▲	▲	▲	▲	▲	▲	▲	▲	▲
Maryland	●	●	●	●		●	●	●	●	●	●	●	●	●	●	●	●	●	●	●	▲	▲	▲	▲	▲	▲	▲	▲	▲	▲	▲	▲	▲
Latvia (LSS)	●	●	●	●	●		●	●	●	●	●	●	●	●	●	●	●	▲	▲	●	▲	▲	▲	▲	▲	▲	▲	▲	▲	▲	▲	▲	▲
England	●	●	●	●	●	●		●	●	●	●	●	●	●	●	●	●	▲	▲	●	▲	▲	▲	▲	▲	▲	▲	▲	▲	▲	▲	▲	▲
Finland	●	●	●	●	●	●	●		●	●	●	●	●	●	▲	●	▲	▲	▲	▲	▲	▲	▲	▲	▲	▲	▲	▲	▲	▲	▲	▲	▲
Delaware Science Coalition, DE	●	●	●	●	●	●	●	●		●	●	●	●	●	●	●	●	●	●	●	●	▲	▲	▲	▲	▲	▲	▲	▲	▲	▲	▲	▲
New Zealand	●	●	●	●	●	●	●	●	●		●	●	●	●	●	●	●	▲	●	●	▲	▲	▲	▲	▲	▲	▲	▲	▲	▲	▲	▲	▲
Jersey City Public Schools, NJ	●	●	●	●	●	●	●	●	●	●		●	●	●	●	●	●	●	●	●	●	▲	▲	▲	▲	▲	▲	▲	▲	▲	▲	▲	▲
Fremont/Lincoln/WestSide PS, NE	●	●	●	●	●	●	●	●	●	●	●		●	●	●	●	●	●	●	●	●	▲	▲	▲	▲	▲	▲	▲	▲	▲	▲	▲	▲
Missouri	●	●	●	●	●	●	●	●	●	●	●	●		●	●	●	●	●	●	●	▲	▲	▲	▲	▲	▲	▲	▲	▲	▲	▲	▲	▲
Lithuania	▼	▼	●	●	●	●	●	●	●	●	●	●	●		●	●	●	●	●	●	●	▲	▲	▲	▲	▲	▲	▲	▲	▲	▲	▲	▲
Italy	▼	▼	▼	●	●	●	●	▼	●	●	●	●	●	●		●	●	●	●	●	●	●	▲	▲	▲	▲	▲	▲	▲	▲	▲	▲	▲
Romania	▼	▼	▼	●	●	●	●	●	●	●	●	●	●	●	●		●	●	●	●	●	●	▲	▲	●	▲	▲	▲	▲	▲	▲	▲	▲
Israel	▼	▼	▼	●	●	●	●	▼	●	●	●	●	●	●	●	●		●	●	●	●	●	▲	▲	●	▲	▲	▲	▲	▲	▲	▲	▲
Cyprus	▼	▼	▼	●	●	▼	▼	▼	●	▼	●	●	●	●	●	●	●		●	●	●	▲	▲	▲	▲	▲	▲	▲	▲	▲	▲	▲	▲
Moldova	▼	▼	▼	●	●	▼	▼	▼	●	●	●	●	●	●	●	●	●	●		●	●	●	▲	▲	●	▲	▲	▲	▲	▲	▲	▲	▲
Chicago Public Schools, IL	▼	▼	▼	●	●	●	●	▼	●	●	●	●	●	●	●	●	●	●	●		●	●	●	●	●	▲	▲	▲	▲	▲	▲	▲	▲
Rochester City Sch. Dist., NY	▼	▼	▼	▼	▼	▼	▼	▼	●	▼	●	●	▼	●	●	●	●	●	●	●		●	●	●	●	●	▲	▲	▲	▲	▲	▲	▲
Macedonia, Rep. of	▼	▼	▼	▼	▼	▼	▼	▼	▼	▼	▼	▼	▼	▼	●	●	●	▼	●	●	●		●	●	●	▲	▲	▲	▲	▲	▲	▲	▲
Thailand	▼	▼	▼	▼	▼	▼	▼	▼	▼	▼	▼	▼	▼	▼	▼	▼	▼	▼	▼	●	●	●		●	●	●	▲	▲	▲	▲	▲	▲	▲
Tunisia	▼	▼	▼	▼	▼	▼	▼	▼	▼	▼	▼	▼	▼	▼	▼	▼	▼	▼	▼	●	●	●	●		●	●	▲	▲	▲	▲	▲	▲	▲
Miami-Dade County PS, FL	▼	▼	▼	▼	▼	▼	▼	▼	▼	▼	▼	▼	▼	▼	▼	●	●	▼	●	●	●	●	●	●		●	●	●	●	▲	▲	▲	▲
Jordan	▼	▼	▼	▼	▼	▼	▼	▼	▼	▼	▼	▼	▼	▼	▼	▼	▼	▼	▼	▼	●	▼	●	●	●		●	●	●	▲	▲	▲	▲
Iran, Islamic Rep.	▼	▼	▼	▼	▼	▼	▼	▼	▼	▼	▼	▼	▼	▼	▼	▼	▼	▼	▼	▼	▼	▼	▼	▼	●	●		●	●	▲	▲	▲	▲
Turkey	▼	▼	▼	▼	▼	▼	▼	▼	▼	▼	▼	▼	▼	▼	▼	▼	▼	▼	▼	▼	▼	▼	▼	▼	●	●	●		●	▲	▲	▲	▲
Indonesia	▼	▼	▼	▼	▼	▼	▼	▼	▼	▼	▼	▼	▼	▼	▼	▼	▼	▼	▼	▼	▼	▼	▼	▼	●	●	●	●		▲	▲	▲	▲
Chile	▼	▼	▼	▼	▼	▼	▼	▼	▼	▼	▼	▼	▼	▼	▼	▼	▼	▼	▼	▼	▼	▼	▼	▼	▼	▼	▼	▼	▼		▲	▲	▲
Morocco	▼	▼	▼	▼	▼	▼	▼	▼	▼	▼	▼	▼	▼	▼	▼	▼	▼	▼	▼	▼	▼	▼	▼	▼	▼	▼	▼	▼	▼	▼		●	▲
Philippines	▼	▼	▼	▼	▼	▼	▼	▼	▼	▼	▼	▼	▼	▼	▼	▼	▼	▼	▼	▼	▼	▼	▼	▼	▼	▼	▼	▼	▼	▼	●		▲
South Africa	▼	▼	▼	▼	▼	▼	▼	▼	▼	▼	▼	▼	▼	▼	▼	▼	▼	▼	▼	▼	▼	▼	▼	▼	▼	▼	▼	▼	▼	▼	▼	▼	

▲ Average achievement significantly higher than comparison participant

● No statistically significant difference from comparison participant

▼ Average achievement significantly lower than comparison participant

Significance tests adjusted for multiple comparisons

SOURCE: IEA Third International Mathematics and Science Study (TIMSS), 1998-1999.

APPENDIX C

Percentiles and Standard Deviations of Mathematics Achievement

8th Grade Mathematics

	5th Percentile	25th Percentile	50th Percentile	75th Percentile	95th Percentile
United States	356 (0.9)	442 (1.9)	504 (0.8)	562 (0.5)	642 (2.3)
Australia	387 (4.1)	472 (1.7)	529 (1.7)	581 (1.9)	648 (3.3)
Belgium (Flemish)	423 (3.4)	511 (1.8)	563 (1.0)	611 (1.6)	675 (3.6)
Bulgaria	367 (4.0)	454 (1.3)	512 (1.7)	567 (2.0)	649 (3.5)
Canada	406 (2.2)	484 (1.6)	533 (0.7)	581 (1.2)	646 (1.7)
Chile	253 (6.5)	336 (2.5)	391 (1.2)	448 (0.8)	533 (1.4)
Chinese Taipei	396 (3.5)	524 (0.6)	595 (1.5)	656 (0.4)	739 (2.2)
Cyprus	335 (5.1)	422 (3.7)	481 (0.9)	534 (0.4)	603 (1.3)
Czech Republic	392 (3.2)	467 (1.5)	517 (1.2)	573 (1.5)	653 (5.8)
England	360 (2.0)	442 (1.5)	496 (0.6)	551 (0.3)	632 (2.4)
Finland	408 (1.5)	479 (0.8)	523 (1.0)	565 (1.1)	623 (2.6)
Hong Kong, SAR	456 (0.8)	538 (0.4)	587 (0.5)	632 (1.5)	693 (1.0)
Hungary	386 (3.9)	476 (1.6)	536 (0.8)	590 (1.0)	667 (2.4)
Indonesia	239 (6.0)	337 (3.5)	401 (2.3)	469 (1.2)	574 (2.3)
Iran, Islamic Rep.	284 (3.9)	367 (1.0)	423 (1.0)	478 (2.2)	556 (3.8)
Israel	300 (6.1)	402 (1.8)	473 (1.5)	534 (0.9)	614 (2.0)
Italy	331 (2.6)	423 (2.6)	482 (1.6)	540 (1.7)	615 (3.2)
Japan	441 (1.4)	529 (1.4)	583 (0.9)	633 (1.5)	702 (2.2)
Jordan	258 (0.7)	357 (2.3)	429 (1.0)	498 (1.8)	596 (1.1)
Korea, Rep. of	448 (2.1)	538 (1.5)	592 (1.1)	640 (1.3)	710 (2.3)
Latvia (LSS)	377 (1.4)	453 (2.9)	505 (1.4)	557 (1.5)	631 (1.8)
Lithuania	354 (2.7)	429 (2.3)	482 (1.6)	534 (2.3)	608 (3.5)
Macedonia, Rep. of	287 (5.2)	386 (0.9)	451 (0.5)	510 (1.2)	594 (2.5)
Malaysia	387 (2.2)	464 (0.6)	519 (0.6)	577 (1.3)	648 (2.5)
Moldova	331 (3.5)	412 (1.9)	468 (1.6)	528 (0.5)	607 (2.0)
Morocco	181 (5.4)	277 (3.3)	340 (1.7)	401 (1.5)	477 (0.8)
Netherlands	410 (4.8)	495 (1.9)	545 (1.5)	590 (1.9)	653 (3.7)
New Zealand	341 (3.5)	430 (1.7)	493 (1.4)	554 (2.3)	632 (4.0)
Philippines	185 (5.0)	278 (2.0)	345 (2.7)	414 (2.5)	504 (4.1)
Romania	312 (7.3)	412 (3.2)	477 (1.6)	537 (0.9)	616 (3.7)
Russian Federation	385 (3.7)	471 (1.0)	526 (0.5)	584 (1.2)	666 (2.9)
Singapore	464 (3.2)	555 (1.3)	608 (2.0)	658 (1.7)	728 (3.7)
Slovak Republic	407 (2.5)	485 (0.5)	534 (1.1)	585 (1.2)	656 (2.2)
Slovenia	392 (2.3)	476 (0.9)	531 (1.5)	587 (1.4)	663 (3.9)
South Africa	113 (8.2)	200 (4.1)	263 (4.7)	337 (3.7)	485 (3.2)
Thailand	328 (4.4)	412 (1.9)	465 (1.0)	524 (1.3)	609 (2.5)
Tunisia	341 (2.9)	406 (1.6)	449 (0.8)	491 (1.1)	551 (1.5)
Turkey	290 (4.7)	371 (2.6)	428 (1.7)	486 (1.0)	572 (1.6)

SOURCE: IEA Third International Mathematics and Science Study (TIMSS), 1998-1999.

( ) Standard errors appear in parentheses.

	5th Percentile	25th Percentile	50th Percentile	75th Percentile	95th Percentile
States					
Connecticut	370 (3.8)	457 (0.2)	516 (1.3)	567 (2.0)	649 (3.0)
Idaho	347 (5.1)	445 (1.9)	502 (1.8)	551 (3.4)	616 (2.8)
Illinois	373 (0.6)	456 (1.6)	510 (2.2)	563 (2.4)	645 (4.2)
Indiana	386 (4.5)	466 (2.5)	516 (1.9)	566 (1.7)	639 (1.0)
Maryland	348 (4.4)	433 (1.6)	496 (2.2)	559 (1.6)	632 (3.2)
Massachusetts	377 (4.0)	461 (1.9)	514 (1.0)	568 (1.3)	642 (3.3)
Michigan	373 (1.3)	465 (0.6)	522 (1.3)	574 (1.5)	641 (2.3)
Missouri	357 (4.2)	441 (2.6)	493 (2.0)	542 (2.1)	610 (4.1)
North Carolina	358 (4.2)	437 (3.7)	496 (1.8)	555 (1.6)	629 (3.7)
Oregon	370 (1.8)	462 (1.7)	519 (2.2)	570 (1.2)	642 (1.1)
Pennsylvania	368 (3.9)	455 (1.9)	511 (1.9)	563 (1.0)	637 (5.6)
South Carolina	352 (3.0)	438 (1.5)	503 (1.5)	567 (2.7)	647 (3.5)
Texas	365 (3.5)	453 (0.9)	522 (1.1)	581 (1.7)	654 (2.2)
Districts and Consortia					
Academy School Dist. #20, CO	400 (2.8)	479 (2.6)	533 (1.0)	579 (1.6)	647 (4.0)
Chicago Public Schools, IL	337 (3.1)	412 (1.2)	462 (2.0)	515 (2.6)	588 (3.6)
Delaware Science Coalition, DE	325 (3.6)	421 (2.6)	482 (2.7)	544 (2.5)	619 (3.4)
First in the World Consort., IL	435 (0.0)	510 (0.1)	560 (0.1)	610 (0.3)	684 (0.8)
Fremont/Lincoln/WestSide PS, NE	334 (5.6)	429 (1.0)	496 (3.7)	550 (3.2)	627 (5.1)
Guilford County, NC	372 (5.3)	456 (2.0)	519 (1.0)	572 (1.1)	644 (4.8)
Jersey City Public Schools, NJ	331 (3.2)	416 (0.8)	474 (2.0)	531 (2.6)	619 (2.4)
Miami-Dade County PS, FL	257 (6.0)	358 (1.7)	420 (3.6)	489 (1.9)	585 (3.5)
Michigan Invitational Group, MI	409 (3.6)	483 (2.3)	537 (1.2)	580 (2.6)	649 (5.6)
Montgomery County, MD	392 (6.3)	485 (1.5)	544 (2.7)	595 (3.2)	665 (1.2)
Naperville Sch. Dist. #203, IL	458 (3.6)	524 (1.2)	569 (2.7)	614 (1.9)	681 (2.4)
Project SMART Consortium, OH	393 (3.6)	467 (1.7)	520 (1.7)	576 (2.4)	642 (1.0)
Rochester City Sch. Dist., NY	312 (7.1)	391 (5.0)	442 (2.9)	497 (1.3)	585 (5.3)
SW Math/Sci. Collaborative, PA	381 (4.9)	462 (0.9)	519 (1.1)	572 (1.0)	649 (3.5)

SOURCE: IEA Third International Mathematics and Science Study (TIMSS), 1998-1999.

8th Grade Mathematics

	Overall		Girls		Boys	
	Mean	Standard Deviation	Mean	Standard Deviation	Mean	Standard Deviation
United States	502 (4.0)	88 (2.4)	498 (3.9)	84 (2.1)	505 (4.8)	91 (3.0)
Australia	525 (4.8)	80 (2.9)	524 (5.7)	77 (3.8)	526 (5.7)	83 (3.3)
Belgium (Flemish)	558 (3.3)	77 (2.8)	560 (7.2)	74 (4.8)	556 (8.3)	79 (5.6)
Bulgaria	511 (5.8)	86 (3.8)	510 (5.9)	84 (3.5)	511 (6.9)	88 (4.7)
Canada	531 (2.5)	73 (1.7)	529 (2.5)	72 (1.9)	533 (3.2)	74 (1.7)
Chile	392 (4.4)	85 (3.5)	388 (4.3)	82 (2.9)	397 (5.8)	89 (4.2)
Chinese Taipei	585 (4.0)	104 (1.8)	583 (3.9)	98 (2.4)	587 (5.3)	110 (2.1)
Cyprus	476 (1.8)	82 (1.7)	479 (2.1)	77 (2.2)	474 (2.7)	85 (1.9)
Czech Republic	520 (4.2)	79 (2.4)	512 (4.0)	78 (2.6)	528 (5.8)	80 (2.9)
England	496 (4.1)	83 (2.2)	487 (5.4)	79 (3.5)	505 (5.0)	86 (2.8)
Finland	520 (2.7)	65 (1.3)	519 (3.0)	63 (1.7)	522 (3.5)	68 (1.7)
Hong Kong, SAR	582 (4.3)	73 (3.0)	583 (4.7)	69 (2.8)	581 (5.9)	77 (4.4)
Hungary	532 (3.7)	85 (2.0)	529 (4.0)	82 (2.3)	535 (4.3)	88 (2.6)
Indonesia	403 (4.9)	101 (2.9)	401 (5.4)	102 (3.4)	405 (5.0)	101 (3.1)
Iran, Islamic Rep.	422 (3.4)	83 (2.3)	408 (4.2)	81 (2.5)	432 (4.8)	83 (2.5)
Israel	466 (3.9)	96 (2.6)	459 (4.2)	90 (2.4)	474 (4.8)	100 (3.8)
Italy	479 (3.8)	87 (2.3)	475 (4.5)	85 (2.8)	484 (4.3)	88 (2.8)
Japan	579 (1.7)	80 (1.1)	575 (2.4)	76 (2.1)	582 (2.3)	82 (1.3)
Jordan	428 (3.6)	103 (1.6)	431 (4.7)	96 (2.3)	425 (5.9)	109 (2.2)
Korea, Rep. of	587 (2.0)	79 (1.0)	585 (3.1)	79 (1.3)	590 (2.2)	80 (1.6)
Latvia (LSS)	505 (3.4)	78 (2.0)	502 (3.8)	75 (2.6)	508 (4.4)	81 (2.4)
Lithuania	482 (4.3)	78 (2.6)	480 (4.7)	76 (3.2)	483 (4.8)	80 (3.2)
Macedonia, Rep. of	447 (4.2)	93 (2.5)	446 (5.3)	92 (3.0)	447 (4.3)	94 (2.8)
Malaysia	519 (4.4)	81 (2.0)	521 (4.7)	79 (2.2)	517 (6.0)	83 (2.6)
Moldova	469 (3.9)	85 (2.1)	468 (4.1)	83 (2.6)	471 (4.7)	87 (2.6)
Morocco	337 (2.6)	91 (2.0)	326 (5.3)	90 (3.5)	344 (4.1)	91 (2.0)
Netherlands	540 (7.1)	73 (4.2)	538 (7.6)	73 (4.4)	542 (7.0)	74 (4.3)
New Zealand	491 (5.2)	89 (2.3)	495 (5.5)	87 (2.9)	487 (7.6)	91 (3.0)
Philippines	345 (6.0)	97 (2.8)	352 (6.9)	96 (4.1)	337 (6.5)	98 (2.8)
Romania	472 (5.8)	93 (3.5)	475 (6.3)	90 (3.8)	470 (6.2)	96 (3.8)
Russian Federation	526 (5.9)	86 (3.0)	526 (6.0)	83 (3.0)	526 (6.4)	90 (3.8)
Singapore	604 (6.3)	79 (2.9)	603 (6.1)	76 (3.0)	606 (7.5)	82 (3.3)
Slovak Republic	534 (4.0)	75 (1.6)	532 (4.2)	72 (1.9)	536 (4.5)	79 (2.3)
Slovenia	530 (2.8)	83 (2.0)	529 (3.0)	79 (1.9)	531 (3.6)	86 (2.8)
South Africa	275 (6.8)	109 (4.7)	267 (7.5)	110 (5.1)	283 (7.3)	108 (4.7)
Thailand	467 (5.1)	85 (2.5)	469 (5.7)	84 (2.8)	465 (5.5)	86 (2.9)
Tunisia	448 (2.4)	64 (0.9)	436 (2.4)	64 (1.2)	460 (2.9)	61 (1.3)
Turkey	429 (4.3)	86 (2.0)	428 (4.7)	83 (2.1)	429 (4.4)	87 (2.4)

SOURCE: IEA Third International Mathematics and Science Study (TIMSS), 1998-1999.

( ) Standard errors appear in parentheses.

Standard Deviations of Achievement in Mathematics – States and Districts/Consortia

8th Grade Mathematics

	Overall		Girls		Boys	
	Mean	Standard Deviation	Mean	Standard Deviation	Mean	Standard Deviation
States						
Connecticut	512 (9.1)	85 (4.8)	506 (8.9)	81 (4.2)	520 (9.8)	89 (5.9)
Idaho	495 (7.4)	82 (3.4)	495 (7.1)	78 (3.0)	495 (8.2)	86 (4.7)
Illinois	509 (6.7)	82 (2.6)	505 (8.0)	81 (3.5)	514 (6.1)	83 (3.2)
Indiana	515 (7.2)	76 (2.4)	510 (6.8)	72 (3.0)	519 (8.0)	80 (3.4)
Maryland	495 (6.2)	88 (3.5)	490 (6.4)	85 (3.7)	499 (6.8)	90 (3.9)
Massachusetts	513 (5.9)	82 (3.2)	510 (6.4)	79 (3.7)	517 (6.0)	84 (3.2)
Michigan	517 (7.5)	81 (3.8)	512 (7.2)	77 (3.6)	522 (8.1)	85 (4.5)
Missouri	490 (5.3)	77 (2.9)	488 (5.9)	74 (3.3)	491 (5.6)	81 (3.4)
North Carolina	495 (7.0)	84 (2.6)	494 (7.9)	82 (2.7)	497 (6.9)	86 (3.4)
Oregon	514 (6.0)	83 (2.6)	514 (6.6)	79 (3.2)	514 (6.9)	86 (2.9)
Pennsylvania	507 (6.3)	82 (3.0)	503 (6.2)	77 (3.3)	512 (7.2)	86 (3.1)
South Carolina	502 (7.4)	90 (3.1)	501 (8.0)	85 (3.0)	502 (7.6)	95 (4.1)
Texas	516 (9.1)	90 (3.6)	513 (8.2)	86 (3.3)	519 (10.7)	95 (4.7)
Districts and Consortia						
Academy School Dist. #20, CO	528 (1.8)	74 (1.7)	526 (2.9)	71 (2.3)	531 (3.4)	77 (2.6)
Chicago Public Schools, IL	462 (6.1)	76 (3.0)	460 (6.3)	74 (3.4)	465 (6.7)	79 (3.5)
Delaware Science Coalition, DE	479 (8.9)	90 (4.3)	475 (8.9)	85 (4.1)	485 (11.1)	94 (5.9)
First in the World Consort., IL	560 (5.8)	77 (4.3)	556 (6.7)	74 (3.5)	564 (6.8)	80 (6.5)
Fremont/Lincoln/WestSide PS, NE	488 (8.2)	89 (6.1)	485 (8.3)	88 (8.9)	491 (10.2)	89 (6.3)
Guilford County, NC	514 (7.7)	85 (3.2)	507 (8.3)	82 (4.2)	521 (8.2)	87 (3.3)
Jersey City Public Schools, NJ	475 (8.6)	87 (4.5)	472 (8.8)	87 (4.7)	478 (9.2)	87 (6.3)
Miami-Dade County PS, FL	421 (9.4)	99 (4.0)	419 (9.3)	93 (5.7)	423 (12.1)	104 (3.8)
Michigan Invitational Group, MI	532 (5.8)	73 (3.6)	535 (5.4)	71 (3.6)	529 (7.4)	74 (4.3)
Montgomery County, MD	537 (3.5)	86 (2.9)	534 (5.5)	86 (6.0)	540 (4.4)	85 (3.4)
Naperville Sch. Dist. #203, IL	569 (2.8)	69 (2.0)	566 (3.3)	68 (3.1)	573 (3.3)	69 (2.6)
Project SMART Consortium, OH	521 (7.5)	77 (3.6)	518 (7.8)	76 (4.1)	523 (8.1)	77 (4.0)
Rochester City Sch. Dist., NY	444 (6.5)	82 (4.5)	439 (7.8)	78 (4.6)	450 (6.6)	85 (5.4)
SW Math/Sci. Collaborative, PA	517 (7.5)	82 (3.7)	509 (7.5)	80 (4.0)	525 (8.5)	83 (3.9)

SOURCE: IEA Third International Mathematics and Science Study (TIMSS), 1998-1999.

APPENDIX D

Descriptions of Mathematics Items at Each Benchmark

D

Lower Quarter Benchmark Items

Fractions and Number Sense

H09 Rounds to estimate the sum of two three-digit numbers.

R07 Subtracts a three-decimal-place number from another with multiple regrouping.

R13 Subtracts a four-digit number from another involving zeroes.

Data Representation, Analysis, and Probability

A06 Calculates and compares the averages of two sets of data.

P16 Reads a thermometer and locates the reading in a table.

Algebra

P09 Selects an expression in exponential notation for repeated multiplication.

Median Benchmark Items

Fractions and Number Sense

B08 Solves a word problem by finding the missing term in a proportion.

C06 Determines which is the most unreasonable estimate for two 3-digit numbers.

D12 Estimates the value, to one decimal place, of a point on a number line marked at whole number intervals.

E04 Arranges four given digits in descending and ascending order and finds the difference between those two numbers.

H08 Selects a figure with shaded parts that represents a familiar fraction.

I05 Solves a word problem involving subtraction of a two-place decimal number from another.

K01 Identifies a circular model of a fraction that best approximates a given rectangular model of the same fraction.

Median Benchmark Items continued

Fractions and Number Sense continued

K02 Solves a word problem by adding numbers with up to three decimal places.

K06 Selects the approximate quantity remaining after an amount is decreased by a given percent.

L09 Given an object of one length, to one decimal place, estimates the length of a second object in a diagram.

L10 Identifies the numerical equivalent of a decimal number given in words.

M04 Selects the smallest fraction from a set of familiar fractions.

N11 Rounds a number less than 100,000 to the nearest hundred.

P13 In a word problem, uses rounding to identify the number sentence that gives the best estimate for the product.

Measurement

D11 Selects appropriate metric unit to measure weight (mass).

G02 Identifies an unlabeled midway point on a number line marked in tenths.

L13 Recognizes the inverse relation between length of nonstandard units and the number of those units required to cover a distance.

Data Representation, Analysis, and Probability

B07 Given two line graphs, identifies the relevant one and determines the interval showing the greatest increase.

C02 Reads and interprets information from a pie graph.

E01 Solves problem by interpreting information from a graph of two intersecting lines.

H07 Reads data from a bar graph to solve a word problem.

J13 Determines how many items are represented by one symbol in a pictogram.

M03 Recognizes that the probability of an outcome of a single event is inversely related to the number of elements in the population of events.

Data Representation, Analysis, and Probability continued

M09 Given a table of values for two variables, selects the graph that could represent the given data.

Q04 Solves a comparison problem by associating elements of a bar graph with a verbal description.

Geometry

C03 Identifies corresponding parts of congruent trapezoids.

J15 Selects the pair of similar triangles from a set of triangles.

J16 Locates the point on a grid with 5-unit divisions when the point lies between the grid lines.

K03 Identifies the diagrammatic representation of a three-dimensional object after rotation.

Algebra

A02 Using properties of a balance, reasons to find an unknown weight (mass).

B12 Identifies the linear equation corresponding to a given verbal statement involving a variable.

D08 Solves for missing number in a proportion.

G04 Solves equation for missing number in a proportion.

H12 Selects the expression that represents a situation involving multiplication.

J17 Finds a missing y value in a table relating x and y values.

V04A Given a sequence of diagrams growing in one-dimension and a partially completed table, finds the next two terms in the table.

Upper Quarter Benchmark Items

Fractions and Number Sense

A01 Finds 4/5 of a region divided into 10 equal parts.

A04 Solves a word problem by finding the missing term in a proportion.

B09 Given two equivalent fractions, identifies the pictorial representation showing they are equivalent.

B10 Selects the smallest of a set of numbers with differing numbers of decimal places.

D09 Selects the smallest fraction from a set of familiar fractions.

E03 Identifies the fraction of an hour representing a time interval.

F09 Identifies a decimal number given in thousandths between two decimal numbers given in hundredths.

F12 Identifies the interval containing the fraction that represents the shaded part of a circle.

G05 Selects a fraction representing the comparison of part to whole, given each of two parts in a word-problem setting.

I02 Solves a multi-step word problem involving multiplication of whole numbers by fractions.

I06 Writes a fraction less than a given fraction.

J12 Divides one fraction by another with unlike denominator.

J18 Uses a map scale to find the approximate distance between two towns.

K09 Adds three fractions with denominators less than 10.

L18 Subtracts fractions with unlike denominators.

N14 Selects a set of equivalent fractions.

N19 Shades squares in a rectangular grid to represent a given fraction.

O04 Rounds a four-place decimal to the nearest hundredth.

O09 Solves a one-step word problem involving division of a whole number by a unit fraction.

Fractions and Number Sense continued

P14 Estimates the product of two whole numbers in a word problem.

Q05 Selects the statement that describes the effect of adding the same amount to both terms of a ratio.

Q06 Estimates the product of a multiple of 1000 and a two-digit number in a word problem involving knowledge of units of time.

Q09 Multiplies and adds fractions with different denominators in the correct order.

R15 Solves a multi-step word problem that involves dividing a quantity in a given ratio.

T04 Solves a word problem that involves multiplying a decimal in thousandth by a multiple of a hundred.

V01 Provides an example of a measure that would round to a given value.

V03 Determines the ratio of part to total in a word problem.

Measurement

A03 Given a length rounded to the nearest centimeter, identifies what the actual length could have been in centimeters to one decimal place.

C01 Compares volume by visualizing and counting cubes.

I07 Finds the area between two rectangles when one is inside the other and their sides are parallel.

M01 Reads the value indicated by an unlabeled tick mark on a circular scale.

N15 Identifies an angle of a given size in a diagram.

O06 Given the start time, and the duration of an event expressed as a fraction of an hour, determines the end time.

S02A Finds a fraction of a given area of an irregular figure composed of squares of equal sides.

T03 Finds the area of a rectangle contained in a parallelogram of given dimensions.

Upper Quarter Benchmark Items continued

Data Representation, Analysis, and Probability

F08 Understands independence in a probability setting.

H11 In a word problem, finds the missing number in a proportion.

I09 Given the set of possible outcomes expressed as fractions of all outcomes, recognizes that probability is associated with the size of a fraction.

K07 In a word problem, when given the possible number of outcomes and the probability of successful outcomes, solves for the number of successful outcomes.

L11 Determines the number of values on the horizontal axis of a line graph that correspond with a given value on the vertical axis.

N18 Identifies the number of successful outcomes of a simple experiment and calculates the probability of success.

O01 On a given graph, interpolates to find a value between graduations on one axis matching a given value on the other axis.

O05 In a word problem, when given the possible number of outcomes and the probability of successful outcomes, solves for the number of successful outcomes.

R09 Solves a word problem by extrapolating a graph of a non-linear relationship.

Geometry

A05 Identifies pairs of congruent triangles.

B11 Visualizes the arrangement of the faces of a cube given its net.

D07 Applies knowledge of symmetry to select the measure of an angle.

E02 Uses properties of congruent triangles to find the measure of an angle.

G03 Solves a problem involving adjacent and vertical angles.

J11 Identifies a false statement about the properties of rectangles.

M07 Uses knowledge of a straight angle to find the measure of an angle.

O03 Given two parallel lines cut by a transversal, selects a pair of supplementary angles.

O08 Selects the center of rotation when shown a diagram of a triangle and its image under a quarter turn.

Geometry continued

R11 Determines the number of triangles of given dimensions needed to cover a given rectangle.

Algebra

C05 Finds a specified term in a sequence given the first three terms pictorially.

D10 Identifies algebraic equation (formula) corresponding to a verbal description involving a constant and two variables.

E05 Identifies the linear relationship between the first and second terms in a set of ordered pairs.

F11 Solves a two-step problem involving multiplication and division of whole numbers and fractions.

G06 Finds the value of an algebraic expression involving multiplication of negative integers.

H10 Identifies the linear equation that describes the relationship between two variables given a table of values.

L12 Translates a word problem into a short finite arithmetic sequence and sum the sequence.

L14 Identifies the linear equation that describes the relationship between two variables given a table of values.

L17 Solves a linear equation involving transposing.

N13 Evaluates a rational expression in one variable for a given value of the variable.

O07 Solves a linear equation involving parentheses.

P11 Selects a multiplicative expression for repeated addition.

Q02 Subtracts algebraic fractions with the same numeric denominator.

Q07 Selects the formula satisfied by the given values of the variables.

R12 Selects a simple, multiplicative expression in one variable that is positive for all negative values of that variable.

S01A Given a sequence of diagrams growing in two dimensions and a partially completed table, finds the next two terms in the table.

V04B Knowing the first five terms of a sequence growing in one dimension, finds the seventh term.

Top 10% Benchmark Items

Fractions and Number Sense

C04	Identifies the pair of numbers satisfying given conditions involving ordering integers, decimals, and common fractions.
F07	Solves a time-distance-rate problem that involves division of decimals and conversion of minutes to seconds.
J14	Identifies the correct position for the decimal point in the quotient in a division of a decimal written in hundredths by a decimal written in thousandths.
M06	Given the total number and the ratio of the two parts, finds the value of one part.
M08	Multiplies a two-place decimal by a three-place decimal.
N16	Solves multi-step problem with fractions requiring analysis of the verbal relations described.
N17	Solves a word problem involving multiplication of two-digit one-place decimals and subtraction of decimals.
O02	Finds the percent change given the original and the new quantities.
P15	Solves a word problem involving both addition and subtraction of familiar fractions.
P17	Writes a decimal expressed in hundredths as a fraction in lowest terms.
Q08	Orders a set of decimals of up to three decimal places.
R08	In a word problem, finds an average by dividing a decimal by a multiple of 100.
R14	Solves a two-step problem involving multiplication of a whole number by a fraction.
T02A	In a multi-step word problem, finds how many of each of two groups of different sizes are required to produce a given number.

Measurement

E06 Identifies the length of a rectangle given its perimeter and width.

F10 Recognizes that precision of measurement is related to the size of the unit of measurement.

I03 Applies knowledge of number of milliliters in a liter to solve a word problem.

J10 Finds the area between two rectangles when one is inside the other and their sides are parallel.

K05 Finds the perimeter of a square given that its area is a square number less than 150.

P12 Estimates the length of a curved piece of string adjacent to a ruler.

Q03 From a set of times expressed variously in days, hours, minutes, and seconds, determines which is least.

S02B Finds the length of a side of a square, given that its area is a square number.

U02A Uses computation with fractions to find the length and width of a rectangle and draws and labels that rectangle on a grid.

U02B Given the dimensions of two rectangles, expresses the ratio of their areas.

U03 Finds the area of a triangle, on the same base and with the same height as a square, when the length of a side of the square is known.

Data Representation, Analysis, and Probability

G01 Reads data from a frequency table to solve a problem.

Geometry

I08 Given only the coordinates of two points on the line, selects the coordinates of a third point on that line.

K08 Uses properties of congruent triangles and the sum of the angles of a triangle to find the measure of an angle.

L16 Identifies the measure of an angle of a quadrilateral given the measures of the other three angles.

Top 10% Benchmark Items continued

Geometry continued

M05 Identifies the image of a triangle under a rotation about a point in the plane.

N12 Locates a point on a number line given its distance from two given points.

P10 Uses properties of similar triangles to find the length of a corresponding side.

Q10 Solves a problem involving measures of overlapping angles.

Algebra

K04 Solves a linear inequality involving a fraction.

L15 Uses proportion to find missing values in a table.

R10 Recognizes properties of operations on real numbers represented in symbolic form.

S01B Knowing the first five terms of a sequence growing in one dimension, finds the seventh term.

T01 Solves a multi-step word problem in which there are two unknowns and displays the method of solution.

V04C Given the initial terms in a sequence and, for example, the 50th term of that sequence, generalizes to find the next term.

Items Above the Top 10% Benchmark

Fractions and Number Sense

S03 Solves a word problem involving multiplication and subtraction of decimals.

T02B Uses information in a word problem to determine numerator and denominator and writes the relevant common fraction.

U01 Estimates the total time in minutes for an event made up of a series of events, each given in minutes and seconds.

Measurement

P08 Finds the ratio of width to perimeter for a rectangle when given the ratio of length to width of a rectangle.

S02C Finds the perimeter of a figure made up of squares with known length of sides.

Data Representation, Analysis, and Probability

V02 Selects relevant information from two advertisements to solve a complex word problem involving decimals.

Algebra

I01 Identifies what the variable represents in an equation for a given situation.

I04 Identifies numbers common to two different arithmetic sequences.

Q01 Selects an algebraic expression to answer a question about a set of linked verbal statements.

S01C Generalizing from the first several terms of a sequence growing in two dimensions, explains a way to find a specified term, e.g. the 50th.

APPENDIX E Acknowledgments

E

TIMSS 1999 and the TIMSS Benchmarking Study were collaborative efforts among hundreds of individuals around the world. Staff from the national research centers in each participating country and from each Benchmarking jurisdiction, the International Association for the Evaluation for Educational Achievement (IEA), the International Study Center (ISC) at Boston College, advisors, and funding agencies worked closely to develop and implement the projects. They would not have been possible without the tireless efforts of all involved. Below, the individuals and organizations are acknowledged for their contributions. Given that implementing the studies has spanned approximately four years and involved so many people and organizations, this list may not pay heed to all who contributed throughout the life of the project. Any omission is inadvertent. TIMSS 1999 and the Benchmarking Study also acknowledge the students, teachers, and school principals who contributed their time and effort to the study. This report would not be possible without them.

Funding Agencies

Funding for the international coordination of TIMSS 1999 was provided by the National Center for Education Statistics (NCES) in the U.S. Department of Education, the U.S. National Science Foundation (NSF), the World Bank, and participating countries. Each participating country was responsible for funding local project costs and implementing TIMSS 1999 in accordance with the international procedures. Funding for the overall design, administration, data management, and quality assurance activities of TIMSS Benchmarking was provided by NCES, NSF, and the Office of Educational Research and Improvement (OERI) in the U.S. Department of Education. Valena Plisko, Eugene Owen, and Patrick Gonzales of NCES; Janice Earle, Larry Suter, and Elizabeth VanderPutten of NSF; Carol Sue Fromboluti and Jill Edwards Staton of OERI, and Maggie McNeely formerly of OERI each played a crucial role in making TIMSS 1999 and the Benchmarking Study possible and for ensuring the quality of the studies. Each Benchmarking participant contracted directly with Boston College to fund data-collection activities in its own jurisdiction.

Management and Operations

TIMSS 1999 was conducted under the auspices of the IEA. TIMSS 1999 was co-directed by Michael O. Martin and Ina V.S. Mullis, and managed centrally by the staff of the International Study Center in the Lynch School of Education at Boston College. Although the study was directed by the International Study Center and its staff members implemented various parts of TIMSS 1999, important activities also were carried out in centers around the world. In the IEA Secretariat in Amsterdam, Hans Wagemaker, Executive Director, was responsible for overseeing fundraising and country participation. The IEA Secretariat also coordinated translation verification and recruiting of international quality control monitors. The data were processed centrally by the IEA Data Processing Center in Hamburg. Statistics Canada in Ottawa was responsible for collecting and evaluating the sampling documentation from each country and for calculating the sampling weights. Educational Testing Service (ETS) in Princeton, New Jersey, conducted the scaling of the achievement data.

For the Benchmarking Study, Westat in Rockville, Maryland, was responsible for sampling, data collection activities, and preliminary data processing. National Computer Systems (NCS) in Iowa City, Iowa, conducted the scoring for Benchmarking jurisdictions along with the national scoring effort. All data were processed in accordance with international standards at the IEA Data Processing Center. Scaling of the achievement data was conducted by Educational Testing Service.

IEA Secretariat

Hans Wagemaker, Executive Director

Barbara Malak, Manager Membership Relations

Leendert Dijkhuizen, Fiscal Officer

International Study Center at Boston College Responsible for TIMSS and PIRLS

Michael O. Martin, Co-Director

Ina V.S. Mullis, Co-Director

Eugenio J. González, Director of Operations and Data Analysis

Kathleen M. O'Connor, TIMSS Benchmarking Coordinator

Kelvin D. Gregory, TIMSS Study Coordinator

Teresa A. Smith, TIMSS Science Coordinator

Robert Garden, TIMSS Mathematics Coordinator

Dana L. Kelly, PIRLS Study Coordinator

Steven Chrostowski, Research Associate

Ce Shen, Research Associate (former)

Julie Miles, Research Associate

Steven Stemler, Research Associate

Ann Kennedy, Research Associate

Maria José Ramirez, Research Assistant

Joseph Galia, Statistician/Programmer

Lana Seliger, Statistician/Programmer (former)

Andrea Pastelis, Database Manager

Kieran Brosnan, Technology Support Specialist

Christine Conley, Publications Design Manager

José Nieto, Publications Manager

Tom Hoffmann, Internet Communications Manager

Mario Pita, Data Graphics Specialist

Betty Hugh, Data Graphics Specialist

Christina Lopez, Data Graphics Specialist (former)

Isaac Li, Data Graphics Assistant

Kathleen Packard, Manager, Finance

Susan Comeau, Manager, Office Administration

Ann Tan, Manager, Conference Administration

Monica Guidi, Administrative Coordinator

Laura Misas, Administrative Coordinator

Rita Holmes, Administrative Coordinator

Statistics Canada

Pierre Foy, Senior Methodologist

Marc Joncas, Senior Methodologist

Andrea Farkas, Junior Methodologist

Salina Park, Cooperative Exchange Student

IEA Data Processing Center

Dirk Hastedt, Senior Researcher

Heiko Sibberns, Senior Researcher

Knut Schwippert, Senior Researcher

Caroline Dupeyrat, Researcher

Oliver Neuschmidt, Researcher

Stephan Petzchen, Research Assistant

Anneke Niemeyer, Research Assistant

Juliane Pickel, Research Assistant

Educational Testing Service

Kentaro Yamamoto, Principal Research Scientist

Ed Kulick, Manager, Research Data Analysis

Westat

Nancy Caldwell, Vice President and Associate Director, Survey Operations Group

Keith Rust, Vice President and Associate Director, Statistical Group

Stephen Roey, Senior Systems Analyst

Project Management Team

Michael Martin, International Study Center

Ina Mullis, International Study Center

Eugenio González, International Study Center

Hans Wagemaker, IEA Secretariat

Dirk Hastedt, IEA Data Processing Center

Pierre Foy, Statistics Canada

Kentaro Yamamoto, Educational Testing Service

Eugene Johnson, American Institutes for Research

Sampling Referees

Pierre Foy, Statistics Canada – TIMSS 1999 Benchmarking

Keith Rust, Westat – TIMSS 1999 International

Benchmarking Participants

Individuals from each Benchmarking jurisdiction were instrumental in conducting the TIMSS Benchmarking Study in their state, district, or consortium. They were responsible for obtaining funding for the project; obtaining cooperation of sampled schools, classes, and students; responding to curriculum questionnaires; reviewing data; contributing to the development of the Benchmarking reports; and coordinating activities with the International Study Center. Jurisdictions would like to acknowledge the following people for their extensive contributions.

States

Connecticut

Patricia Brandt
Connecticut Department of Education
165 Capital Avenue
Hartford CT 06145-2219

Abigail L. Hughes
Connecticut Department of Education
165 Capital Avenue
Hartford CT 06145-2219

Douglas Rindone
Connecticut Department of Education
165 Capital Avenue
Hartford CT 06145-2219

Theodore S. Sergi
Connecticut Department of Education
165 Capital Avenue
Hartford CT 06145-2219

Idaho

Tom Farley
Idaho Department of Education
P.O. Box 83720
Boise ID 83720-0027

Susan Harrington
Idaho Department of Education
P.O. Box 83720
Boise ID 83720-0027

Sally Tiel
Idaho Department of Education
P.O. Box 83720
Boise ID 83720-0027

Illinois

Mervin Brennan
Illinois State Board of Education
100 North First Street
Springfield IL 62777

Carmen Chapman
Illinois State Board of Education
100 North First Street
Springfield IL 62777

Megan Forness
Illinois State Board of Education
Assessment E216
100 North First Street
Springfield IL 62777

Andy Metcalf
Illinois State Board of Education
100 North First Street
Springfield IL 62777

Pam Stanko
Illinois State Board of Education
100 North First Street
Springfield IL 62777

Indiana

Larry Grau
Office of the Governor
State House
200 West Washington Street, Room 206
Indianapolis IN 46204-2797

Dwayne James
Indiana Department of Education
Room 229, State House
Indianapolis IN 46204

Stan Jones
Commissioner for Higher Education
101 West Ohio Street - Suite 550
Indianapolis IN 46204

Cheryl Orr
Indiana's Education Roundtable
101 West Ohio Street - Suite 550
Indianapolis IN 46204

Suellen Reed
Superintendent of Public Instruction
Room 229, State House
Indianapolis IN 46204-2797

Cynthia Roach
Indiana Department of Education
Division of Assessment
Room 229, State House
Indianapolis IN 46204-2797

Maryland

Diane Householder
Maryland State Department of Education
200 West Baltimore Street
Baltimore MD 21201-2595

Mark Moody
Maryland State Department of Education
200 West Baltimore Street
Baltimore MD 21201-2595

Kathy Rosenberg
Maryland State Department of Education
200 West Baltimore Street
Baltimore MD 21201-2595

Massachusetts

Jeffrey Nellhaus
Massachusetts Department of Education
350 Main Street
Malden MA 02148-5023

Sheldon Rothman
Massachusetts Department of Education
350 Main Street
Malden MA 02148-5023

Kit Viator
Massachusetts Department of Education
350 Main Street
Malden MA 02148-5023

Lori Wright
Massachusetts Department of Education
350 Main Street
Malden MA 02148-5023

Michigan

Charles Allan
Michigan Department of Education
Curriculum Development Program
P.O. Box 30008
Lansing MI 48909

Missouri

James Friedebach
Missouri Department of Education
205 Jefferson
P.O. Box 480
Jefferson City MO 65102-0480

North Carolina

Louis Fabrizio
North Carolina Department
of Public Instruction
301 North Wilmington Street
Raleigh NC 27601-2825

Tammy Howard
North Carolina Department
of Public Instruction
301 North Wilmington Street
Raleigh NC 27601-2825

Oregon

Joanne Flint
Oregon Department of Education
255 Capital Street NE
Salem OR 97310-0203

Wayne Neuberger
Oregon Department of Education
255 Capital Street NE
Salem OR 97310-0203

Pennsylvania

R. Jay Gift
Pennsylvania Department of Education
333 Market Street, 8th Floor
Harrisburg PA 17126-0333

Frank Marburger
Pennsylvania Department of Education
333 Market Street, 8th Floor
Harrisburg PA 17126-0333

Lee Plempel
Pennsylvania Department of Education
333 Market Street, 8th Floor
Harrisburg PA 17126-0333

Charlie Wayne
Pennsylvania Department of Education
333 Market Street, 8th Floor
Harrisburg PA 17126-0333

South Carolina

Karen Horne
South Carolina Department of Education
1429 Senate Street
Columbia SC 29201

Susan Agruso
South Carolina Department of Education
1429 Senate Street
Columbia SC 29201

Lane Peeler
South Carolina Department of Education
611-B Rutledge Building
1429 Senate Street
Columbia SC 29201

Paul Sandifer
South Carolina Department of Education
607 Rutledge Building
1429 Senate Street
Columbia SC 29201

Teri Siskind
South Carolina Department of Education
607 Rutledge Building
1429 Senate Street
Columbia SC 29201

Texas

Chris Castillo Comer
Texas Education Agency
1701 North Congress Avenue
Austin TX 78701

Ed Miller
Texas Education Agency
1701 North Congress Avenue
Austin TX 78701-1494

Phyllis Stolp
Texas Education Agency
1700 North Congress Avenue
Austin TX 78701

Districts and Consortia

Academy School District #20

Wendy Crist
Academy School District #20
7610 North Union Boulevard
Colorado Springs CO 80920

Alisabeth Hohn
Academy School District #20
7610 North Union Boulevard
Colorado Springs CO 80920

Chicago Public Schools

Gery Chico
Chicago Public Schools
125 South Clark Street
Chicago IL 60603

Richard Daley
City Hall
121 North LaSalle Street
Chicago IL 60603

Joseph Hahn
Chicago Public Schools
125 South Clark Street 11th Floor
Chicago IL 60603

Phil Hansen
Chicago Public Schools
125 South Clark Street
Chicago IL 60603

Paul Vallas
Chicago Public Schools
125 South Clark Street
Chicago IL 60603

Melanie Wojtulewicz
Chicago Public Schools
1326 West 14th Place
Room 315A
Chicago IL 60608

Delaware Science Coalition

Gail Ames
Delaware Science Coalition
2916 Duncan Road
Wilmington DE 19808

John Collette
Delaware Science Coalition
309 Brockton Road
Wilmington DE 19803

Julie Cwikla Banks
University of Delaware
305 M Willard Hall
Newark DE 19716

Valerie Maxwell
Appoquinimink School District
118 South Sixth Street
Odessa DE 19730

First in the World Consortium

Elaine Aumiller
North Central Regional
Education Lab (NCREL)
1120 East Diehl Road, Suite 200
Naperville IL 60563

Blase Masini
North Central Regional
Education Lab (NCREL)
1120 East Diehl Road, Suite 200
Naperville IL 60563

Paul Kimmelman
1306 Hidden Lake Drive
Buffalo Grove IL 60089

David Kroeze
First in the World Consortium
Northbrook School District #27
1250 Sanders Road
Northbrook IL 60062

Fremont/Lincoln/Westside Public Schools

James Findley
Westside Public Schools
909 South 76th Street
Omaha NE 68114-4599

Marilyn Moore
Lincoln Public Schools
Box 82889
Lincoln NE 68501-2889

Stephen Sexton
Fremont Public Schools
957 North Pierce Street
Fremont NE 68025

Terry Snyder
Fremont Public Schools
957 North Pierce Street
Fremont NE 68025

Guilford County Schools

Lynne Johnson
Guilford County Schools
120 Franklin Boulevard
Greensboro NC 27401

Diane Spencer
Guilford County Schools
120 Franklin Boulevard
Greensboro NC 27401

Sadie Bryant Woods
Guilford County Schools
134 Franklin Boulevard
Greensboro NC 27401

Jersey City Public Schools

Richard DiPatri
Jersey City Public Schools
State District Superintendent
346 Claremont Avenue
Jersey City NJ 07305

Adele Macula
Jersey City Board of Education
346 Claremont Avenue
Jersey City NJ 07305

Aldo Sanchez-Abreu
Jersey City Board of Education
346 Claremont Avenue
Jersey City NJ 07305

Patsy Wang-Iverson
Mid-Atlantic Eisenhower Consortium
Research for Better Schools
444 North Third Street
Philadelphia PA 19123

Miami-Dade County Public Schools

Joseph Burke
Miami-Dade County Public Schools
1500 Biscayne Boulevard
Room 327T
Miami FL 33132

Gisela Feild
Miami-Dade County Public Schools
1500 Biscayne Boulevard
Suite 225
Miami FL 33132

Joseph Mathos
Miami-Dade County Public Schools
1450 Northeast 2nd Avenue #931
Miami FL 33132

Vilma Rubiera
Miami-Dade County Public Schools
1500 Biscayne Boulevard
Suite 225
Miami FL 33132

Alex Shneyderman
Miami-Dade County Public Schools
1500 Biscayne Boulevard
Suite 225
Miami FL 33132

Constance Thornton
Miami-Dade County Public Schools
1500 Biscayne Boulevard
Suite 327
Miami FL 33132

Michigan Invitational Group

Robert Dunn
Michigan Invitational Group
Michigan Department of Education
658 Grat Strasse
Manchester MI 48158

Montgomery County Public Schools

Marlaine Hartzman
Montgomery County Public Schools
850 Hungerford Drive, Room 11
Rockville MD 20850

John Larson
Montgomery County Public Schools
850 Hungerford Drive, Room 11
Rockville MD 20850

Naperville Community School District 203

Russ Bryan
Naperville Community School District 203
203 West Hillside Road
Naperville IL 60540

Lenore Johnson
Naperville Community School District 203
203 West Hillside Road
Naperville IL 60540

Jack Hinterlong
Naperville Community School District 203
203 Hillside Road
Naperville IL 60540

Donald E. Weber, Ed.D
Naperville Community School District 203
203 West Hillside Road
Naperville IL 60540

Jodi Wirt
Naperville Community School District 203
203 Hillside Road
Naperville IL 60540

Project SMART Consortium

Dennis Kowalski
Strongville City School
13200 Pearl Road
Stongsville OH 44136

Terry Krivak
c/o Ohio Aerospace Institute
22800 Cedar Point Road
Cleveland OH 44142

Anne Mikesell
Ohio Department of Education
25 South Front Street, 5th Floor
Columbus OH 43215

Linda Williams
Mentor Exempted Village
6451 Center Street
Mentor OH 44060

Paul R. Williams
Project SMART Consortium
Beachwood City School District
24601 Fairmount Boulevard
Beachwood OH 44122

Rochester City School District

Ann Pinnella Brown
Rochester City School District
131 West Broad Street
Rochester NY 14614

Cecilia Golden
Rochester City School District
131 West Broad Street
Rochester NY 14614

Corinthia Sims
Rochester City School District
131 West Broad Street
Rochester NY 14614

Southwest Pennsylvania Math and Science Collaborative

Nancy Bunt
2650 Regional Enterprise Tower
425 Sixth Avenue
Pittsburgh PA 15219

Marcia Seeley
2650 Regional Enterprise Tower
425 Sixth Avenue
Pittsburgh PA 15219

Lou Tamler
2650 Regional Enterprise Tower
425 Sixth Avenue
Pittsburgh PA 15219

Cynthia A. Tananis
University of Pittsburgh
5P26 WWPH School of Education
Pittsburgh PA 15260

National Research Coordinators

The TIMSS 1999 National Research Coordinators and their staff had the enormous task of implementing the TIMSS 1999 design. This required obtaining funding for the project; participating in the development of the instruments and procedures; conducting field tests; participating in and conducting training sessions; translating the instruments and procedural manuals into the local language; selecting the sample of schools and students; working with the schools to arrange for the testing; arranging for data collection, coding, and data entry; preparing the data files for submission to the IEA Data Processing Center; contributing to the development of the international reports; and preparing national reports. The way in which the national centers operated and the resources that were available varied considerably across the TIMSS 1999 countries. In some countries, the tasks were conducted centrally, while in others, various components were subcontracted to other organizations. In some countries, resources were more than adequate, while in some cases, the national centers were operating with limited resources. Of course, across the life of the project, some NRCs have changed. This list attempts to include all past NRCs who served for a significant period of time as well as all the present NRCs. All of the TIMSS 1999 National Research Coordinators and their staff members are to be commended for their professionalism and their dedication in conducting all aspects of TIMSS.

Australia

Susan Zammit
Australian Council for Educational Research (ACER)
19 Prospect Hill Road
Private Bag 55
Camberwell, Victoria 3124

Belgium (Flemish)

Jan Van Damme
Afd. Didactiek
Vesaliusstraat 2
B-3000 Leuven

Christiane Brusselmans-Dehairs
Jean-Pierre Verhaeghe
Vakgroep Onderwijskunde Universiteit Gent
Henri Dunantlaan 2
B-9000 Gent

Ann Van Den Broeck
Dekenstraat 2
Afd. Didactiek
B-3000 Leuven

Bulgaria

Kiril Bankov
Faculty of Mathematics and Informatics
University of Sofia
1164 Sophia

Canada

Alan Taylor
Applied Research and Evaluation Services (ARES)
University of British Columbia
6058 Pearl Avenue,
Burnaby, BC V5H 3P9

Richard Jones
Education Quality & Accountability Office (EQAO)
2 Carlton Street, Suite 1200
Toronto, ON M5B2M9

Jean-Louis Lebel
Direction de la sanction des etudes
1035 rue De La Chevrotiere
26 etage
Quebec GIR 5A5

Michael Marshall
University of British Columbia
Faculty of Education, Room 6
2125 Main Mall
Vancouver, BC V6T1Z4

Chile

Maria Inès Alvarez
Unidad de Curriculum y Evaluación
Ministerio de Educación
Alameda 1146
Sector B, Piso 8

Chinese Taipei

Jau-D Chen
Dean of General Affairs
National Taiwan Normal University
162, East Hoping Road Section 1
Taipei, Taiwan 117

Cyprus

Constantinos Papanastasiou
Dept. of Education
University of Cyprus
P.O. Box 20537
Nicosia CY-1678

Czech Republic

Jana Paleckova
Institute for Information of Education (UIV)
Senovazne nam.26
111 21 Praha 1

England

Graham Ruddock
National Foundation for Educational Research (NFER)
The Mere, Upton Park
Slough, Berkshire SL1 2DQ

Finland

Pekka Kupari
University of Jyvaskyla
Institute for the Educational Research
P.O. Box 35
SF – 40351 Jyvaskyla

Hong Kong, SAR

Frederick Leung
The University of Hong Kong – Dept. of Curriculum
Faculty of Education, Room 219
Pokfulam Road
Hong Kong, SAR

Hungary

Péter Vari
National Institute of Public Education
Centre for Evaluation Studies
Dorottya u.8, Pf 701/420
1051 Budapest

Indonesia

Jahja Umar
Examiniation Development Center
Jalan Gunung Sahari Raya – 4
Jakarta Pusat
Jakarta

Iran, Islamic Republic

Ali Reza Kiamanesh
Ministry of Education
196, Institute for Education Research
Keshavaraz Boulevard
Tehran, 14166

Israel

Ruth Zuzovsky
Tel Aviv University
School of Education
Center for Science and Technology Education
Ramat Aviv 69978

Italy

Anna Maria Caputo
Ministerio della Pubblica Istruzione
Centro Europeo Dell 'Educazione (CEDE)
5- Villa Falconieri
Frascati (Roma) 00044

Japan

Yuji Saruta
Hanako Senuma
National Institute for Educational Research (NIER)
6-5-22 Shimomeguro
Meguro-ku, Tokyo 153-8681

Jordan

Tayseer Al-Nhar
National Center for Human Resources Development
P.O. Box 560
Amman, Jordan 11941

Korea, Republic of

Sungsook Kim
Chung Park
Korea Institute of Curriculum & Evaluation(KICE)
25-1 Samchung-dong
GhongRo-Gu, Seoul 110-230

Latvia

Andrejs Geske
University of Latvia
IEA National Research Center
Jurmalas Gatve 74/76, Room 204A
Riga LV-1083

Lithuania

Algirdas Zabulionis
National Examinations Center
Ministry of Education and Science
M. Katkaus 44
Vilnius LT2051

Macedonia, Republic of

Anica Aleksova
Ministry of Education and Science
Bureau for Development of Education
Ruder Boskovic St. bb.
1 000 Skopje

Malaysia

Azmi Zakaria
Ministry of Education
Level 2,3 &5 Block J South
Pusat Bandar Damansara, Kuala Lumpur 50604

Moldova, Republic of

Ilie Nasu
Ministry of Education and Science
University "A. Russo"
Str. Puschin 38
Balti 3100

Lidia Costiuc
1 Piata Mazzi Adunazi Nationale
Chisinau

Morocco

Mohamed Fatihi
Direction de l'Evaluation du Systeme Educatif
Innovations Pedagogiques
32 Boulevard Ibn Toumert
Place Bob Rouah, Rabat

Netherlands

Klaas Bos
University of Twente
Centre for Applied Research in Education(OCTO)
P.O. Box 217
7500 AE Enschede

New Zealand

Megan Chamberlain
Ministry of Education
CER Unit-Research Division
45-47 Pipitea Street
Thorndon, Wellington

Philippines

Ester Ogena
DOST-Science Education Institute
3F PTRI Blg
Bicutan, Taguig
Metro Manila 1604

Vivien Talisayon
Institute Of Science & Mathematics Education Development
University of the Philippines UPISMED
Diliman, Quezon City

Romania

Gabriela Noveanu
Institute for Educational Sciences
Evaluation and Forecasting Division
Str. Stirbei Voda 37
Bucharest Ro-70732

Russian Federation

Galina Kovalyova
Center for Evaluating the Quality of Education
Institute of General Secondary Education
ul. Pogodinskaya 8
Moscow 119905

Singapore

Cheow Cher Wong
Research and Evaluation Branch
Ministry of Education
1 North Buona Vista Dr /MOE Building
Singapore, Singapore 138675

Slovak Republic

Olga Zelmanova
Maria Berova
SPU-National Institute for Education
Pluhova 8, P. O. Box 26
Brastislava 830 00

Slovenia

Barbara Japelj
Educational Research Institute Ljubljana
Gerbiceva 62
Ljubljana 1000

South Africa

Sarah Howie
Human Sciences Research Council
134 Pretorius Street
Private Bag X41
Pretoria 0001

Thailand

Precharn Dechsri
Institute For the Promotion of Teaching Science & Technology (IPST)
924 Sukhumvit Rd. Ekamai
Bangkok 10100

Tunisia

Ktari Mohsen
Ministere de l'Education
Boulevard Bab-Bnet
Tunis

Turkey

Yurdanur Atlioglu
Educational Research and Development Directorate
Gazi Mustafa Kemal Bulvani
No 109/5-6-7
Maltepe, Ankara 06570

United States

Patrick Gonzales
National Center for Education Statistics
United States Dept. of Education
1990 K Street, NW Room 9071
Washington, DC 20006

TIMSS 1999 Advisory Committees

The International Study Center at Boston College was supported in its work by advisory committees. The Subject Matter Item Replacement Committee was instrumental in developing the TIMSS 1999 tests, and the Questionnaire Item Review Committee revised the TIMSS questionnaires. The Scale Anchoring Panel developed the descriptions of the international benchmarks in mathematics and science.

Subject Matter Item Replacement Committee

Mathematics

Antoine Bodin, France
Anna-Maria Caputo, Italy
Nobert Delagrange, Belgium (Flemish)
Jan de Lange, Netherlands
Hee-Chan Lew, Republic of Korea
Mary Lindquist, United States
David Robitaille, Canada

Science

Hans Ernst Fischer, Germany
Galina Kovalyova, Russian Federation
Svein Lie, Norway
Masao Miyake, Japan
Graham Orpwood, Canada
Jana Strakova, Czech Republic
Carolyn Swain, England

Special Consultants

Chancey Jones, Mathematics
Christine O'Sullivan, Science

Questionnaire Item Review Committee

Im Hyung, Republic of Korea
Barbara Japelj, Slovenia
Trevor Williams, United States
Graham Ruddock, England
Klaas Bos, Netherlands

Scale Anchoring Committees

Mathematics

Anica Aleksova, Republic of Macedonia
Lillie Albert, United States
Kiril Bankov, Bulgaria
Jau-D Chen, Chinese Taipei
John Dossey, United States
Barbara Japelj, Slovenia
Mary Lindquist, United States
David Robitaille, Canada
Graham Ruddock, United Kingdom
Hanako Senuma, Japan
Pauline Vos, The Netherlands

Science

Audrey Champagne, United States
Galina Kovalyova, Russian Federation
Jan Lokan, Australia
Jana Paleckova, Czech Republic
Senta Raizen, United States
Vivien Talisayon, Philippines
Hong Kim Tan, Singapore

Typography
This book was set in ITC New Baskerville, designed by George W. Jones, and Frutiger, designed by Adrian Frutiger for Linotype-Hell, released by Adobe.

Book Design and Illustrations
José R. Nieto

Layout and Production
Betty Hugh
Mario Pita
Christina López

Production Assistants
Nathan Pyritz
Susan Messner

Cover Design
Christine Conley